Forward Osmosis

Fundamentals and Applications

Edited by
Ho Kyong Shon
Sherub Phuntsho
Tian C. Zhang
Rao Y. Surampalli

Sponsored by the
Hazardous, Toxic, and Radioactive Waste Engineering Committee of the
Environmental Council of the
Environmental and Water Resources Institute of the
American Society of Civil Engineers

ASCE

EWRI ENVIRONMENTAL & WATER RESOURCES INSTITUTE

Published by the American Society of Civil Engineers

Cataloging-in-Publication Data on file with Library of Congress

1801 Alexander Bell Drive
Reston, Virginia, 20191-4382
www.asce.org/bookstore | ascelibrary.org

Errata: Errata, if any, can be found at http://dx.doi.org/10.1061/9780784414071

ISBN 978-0-7844-1407-1 (print)
ISBN 978-0-7844-7923-0 (PDF)
Manufactured in the United States of America.

22 21 20 19 18 17 16 15 1 2 3 4 5

Contents

Preface

The current world population is 7 billion and is expected to reach 9 billion by 2050, creating a significant challenge in meeting the increasing demand for potable water. The fresh water resources have already become limited and the human encroachment on water catchments, unsustainable use of water and human induced climate change is further worsening the water scarcity issues in many parts of the world. To secure sustainable water resources from nonconventional sources, the projection of scientific solutions plays an essential role and has been widely recognised. Reverse osmosis (RO) is one such technology that can play a leading role in resolving water scarcity issues. The state-of-the-art RO technology has significantly improved the scope for the use of saline water and impaired wastewater effluent as an alternate source of water to augment fresh water or to reduce pressure on freshwater resources. The state-of-the-art seawater RO desalination plant now consumes energy by several orders of magnitude lower than the seawater RO about three decades ago. However, the energy required for seawater RO has almost reached a plateau, and any more efforts towards reducing energy consumption requires additional processes thereby increasing the total cost of the final water. Moreover, the law of thermodynamics have set a limit to the theoretical and practical energy for desalination of seawater. Membrane fouling particularly bio-fouling still poses a significant challenge to the RO desalination process further increasing the energy consumption. Therefore, RO still remains energy intensive in nature. Energy is an important component of a desalination plant given its role on global warming and climate change. Therefore, any low energy desalination technologies could make desalination technology more affordable and have a significant impact in meeting the increasing water demand.

Forward osmosis (FO) is now an emerging technology for desalination. The FO process works on the principle of the natural osmotic process, driven by the osmotic gradient between two solutions of different osmolar concentrations when they are separated by a semi-permeable membrane. When a semi-permeable membrane separates saline feed water and the highly concentrated solution (referred to as draw solution or DS), water moves from the saline water (lower solute concentration) to the concentrated DS (higher solute concentration), while retaining the solutes on both sides of the membrane. The main feature of this process is that the water transport across a semi-permeable membrane in the FO process is driven by the osmotic gradient and does not require hydraulic pressure therefore consuming significantly less energy than the RO process where direct hydraulic pressure is required. The other advantage of FO process is the low fouling potential of the process because of the absence of the hydraulic pressure. In most cases, fouling is physically reversible that reduces the need for chemical cleaning like the RO process.

However, FO technology still suffers from some major technological barriers. The first barrier is the lack of a membrane suitable for the FO process. The conventional salt rejecting membranes such as RO membranes are asymmetric and not suitable for FO process as it causes concentration polarisation effects, some of which are not only unique to the FO process but also pose a significant decrease in process efficiency. However, significant progress has been made in FO membrane fabrication, with thin film composites having comparatively higher water flux than the existing commercial FO membranes. The second drawback with the FO process is the lack of suitable DS especially when the FO process is used for drinking water purposes and where draw solute separation is essential. The separation and recovery of the DS require an additional processing unit, which consumes energy and therefore still remains a significant challenge for drinking water applications. The success of FO desalination for potable purpose will depend on how easily and efficiently the draw solute can be separated from the water.

Although FO technology has recently drawn a lot of attention, information on the FO process and its application is currently not available. Therefore, this book geared to provide the essential insight and information, background materials and application on various aspects of FO technology. This book explains the concept of FO technology and evaluates the scope and limitations of the technology. It begins with a brief explanation on the concept of FO technology, followed by a discussion on the values, delineation, fundamental of FO technology and FO membrane preparation, draw solution, modeling, process design parameter and fouling. Different applications in desalination, membrane bioreactor, water reclamation, fertigation and landfill leachate are envisaged. Furthermore, the removal of various pollutants such as emerging contaminants, boron and heavy metals is also addressed. In the final chapter, energy balance is covered with osmotic power by pressure retarded osmosis. The book also assesses the limitations of different FO processes and suggests alternatives in overcoming these specific limitations.

This book will be of interest to students, scientists, engineers, government officers, process managers and practicing professionals. The book will provide a good reference for in-depth and broad understanding and comprehensive information on the principles of emerging FO technology, different applications, recent development, and critical interpretation of new directions toward the future.

—HKS, SP, TCZ, RYS

Contributing Authors

S. Azari, *University of South Australia, Adelaide, Australia*
H. W. Bai, *Nanyang Technological University, Singapore*
G. Blandin, *University of New South Wales, Australia*
L. Chekli, *University of Technology, Sydney, Australia*
D. Chen, *Nanyang Technological University, Singapore*
G. Chen, *X Shanghai Advanced Research Institute, China*
J. Cho, *Yonsei University, Seoul, South Korea*
C. Cote, *Commonwealth Scientific and Industrial Research Organisation (CSIRO), Brisbane, Australia*
M. Cunnington, *Commonwealth Scientific and Industrial Research Organisation (CSIRO), Brisbane, Australia*
T.-S. Chung (Neal), *University of Singapore, Singapore*
M.J. Francis, *Commonwealth Scientific and Industrial Research Organisation (CSIRO), Brisbane, Australia*
G. Han, *National University of Singapore, Singapore*
T. He, *Shanghai Advanced Research Institute, China*
J. Herrron, *Hydration Technology Inc., USA*
S. Hong, *Korea University, Korea*
J. H. Kim, *Gwangju Institute of Science and Technology (GIST), South Korea*
J. E. Kim, *University of Technology, Sydney, Australia*
K.-Y. Kim, *Gwangju Institute of Science and Technology (GIST), South Korea*
L. H. Kim, *Gwangju Institute of Science and Technology (GIST), South Korea*
S.-J. Kim, *Gwangju Institute of Science and Technology (GIST), South Korea*
S. Kim, *Gwangju Institute of Science and Technology (GIST), South Korea*
I. S. Kim, *Gwangju Institute of Science and Technology (GIST), South Korea*
P. Le-Clech, *University of New South Wales, Australia*
J. Lee, *Gwangju Institute of Science and Technology (GIST), South Korea*
S. Lee, *Kookmin University, South Korea*
X. Li, *National University of Singapore, Singapore*
F. Lotfi, *University of Technology, Sydney, Australia*
T. Majeed, *University of Technology, Sydney, Australia*
Z. Y. Liu, *Nanyang Technological University, Singapore*
L. D. Nghiem, *University of Wollongong, Australia*
H. Y. Ng, *National University of Singapore, Singapore*
A. Nguyen, *University of South Australia, Adelaide, Australia*
R. C. Ong, *University of Singapore, Singapore*
M. Park, *Gwangju Institute of Science and Technology (GIST), South Korea*
D. H. H. Phuoc, *National University of Singapore, Singapore*

W. E. Price, *University of Wollongong, Australia*
M. J. Park ***University of Technology, Sydney, Australia***
S. Phuntsho, *University of Technology, Sydney, Australia*
A. Razmjou, *Monash University, Melbourne, Australia*
S. Sahebi, *University of Technology, Sydney, Australia*
X.-M. Li, *Shanghai Advanced Research Institute, China*
H. K. Shon, *University of Technology, Sydney, Australia*
G. P. Simon, *Monash University, Melbourne, Australia*
X. X. Song, *Nanyang Technological University, Singapore*
S. Su, *Commonwealth Scientific and Industrial Research Organisation (CSIRO), Brisbane, Australia*
D. D. Sun, *Nanyang Technological University, Singapore*
P. Sukitpaneenit, *National University of Singapore, Singapore*
R. Y. Surampalli, *Global Institute for Energy, Environment and Sustainability*
C. Y. Tang, *Nanyang Technological University, Singapore*
M. Tang, *National University of Singapore, Singapore*
R. Thiruvenkatachari, *Commonwealth Scientific and Industrial Research Organisation (CSIRO), Brisbane, Australia*
H. Wang, *National University of Singapore, Singapore*
H. Wang, *Monash University, Melbourne, Australia*
J. Wie, *Nanyang Technological University, Singapore*
L. Xiao, *National University of Singapore, Singapore*
M. Xie, *University of Wollongong, Australia*
Y. Yuan, *University of Nebraska-Lincoln, Lincoln, NE, USA*
T. C. Zhang, *University of Nebraska-Lincoln, Lincoln, NE, USA*
S. Zhang, *University of Singapore, Singapore*
L. Zou, *University of South Australia, Adelaide, Australia*

CHAPTER 1

Introduction: Role of Membrane Science and Technology and Forward Osmosis Processes

Sherub Phuntsho
Ho Kyong Shon
Tian Zhang
Rao Surampalli

1.1 BACKGROUND

An estimated 1,386,000,000 km^3 of water is found on the Earth covering about 70% of our planet. Water exists in three major forms: saline water (97.5%), fresh water (0.5%) and others such as glacial ice/snow (2.5%) (Khawaji et al. 2008). Water is in fact the very reason that various forms of life thrive on this planet. Most ancient civilizations prospered along the rivers and lakes because of the availability and accessibility of abundant sources of fresh water. The same civilizations also perished once water sources dried out. While any form of water supports life on earth, it's the availability of the fresh water that is important for the survival and livelihood of the human on this planet. Unfortunately, only a very small fraction of water on this earth is available as fresh water for direct use by living beings.

Recently, water scarcity issues have been steadily growing in many parts of the world. According to the current estimate, more than one third of the world's population now lives in water-stressed countries and may rise to nearly two thirds by 2025 (Service 2006). Water issues are now experienced even in those countries once considered water-rich nations (Elimelech and Phillip 2011). In modern society, water is needed for various purposes: domestic consumption (drinking, cooking and washing), in growing food (irrigation) and industrial consumption (support economic growth and improve livelihood). The demand for fresh water for these uses have been expanding very rapidly driven by the rapid population growth, rapid economic growth for improving the living standards in the developing countries and increased food demand. However, the existing fresh water sources are mostly declining due to loss of vital catchment areas to human encroachment such as expanding cities, clearing forests for agriculture and in many areas due to unsustainable exploitation of forest resources such as commercial logging. Pollution of existing fresh water resources from domestic and industrial discharges has also rendered many rivers, fresh water lakes and groundwater unsuitable for human consumption without adequate treatment. Water mismanagement is another critical factor that results in water issues in many parts of the world. Global warming and climate change is another recent factor that has been responsible for exacerbating the water issues. Climate change has resulted in frequent drought and unpredictable rainfall events in many parts of the world, creating further uncertainty about water availability and food security (Whetton et al. 1993; Mearns et al. 1996; Hughes 2003; McDonald et al. 2011).

Given that the world population would grow from 7 billion in 2011 (UNFPA 2011) to 9 billion by 2050 (UN 2009b; UN 2009a) and the decline of available fresh water resources, there is no doubt that one of the most crucial challenges for this century will be dealing with the pressure from the significant increase in the demand for food, water and energy to sustain this enormous population growth (Ward and Pulido-Velazquez 2008). The growing water scarcity have also led to increased competition for the limited fresh water resources between the various water users, between different regions and even between different countries. Someone even termed water as the next black oil, which could be a reality if the demands for water continue to rise and the fresh water resources continue to decline in the future. It is even envisaged that the future source of dispute between the communities or between the different nations would be due to the water sources, especially those that share the same water resources for their livelihood that could potentially lead to violent conflicts.

Recently, there has been a significant increase in the number of publications on the FO process indicating the worldwide interest in this particular subject. Because of the early stages of the FO technology, comprehensive literature on the FO process is lacking other than the few review papers recently published as journal articles. There is a need for a book that provides comprehensive information on the forward osmosis process and the status of the technology. Therefore, the main objective of this book is to introduce the principles and concept of the FO process and the potential for various applications such as desalination, fertigation, wastewater reuse and mine water treatment. The book also covers other important aspects of the FO process such as membrane fabrication and modeling. In the later part of the sections in this chapter, a brief outline of each chapter is presented.

1.2 ROLE OF MEMBRANE SCIENCE AND TECHNOLOGY IN SOLVING THE WATER SCARCITY ISSUES

Several measures have been adopted to solve the water stress including water conservation, infrastructure improvement/repair, conservation and improvement of catchment areas and improvement of distribution system. Although these measures do not create any new water resources, it helps in alleviating water stress to a certain degree by improving the management efficiency of the existing water resources (Elimelech and Phillip 2011). Given that the water demand will continue to rise with the rise in the worlds' population, the scientific and engineering solutions is expected to play a crucial role in solving some of the water issues on this planet (Jury and Vaux 2005). The development of several water treatment technologies has in fact rendered most water sources safe for various uses including water from nonconventional water sources.

There are various water treatment technologies available on the market. One of the most advanced and reliable water treatment technologies is the membrane process. Depending on the membrane pore size and their removal capacity, the membrane process is classified into four different types: microfiltration (MF), ultrafiltration (UF), nanofiltration (NF) and reverse osmosis (RO). MF has the largest pore size while RO has the smallest pore size. MF, UF and RO are the most widely used membrane processes while the application of NF is also gaining popularity in some industries.

However, recognizing the limitations with the conventional water resources, many countries are now turning towards the non-conventional sources of water such as wastewater reuse and saline water to augment the limited fresh water resources. In fact, many cities in the Middle East (such as Saudi Arabia, Israel, etc.) have built seawater desalination plants as the main source of water supply systems. Wastewater reclamation and reuse is gaining popularity in many countries. For example, Israel reclaims and recycles almost 100% of its wastewater for non-potable uses such as irrigation. Many other cities have built desalination plants to augment water supply during the time of drought such as Sydney, Melbourne, Brisbane and Adelaide in Australia. These nonconventional water sources create new water and not only in meeting water demand but also help enhance water security during the time of drought, which has become quite frequent. Many countries are now considering water reuse and desalination, as reliable alternatives for water management options, and this has been made possible due to rapid development in desalination technology that is robust and reliable.

There are several technologies available in the market for desalination: multi-stage flash (MSF), multiple-effect distillation (MED), mechanical vapor compression (MVC) and RO. While thermal-based processes such as MED and MSF are popular in the oil-rich countries, the membrane-based RO desalination technology has been gaining popularity due to rapid improvement and development in the membrane technology. RO now accounts for up to 50% of the entire desalination market in the world (Ettouney and Wilf 2009). The RO desalination process continues to grow quickly because of advances in the technology, especially in the development of more efficient and less expensive membranes, and also due to significant reductions in the energy requirement compared to three or four decades ago (Fritzmann et al. 2007). In some countries such as the USA, Spain, Cyprus and Malta, the RO process is the only technology used for desalination (Ettouney and Wilf 2009). Besides desalination of saline water, the RO process has been used for wastewater reclamation and reuse. RO has a high rejection of many dissolved solutes, and its application for wastewater provides very high quality of water for reuse although direct potable use is still not gaining wide social acceptance.

1.3 ISSUES AND CHALLENGES WITH CURRENT REVERSE OSMOSIS DESALINATION PROCESSES

There has been a significant investment in research and development in the membrane process since the 1960s, and the result has been a drastic improvement and advancement in membrane technology. However, some of the key issues and challenges that continue to plague the membrane processes, particularly with the RO process are:

- High energy consumption;
- High capital cost; and
- Membrane fouling and in particular biofouling

1.3.1 Energy Consumption of RO Processes

The performance of the RO process has significantly improved since the 1970s. Today, the state-of-the-art seawater RO (SWRO) desalination plant consumes only about 3.0 kWh/m^3 of energy compared to 10 kWh/m^3 in the 1970s, which is a tremendous reduction in energy consumption. However, SWRO desalination is still seen or remains as an energy intensive process (McGinnis and Elimelech 2007; Greenlee et al. 2009). Seawater has an osmotic pressure of about 27 bar and therefore SWRO has to be operated at hydraulic pressure higher than 27 bar. The pressure requirement further rises with the increase in the feed recovery rate. For example, SWRO is usually operated at a maximum of about 50% recovery rate, and as such, desalination plants are generally operated at hydraulic pressure ranging from 60 to 80 bar.

Moreover, the law of thermodynamics sets a minimum energy of 0.8 kWh/m^3 to produce fresh water from the seawater at zero recovery irrespective of the process used and of 1.14 kWh/m^3 at the 50% recovery rate (Elimelech and Phillip 2011). However, practically, the most efficient RO desalination plant equipped with the energy recovery system has been reported to consume about up to 3.0 kWh/m^3 excluding the energy consumed by the pre-treatment processes, and this unit energy consumption for seawater RO desalination increases with recovery rates (Elimelech and Phillip 2011; Subramani et al. 2011). Although this is a significant reduction in energy consumption by the SWRO desalination process, this energy is still about three to four times higher than the energy required for the treatment and supply of freshwater from the conventional sources. For this reason, it is often difficult for engineers to convince the conservative environmental groups about the benefits of RO desalination technology. Since energy and climate change issues are strongly inter-related (Crowley 2000; Semiat 2008), addressing global water scarcity problems requires robust and innovative methods of purifying water at minimum energy and cost (McGinnis and Elimelech 2008; Shannon et al. 2008; Elimelech and Phillip 2011). Low energy and low cost desalination technologies could have a significant impact on humanity by utilizing the unlimited saline water sources available on this planet.

1.3.2 High Capital Cost

SWRO plants operate at a hydraulic pressure of 80 bar or more. To withstand such high pressures, all the membrane accessories such as pressure vessels, pipes and fittings have to be made of materials with high pressure ratings. Pipes and fittings are usually made of stainless steel and pressure vessels of thick PVC housing. The high-pressure rating of the RO membrane accessories makes the SWRO desalination plants highly capital intensive, which is a very significant issue in terms of the affordability by most developing societies in the world.

1.3.3 Challenges with Membrane Fouling

Fouling still proves to be a major challenge for the membrane processes (Greenlee et al. 2009; Phuntsho et al. 2011a; Phuntsho et al. 2011b). Although various strategies are adopted to reduce the impact of membrane fouling such as multiple pre-treatments, physical and chemical cleaning, fouling, particularly biofouling continue to pose a major challenge to the SWRO desalination process. Membrane fouling reduces membrane efficiency and increases the energy consumption. Moreover, cleaning increases the plant down time, thereby hampering the production efficiency and treatment cost. Besides, most pre-treatment strategies involve additional processes that proportionately increases the capital and operational costs of the plant (Semiat 2008). People are working on identifying the main component responsible for biofouling as well as in the early detection of biofouling.

1.4 FORWARD OSMOSIS: FUNDAMENTALS AND PRINCIPLES

1.4.1 Osmosis

Osmosis refers to the phenomenon that describes the spontaneous passage or diffusion of solvents or water through a semipermeable membrane while preventing the passage of solutes. Osmosis is therefore ubiquitous in all living cells. Although osmotic phenomenon was first studied in 1877 by a German plant physiologist, Wilhelm Pfeffer, the general term 'osmose' (now osmosis) was introduced earlier in 1854 by a British chemist, Thomas Graham (Britannica 2012). When a semipermeable membrane separates the solution and a pure solvent, the solution tends to become more diluted by absorbing the solvent through the membrane. If hydraulic pressure is applied on the solution to stop the movement of pure solvent across the membrane and to maintain a condition of equilibrium (no flow of solvent), this equivalent pressure has been termed as osmotic pressure (Robinson and Stokes 1959). Osmotic pressure is a colligative property and therefore refers to its chemical potential of the solvent in the solution, or alternatively it includes vapor pressure lowering, boiling point elevation, freezing point depression and osmotic pressure (Rudin 1999).

Osmosis therefore describes the natural diffusion of water through a semi-permeable membrane from a solution containing a lower salt concentration to a solution containing a higher salt concentration (Cath et al. 2006). The osmotic pressure (π) of an ideal dilute solution is given by Van't Hoff's (1887) equation shown below.

$$\pi = nMRT \tag{1.1}$$

where n stands for the Van't Hoff factor (refers to the number of individual particles of compounds dissolved in the solution, for example $n = 2$ for NaCl, $n = 1$ for glucose); M is the molar concentration (molarity) of the solution; R is the gas constant ($R = 0.0821\ \mathrm{L \cdot atm \cdot mol^{-1} \cdot K^{-1}}$); and T is the absolute temperature (in K) of the solution. The osmotic coefficients of a few commonly available salts such as NaCl and KCl, etc. are provided in the work by Robinson and Stokes (1959). Osmotic pressure of most solutions can also be readily modeled using thermodynamic software such as OLI Stream Analyser (OLI Systems Inc., Morris Plains, NJ, US).

The Van't Hoff equation is, however, applicable only to ideal and dilute solutions where ions behave independently of one another, and it is generally used for the determination of the molecular weight (MW) of large molecules (Yokozeki 2006). For concentrated solutions, the interactions between the solutes increase and the solution becomes a non-ideal one, ultimately altering the activity coefficient of each solute/ion and the osmotic pressure of the solution (Snoeyink and Jenkins 1980). For general solutions, the osmotic pressure can be given by the concentration dependence osmotic equation (Stigter and Hill 1959; Rudin 1999), also known as the Virial equation:

$$\frac{\pi}{c} = RT\left[\frac{1}{M_w} + Bc + Cc^2 + Dc^3 + \cdots\cdots\right] \tag{1.2}$$

where c is the concentration in terms of grams of solute per liter (g/L); B, C and D are the osmotic viral coefficients which are functions of the temperature and the chemical potentials of the species in the salt solution (Stigter and Hill 1959). The viral coefficient can be determined empirically by fitting experimental osmotic pressure data and generally the determination of B and C is sufficient to reproduce observed data (Yokozeki 2006). A solute with a small MW combined with high water solubility can generate a higher osmotic pressure (on an equal mass basis) and therefore can lead to higher water fluxes (McCutcheon et al. 2005).

The alternative way of determining the osmotic coefficient of a solution is by measuring the osmolality of the solution using an osmometer. An osmometer measures osmolality in osmoles/kg of water based on the freezing point depression method, but the application of osmometers is limited to solutions with infinite dilutions and therefore is not suitable at higher concentrations. Osmolality can be converted into osmotic pressure (π) with Eq. 1.3 (Zhang and Song 2000):

$$\pi = 19{,}300 \times \text{Osmolality (in mm of Hg)} \tag{1.3}$$

1.4.2 Forward Osmosis Processes

The FO process is in fact an engineered osmotic process in which an artificially high concentrated solution, termed a draw solution (DS), is used on one side of the semi-permeable membrane and the water to be treated is on the other side of the same membrane. Although FO is based on the principle of osmosis, the term 'forward osmosis' (FO) has been probably coined to distinguish it from RO, which is the term that has been used for the membrane desalination process for many decades. The semi-permeable membrane, usually made from polymeric materials, acts as a barrier that allows small molecules such as water to pass through while blocking larger molecules such as salts, sugars, starches, proteins, viruses, bacteria, and parasites (Xu et al. 2010). Both RO and FO processes use a semi-permeable membrane to separate water from dissolved solutes effectively, although their driving forces are different. The main difference between the two processes is that the driving force in the RO process is created by hydraulic pressure, while the driving force in the FO process is created by the concentration or osmotic difference.

Pressure-retarded osmosis (PRO) is a process closely related with the FO process. PRO has been tested and evaluated since the 1960s as a potential process for power generation using the salinity gradient (Loeb 1976; Loeb et al. 1976; Loeb and Mehta 1979; Loeb 1998; Loeb 2001; Loeb 2002a; Loeb 2002b; Cath et al. 2006; Achilli and Childress 2010; Yip and Elimelech 2012). PRO uses the osmotic pressure difference between seawater, or concentrated brine, and fresh water to pressurize the saline stream, thereby converting the osmotic pressure of seawater into a hydrostatic pressure that can be used to produce electricity. PRO is viewed as an intermediate process between FO and RO (Cath et al. 2006); however, the net water flux is still from the feed solution to the concentrated draw solution (similar to FO). The general equation describing water transport in FO, RO, and PRO is given by the following equation (Cath et al. 2006).

$$J_w = A(\sigma\Delta\pi - \Delta P) \tag{1.4}$$

where, J_w is the water flux, A is the water permeability constant of the membrane, σ is the reflection coefficient defined as the ratio between the negative solute-water phenomenological coefficient divided by the pure water permeability

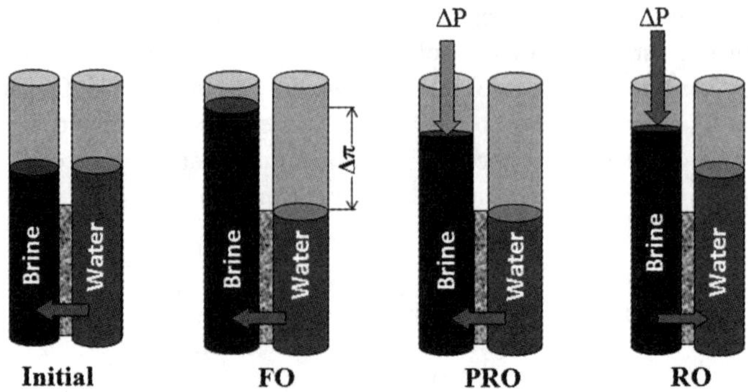

Figure 1.1. Principles of osmotic processes: initial state of the solutions, forward osmosis (FO), pressure retarded osmosis (PRO) and reverse osmosis (RO)

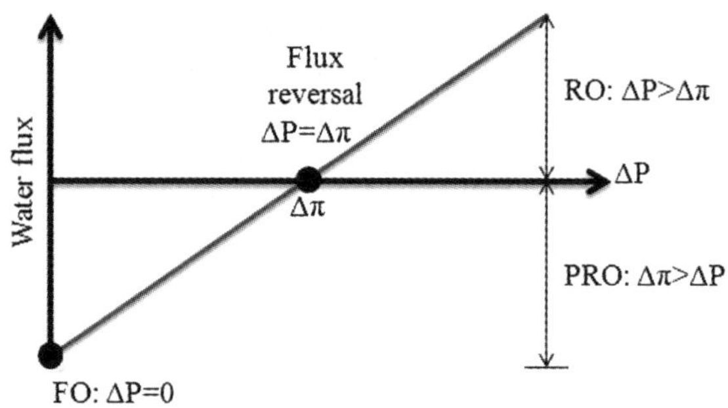

Figure 1.2. The direction and magnitude of water flux as a function of applied pressure in FO, PRO and RO (adapted from Lee et al. 1981)

(Mulder 1997; Hancock and Cath 2009) and ΔP is the applied hydraulic pressure. The following conditions describe each process of the equation:

For RO process, $\Delta P > \Delta\pi$

For FO process, $\Delta P = \text{zero}$

For PRO process, $\Delta\pi > \Delta P$

Figure 1.1 describes the flux directions of the permeating water in the RO, FO and PRO processes. Figure 1.2 illustrates the different points, describing the RO zone, point of FO process and PRO zone along with the flux reversal point as characterized by Lee et al. (1981).

1.4.3 Forward Osmosis: Its Drivers and the Current Status

One of the two major challenges of this millennium will no doubt be energy and water, and the need for these two resources is increasing every year to support the rapid population growth and the growing economy as discussed earlier. As the demand for water increases, the issue of water scarcity will continue to grow unless new sources of water are available. Under such circumstances, desalination is expected to play a major role in helping to create new sources of water using the saline water source, which is abundant on the earth. However, as acknowledged earlier, the issues and challenges remain with the current desalination technologies such as RO as they are energy and capital intensive. Researchers have been therefore exploring for low energy and affordable desalination technologies as an alternative to the RO process. Of the several new potential desalination processes identified, FO has been identified as one of promising emerging technologies for desalination and water reuse applications.

The FO process operates without the use of any hydraulic energy, and hence, this is one of the main drivers of the FO technology. As the FO process operates at low hydraulic pressure, the cost of the pumps and other membrane accessories would be much lower and ultimately lower capital cost than the RO process. The other perceived advantage of the FO process has been the low fouling issues associated with it. It has been observed through several studies that, fouling issues and challenges are less problematic than the RO process because fouling in the FO process is generally physically reversible thereby avoiding the need for expensive chemical cleaning. The other driving factor of the FO process is that, the RO technology process has significantly improved its energy efficiency during the last four decades and any attempt to further decrease the energy is likely to increase the total cost of the water. Hence, it is now understood within the research circles that, the opportunity for further significant reduction in energy consumption for the SWRO desalination process is slim. Hence, researchers are now putting in more research efforts on some of these emerging desalination technologies.

1.5 BOOK OUTLINE

This book comprises 20 chapters including this chapter (Introduction). The following paragraph provides a brief outline of each chapter.

Chapter 2. In this chapter, modeling of the FO process is introduced. The chapter covers the basic concepts of FO process, mass transport in the FO process, flux modeling of single-skinned FO membranes and advance double-skinned FO membranes, modeling of salt accumulation in the osmotic membrane bioreactor, and modeling of a related PRO process. Additionally, the computational fluid dynamics approach is also introduced as a useful tool for FO modeling. The chapter provides a systematic insight into the FO mass transport, to understand the governing mechanisms over flux behavior on a quantitative basis, and finally to achieve clear concepts about how to improve the efficiency of the FO process.

Chapter 3. In this chapter, the impacts of spacers in the FO process are discussed including the complexity of concentration polarization (CP) effects. In addition, CPI (Concentration Polarization Index) is suggested as a CP indicator pertaining to the degree of concentration polarization and permeate flux in FO processes. The phenomena of boundary layer compression and boundary layer disruption are compared between RO and FO processes. Lastly, sensitivity analysis is included in order to gain insights into FO separation characteristics and to suggest further strategy for FO process development.

Chapter 4. This chapter discusses the modifications made to the design of standard pressure filtration elements to transform them into FO element designs. Primarily this entails redesigning the permeate channels in the element to accept a draw solution pumped separately from the permeate port.

Chapter 5. This chapter reviews different types of DS available for various FO applications. Classification of DS into different categories has been suggested. The characteristics of different DS affecting the performance of the FO process have been discussed. The chapter also provides a valuable insight on the selection of suitable draw solution (DS) for a particular type of FO application.

Chapter 6. In this chapter, a new kind of draw solute, dextran coated Fe_3O_4 magnetic nanoparticles (DF-MNPs) is introduced and its performance in the FO process evaluated. DF-MNP combines the advantages of high osmotic pressure of dextran and strong magnetism of Fe_3O_4 MNPs. DF-MNPs show good performance for the brackish water desalination with high water flux and easy recovery using an external magnet, which is favorable for significantly reducing the operation cost in the FO process.

Chapter 7. In this chapter, thermo-sensitive polymer hydrogels are introduced as a new class of draw agent for forward osmosis (FO) desalination. The effect of chemical alteration of hydrogels including the addition of ionic groups, light-absorbing carbon and graphene moieties on the FO desalination performance is reviewed. The key parameters for the FO and water recovery step are examined; in particular, influential parameters such as hydrogel particle size, membrane type, feed solution, particle-particle contact areas, particle-membrane contact zones and hydrogel thickness on the membrane surface are explored.

Chapter 8. This chapter introduces the state-of-the-art technologies for the fabrication of FO membranes including analysis on the mass transport phenomena and strategies for the design of effective FO membranes. The chapter mainly focuses on the fabrication of asymmetric membranes by both phase inversion (cellulose acetate, polybenzimidazole and polyamide-imide) and interfacial polymerization (thin film composite, layer-by-layer and biomimetic membranes). The chapter also introduces some of the methods for the characterization of FO membranes. Each type of membrane is discussed in terms of material, structure, preparation methods and FO performance.

Chapter 9. This chapter introduces a novel nanofiber FO membrane (NFM) made of electrospun nanofiber support layer and a thin active layer. The key features of the electrospun nanofiber support layer are the low tortuosity, high porosity and very thin support layer, which are critical for the internal concentration polarization that restrains the performance of the conventional FO membranes. This chapter also suggests the methods for characterizing the NFM membrane in terms of structures and water permeability. This study paves the new way for the design and development of next-generation FO membranes with high water productivity.

Chapter 10. Like any other membrane processes, FO also is not immune to fouling although the advantage of FO process is the physically reversible nature of the fouling. This chapter discusses the various options for the surface modifications of the FO membrane to improve its performance in terms of fouling resistance. The three coating methods to improve the anti-fouling properties of the FO membrane such as anionic, hydrophilic polymer and bio-inspired zwitterionic coating are discussed in details. The chapter also introduces some of the methods of characterization of surface modified FO membranes.

Chapter 11. In order to understand fouling in the FO process, background knowledge on fouling is needed. Since numerous cases of fouling have been studied in the pressure-driven membrane process, understanding fouling in hydraulic pressure-driven membrane processes is needed first. Then, moving to fouling in the osmotic pressure-driven membrane process would be helpful even though the two processes have different driving forces for separation.

Chapter 12. This chapter reviews the important aspects of biofouling in the FO membrane process. The distinction between biofouling phenomena in RO and FO membranes are discussed.

Chapter 13. With diminishing supplies and an escalating demand for drinking water, motivation for societies to recycle wastewater and saline water becomes stronger over the years. This chapter introduces the application of FO process for the wastewater reclamation and reuse. The chapter provides an overview of the fundamental terminologies, concepts, performance discussions and concerns of FO hybrid systems for water reclamation and reuse.

Chapter 14. This chapter introduces the principles of the FO process for seawater desalination and historical evolution of desalination technologies from RO to FO, followed by the in-depth presentation of different case studies on FO desalination processes. The chapter presents a comprehensive review of the FO operation conditions, performance parameters, and key issues related to design/management, future trend and research directions of FO for seawater desalination.

Chapter 15. This chapter reviews the state-of-the-art treatment of shale gas produced water by the FO process with the focus on the treatment of shale gas flow-back water (SGW). A brief introduction of origin and chemical/physical characteristics of the SGW are given, and the advantages and limitations of potential treatments methods are analyzed. The process parameters, selection of membrane and DS were summarized. Finally, the potential of utilization of FO processes for the treatment of SGW in a large scale are discussed.

Chapter 16. This chapter provides a comprehensive review on the current state of knowledge on the rejection of emerging trace organic chemicals by FO processes. The review begins with a brief introduction on the occurrence of emerging trace organic chemicals in municipal wastewater effluent. Three major aspects are outlined to delineate their impacts on the rejections of trace organic chemicals: process parameters including properties of membrane and draw solution and operating conditions; membrane fouling; and FO-based hybrid processes. This review ends with several concluding remarks that will further strengthen the potential of FO in effectively dealing with emerging trace organic chemicals.

Chapter 17. The FO process simply converts from one form of solution to another form of solution with similar osmotic pressure. One of the challenges for FO process therefore is the need to separate the draw solute and the pure water through a post-treatment process and hence, FO cannot be used as a standalone process. However, the FO process becomes ideal for those applications where the presence of draw solutes adds value to the FO processed water (diluted DS), and therefore, the separation of draw solutes and water does not become essential. One such application is the fertilizer driven FO (FDFO) process in which the fertilizer solution is used as a DS to desalinate or treat any impaired water by FO process. The chapter explains the concept of the FDFO process and examines the performances of some of the commonly used fertilisers as DS. The limitations of the FDFO process are identified, and their options suggested to overcome the limitations.

Chapter 18. In this chapter, the integration of forward osmosis (FO) with other water treatment methods has been examined with a particular focus on the integration of FO and RO technologies. The limitations of RO in relation to the quality of the feed water are discussed; in particular for the treatment of mine-affected water. Different modes of integration have been presented, and the advantage and limitations of different hybrid configurations examined.

Chapter 19. This chapter presents a brief overview of the principles of pressure assisted osmosis (PAO) processes and the recent approaches aiming at the enhancement of water flux and reduction of reverse solute flux. The potential of PAO for desalination and wastewater reclamation is also highlighted.

Chapter 20. Salinity gradient energy (or osmotic power) has huge potential as a sustainable blue energy source. This chapter focuses on development of special membranes for osmotic power generation via PRO processes. The concept of osmotic power generation and the fundamentals of PRO are presented in terms of osmotic processes, water and salt permeabilities, concentration polarization and power density. In addition, preferential types of PRO membranes and some recent exciting developments are reviewed.

References

Achilli, A. and Childress, A. E. (2010). "Pressure retarded osmosis: From the vision of Sidney Loeb to the first prototype installation - Review." *Desalination*, 261(3), 205–211.

Britannica. (2012). "Osmosis." from http://www.britannica.com/.

Cath, T. Y., Childress, A. E., and Elimelech, M. (2006). "Forward osmosis: Principles, applications, and recent developments: Review." *Journal of Membrane Science*, 281 (2006), 70–87.

Crowley, T. J. (2000). "Causes of Climate Change Over the Past 1000 Years." *Science*, 289(5477), 270–277.

Elimelech, M. and Phillip, W. A. (2011). "The Future of Seawater Desalination: Energy, Technology, and the Environment." *Science*, 333(6043), 712–717.

Ettouney, H. and Wilf, M. (2009). *Commercial Desalination Technologies: Seawater Desalination*, G. Micale, L. Rizzuti and A. Cipollina, Springer Berlin Heidelberg: 77–107.

Fritzmann, C., Löwenberg, J., Wintgens, T., and Melin, T. (2007). "State-of-the-art of reverse osmosis desalination." *Desalination*, 216 (1–3), 1–76.

Greenlee, L. F., Lawler, D. F., Freeman, B. D., Marrot, B., and Moulin, P. (2009). "Reverse osmosis desalination: Water sources, technology, and today's challenges." *Water Research*, 43(9), 2317–2348.

Hancock, N. T. and Cath, T. Y. (2009). "Solute Coupled Diffusion in Osmotically Driven Membrane Processes." *Environmental Science & Technology*, 43(17), 6769–6775.

Hughes, L. (2003). "Climate change and Australia: Trends, projections and impacts." *Blackwell Publishing Ltd.*, 28: 423–443.

Jury, W. A. and Vaux, H. (2005). "The role of science in solving the world's emerging water problems." *Proceedings of the National Academy of Sciences*, 102(44), 15715–15720.

Khawaji, A. D., Kutubkhanah, I. K., and Wie, J.-M. (2008). "Advances in seawater desalination technologies." *Desalination*, 221(1–3), 47–69.

Lee, K. L., Baker, R. W., and Lonsdale, H. K. (1981). "Membranes for power generation by pressure-retarded osmosis." *Journal of Membrane Science*, 8(2), 141–171.

Loeb, S. (1976). "Production of energy from concentrated brines by pressure-reterded osmosis 1. Preliminary technical and economic correlations." *Journal of Membrane Science*, 1(1), 49–63.

Loeb, S. (1998). "Energy production at the Dead Sea by pressure-retarded osmosis: challenge or chimera?" *Desalination*, 120(3), 247–262.

Loeb, S. (2001). "One hundred and thirty benign and renewable megawatts from Great Salt Lake? The possibilities of hydroelectric power by pressure-retarded osmosis." *Desalination*, 141(1), 85–91.

Loeb, S. (2002a). "Large-scale power production by pressure-retarded osmosis, using river water and sea water passing through spiral modules." *Desalination*, 150(2), 205–205.

Loeb, S. (2002b). "One hundred and thirty benign and renewable megawatts from Great Salt Lake? The possibilities of hydroelectric power by pressure-retarded osmosis with spiral module membranes." *Desalination*, 142(2), 207–207.

Loeb, S. and Mehta, G. D. (1979). "2-coefficient water transport-equation for pressure retarded osmosis." *Journal of Membrane Science*, 4(3), 351–362.

Loeb, S., Vanhessen, F., and Shahaf, D. (1976). "Production of energy from concentrated brines by pressure-reterded osmosis 2. experimental results and projected energy costs." *Journal of Membrane Science*, 1(3), 249–269.

McCutcheon, J. R., McGinnis, R. L., and Elimelech, M. (2005). "A novel ammonia-carbon dioxide forward (direct) osmosis desalination process." *Desalination*, 174(2005), 1–11.

McDonald, R. I., Green, P., Balk, D., Fekete, B. M., Revenga, C., Todd, M., and Montgomery, M. (2011). "Urban growth, climate change, and freshwater availability." *Proceedings of the National Academy of Sciences*, 108(15), 6312–6317.

McGinnis, R. L. and Elimelech, M. (2007). "Energy requirements of ammonia-carbon dioxide forward osmosis desalination." *Desalination*, 207(1–3), 370–382.

McGinnis, R. L. and Elimelech, M. (2008). "Global Challenges in Energy and Water Supply: The Promise of Engineered Osmosis." *Environmental Science & Technology*, 42(23), 8625–8629.

Mearns, L. O., Rosenzweig, C., and Goldberg, R. (1996). "The effect of changes in daily and interannual climatic variability on CERES-Wheat: A sensitivity study." *Climatic Change*, 32(3), 257–292.

Mulder, M. (1997). *Basic principles in membrane technology*, Kluwer Academic Publishers.

Phuntsho, S., Listowski, A., Shon, H. K., Le-Clech, P., and Vigneswaran, S. (2011a). "Membrane autopsy of a 10 year old hollow fibre membrane from Sydney Olympic Park water reclamation plant." *Desalination*, 271(1–3), 241–247.

Phuntsho, S., Shon, H. K., Vigneswaran, S., and Cho, J. (2011b). "Assessing membrane fouling potential of humic acid using flow field-flow fractionation." *J. Membr. Sci.*, 373(1–2), 64–73.

Robinson, R. A. and Stokes, R. H. (1959). *Electrolyte Solutions*, 2nd ed., Reprinted by Courier Dover Publications, New York (2002).

Rudin, A. (1999). *The elements of polymer science and engineering: an introductory text and reference for engineers and chemists*, San Diego, CA London: Academic Press.

Semiat, R. (2008). "Energy Issues in Desalination Processes." *Environmental Science & Technology*, 42(22), 8193–8201.

Service, R. F. (2006). "Desalination Freshens Up." *Science*, 313(5790), 1088–1090.

Shannon, M. A., Bohn, P. W., Elimelech, M., Georgiadis, J. G., Marinas, B. J., and Mayes, A. M. (2008). "Science and technology for water purification in the coming decades." *Nature*, 452(7185), 301–310.

Snoeyink, V. L. and Jenkins, D. (1980) *Water chemistry*, John Wiley & Sons.

Stigter, D., and Hill, T. L. (1959). "Theory of the Donnan membrane equilibrium. II. Calculation of the osmotic pressure and the salt distribution in a Donnan system with highly-charged colloid particles." *Journal of Physical Chemistry*, 63, 551–556.

Subramani, A., Badruzzaman, M., Oppenheimer, J., and Jacangelo, J. G. (2011). "Energy minimization strategies and renewable energy utilization for desalination: A review." *Water Research*, 45(5), 1907–1920.

UN. (2009a) *Role of desalination in addressing water scarcity/Economic and Social Commission for Western Asia*, United Nations, New York.

UN. (2009b) "World population prospects: The 2008 revision." Population Newsletter 87.

UNFPA. (2011). "State of world population 2011." *Information and External Relations Division*, United Nations Population Fund.

Van't Hoff, J. H. (1887). "Die Rolle der osmotischen Druckes in der Analogie zwischen Lösungen und Gasen." *Z. Phys. Chem.*, 1, 481–508.

Ward, F. A. and Pulido-Velazquez, M. (2008). "Water conservation in irrigation can increase water use." *Proceedings of the National Academy of Sciences*, 105(47), 18215–18220.

Whetton, P. H., Fowler, A. M., Haylock, M. R., and Pittock, A. B. (1993). "Implications of climate change due to the enhanced greenhouse effect on floods and droughts in Australia." *Climatic Change*, 25(3), 289–317.

Xu, Y., Peng, X., Tang, C. Y., Fu, Q. S., and Nie, S. (2010). "Effect of draw solution concentration and operating conditions on forward osmosis and pressure retarded osmosis performance in a spiral wound module." *Journal of Membrane Science*, 348 (1–2), 298–309.

Yip, N. Y. and Elimelech, M. (2012). "Thermodynamic and Energy Efficiency Analysis of Power Generation from Natural Salinity Gradients by Pressure Retarded Osmosis." *Environmental Science & Technology*, 46(9), 5230–5239.

Yokozeki, A. (2006). "Osmotic pressures studied using a simple equation-of-state and its applications." *Applied Energy*, 83(1), 15–41.

Zhang, M. and Song, L. (2000). "Mechanisms and parameters affecting flux decline in cross-flow microfiltration and ultrafiltration of colloids." *Environmental Science & Technology*, 34(17), 3767–3773.

CHAPTER 2

Modeling of Forward Osmosis Processes

Jing Wei
Chuyang Y. Tang

2.1 INTRODUCTION

Forward osmosis (FO) is an emerging membrane technology that utilizes an osmotic pressure difference to drive the water transport through a semi-permeable membrane from the lower osmotic pressure side (referred to as feed solution) to the higher osmotic pressure side (referred to as draw solution) (Cath et al. 2006). Compared to the pressure-driven membrane processes such as nanofiltration (NF) and reverse osmosis (RO), FO can have high rejection to a wide range of solutes and contaminants and requires minimal hydraulic pressure (Holloway et al. 2007; Wang et al. 2010b; Yip et al. 2010; Jin et al. 2012; Xie et al. 2012). This advantage makes FO a very attractive alternative to the conventional pressure-driven membrane processes to overcome the high energy consumption of pumping system, when the high-osmotic-pressure draw solutions are naturally available (e.g., seawater and RO brine) or the draw solutions can be easily reconcentrated (Achilli et al. 2010; Chekli et al. 2012).

Although FO technology has demonstrated great potential and advantages in a variety of applications such as wastewater treatment, desalination and power generation (Cath et al. 2006; Zhao et al. 2012), there are still a few technical challenges to be overcome. One of the most critical problems is the significantly lower water flux obtained in experimental study than that estimated with the osmotic pressure difference (Lee et al. 1981; Cath et al. 2006; McCutcheon et al. 2006). One main reason to this discrepancy is the presence of concentration polarization. Similar to pressure-driven membrane processes, there is concentration polarization at the solution-membrane interface (referred to as external concentration polarization or ECP) in the FO process due to the unequal convection and diffusion of solutes between the bulk solution and the membrane surface. The ECP in the FO process can be mitigated by improving the hydraulic condition near the membrane surface (Wang et al. 2010c). A unique and critical problem in FO as well as other osmotically driven membrane processes is the concentration polarization inside the membrane due to the opposite directions of water flux and solute flux. In the pressure-driven membrane processes, solute flux is in the same direction as water flux by convection. The special flow pattern in the FO process leads to a concentration polarization layer build-up within the membrane support, i.e., either (1) a concentrative internal concentration polarization (concentrative ICP) when the membrane active layer faces the draw solution (the AL-DS orientation), or (2) a dilutive internal concentration polarization (dilutive ICP) when the membrane active layer faces the feed solution (the AL-FS orientation) (McCutcheon and Elimelech 2006). The ICP causes drastic loss of effective osmotic pressure difference across the active layer and reduces the available water flux (Loeb et al. 1997; Gray et al. 2006; McCutcheon and Elimelech 2006; Tang et al. 2010).

Unlike the pressure-driven membrane processes where water flux and solute flux are approximately linear to the driving force, mass transport in FO process shows highly non-linear behavior with respect to membrane permeability and osmotic pressure difference due to the ICP phenomenon (Tang et al. 2010; Wei et al. 2013). ICP is one of the major obstacles to the FO membrane optimization and hence on the efficiency of the FO process. In this chapter, modeling of the FO process is introduced. The objective of this chapter is to provide a systematic insight into the FO mass transport, to understand the governing mechanisms over flux behavior on a quantitative basis, and finally to achieve clear concept about how to improve the efficiency the FO process. The contents include the basic concepts of FO process, mass transport in the FO process, flux modeling of single-skinned FO membranes and advance double-skinned FO membranes, modeling of salt accumulation in the osmotic membrane bioreactor, and modeling of a related pressure-retarded osmosis process. Additionally, the computational fluid dynamics approach is also introduced as a useful tool for FO modeling.

2.2 MASS TRANSPORT IN FO PROCESSES

When a high osmotic pressure draw solution and a lower osmotic pressure feed solution are separated by a semi-permeable membrane, dilution of the draw solution and concentration of the feed solution will take place simultaneously due to transfer of solvent under the influence of driving force created by the osmotic pressure difference ($\Delta\pi$) (Moody and Kessler 1976; Lee et al. 1981). An osmotic pressure difference is generated due to the chemical potential difference of solvent in the draw and feed solutions. With identical solutes in both solutions, the chemical potential of solvent in the low concentration solution is higher and thus drives the solvent transport into the high concentration solution until equal chemical potentials are reached, i.e., identical solute concentrations in both solutions, or $\Delta\pi$ is counterbalanced by a hydrostatic pressure difference, etc. (Mulder 1996). The physical meaning of osmotic pressure (π) can be interpreted as the pressure needed to prevent the transport of pure water through an ideal semi-permeable membrane into this solution (Mulder 1996).

A commonly used model for the mass transport across an FO membrane is the solution-diffusion model. In the FO process, the water flux (J_v) and solute flux (J_s) across the membrane can be calculated by (Lee et al. 1981):

$$J_v = A(\sigma\Delta\pi - \Delta P) \tag{2-1}$$

$$J_s = B\Delta C \tag{2-2}$$

where A and B are the membrane water permeability coefficient and solute permeability coefficient, respectively; σ is the reflection coefficient, which is usually assumed to be unity; ΔP is the applied hydraulic pressure. In the FO process, the hydraulic pressure applied on the system is zero ($\Delta P = 0$), and thus, water flux is equal to the product of membrane water permeability and osmotic pressure difference:

$$J_v = A(\pi_{draw} - \pi_{feed}) \tag{2-3}$$

where π_{draw} and π_{feed} are the osmotic pressures of draw solution and feed solution, respectively.

It should be noticed that the FO water flux represented by Eq. (2-3) is applicable only under ideal condition (Moody and Kessler 1976). To be specific, the membrane is assumed to be a symmetric thin active layer that is permeable to water but completely impermeable to solutes. Moreover, perfect mixing is provided at both sides of the membrane to maintain identical concentrations between the bulk phase and the solution-membrane interface, i.e., there is no concentration polarization. The osmotic pressure profile across the membrane is shown in Figure 2.1.

However, the experimental FO water flux is typically much lower than the product of the membrane water permeability and the osmotic pressure difference of the bulk solutions given by Eq. (2-3) (Lee et al. 1981; Cath et al. 2006; McCutcheon et al. 2006). One main reason is the concentration polarization phenomenon (Moody and Kessler 1976). Concentration polarization is a universal problem in all membrane processes for liquid separation (Sablani et al. 2001). With a driving force

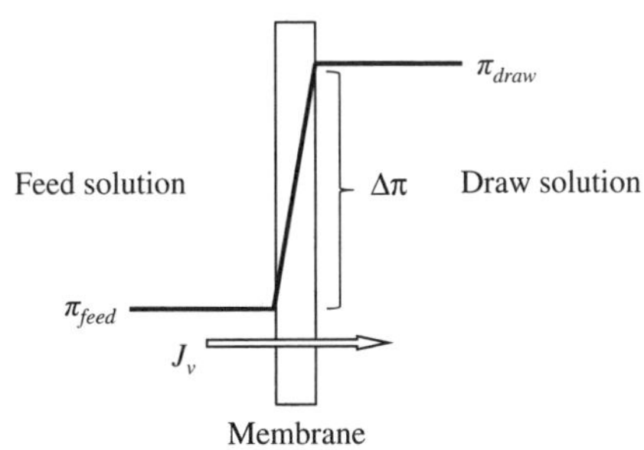

Figure 2.1. Osmotic pressure profile across the membrane in ideal FO processes
SOURCE: Reproduced from Cath et al. (2006) with permission from Elsevier

applied, i.e., hydraulic pressure for pressure-driven membrane processes or osmotic pressure difference for FO, water permeates through the membrane while the transport of feed solutes is hindered. The retained solutes will accumulate at the membrane surface and forms a boundary layer (Mulder 1996). This phenomenon of retained components accumulating and permeating components being depleted at the membrane surface is referred to as concentration polarization (Mulder 1996; Strathmann 1981). In the FO process, concentration polarization would exist externally on both sides of the membrane if solutes are present in both the feed and draw solutions. The accumulation of solutes at the feed side is referred to as concentrative external concentration polarization (concentrative ECP) as in pressure-driven membrane processes. Meanwhile, the solutes at the draw side are flushed away from membrane surface due to the convection flow of water permeate. The depletion of solutes at the draw side is referred to as dilutive external concentration polarization (dilutive ECP) (McCutcheon and Elimelech 2006).

Figure 2.2 illustrates the osmotic pressure profile when ECP exists on both sides of the symmetric membrane. The ECP increases the osmotic pressure at the feed side (for the concentrative ECP) and reduces the osmotic pressure at the draw side (for the dilutive ECP). As a result, the effective driving force $\Delta\pi_{eff}$ is reduced and leads to a lower water flux than that estimated with $\Delta\pi$ in the bulk phases shown by Equation 2-3. In the pressure-driven membrane processes, the influence of ECP on membrane performance has been well studied by prior studies (Strathmann 1981; Mulder 1996; Elimelech and Bhattacharjee 1998; Sablani et al. 2001), which may have a similar effect on the FO process. Nevertheless, it is found that the influence of ECP on the FO flux is relatively small, which could be mitigated by controlling the flux and mass transfer coefficient. Effective approaches to reduce ECP may include cross-flow velocity adjustment, using spacers, vibrations, ultrasound, etc. (Lee et al. 1981; Sablani et al. 2001; Cath et al. 2006).

Compared to the ECP, a more severe problem contributing to the low FO water flux is ICP. In the ideal case, perfect mixing is provided on the membrane surface to homogenize solutions and to mitigate the effect of concentration polarization. Under FO conditions, however, concentration polarization also takes place within the membrane. The semi-permeable membranes typically have

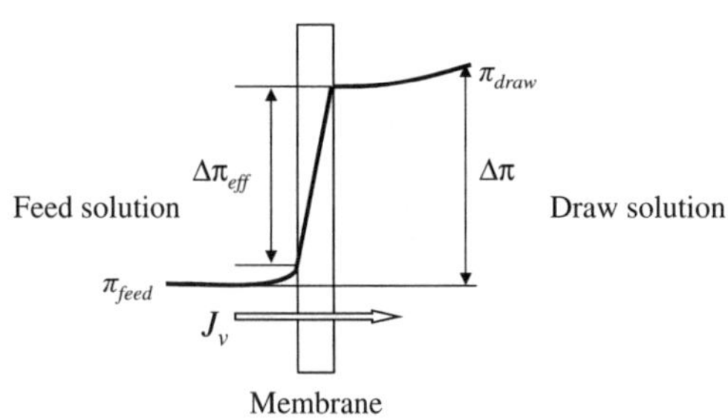

Figure 2.2. Osmotic pressure profile across the FO membrane in the presence of ECP
SOURCE: Adapted from McCutcheon and Elimelech (2006) with permission from Elsevier

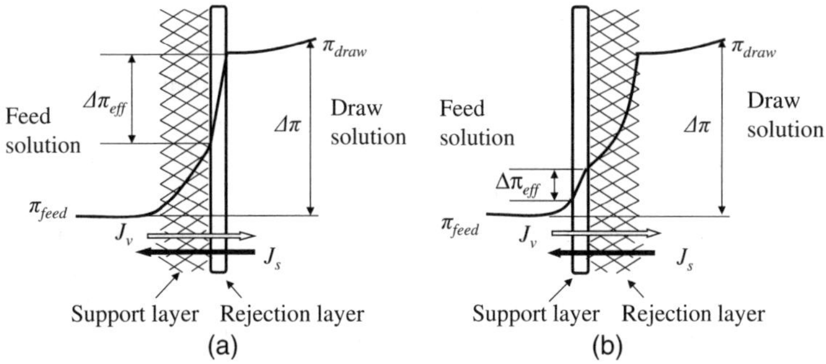

Figure 2.3. Osmotic pressure profiles across a FO membrane in the (a) AL-DS and (b) AL-FS orientations.
SOURCE: Adapted from Cath et al. (2006) with permission from Elsevier

a relatively thicker and porous support layer beneath the thin selective layer for sufficient handling strength (McCutcheon et al. 2005). This support layer tends to resist mass diffusion and thus accumulates a concentration polarization layer within it. This phenomenon is observed uniquely in osmotically driven membrane processes and is referred to as ICP (Mehta and Loeb 1978; Loeb et al. 1997; McCutcheon and Elimelech 2006).

In the FO process, the asymmetric semi-permeable membrane can be loaded with the active rejection layer facing either the draw solution (Figure 2.3(a)) or the feed solution (Figure 2.3(b)). In the AL-DS orientation, water permeates from the feed solution through the membrane rejection layer while the feed solutes transport into the porous support due to convection. As the feed solutes are retained by the semi-permeable rejection layer, a boundary layer is formed within the support. On the other hand, the diffusion of solutes back to the bulk phase will be hindered by the support layer. This phenomenon is referred to as concentrative ICP (McCutcheon et al. 2005; McCutcheon and Elimelech 2006). The ICP also takes place in another orientation. In the AL-FS orientation, water permeates through the rejection layer and dilutes the draw solution in the support layer. Draw solutes have to diffuse towards the rejection-layer-support interface to restore the osmotic driving force. However, the diffusion will be hindered by the support layer, leading to depletion of draw solutes near the rejection-layer-support interface, which is referred to as dilutive ICP (McCutcheon et al. 2005; McCutcheon and Elimelech 2006). Both the concentrative and dilutive ICP could lead to a severe reduction in the $\Delta\pi_{eff}$ across the rejection layer (i.e., $\Delta\pi_{eff} < \Delta\pi$ as in Figure 2.3). According to Eq. (2-3), the water flux should increase linearly with the osmotic pressure difference when the ECP is minimized. However, many studies showed a non-linear dependence of flux on $\Delta\pi$, i.e., the discrepancy between experimental and predicted flux appeared to increase at a higher osmotic driving force (McCutcheon et al. 2005; Gray et al. 2006; Tang et al. 2010; Xu et al. 2010; Chanukya et al. 2012). This is because the influence of ICP is enhanced at high water flux levels. The ICP phenomenon can be considered as hindered solute diffusion by the membrane support layer (Lee et al. 1981; Loeb et al. 1997; Loeb 2002; Gray et al. 2006; Tang et al. 2010).

The FO membranes are usually not ideally impermeable to solutes. The resulting reverse diffusion of draw solutes through the rejection layer can enhance the ICP by increasing the osmotic pressure of feed (in the AL-DS orientation) and reducing the osmotic pressure of draw (in the AL-FS orientation) near the interface between the support and rejection layer, and thus, further reduces $\Delta\pi_{eff}$ (Lee et al. 1981; Loeb et al. 1997; Tang et al. 2010). Even when pure water is used as the feed and the AL-DS orientation is adapted, which was expected to have insignificant ICP, concentrative ICP still can take place due to reverse diffusion of draw solutes. Therefore, solute diffusion is also considered as one of the most important problems in the osmotically driven membrane processes.

Studies on the ICP have proved that ICP contributes to dramatic FO water flux reduction, especially under conditions of high flux and/or high draw solution concentration (Loeb et al. 1997; McCutcheon and Elimelech 2006; McCutcheon and Elimelech 2007; Tan and Ng 2008; Tang et al. 2010; Wei et al. 2011a). Moreover, unlike ECP that can be mitigated by optimizing external hydraulic conditions, ICP is less affected by external shear or turbulence due to the protection of the support layer (Lee et al. 1981; Qin et al. 2009).

2.3 SIMULATION OF INFLUENCE OF MEMBRANE PARAMETERS ON FO WATER FLUX

The water flux in the FO process can be predicted by a classic ICP model (Lee et al. 1981; Loeb et al. 1997). The details of modeling development can refer to Appendix A. In this model, water flux in the two membrane orientations is:

Concentrative ICP (AL-DS):

$$J_v = K_m \ln\left(\frac{A\pi_{draw} - J_v + B}{A\pi_{feed} + B}\right) \tag{2-4}$$

Dilutive ICP (AL-FS):

$$J_v = K_m \ln\left(\frac{A\pi_{draw} + B}{A\pi_{feed} + J_v + B}\right) \tag{2-5}$$

where A is water permeability coefficient; B is solute permeability coefficient; and K_m is the mass transfer coefficient, defined as:

$$K_m = \frac{D}{S} \tag{2-6}$$

where D is the solute diffusion coefficient; S is the structural parameter of membrane, defined as:

$$S = \frac{\tau l}{\varepsilon} \tag{2-7}$$

where τ, l and ε are the tortuosity, the actual thickness and porosity of the porous support layer, respectively. The structural parameter is an important property of FO membrane, which is analogous to the boundary layer thickness of ECP in the pressure-driven membrane processes. The FO solute flux can be determined by (Tang et al. 2010):

$$J_s = \frac{B}{A \cdot \beta R_g T} J_v \tag{2-8}$$

where β is the van't Hoff coefficient; R_g is the universal gas constant; T is the absolute temperature.

The ICP model shows that the FO water flux could be controlled by the membrane properties, i.e., structural parameter (S), water permeability (A) and solute permeability (B). One of the key solutions to overcome the ICP problem is using an efficient membrane that is specially tailored for osmotically driven membrane processes. Up to the present there have been a handful of FO membranes reported (Zhao et al. 2012), however, the experimental FO water flux is still far lower than that expected with the $\Delta\pi$ applied. More efficient membranes have yet to be developed for the practical application of FO technology. On the other hand, the water flux in the FO process shows a highly non-linear relationship with the membrane properties and the bulk osmotic pressure difference (McCutcheon and Elimelech 2006; Tang et al. 2010; Wei et al. 2011b; Wei et al. 2013). Simulations using the ICP model can help evaluate the effectiveness of parameter optimization on the FO water flux, providing a clear insight into membrane customization.

In this section, simulation was performed using the ICP model to observe the relationship between FO water flux and the membrane properties (S, A and B values), and to obtain implications for membrane optimization. Reference values of the modeling parameters were determined according to current membrane fabrication techniques, as listed in Table 2.1.

2.3.1 Influence of Structural Parameter

The predicted FO water flux with varying membrane structural parameters is plotted in Figure 2.4. The simulation result shows a pronounced influence of structural parameter on the FO water flux. A water flux of 179 L/m^2-h can be achieved in both the AL-DS and AL-FS orientations if the membrane is support-free ($S = 0$). This approximates to the water flux under equivalent hydraulic pressure as the applied osmotic pressure difference. Increasing the S value results

Table 2.1. Reference values of modeling parameters

Parameter	*Reference value*
Solute diffusion coefficient, D	$1.3 \times 10^{-9} m^2/s$
Osmotic pressure of draw solution, π_{draw}	99.9 bar for 2 M NaCl
Osmotic pressure of feed solution, π_{feed}	0.45 bar for 10 mM NaCl
Water permeability coefficient, A	5.0×10^{-12} m/s Pa
Solute permeability coefficient, B	1.0×10^{-7} m/s
Structural parameter, S	0.7 mm

in a higher resistance to mass transport. The diffusion of solute from membrane support toward bulk phase of feed solution (in AL-DS) or the replenishment of draw solute in support layer (in AL-FS) will be retarded, causing a significant loss of osmotic driving force and correspondingly water flux reduction. The FO water flux is sensitive to S value variation in an exponential manner (see Eqs. (2-4) and (2-5)). In particular, the dilution of draw solution will cause a severer osmotic pressure difference loss than the concentration of feed solution, when relatively diluted feed solution is used. Thus a sharp decrease of water flux with larger S value is observed in the AL-FS orientation.

According to the definition of structural parameter (Eq. (2-7)), reducing the support layer thickness, tortuosity and increasing the support porosity can have a smaller S value. For example, the support layer of a conventional RO membrane typically has dense and sponge-like pore structure. As a result, such a membrane usually achieves poor FO performance, despite of its superior water permeability and salt rejection in pressure-driven mode (McCutcheon et al. 2005; Tiraferri et al. 2011; Wei et al. 2011b). The commercial FO membranes as well as recently reported FO membranes have been greatly improved in term of support layer

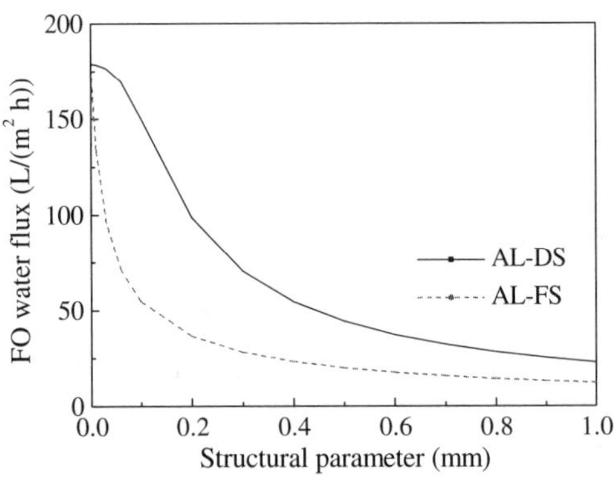

Figure 2.4. Predicted FO water flux as a function of membrane structural parameter. Simulation conditions: $A = 5.0 \times 10^{-12}$ m/s Pa, $B = 1.0 \times 10^{-7}$ m/s, $D = 1.3 \times 10^{-9}$ m²/s, $\pi_{draw} = 99.9$ bar, and $\pi_{feed} = 0.45$ bar

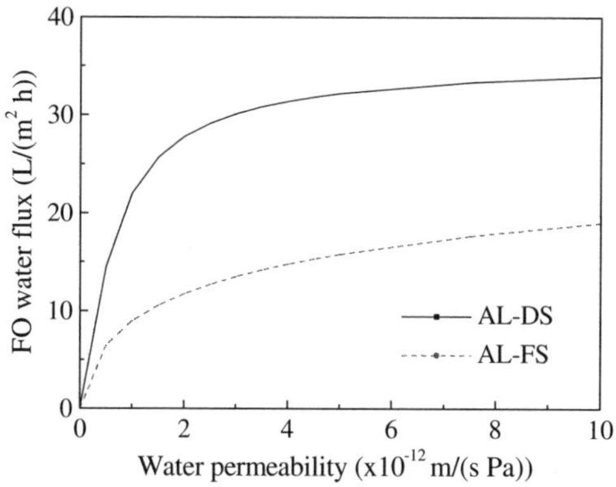

Figure 2.5. Predicted FO water flux as a function of membrane water permeability. Simulation conditions: $B = 1.0 \times 10^{-7}$ *m/s,* $S = 0.7$ *mm,* $D = 1.3 \times 10^{-9}$ m^2/s, $\pi_{draw} = 99.9$ *bar, and* $\pi_{feed} = 0.45$ *bar*

thickness and pore structure (Chou et al. 2010; Wang et al. 2010b; Yip et al. 2010; Saren et al. 2011; Wei et al. 2011b). Enhanced FO water flux has been observed with these membranes. Yet further improvement of the support layer is still necessary in future FO membrane fabrication, of which the main concerns would be reduction of *S* value to as low as possible while reinforcing the mechanical strength of membrane for module assembly.

2.3.2 Influence of Water Permeability

The effect of membrane water permeability on FO water flux is shown in Figure 2.5. When the *A* value is small, J_v increases almost proportionally to *A* value when other parameters are fixed. According to Eqs. (2-4) and (2-5), ICP effect is exponentially enhanced with water flux. Thus ICP at a relatively low flux level is likely less important, whereas the membrane frictional loss tends to be more critical. In this case, a high water permeability coefficient is more important to increase the FO water flux. However, less improvement of water flux can be achieved by further increasing the *A* value after a certain region. Again, this can be explained by the enhanced ICP effect. Any attempt to increase the water flux will be counter-balanced by the corresponding severer ICP.

On the other hand, both RO and FO literature have shown a strong trade-off between the water permeability and solute rejection of membranes (Tang et al. 2009; Geise et al. 2011; Wei et al. 2011a; Yip et al. 2011). Membranes with higher water permeability tend to have a lower salt rejection. This trade-off imposes a critical constraint for optimizing FO membrane performance. While a high permeability membrane tends to have lower hydraulic frictional loss, its lower solute rejection could induce a higher solute flux and causes severe ICP. The net effect of this two opposed mechanisms on water flux depends on the operating conditions. Generally, the higher water flux level is, the more significant impact of

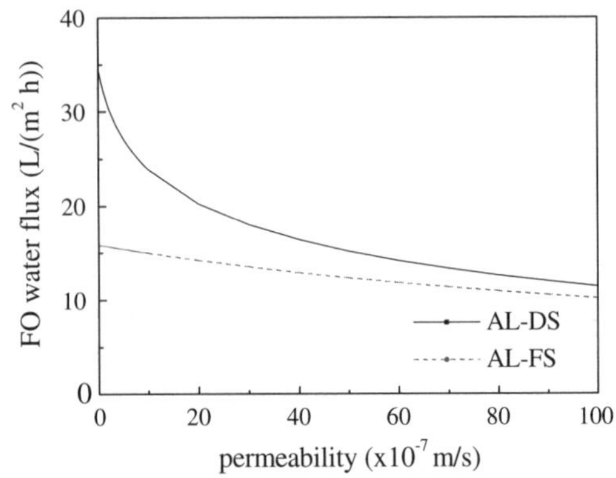

Figure 2.6. Predicted FO water flux as a function of solute permeability. Simulation conditions: $A = 5.0 \times 10^{-12}$ *m/s Pa,* $S = 0.7$ *mm,* $D = 1.3 \times 10^{-9}$ m^2/s, $\pi_{draw} = 99.9$ *bar, and* $\pi_{feed} = 0.45$ *bar*

ICP will be. According to Figure 2.5, FO membranes with A values nearly reaching the plateau region are not suggested to further improve the permeability. In this region, membrane frictional resistance is no longer a governing mechanism, increasing the A value has a minor benefit to water flux but have the negative effect of lower salt rejection.

2.3.3 Influence of Salt Permeability

The effect of salt permeability on FO water flux is shown in Figure 2.6. Generally, lower water flux is achieved with higher membrane salt permeability, i.e., by using a lower rejection membrane. In the simulation of Figure 2.6, the membrane water permeability is fixed. The increasing salt permeability means a higher B/A ratio (i.e., lower membrane selectivity). According to Eq. 2-8, the salt flux/water flux ratio (J_s/J_v) is directly determined by the membrane selectivity (Tang et al. 2010). A larger B/A value induces severe reverse diffusion of draw solute and leads to lower water flux, which is referred to as a J_s-induced-ICP mechanism. When the FO membrane has a larger structural parameter, the effect of J_s-induced-ICP will be further enhanced.

In comparison, the FO water flux is more sensitive to the change of B value in the AL-DS orientation. In this orientation, ICP in the feed side is contributed by accumulation of feed solute and reverse diffusion of draw solute. With a low concentration feed solution ($\pi_{feed} \ll B/A$), the influence of J_s on salt accumulation will be even more prominent. In the AL-FS orientation, the effect of B value is less pronounced, possibly due to the self-compensation mechanism in this orientation and the dominant dilution effect by the permeate. Optimization of the B value would be more effective when the B/A value is a governing parameter to water flux. In addition, modification of the membrane rejection layer may result in lower the B value but simultaneously lower the A value. Thus optimal combination of

A and *B* values should be chosen according to membrane properties and operating conditions.

2.3.4 Implications for FO Membrane Optimization

The simulating effect of the membrane properties on FO water flux suggests that an efficient FO membrane should have small structural parameter, high water permeability as well as low solute permeability (McCutcheon and Elimelech 2006; Hancock and Cath 2009; Wei et al. 2011b). In view of the constraint by membrane fabrication technology, however, it is challenging to improve the membranes in all aspects due to the trade-off of these parameters. Additionally, the relationships between FO water flux and these membrane properties are highly non-linear as a result of the competing effect of different mechanisms. For example, a highly water-permeable membrane has less membrane frictional loss but the correspondingly high solute permeability would increase the risk of J_s-induced-ICP. Although using a draw solute with larger molecular weight could reduce the solute permeability but its lower diffusion coefficient will reduce the mass transfer coefficient in the support layer and enhance ICP. Therefore, the governing mechanisms should be identified before the selection of FO membranes and draw solution to achieve the optimal FO performance in an individual application.

The competing mechanisms governing the FO flux performance are summarized in Table 2.2. In general, the frictional loss due to the hydraulic resistance of membrane (M_R) and the J_s-induced ICP ($M_{ICP\text{-}Js}$) occurs in both the AL-DS and the AL-FS orientations. The concentrative ICP contributed by feed solute ($M_{ICP\text{-}feed}$) is only applicable to the AL-DS orientation, whereas the dilutive ICP of draw solute ($M_{ICP\text{-}draw}$) is only applicable to the AL-FS orientation. These mechanisms are related to properties of the membrane, draw solution and feed solution. In the potential FO applications such as FO desalination and wastewater treatment, the properties of feed solution are likely to be given. Thus, the improvement of FO performance calls for careful selection of membranes, as well as draw solution if allowed in some cases.

A set of systematic criteria for selection of membrane and draw solution in the FO process is proposed based on the governing mechanisms listed in Table 2.2. First of all, a critical property of high performance FO membranes is small *S* value to reduce the ICP propensity (see Section 2.3.1 and Table 2.2). As far as the rejection layer is concerned, high selectivity to draw solute ($B_{draw\,solute}/A \ll \pi_{feed}$) is of critical importance as well (Xiao et al. 2011), in order to minimize the loss of osmotic driving force due to $M_{ICP\text{-}J_s}$. This can be done by selecting a suitable combination of membrane and draw solute. For example, a membrane with a RO-type rejection layer is required for the NaCl-based process whereas a membrane with NF-type rejection layer can be used for the $MgCl_2$ based process. In applications that allow certain freedom of draw solute type, solutes with high diffusion coefficient (D_{draw}) are preferred as long as $M_{ICP\text{-}J_s}$ does not dominate the flux performance. Additional criteria of the draw solutes also include high solubility, high osmotic pressure, no toxicity, acceptable cost of dosing, available

Table 2.2. Major mechanisms and important parameters governing the water flux of FO membranes

Orientation	Major mechanisms	Important parameters: Membrane related	Important parameters: Draw solution related	Important parameters: Feed solution related
AL-DS (higher flux, less fouling resistant)	Frictional loss (M_R)	A	–	–
	J_s-induced ICP ($M_{ICP\text{-}Js}$)	A, $B_{draw\,solute}$, S	$B_{draw\,solute}$, D_{draw}	–
	Concentrative ICP of feed solutes ($M_{ICP\text{-}feed}$)	$B_{feed\,solute}$, S	–	$B_{feed\,solute}$, π_{feed}, D_{feed}
AL-FS (lower flux, more fouling resistant)	Frictional loss (M_R)	A	–	–
	J_s-induced ICP ($M_{ICP\text{-}Js}$)	A, $B_{draw\,solute}$, S	$B_{draw\,solute}$, D_{draw}	–
	Dilutive ICP of draw solutes ($M_{ICP\text{-}draw}$)	S	D_{draw}	–

technique of draw solution regeneration (Achilli et al. 2010), low potential of interfering membrane fouling (Zou et al. 2011; She et al. 2012a), etc. Furthermore, the membrane should have a water permeability A as high as possible so long as the resultant membrane selectivity satisfies $B_{draw\,solute}/A \ll \pi_{feed}$. The propensity to fouling (particularly the AL-DS orientation) and required feed solution rejection should be considered as well.

2.4 MODELING OF DOUBLE-SKINNED FO MEMBRANES

The presence of ICP in the FO process causes drastic reduction of effective osmotic driving force and obtainable water flux. Prior experimental and theoretical studies demonstrated that ICP exists in both the AL-DS and AL-FS orientations. In particular, significantly lower water flux is usually observed in the AL-FS orientation due to the severe dilution effect of draw solution within the support layer (Gray et al. 2006; McCutcheon and Elimelech 2006; Tang et al. 2010). On the other hand, operation in the AL-DS orientation may not be a superior choice as it directly exposes the porous surface of FO membrane towards feed solution. When feed solution contains foulant, the membrane is susceptible to fouling (Zou et al.

2011). Fouling in the AL-DS orientation not only increases the hydraulic resistance of membrane as that in the pressure-driven membrane process, but also leads to a unique problem referred to as pore-clogging-enhanced-ICP as the foulant entering the support layer can reduce support layer porosity and increase the tortuosity of water flow path (Tang et al. 2010). By comparison, the AL-FS orientation is less prone to fouling-induced flux reduction because of the protection by the dense active layer and the self-compensation flux behavior, i.e., the tendency of flux reduction can be compensated by a correspondingly reduced ICP degree. Thus, there is a dilemma faced by the conventional asymmetric FO membranes whether to choose the AL-FS orientation with relatively stable but a lower water flux level due to severe ICP, or the AL-DS orientation with high initial water flux but potential flux reduction caused by fouling (Tang et al. 2011).

To solve the above dilemma, an innovative concept of double-skinned FO membrane is proposed. This type of membranes contains a dense skin layer facing the draw solution for solute rejection and a second skin layer facing the feed solution to protect the porous support from fouling. With this novel design, the membrane is able to operate in an orientation similar to AL-DS with minimized ICP as well as high foulant resistance. A handful of such double-skinned FO membranes have been reported recently, e.g., double-skinned layer-by-layer assembly FO membranes (Qi et al. 2012), double-skinned thin film composite FO membranes (Fang et al. 2012), double-skinned asymmetric cellulose acetate membranes (Wang et al. 2010a; Su et al. 2012), etc. As increasing attention is paid to the FO fouling issue, the double-skinned design is expected to have a larger market in the future practical application. In this section, the concepts and modeling of double-skinned FO membranes are elucidated to provide knowledge about mass transport in these unique membranes as well as implications for membrane optimization.

2.4.1 Modeling Double-Skinned FO Membranes

The design of a double-skinned FO membrane is illustrated in Figure 2.7. The two skin layers of double-skinned membrane can be customized to have different water permeability and solute permeability. A porous support layer is formed between the two skin layers. An analytical model of double-skinned FO membranes has been developed by Tang and coworkers (Tang et al. 2011) to understand the mass transport through the membrane. Derivation of the model can be referred to Appendix B. Water flux of a double-skinned FO membrane can be determined by:

$$J_v = K_m \ln\left(\frac{A\pi_{draw} - \frac{B}{B_{draw}} J_v + B}{A\pi_{feed} + \frac{B}{B_{feed}} J_v + B}\right) \tag{2-9}$$

where A and B are the overall water permeability coefficient and overall solute permeability coefficient of the membrane, respectively; B_{draw} and B_{feed} are the solute permeability coefficients of the draw skin and feed skin, respectively.

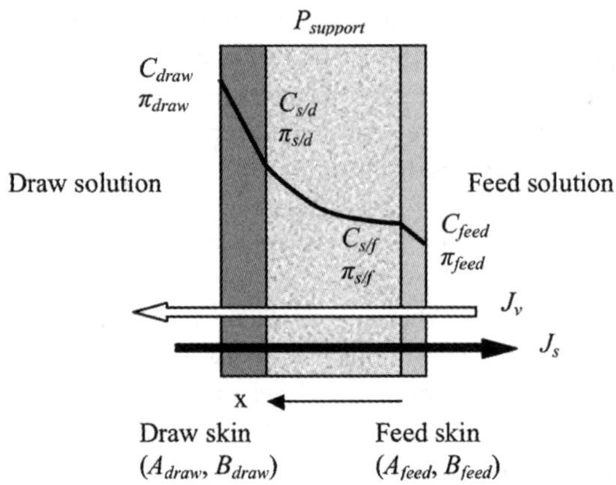

Figure 2.7. Schematic drawing of a double-skinned FO membrane
SOURCE: Tang et al. (2011) with permission from Elsevier

The A and B values can be determined by pressure-driven filtration tests. The contribution of feed skin and draw skin to overall membrane permeability is in a resistance-in-series mode:

$$\frac{1}{A} = \frac{1}{A_{draw}} + \frac{1}{A_{feed}} \tag{2-10}$$

$$\frac{1}{B} = \frac{1}{B_{draw}} + \frac{1}{B_{feed}} \tag{2-11}$$

where A_{draw} and A_{feed} are the water permeability coefficients of the draw skin and feed skin, respectively.

Eq. (2-9) is not only applicable to double-skinned membranes but also usable for single-skinned membranes. To be specific, when the feed skin of the double-skinned membrane is completely permeable to water and solute ($A_{feed} \rightarrow \infty$ and $B_{feed} \rightarrow \infty$), Eq. (2-9) can be simplified to:

$$J_v = K_m \ln\left(\frac{A_{draw}\pi_{draw} - J_v + B_{draw}}{A_{draw}\pi_{feed} + B_{draw}}\right) \tag{2-12}$$

On the other hand, if the draw skin of double-skinned membrane is completely permeable to water and solute ($A_{draw} \rightarrow \infty$ and $B_{draw} \rightarrow \infty$), Eq. (2-9) can be simplified to:

$$J_v = K_m \ln\left(\frac{A_{feed}\pi_{draw} + B_{feed}}{A_{feed}\pi_{feed} + J_v + B_{feed}}\right) \tag{2-13}$$

It shows that Eqs. (2-12) and (2-13) are actually the governing equations for single-skinned membrane in the AL-DS and AL-FS orientations, respectively. Therefore, Eq. (2-9) can work as a general equation for both single and double skinned FO membranes.

The solute flux of double-skinned membrane can be determined by:

$$J_v\left(\frac{1}{A_{draw}}+\frac{1}{A_{feed}}\right)=\beta R_g T\cdot J_s\left(\frac{1}{B_{draw}}+\frac{1}{B_{feed}}\right) \tag{2-14}$$

$$\frac{J_s}{J_v}=\frac{B}{A\cdot\beta R_g T} \tag{2-15}$$

According to Eq. (2-15), the ratio of J_s/J_v is determined by the membrane selectivity (B/A) and solute properties (β) for the double-skinned membranes, which is similar to that for single-skinned membranes. This ratio is not affected by the membrane orientation or solution concentration.

A unique term in the double-skinned FO membrane model is the hydraulic pressure in the support layer ($P_{support}$). It is generated due to the unbalanced water/solute transport through the feed and draw skins. This pressure is nearly uniform within the porous support layer due to the relatively low hydraulic resistance of the support layer, and can be estimated by:

$$P_{support}=\left(1-\frac{B/A}{B_{draw}/A_{draw}}\right)\cdot\frac{J_v}{A_{draw}} \tag{2-16}$$

or

$$P_{support}=\left(\frac{B/A}{B_{feed}/A_{feed}}-1\right)\cdot\frac{J_v}{A_{feed}} \tag{2-17}$$

It is suggested by Eqs. (2-16) and (2-17) that a hydraulic pressure will be developed within the double-skinned membrane when selectivity of the draw and feed skins are different. From the membrane fabrication point of view, proper combination of B_{draw}/A_{draw} and B_{feed}/A_{feed} values should be designed to avoid pressure accumulation in the support layer and to prevent the mechanical damage of membrane.

2.4.2 Implications to Double-Skinned FO Membrane Design

With the double-skinned FO membrane model, one can better understand the effects of feed skin and draw skin properties on the FO flux performance. The main function of feed skin is to prevent foulant entering the porous support and to

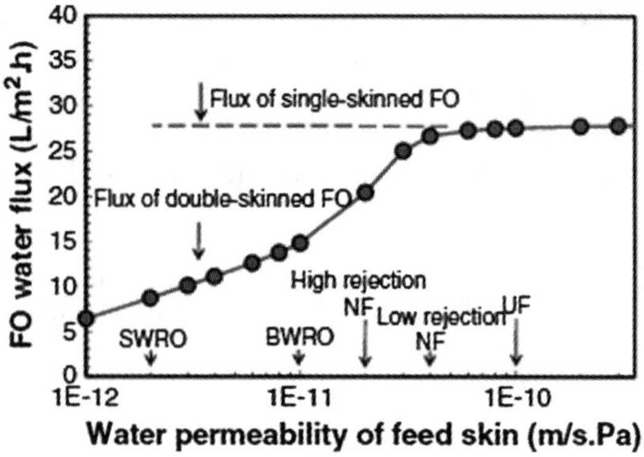

Figure 2.8. Effects of feed skin properties on the FO water flux of double-skinned membranes. Simulation conditions: BWRO-type draw skin ($A_{draw} = 1.0 \times 10^{-11}$ m/sPa, $B_{draw} = 3.0 \times 10^{-7}$ m/s), $K_m = 2.0 \times 10^{-6}$ m/s, 1 M NaCl draw solution and 10 mM NaCl feed solution

SOURCE: Adapted from Tang et al. (2011) with permission from Elsevier

achieve a long-term stable water flux. However, the feed skin would also contribute to the overall membrane permeability in a resistance-in-series mode according to Eq. (2-10). In a previous simulation study (Tang et al. 2011), effect of the feed skin water permeability on FO water flux was investigated. With a brackish-water-RO (BWRO)-type rejection layer as the draw skin and NaCl draw solutions, higher FO water flux ($J_{v,double}$) can be achieved by increasing the water permeability of feed skin (Figure 2.8). For example, simulation results with a seawater-RO (SWRO)-type feed skin shows that the corresponding overall membrane water permeability is only 16.7% as that of the BWRO-type draw skin, and thus results in a significantly lower water flux compared to that of the single-skinned membrane with a BWRO-type draw skin ($J_{v,single}$). A less permeable feed skin ($A_{feed} < A_{draw}$) will dominate the overall hydraulic resistance of membrane. Modifying the feed skin from SWRO-type to BWRO-type (i.e., $A_{feed} = A_{draw}$) can effectively increase the overall membrane permeability to 50% as that of the BWRO-type draw skin. Nevertheless, there is still 50% of FO water flux loss compared to $J_{v,single}$. Moreover, the high solute rejection feed skin will lead to a higher $C_{s/f}$ and enhance ICP. Therefore, a more permeable feed skin ($A_{feed} \gg A_{draw}$) would be preferred. According to the simulation results (Figure 2.8), membrane with a low-rejection-NF (NFlr)-type skin layer can have a $J_{v,double}/J_{v,single}$ ratio up to 95%, which means that this feed skin has only marginal contribution to the overall membrane resistance. Further modification of the skin layer to ultrafiltration (UF)-type achieves relatively minor flux enhancement ($J_{v,double}/J_{v,single} \approx 99\%$). Additionally, a non-porous NFlr-type skin is superior over the porous UF-type skin layer with respect to anti-fouling performance.

The draw skin of a double-skinned FO membrane works as a main barrier layer to draw solute while allows water passage. According to a prior simulation

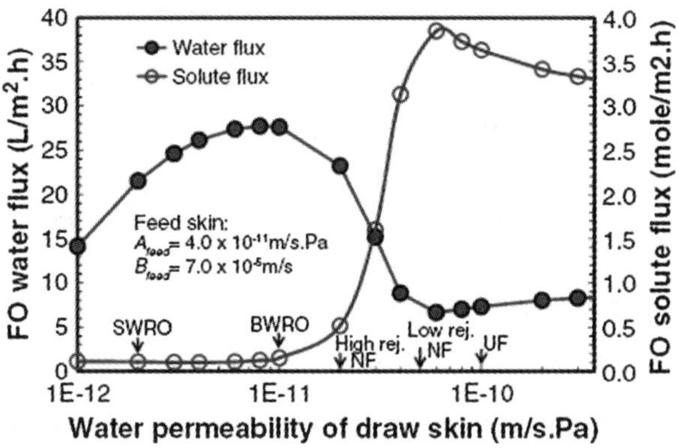

Figure 2.9. Effects of draw skin properties on the FO water flux and solute flux of double-skinned membranes. Simulation conditions: low rejection NF-type feed skin ($A_{feed} = 4.0 \times 10^{-11}$ m/s Pa, $B_{feed} = 7.0 \times 10^{-5}$ m/s), $K_m = 2.0 \times 10^{-6}$ m/s, 1 M NaCl draw solution and 10 mM NaCl feed solution

SOURCE: Adapted from Tang et al. (2011) with permission from Elsevier

work (Tang et al. 2011), draw skin with superior solute rejection (SWRO-type) may dramatically increase the membrane resistance as a result of the permeability-selectivity trade-off. On the other hand, a highly permeable draw skin may enhance the overall membrane solute permeability, which results in high solute flux and severe J_s-induced ICP. In both cases, drastic reduction of FO water flux is exhibited. A balance between draw skin water permeability and selectivity is necessary to achieve high water flux with a reasonable solute flux. For example, maximum FO water flux is shown by membrane with a BWRO-type draw skin when low rejection NF-type feed skin and NaCl solutions are used in the simulation (Figure 2.9).

2.5 MODELING OF SALT ACCUMULATION IN FO PROCESSES

Osmotic membrane bioreactor (OMBR) is an innovative water reclamation technology that combines biological degradation and osmotically driven membrane separation process (Cornelissen et al. 2008; Achilli et al. 2009b; Qin et al. 2010; Cornelissen et al. 2011; Lay et al. 2011; Yap et al. 2012). Compared to traditional MBR technologies based on microfiltration (MF) or UF separation, OMBR can have higher contaminant removal efficiency due to the high-rejection FO membranes (Achilli et al. 2009b; Alturki et al. 2012; Xie et al. 2012). In addition, potentially lower fouling tendency and correspondingly higher membrane cleaning efficiency have been reported for OMBR, which might be attributed to its mild flux condition and special membrane characteristics (Cornelissen et al. 2008; Achilli et al. 2009b; Lay et al. 2010, 2012).

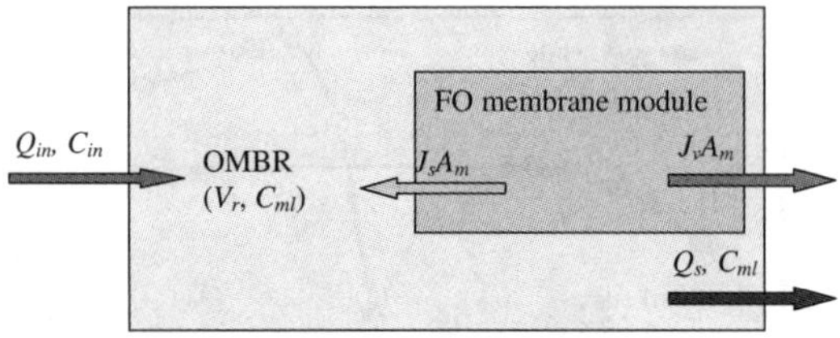

Figure 2.10. Mass balance in an OMBR
SOURCE: Xiao et al. (2011) with permission from Elsevier

A critical problem in the operation of OMBR is salt accumulation in the bioreactor, which is contributed by two factors: (1) influent solute retained by the FO membrane, and (2) reverse diffusion of draw solute into the reactor (Xiao et al. 2011). Particularly, reverse diffusion of the draw solute is a unique problem in OMBR, which does not present in conventional pressure-driven MBR. It can significantly enhance the salinity in the reactor if a low rejection membrane is used. Salt accumulation in bioreactor reduces the effective osmotic driving force and results in decreasing water flux. Moreover, increasing salinity could inhibit and even have toxicity effect on the microbial communities, and thus deteriorate the biological activity (Alturki et al. 2012).

Salt accumulation in an OMBR (Figure 2.10) can be estimated according to the mass balance of salt and water in the reactor. Details of modeling development can be found in Appendix C. At steady state, salt concentration (C_{ml}) and osmotic pressure (π_{ml}) of the mixed liquor in an OMBR are:

$$C_{ml}^{steady} = \frac{\theta_{SRT}}{\theta_{HRT}} C_{in} + \left(\frac{\theta_{SRT}}{\theta_{HRT}} - 1 \right) \frac{B}{A\beta R_g T} \tag{2-18}$$

$$\pi_{ml}^{steady} = \frac{\theta_{SRT}}{\theta_{HRT}} \pi_{in} + \left(\frac{\theta_{SRT}}{\theta_{HRT}} - 1 \right) \frac{B}{A} \tag{2-19}$$

where θ_{HRT} and θ_{SRT} are the hydraulic retention time and sludge retention time, respectively; C_{in} and π_{in} are the solute concentration and osmotic pressure of influent sewage, respectively. Water flux (J_v) of OMBR at steady state can be estimated by:

$$J_v^{steady} = K_m \left(\ln \left(\frac{\pi_{draw}}{\pi_{in} + B/A} \right) - \ln \left(\frac{\theta_{SRT}}{\theta_{HRT}} \right) \right) \tag{2-20}$$

Eqs. (2-18)–(2-20) illustrate the contribution of influent salt and draw solute reverse diffusion to salt accumulation and water flux in OMBR. OMBR performance is governed by the operating conditions ($\theta_{SRT}/\theta_{HRT}$ ratio, concentrations

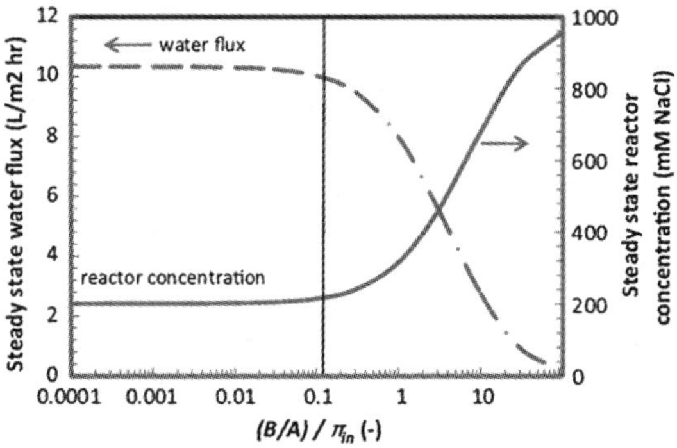

Figure 2.11. Influence of membrane selectivity (B/A) on solute concentration and water flux of an OMBR. Simulation conditions: 10 mM NaCl initial feed concentration, 1 M NaCl draw solution, initial hydraulic retention time 8 hours, and sludge retention time 10 days in the AL-FS orientation; A and B values were chosen to keep the initial FO water flux of 15.3 L/m²-h
SOURCE: Adapted from Xiao et al. (2011) with permission from Elsevier

of influent wastewater and draw solution), as well as the membrane properties (mass diffusion coefficient and membrane selectivity). This model provides important implications for OMBR optimization. Firstly, within the operating cost limit adjusting the $\theta_{SRT}/\theta_{HRT}$ ratio can effectively control the salt accumulation problem and available water flux. Secondly, low membrane selectivity can cause severe draw solute leakage into the reactor, which not only leads to water flux reduction but also has potentially adverse effects on the biological activity. In this case, a more selective membrane is required to minimize the salt accumulation attributed to the reverse diffusion. However, overly increasing the membrane selectivity may have minor effect on salt accumulation when the retention of influent solute plays a dominant role (i.e., $B/A \ll \pi_{in}$, Figure 2.11) (Xiao et al. 2011). It is suggested by prior simulation work (Xiao et al. 2011) that a reasonable membrane selectivity should be chosen at $B/A \ll 0.1\pi_{in}$. Further reducing the B/A ratio might lead to unnecessarily water flux reduction due to compromised membrane water permeability.

2.6 MODELING OF PRESSURE-RETARDED OSMOSIS PROCESSES

Owing to the recurring surge of FO study, another related osmotically driven membrane process referred to as pressure-retarded osmosis (PRO) has also become increasingly popular in the past few years. In the PRO process, feed solution and draw solution are placed against opposite side of a semi-permeable membrane. Different from the FO process, the draw solution is pressurized in the PRO process, i.e. $\Delta P > 0$ in Eq. 2-1. As water permeates into the draw solution, the diluted seawater will be depressurized by a hydroturbine to harvest the osmotic power. With the PRO technology, the widely available osmotic energy sources (e.g., brackish water, seawater and RO brine) can be utilized to produce electric power (Aaberg 2003; Gerstandt et al. 2008; Skilhagen et al. 2008Achilli et al. 2009a; Achilli and Childress 2010; Patel 2010; She et al. 2012b).

In order to prevent the rejection layer from mechanical damage, the membrane is typically loaded with the rejection layer facing the pressurized draw solution (the AL-DS orientation) in the PRO process. The modeling of PRO process has been conducted to evaluate the potential power production efficiency (Yip and Elimelech 2011; Yip et al. 2011; She et al. 2012b). In a model developed by Yip and coworkers (Yip and Elimelech 2011; Yip et al. 2011), water flux in the PRO process can be determined by:

$$J_v = A\left(\frac{\pi_{draw}\exp(-\frac{J_v}{k}) - \pi_{feed}\exp(\frac{J_v}{K_m})}{1+\frac{B}{J_v}(\exp(\frac{J_v}{K_m}) - \exp(-\frac{J_v}{k}))} - \Delta P\right) \tag{2-21}$$

where k is the mass transfer coefficient of the dilutive ECP boundary layer. The power density in the PRO process can be determined by (Yip et al. 2011; She et al. 2012b):

$$W = J_v \Delta P \tag{2-22}$$

Three influencing phenomena are considered in Eq. (2-21), including (1) the dilutive ECP represented by a term of $\exp(-J_v/k)$, (2) the concentrative ICP represented by a term of $\exp(J_v/K_m)$, and (3) the reverse diffusion of draw solute represented by a term $(1 + B/J_v \exp(J_v/K_m) - \exp(-J_v/k))$. In a prior simulation work based on Eq. (2-21) and (2-22) (Figure 2.12), the effect of each factor was predicted by diminishing the corresponding term (Yip and Elimelech 2011). The results demonstrated a dominant influence of ICP on PRO water flux. Typically, the range of the ECP factor is 0–1, whereas the ICP factor can range from 1 to infinite. Thus, the influence of ICP is more pronounced than that of the ECP. Additionally, although the PRO process has a less strict requirement on solute rejection compared to the FO process, decent membrane selectivity is still preferred to minimize the loss of osmotic driving force. In all of the simulating scenarios, peak power density is achieved when the hydraulic pressure applied on draw side is about half of the osmotic pressure difference ($\Delta P = \Delta\pi/2$).

2.7 COMPUTATIONAL FLUID DYNAMICS MODELING OF FO PROCESSES

Computational fluid dynamics (CFD) is a branch of fluid mechanics that uses numerical methods and algorithms to analyze the fluid flow problems. This method has been employed to investigate the concentration polarization problems

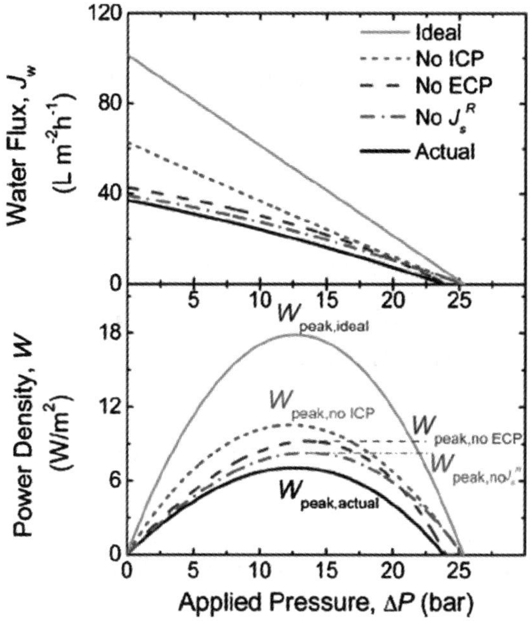

Figure 2.12 Predicted PRO water flux and power density as a function of applied hydraulic pressure. Simulation conditions: $\pi_{draw} = 26.14$ *bar,* $\pi_{feed} = 0.789$ *bar,* $A = 1.11 \times 10^{-11}$ *m/s Pa,* $B = 2.36 \times 10^{-7}$ *m/s,* $S = 0.35$ *mm,* $k = 3.85 \times 10^{-5}$ *m/s*
SOURCE: Adapted from Yip and Elimelech (2011) with permission from American Chemical Society

(both ICP and ECP) in the FO process (Gruber et al. 2011, 2012). In the FO CFD modeling, the fluid flow through a membrane is governed by conservation of mass and momentum as well as the convection-diffusion for the solute mass fraction, which can be described by a weakly compressible formulation of the governing equations as below (Gruber et al. 2011):

$$\frac{\partial \rho}{\partial t} + \nabla \cdot (\rho U) = 0 \tag{2-23}$$

$$\frac{\partial \rho U}{\partial t} + \nabla \cdot (\rho U U) = \nabla \cdot [\mu(\nabla U + \nabla U^T)] - \nabla p + \rho g \tag{2-24}$$

$$\frac{\partial \rho m_A}{\partial t} + \nabla \cdot (\rho U m_A) - \nabla \cdot (\rho D_{AB} \nabla m_A) = 0 \tag{2-25}$$

where ρ is the fluid density; U is the fluid velocity vector; μ is the viscosity of fluid; p is the pressure; g is the gravitational acceleration; m_A is the solute mass fraction; and D_{AB} is the solute diffusion coefficient. Another essential step of CFD modeling is defining the boundary conditions. In a CFD model of FO membrane system, a slip velocity boundary condition in the tangential direction is assumed to parameterize the effects of flow dynamics near membrane:

$$U_{\text{slip}} = \frac{\sqrt{\kappa}}{\alpha}\frac{\partial U}{\partial n_d} \tag{2-26}$$

The ICP model (Eqs. (2-4) and (2-5)) for water flux through the membrane is incorporated as a velocity boundary condition and the convective-diffusive fluxes balance (Eq. (2-27)) is adapted as the boundary conditions for the solute mass fraction on the membrane (Gruber et al. 2011).

$$-\rho_m D_{AB}\frac{\partial m_A}{\partial n_d} n_d + \rho_m m_{A,m} J_v = J_s \tag{2-27}$$

where U_{slip} is the surface slip velocity; κ is the permeability; α is the slip coefficient; $\partial U/\partial n_d$ is the shear rate at the boundary; ρ_m is the fluid density at the membrane surface; n_d is the direction towards the draw which is normal to the membranes; $m_{A,m}$ is the solute mass fraction at the membrane surface. Comparing to semi-analytical methods, the CFD approach is capable of simultaneously modeling the effects of various parameters, e.g., hydrodynamic conditions, spacer, module geometry, and fluid physical properties, even in a three-dimensional scale. Hence it can be used for systematic study of the complex CP phenomenon and the FO process optimization.

2.8 APPENDIXES

2.8.1 Appendix A: Development of ICP Model for FO Water Flux and Solute Flux

The ICP model of the FO process is derived by integrating the solution-diffusion model and the diffusion-convection transport in the porous support of membrane (Lee et al. 1981; Loeb et al. 1997; McCutcheon and Elimelech 2006; Tang et al. 2010). According to the solution-diffusion model, the water flux J_v and solute flux J_s through the rejection layer can be determined by:

$$\text{AL-DS}: J_v = A(\pi_{draw} - \pi_{support}) \tag{A1}$$

$$J_s = B(C_{draw} - C_{support}) \tag{A2}$$

$$\text{AL-FS}: J_v = A(\pi_{support} - \pi_{feed}) \tag{A3}$$

$$J_s = B(C_{support} - C_{feed}) \tag{A4}$$

where A and B are the water permeability coefficient and the solute permeability coefficient of the membrane, respectively; π_{draw} and C_{draw} are the osmotic pressure and the solute concentration of the draw solution, respectively; $\pi_{support}$ and $C_{support}$ are the osmotic pressure and the solute concentration at the interface of rejection layer and support layer; π_{feed} and C_{feed} are the osmotic pressure and solute concentration of the feed solution, respectively. The solute transport in the membrane support layer can be modeled according to the diffusion-convection flow. In the AL-DS orientation, the solute transport into the support by convection (J_vC) and that by solute reverse diffusion from the draw (J_s) have to be balanced by the solute diffusion out of the support:

$$J_vC + J_s = D_{eff}\frac{dC}{dx} \tag{A5}$$

where C is the solute concentration in the support layer at a distance x away from the interface between the rejection layer and support layer; D_{eff} is the effective diffusion coefficient of solute, which is the product of the support layer porosity ε and the solute diffusion coefficient D ($D_{eff} = \varepsilon D$). The boundary conditions for Eq. (A5) are defined as:

$$C = C_{support} \quad \text{at } x = 0 \tag{A6}$$

$$C = C_{feed} \quad \text{at } x = l_{\text{eff}} \tag{A7}$$

where l_{eff} is the effective thickness of the support layer, equal to the product of the actual thickness of support layer (l) and the tortuosity of support layer (τ). With the boundary conditions, Eq. (A1), (A2) and (A5) can be solved to yield:

$$\ln\left(\frac{C_{support} + B(C_{draw} - C_{support})/A(\pi_{draw} - \pi_{support})}{C_{feed} + B(C_{draw} - C_{support})/A(\pi_{draw} - \pi_{support})}\right) = \frac{J_v}{K_m} \tag{A8}$$

where K_m is the mass transfer coefficient of membrane, defined as:

$$K_m = \frac{D_{eff}}{l_{eff}} = \frac{\varepsilon D}{\tau l} = \frac{D}{\tau l/\varepsilon} = \frac{D}{S} \tag{A9}$$

where S is a structural parameter of membrane defined as:

$$S = \frac{\tau l}{\varepsilon} \tag{A10}$$

Assume that the solutions obey the van't Hoff equation and have a linear osmotic-pressure-concentration relationship:

$$\pi = \beta \cdot C R_g T \tag{A11}$$

where β is the vant't Hoff coefficient; R_g is the universal gas constant; T is the absolute temperature. Eq. (A8) can be simplified to:

$$\ln\left(\frac{\pi_{support} + B/A}{\pi_{feed} + B/A}\right) = \frac{J_v}{K_m} \tag{A12}$$

Substituting Eq. (A1) into Eq. (A12), water flux in the AL-DS orientation becomes:

$$J_v = K_m \ln\left(\frac{A\pi_{draw} - J_v + B}{A\pi_{feed} + B}\right) \tag{A13}$$

Similarly, the FO water flux in the AL-FS orientation can be derived:

$$\ln\left(\frac{C_{draw} + B(C_{support} - C_{feed})/A(\pi_{support} - \pi_{feed})}{C_{support} + B(C_{support} - C_{feed})/A(\pi_{support} - \pi_{feed})}\right) = \frac{J_v}{K_m} \tag{A14}$$

and

$$J_v = K_m \ln\left(\frac{A\pi_{draw} + B}{A\pi_{feed} + J_v + B}\right) \tag{A15}$$

For solutions that do not follow the van't Hoff equation, the osmotic pressure and concentration may show a non-linear relationship. Thus Eqs. (A13) and (A15) are not applicable. In this case, water flux can be solved iteratively using Eq. (A1), (A3), (A8) and (A14) with the osmotic pressure-concentration curves of the solutes (Tang et al. 2010).

The FO solute flux can be derived from Eqs. (A1)–(A4):

$$\text{AL-DS:} \; \frac{J_s}{J_v} = \frac{B(C_{draw} - C_{support})}{A(\pi_{draw} - \pi_{support})} \tag{A16}$$

$$\text{AL-FS:} \; \frac{J_s}{J_v} = \frac{B(C_{support} - C_{feed})}{A(\pi_{support} - \pi_{feed})} \tag{A17}$$

When the solutions follow the van't Hoff equation (Eq. (A11)), Eqs. (A16) and (A17) can be simplified to the same equation (Tang et al. 2010):

$$J_s = \frac{B}{A \cdot \beta R_g T} J_v \tag{A18}$$

If the van't Hoff equation is not applicable to the solutions, the FO solute flux can be derived from Eqs. (A2) and (A4) after $C_{support}$ is determined.

2.8.2 Appendix B: Model Development for Double-Skinned FO Membranes

According to the solution-diffusion model, FO water flux and solute flux of a double-skinned FO membrane (Figure 2.7) can be determined by (Tang et al. 2011):

$$J_v = A_{draw}(\pi_{draw} - \pi_{s/d} + P_{support}) \tag{B1}$$

$$J_s = B_{draw} - (C_{draw} - C_{s/d}) \tag{B2}$$

or

$$J_v = A_{feed}(\pi_{s/f} - \pi_{feed} - P_{support}) \tag{B3}$$

$$J_s = B_{feed}(C_{s/f} - C_{feed}) \tag{B4}$$

where A_{draw} and A_{feed} are the water permeability coefficients of the draw skin and the feed skin, respectively; B_{draw} and B_{feed} are the solute permeability coefficients of the draw skin and the feed skin, respectively; $\pi_{s/d}$ and $C_{s/d}$ are the osmotic pressure and the solute concentration at the draw-skin-support interface, respectively; $\pi_{s/f}$ and $C_{s/f}$ are the osmotic pressure and solute concentration at the feed-skin-support interface, respectively; $P_{support}$ is the hydraulic pressure within the support layer. Based on the convection-diffusion mechanism, the solute transport in the support layer is:

$$J_s = D_{eff} \frac{dC}{dx} J_v C \tag{B5}$$

where C is the solute concentration at a distance x away from the feed skin-support interface. Substituting the two boundary conditions:

$$C = C_{s/f} \quad \text{at } x = 0 \tag{B6}$$

$$C = C_{s/d} \quad \text{at } x = l_{eff} \tag{B7}$$

into Eq. (B5), one can obtain:

$$\ln\left(\frac{C_{s/d} + J_s/J_v}{C_{s/f} + J_s/J_v}\right) = \frac{J_v l_{\text{eff}}}{D_{\text{eff}}} = \frac{J_v}{D/S} = \frac{J_v}{K_m} \tag{B8}$$

Further substituting Eqs. (B2) and (B4) into Eq. (B8), one can obtain:

$$J_v = K_m \ln\left(\frac{C_{draw} - J_v/B_{draw} + J_s/J_v}{C_{feed} + J_v/B_{feed} + J_s/J_v}\right) \tag{B9}$$

Assuming that the feed and draw solutions follow the van't Hoff equation, Eqs. (B1)–(B4) can be integrated as:

$$J_v\left(\frac{1}{A_{draw}} + \frac{1}{A_{feed}}\right) = \beta R_g T \cdot J_s\left(\frac{1}{B_{draw}} + \frac{1}{B_{feed}}\right) \tag{B10}$$

The overall water permeability (A) and solute permeability (B) of a double-skinned membrane can be derived from the permeability of the feed skin and the draw skin by assuming a resistance-in-series model:

$$\frac{1}{A} = \frac{1}{A_{draw}} + \frac{1}{A_{feed}} \tag{B11}$$

$$\frac{1}{B} = \frac{1}{B_{draw}} + \frac{1}{B_{feed}} \tag{B12}$$

Substituting Eqs. (B11) and (B12) into Eq. (B10) gives:

$$\frac{J_s}{J_v} = \frac{B}{A \cdot \beta R_g T} \tag{B13}$$

The water flux of double-skinned FO membrane can be solved by substituting Eq. (B13) into Eq. (B9):

$$J_v = K_m \ln\left(\frac{A\pi_{draw} - \frac{B}{B_{draw}} J_v + B}{A\pi_{feed} + \frac{B}{B_{feed}} J_v + B}\right) \tag{B14}$$

Combining Eqs. (B1), (B2) and the van't Hoff equation Eq. (A11), one can determine the hydraulic pressure in the support layer:

$$P_{support} = \frac{J_v}{A_{draw}} - \beta R_g T \frac{J_s}{B_{draw}} \tag{B15}$$

Further substituting Eq. (B13) into Eq. (B15), one can eliminate J_s and obtain:

$$P_{support} = \left(1 - \frac{B/A}{B_{draw}/A_{draw}}\right) \cdot \frac{J_v}{A_{draw}} \tag{B16}$$

Similarly, another form of $P_{support}$ can be written as:

$$P_{support} = \left(\frac{B/A}{B_{feed}/A_{feed}} - 1\right) \cdot \frac{J_v}{A_{feed}} \tag{B17}$$

2.8.3 Appendix C: Model Development for Salt Accumulation in OMBR

In a submerged OMBR (Figure 2.10) where the salt concentrations of the feed sewage and the draw solution are constant, the time-dependent salt concentration of the mixed liquor (C_{ml}) can be determined as (Xiao et al. 2011):

$$V_r \frac{dC_{ml}}{dt} = Q_{in}C_{in} + J_s A_m - Q_s C_{ml} \tag{C1}$$

$$Q_{in} = J_v A_m + Q_s \tag{C2}$$

where V_r is the volume of bioreactor; dC_{ml}/dt is the change of mixed liquor salt concentration over time; Q_{in} an Q_s are the volumetric flow of the influent sewage and the waste sludge, respectively; C_{in} is the solute concentration of the influent sewage; A_m is the FO membrane area. The hydraulic retention time (θ_{HRT}) and the solids retention time (θ_{SRT}) are defined as:

$$\theta_{HRT} = \frac{V_r}{J_v A_m + Q_s} \tag{C3}$$

$$\theta_{SRT} = \frac{V_r}{Q_s} \tag{C4}$$

Assuming that the solute follows the van't Hoff equation (Eq. (A11)), increases of the salt concentration and the osmotic pressure (π_{ml}) in the reactor can be calculated by:

$$\frac{dC_{ml}}{dt} = \frac{1}{\theta_{HRT}}\left(C_{in} + \frac{B}{A\beta R_g T}\right) - \frac{1}{\theta_{SRT}}\left(C_{ml} + \frac{B}{A\beta R_g T}\right) \quad \text{(C5)}$$

$$\frac{d\pi_{ml}}{dt} = \frac{1}{\theta_{HRT}}\left(\pi_{in} + \frac{B}{A}\right) - \frac{1}{\theta_{SRT}}\left(\pi_{ml} + \frac{B}{A}\right) \quad \text{(C6)}$$

where π_{in} and π_{ml} are the osmotic pressures of the influent sewage and the mixed liquor, respectively. At the steady state ($d\pi_{ml}/dt = 0$), the salt concentration and the osmotic pressure in the OMBR are:

$$C_{ml}^{steady} = \frac{\theta_{SRT}}{\theta_{HRT}} C_{in} + \left(\frac{\theta_{SRT}}{\theta_{HRT}} - 1\right)\frac{B}{A\beta R_g T} \quad \text{(C7)}$$

$$\pi_{ml}^{steady} = \frac{\theta_{SRT}}{\theta_{HRT}} \pi_{in} + \left(\frac{\theta_{SRT}}{\theta_{HRT}} - 1\right)\frac{B}{A} \quad \text{(C8)}$$

The water flux of OMBR at the steady state can be derived by integrated Eq. (C8) into the ICP model (Eqs. (A13) and (A15)). As J_v/A and B/A values are usually substantially lower that π_{draw} and π_{ml}, water flux at steady state can be determined:

$$J_v^{steady} = K_m \cdot \ln\left(\frac{\pi_{draw}}{(\theta_{SRT}/\theta_{HRT})(\pi_{in} + B/A)}\right) \quad \text{(C9)}$$

2.9 NOMENCLATURE

Symbols

A	water permeability coefficient of membrane
A_{draw}	water permeability coefficient of the draw skin
A_{feed}	water permeability coefficient of the feed skin
A_m	membrane area
B	solute permeability coefficient of membrane
B_{draw}	solute permeability coefficient of the draw skin
$B_{draw\,solute}$	draw solute permeability coefficient of membrane
B_{feed}	solute permeability coefficient of the feed skin

$B_{feed\,solute}$	feed solute permeability coefficient of membrane
C	solute concentration
C_{draw}	solute concentration of the draw solution
C_{feed}	solute concentration of the feed solution
C_{in}	solute concentration of influent sewage
C_{ml}	solute concentration of the mixed liquor
$C_{s/d}$	solute concentration at the interface of draw skin and support layer
$C_{s/f}$	solute concentration at the interface of feed skin and support layer
$C_{support}$	solute concentration at the interface of rejection layer and support layer
D	solute diffusion coefficient
D_{AB}	diffusion coefficient of the molecule A in B
D_{draw}	draw solute diffusion coefficient
D_{eff}	effective solute diffusion coefficient
D_{feed}	feed solute diffusion coefficient
g	gravitational acceleration
J_v	water flux
$J_{v,double}$	water flux of double-skinned membrane
$J_{v,single}$	water flux of single-skinned membrane
J_s	solute flux
k	mass transfer coefficient of the ECP boundary layer
K_m	mass transfer coefficient of the ICP boundary layer
l	thickness of the membrane support layer
l_{eff}	effective thickness of the support layer
m_A	solute mass fraction
$m_{A,m}$	solute mass fraction at the membrane surface
n_d	direction towards the draw
p	pressure
$P_{support}$	hydraulic pressure in the support layer of double-skinned FO membrane
Q_{in}	volumetric flow of the influent sewage
Q_s	volumetric flow of the waste sludge
R_g	gas constant
S	structural parameter
t	time
T	absolute temperature
U	fluid velocity vector
U_{slip}	surface slip velocity
V_r	volume of bioreactor
W	power density
x	distance away from the interface of rejection layer and support layer
α	slip coefficient

β	van't Hoff coefficient
$\Delta\pi_{eff}$	effective osmotic pressure difference
ΔP	transmembrane hydraulic pressure
$\Delta\pi$	osmotic pressure difference
ε	porosity of the membrane support layer
θ_{HRT}	hydraulic retention time
θ_{SRT}	sludge retention time
κ	permeability
μ	viscosity of fluid
π	osmotic pressure
π_{draw}	osmotic pressure of draw solution
π_{feed}	osmotic pressure of feed solution
π_{in}	osmotic pressure of influent sewage
π_{ml}	osmotic pressure of the mixed liquor
$\pi_{s/d}$	osmotic pressure at the interface of draw skin and support layer
$\pi_{s/f}$	osmotic pressure at the interface of feed skin and support layer
$\pi_{support}$	osmotic pressure at the interface of rejection layer and support layer
ρ	fluid density
ρ_m	fluid density at the membrane surface
σ	reflection coefficient
τ	tortuosity

Abbreviations

AL-DS	membrane orientation with active rejection layer facing draw solution
AL-FS	membrane orientation with active rejection layer facing feed solution
CFD	computational fluid dynamics
CP	concentration polarization
ECP	external concentration polarization
FO	forward osmosis
ICP	internal concentration polarization
MF	microfiltration
$M_{\text{ICP-draw}}$	dilutive internal concentration polarization of draw solution
$M_{\text{ICP-feed}}$	concentrative internal concentration polarization of feed solution
$M_{\text{ICP-Js}}$	internal concentration polarization induced by solute reverse diffusion
M_{R}	membrane frictional resistance loss
NF	nanofiltration
OMBR	osmotic membrane bioreactor
PRO	pressure-retarded osmosis
RO	reverse osmosis
UF	ultrafiltration

References

Aaberg, R. J. (2003). "Osmotic Power: A new and powerful renewable energy source?" *Refocus*, 4(6), 48–50.

Achilli, A., Cath, T. Y., and Childress, A. E. (2009a). "Power generation with pressure retarded osmosis: An experimental and theoretical investigation." *Journal of Membrane Science*, 343(1–2), 42–52.

Achilli, A., Cath, T. Y., Marchand, E. A., and Childress, A. E. (2009b). "The forward osmosis membrane bioreactor: A low fouling alternative to MBR processes." *Desalination*, 239(1–3), 10–21.

Achilli, A., Cath, T. Y., and Childress, A. E. (2010). "Selection of inorganic-based draw solutions for forward osmosis applications." *Journal of Membrane Science*, 364(1–2), 233–241.

Achilli, A. and Childress, A. E. (2010). "Pressure retarded osmosis: From the vision of Sidney Loeb to the first prototype installation—Review." *Desalination*, 261(3), 205–211.

Alturki, A., McDonald, J., Khan, S. J., Hai, F. I., Price, W. E., and Nghiem, L. D. (2012). "Performance of a novel osmotic membrane bioreactor (OMBR) system: Flux stability and removal of trace organics." *Bioresource Technology*, 113, 201–206.

Cath, T. Y., Childress, A. E., and Elimelech, M. (2006). "Forward osmosis: Principles, applications, and recent developments." *Journal of Membrane Science*, 281(1–2), 70–82.

Chanukya, B. S., Patil, S., and Rastogi, N. K. (2012). "Influence of concentration polarization on flux behavior in forward osmosis during desalination using ammonium bicarbonate." *Desalination*, 312, 39–44.

Chekli, L., Phuntsho, S., Shon, H. K., Vigneswaran, S., Kandasamy, J., and Chanan, A. (2012). "A review of draw solutes in forward osmosis process and their use in modern applications." *Desalination and Water Treatment*, 43(1–3), 162–184.

Chou, S., Shi, L., Wang, R., Tang, C. Y., Qiu, C., and Fane, A. G. (2010). "Characteristics and potential applications of a novel forward osmosis hollow fiber membrane." *Desalination*, 261(3), 365–372.

Cornelissen, E. R., Harmsen, D., de Korte, K. F., Ruiken, C. J., Qin, J. J., Oo, H., and Wessels, L. P. (2008). "Membrane fouling and process performance of forward osmosis membranes on activated sludge." *Journal of Membrane Science*, 319(1–2), 158–168.

Cornelissen, E. R., Harmsen, D., Beerendonk, E. F., Qin, J. J., Oo, H., De Korte, K. F., and Kappelhof, J. W. M. N. (2011). "The innovative Osmotic Membrane Bioreactor (OMBR) for reuse of wastewater." *Water Science and Technology*, 63(8), 1552–1565.

Elimelech, M. and Bhattacharjee, S. (1998). "A novel approach for modeling concentration polarization in crossflow membrane filtration based on the equivalence of osmotic pressure model and filtration theory." *Journal of Membrane Science*, 145(2), 223–241.

Fang, W., Wang, R., Chou, S., Setiawan, L., and Fane, A. G. (2012). "Composite forward osmosis hollow fiber membranes: Integration of RO- and NF-like selective layers to enhance membrane properties of anti-scaling and anti-internal concentration polarization." *Journal of Membrane Science*, 394–395, 140–150.

Geise, G. M., Park, H. B., Sagle, A. C., Freeman, B. D., and McGrath, J. E. (2011). "Water permeability and water/salt selectivity tradeoff in polymers for desalination." *Journal of Membrane Science*, 369(1–2), 130–138.

Gerstandt, K., Peinemann, K. V., Skilhagen, S. E., Thorsen, T., and Holt, T. (2008). "Membrane processes in energy supply for an osmotic power plant." *Desalination*, 224(1–3), 64–70.

Gray, G. T., McCutcheon, J. R., and Elimelech, M. (2006). "Internal concentration polarization in forward osmosis: role of membrane orientation." *Desalination*, 197(1–3), 1–8.

Gruber, M. F., Johnson, C. J., Tang, C. Y., Jensen, M. H., Yde, L., and Helix-Nielsen, C. (2011). "Computational fluid dynamics simulations of flow and concentration polarization in forward osmosis membrane systems." *Journal of Membrane Science*, 379(1–2), 488–495.

Gruber, M. F., Johnson, C. J., Tang, C., Jensen, M. H., Yde, L., and Helix-Nielsen, C. (2012). "Validation and analysis of forward osmosis CFD model in complex 3D geometries." *Membranes*, 2(4), 764–782.

Hancock, N. T. and Cath, T. Y. (2009). "Solute coupled diffusion in osmotically driven membrane processes." *Environmental Science and Technology*, 43(17), 6769–6775.

Holloway, R. W., Childress, A. E., Dennett, K. E., and Cath, T. Y. (2007). "Forward osmosis for concentration of anaerobic digester centrate." *Water Research*, 41(17), 4005–4014.

Jin, X., Shan, J., Wang, C., Wei, J., and Tang, C. Y. (2012). "Rejection of pharmaceuticals by forward osmosis membranes." *Journal of Hazardous Materials*, 222–228, 55–61.

Lay, W. C. L., Chong, T. H., Tang, C. Y., Fane, A. G., Zhang, J., and Liu, Y. (2010). "Fouling propensity of forward osmosis: Investigation of the slower flux decline phenomenon." *Water Science and Technology*, 61(4), 922–936.

Lay, W. C. L., Zhang, Q., Zhang, J., McDougald, D., Tang, C., Wang, R., Liu, Y., and Fane, A. G. (2011). "Study of integration of forward osmosis and biological process: Membrane performance under elevated salt environment." *Desalination*, 283, 123–130.

Lay, W. C. L., Zhang, J., Tang, C., Wang, R., Liu, Y., and Fane, A. G. (2012). "Factors affecting flux performance of forward osmosis systems." *Journal of Membrane Science*, 394–395, 151–168.

Lee, K. L., Baker, R. W., and Lonsdale, H. K. (1981). "Membranes for power generation by pressure-retarded osmosis." *Journal of Membrane Science*, 8(2), 141–171.

Loeb, S., Titelman, L., Korngold, E., and Freiman, J. (1997). "Effect of porous support fabric on osmosis through a Loeb-Sourirajan type asymmetric membrane." *Journal of Membrane Science*, 129(2), 243–249.

Loeb, S. (2002). "Large-scale power production by pressure-retarded osmosis, using river water and sea water passing through spiral modules." *Desalination*, 143(2), 115–122.

McCutcheon, J. R., McGinnis, R. L., and Elimelech, M. (2005). "A novel ammonia–carbon dioxide forward (direct) osmosis desalination process." *Desalination*, 174(1), 1–11.

McCutcheon, J. R. and Elimelech, M. (2006). "Influence of concentrative and dilutive internal concentration polarization on flux behavior in forward osmosis." *Journal of Membrane Science*, 284(1–2), 232–242.

McCutcheon, J. R., McGinnis, R. L., and Elimelech, M. (2006). "Desalination by ammonia – carbon dioxide forward osmosis: Influence of draw and feed solution concentrations on process performance." *Journal of Membrane Science*, 278(1–2), 114–123.

McCutcheon, J. R. and Elimelech, M. (2007). "Modeling water flux in forward osmosis: Implications for improved membrane design." *AIChE Journal*, 53(7), 1736–1744.

Mehta, G. D. and Loeb, S. (1978). "Internal polarization in the porous substructure of a semipermeable membrane under pressure-retarded osmosis." *Journal of Membrane Science*, 4, 261–265.

Moody, C. D. and Kessler, J. O. (1976). "Forward osmosis extractors." *Desalination*, 18(3), 283–295.

Mulder, M. (1996). *Basic principles of membrane technology*, Kluwer Academic Publishers.

Patel, S. (2010). "Norway inaugurates osmotic power plant." *Power*, 154(2).

Qi, S., Qiu, C. Q., Zhao, Y., and Tang, C. Y. (2012). "Double-skinned forward osmosis membranes based on layer-by-layer assembly-FO performance and fouling behavior." *Journal of Membrane Science*, 405–406, 20–29.

Qin, J. J., Oo, M. H., Tao, G., Cornelissen, E. R., Ruiken, C. J., de Korte, K. F., Wessels, L. P., and Kekre, K. A. (2009). "Optimization of operating conditions in forward osmosis for osmotic membrane bioreactor." *Open Chemical Engineering Journal*, 3, 22–32.

Qin, J. J., Kekre, K. A., Oo, M. H., Tao, G., Lay, C. L., Lew, C. H., Cornelissen, E. R., and Ruiken, C. J. (2010). "Preliminary study of osmotic membrane bioreactor: Effects of draw solution on water flux and air scouring on fouling." *Water Science and Technology*, 62(6), 1353–1360.

Sablani, S., Goosen, M., Al-Belushi, R., and Wilf, M. (2001). "Concentration polarization in ultrafiltration and reverse osmosis: A critical review." *Desalination*, 141(3), 269–289.

Saren, Q., Qiu, C. Q., and Tang, C. Y. (2011). "Synthesis and characterization of novel forward osmosis membranes based on layer-by-layer assembly." *Environmental Science and Technology*, 45(12), 5201–5208.

She, Q., Jin, X., Li, Q., and Tang, C. Y. (2012a). "Relating reverse and forward solute diffusion to membrane fouling in osmotically driven membrane processes." *Water Research*, 46(7), 2478–2486.

She, Q., Jin, X. and Tang, C. Y. (2012b). "Osmotic power production from salinity gradient resource by pressure retarded osmosis: Effects of operating conditions and reverse solute diffusion." *Journal of Membrane Science*, 401–402, 262–273.

Skilhagen, S. E., Dugstad, J. E., and Aaberg, R. J. (2008). "Osmotic power–power production based on the osmotic pressure difference between waters with varying salt gradients." *Desalination*, 220(1–3), 476–482.

Strathmann, H. (1981). "Membrane separation processes." *Journal of Membrane Science*, 9(1–2), 121–189.

Su, J., Chung, T. S., Helmer, B. J., and de Wit, J. S. (2012). "Enhanced double-skinned FO membranes with inner dense layer for wastewater treatment and macromolecule recycle using Sucrose as draw solute." *Journal of Membrane Science*, 396, 92–100.

Tan, C. H. and Ng, H. Y. (2008). "Modified models to predict flux behavior in forward osmosis in consideration of external and internal concentration polarizations." *Journal of Membrane Science*, 324(1–2), 209–219.

Tang, C. Y., Kwon, Y. N., and Leckie, J. O. (2009). "Effect of membrane chemistry and coating layer on physiochemical properties of thin film composite polyamide RO and NF membranes. II. Membrane physiochemical properties and their dependence on polyamide and coating layers." *Desalination*, 242(1–3), 168–182.

Tang, C. Y., She, Q., Lay, W. C. L., Wang, R., and Fane, A. G. (2010). "Coupled effects of internal concentration polarization and fouling on flux behavior of forward osmosis membranes during humic acid filtration." *Journal of Membrane Science*, 354(1–2), 123–133.

Tang, C. Y., She, Q., Lay, W. C. L., Wang, R., Field, R., and Fane, A. G. (2011). "Modeling double-skinned FO membranes." *Desalination*, 283, 178–186.

Tiraferri, A., Yip, N. Y., Phillip, W. A., Schiffman, J. D., and Elimelech, M. (2011). "Relating performance of thin-film composite forward osmosis membranes to support layer formation and structure." *Journal of Membrane Science*, 367(1–2), 340–352.

Wang, K. Y., Ong, R. C., and Chung, T. S. (2010a). "Double-skinned forward osmosis membranes for reducing internal concentration polarization within the porous sublayer." *Industrial and Engineering Chemistry Research*, 49(10), 4824–4831.

Wang, R., Shi, L., Tang, C. Y., Chou, S., Qiu, C., and Fane, A. G. (2010b). "Characterization of novel forward osmosis hollow fiber membranes." *Journal of Membrane Science*, 355(1–2), 158–162.

Wang, Y., Wicaksana, F., Tang, C. Y., and Fane, A. G. (2010c). "Direct microscopic observation of forward osmosis membrane fouling." *Environmental Science and Technology*, 44(18), 7102–7109.

Wei, J., Liu, X., Qiu, C., Wang, R., and Tang, C. Y. (2011a). "Influence of monomer concentrations on the performance of polyamide-based thin film composite forward osmosis membranes." *Journal of Membrane Science*, 381(1–2), 110–112.

Wei, J., Qiu, C., Tang, C. Y., Wang, R., and Fane, A. G. (2011b). "Synthesis and characterization of flat-sheet thin film composite forward osmosis membranes." *Journal of Membrane Science*, 372(1–2), 292–302.

Wei, J., Qiu, C., Wang, Y.-N., Wang, R., and Tang, C. Y. (2013). "Comparison of NF-like and RO-like thin film composite osmotically-driven membranes—Implications for membrane selection and process optimization." *Journal of Membrane Science*, 427, 460–471.

Xiao, D., Tang, C. Y., Zhang, J., Lay, W. C. L., Wang, R., and Fane, A. G. (2011). "Modeling salt accumulation in osmotic membrane bioreactors: Implications for FO membrane selection and system operation." *Journal of Membrane Science*, 366(1–2), 314–324.

Xie, M., Nghiem, L. D., Price, W. E., and Elimelech, M. (2012). "Comparison of the removal of hydrophobic trace organic contaminants by forward osmosis and reverse osmosis." *Water Research*, 46(8), 2683–2692.

Xu, Y., Peng, X., Tang, C. Y., Fu, Q. S., and Nie, S. (2010). "Effect of draw solution concentration and operating conditions on forward osmosis and pressure retarded osmosis performance in a spiral wound module." *Journal of Membrane Science*, 348 (1–2), 298–309.

Yap, W. J., Zhang, J., Lay, W. C. L., Cao, B., Fane, A. G., and Liu, Y. (2012). "State of the art of osmotic membrane bioreactors for water reclamation." *Bioresource Technology*, 122, 212–222.

Yip, N. Y., Tiraferri, A., Phillip, W. A., Schiffman, J. D., and Elimelech, M. (2010). "High performance thin-film composite forward osmosis membrane." *Environmental Science and Technology*, 44(10), 3812–3818.

Yip, N. Y. and Elimelech, M. (2011). "Performance limiting effects in power generation from salinity gradients by pressure retarded osmosis." *Environmental Science and Technology*, 45(23), 10273–10282.

Yip, N. Y., Tiraferri, A., Phillip, W. A., Schiffman, J. D., Hoover, L. A., Kim, Y. C., and Elimelech, M. (2011). "Thin-film composite pressure retarded osmosis membranes for sustainable power generation from salinity gradients." *Environmental Science and Technology*, 45(10), 4360–4369.

Zhao, S., Zou, L., Tang, C. Y., and Mulcahy, D. (2012). "Recent developments in forward osmosis: Opportunities and challenges." *Journal of Membrane Science*, 396, 1–21.

Zou, S., Gu, Y., Xiao, D., and Tang, C. Y. (2011). "The role of physical and chemical parameters on forward osmosis membrane fouling during algae separation." *Journal of Membrane Science*, 366(1–2), 356–362.

CHAPTER 3

Impacts of Spacers on Forward Osmosis Processes

Joon Ha Kim
Minkyu Park
Jijung Lee

Spacer is usually employed in spiral wound membrane modules in membrane processes to enhance mass transfer in the vicinity of the membrane. In this chapter, the impacts of spacers on FO separation are dealt with numerically taking the complexity of concentration polarization into consideration. In addition, CPI (Concentration Polarization Index) is suggested as a CP indicator pertaining to the degree of concentration polarization and permeate flux in FO processes. FO spacers generally reduce the degree of CP although spacers under a certain configuration have adverse effects on permeate flux. The phenomena of boundary layer compression and boundary layer disruption are compared between RO and FO separation. Lastly, sensitivity analysis is brought up in order to gain insights into FO separation characteristics and to suggest further strategy for FO process development.

3.1 CONCENTRATION POLARIZATION IN FORWARD OSMOSIS

Concentration polarization (CP) is an important factor for analysing the flux performance in membrane processes (Winzeler and Belfort, 1993). CP usually diminishes the driving force for water permeation during membrane processes. In particular, FO membrane process is affected by CP because it is run by concentration difference itself. The concentration profile in and out of membrane is elevated or declined, which is different from the concentration of bulk solutions. The effect of CP on the flux performance in FO separation is much bigger than that of RO separation. Therefore, the characteristics of CP in FO need to be critically considered since its developing status is rather complex.

Figure 3.1 illustrates the comparison of CP between RO and FO membrane processes. External concentration polarization (ECP) and internal concentration polarization (ICP) are separately visualized. The region colored in red and blue

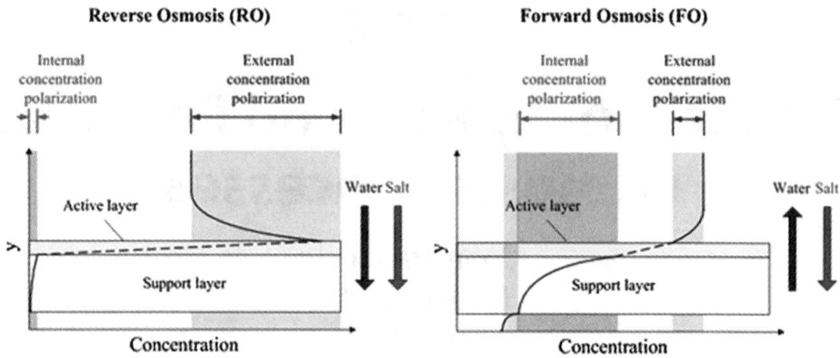

Figure 3.1. Concentration polarization phenomena of RO and FO. ECP plays more important role than ICP in RO, while the degree of ICP is more severe than ECP in FO

shows the extent of ICP and ECP respectively. The CP in membrane separation can be classified into ECP and ICP according to the place where it happens, outside the membrane or inside the membrane. ECP refers the CP phenomenon that occurs outside the membrane externally in the vicinity of membrane. ICP means the CP generated inside the membrane such as within the porous support layer. In addition, depending on the concentration profile, dilutive CP and concentrative CP exist.

We primarily focus on concentrative external concentration polarization (CECP) for RO separation. However, we need to think about total four cases for FO separation—concentrative ECP (CECP), dilutive ECP (DECP), concentrative ICP (CICP), and dilutive ICP (DICP). Since water penetrates from lower concentration of feed solution to higher concentration of draw solution, feed solution becomes gradually concentrated thereby suffering concentrative CP whereas draw side continually get diluted thereby undergoing dilutive CP.

ICP is hardly brought up in RO separation. As seen in Figure 3.1, degree of ICP is negligible compared to ECP. In RO, water moves from concentrated solution (feed side) to diluted solution (permeate side) by applying hydraulic pressure exceeding osmotic pressure difference. Solutes and other species in the feed solution are transported toward the active layer surface by convective flow at a velocity equivalent to that of water permeation. Then, solutes including monovalent, divalent, and whatnot are rejected due to the semipermeable nature of membrane, and solutes are retained on the membrane surface externally. The concentration near the active layer is sharply increased and the permeate concentration is very low. The concentration gradation within the support layer is also low.

On the other hand, the degree of ICP is significant compared to ECP in FO (Figure 3.1 right). In FO separation, water permeates through the active layer, which is an actual selective barrier that allow only water passes through, until when the difference in chemical potential becomes equilibrated to the point where the concentration of feed and draw side is identical. As water flux continues, feed solution is concentrated and draw solution is diluted. During the separation process, ICP within the porous support layer in FO separation is unavoidable. Steep concentration gradient (i.e. ICP) is necessarily generated inside the porous layer where solutes move relatively freely along the concentration gradient. Moreover, it is not easy to disturb CP layer because hydrodynamic mixing is not applicable to the porous support layer. Instead, diffusion phenomena inside the porous support layer are crucial for the FO performance.

According to the membrane orientation, the aspect of CP becomes differentiated. Figure 3.2 depicts two cases of membrane orientation. The terminology, PRO mode and FO mode, is widely used in literatures to distinguish merely the membrane orientation (McCutcheon and Elimelech (2006), Chou et al. (2010), Mi and Elimelech (2008)). The orientation in which active layer faces feed solution and porous support layer faces draw solution is FO mode; PRO mode refers the opposite. In terms of application, FO mode is for water production guaranteeing higher water quality and PRO mode aims power generation assuring higher water flux. To avoid the confusion arising from the terminology, AL-facing DS and AL-facing-FS can be used instead of PRO mode and FO mode respectively.

From Figure 3.2, we can know that dilutive ECP pairs with concentrative ICP in PRO mode and dilutive ICP pairs with concentrative ECP in FO mode. ECP at the support layer surface is normally negligible with the assumption that solute freely moves at support layer interface. There are no combinations other than above-mentioned cases that we need to consider.

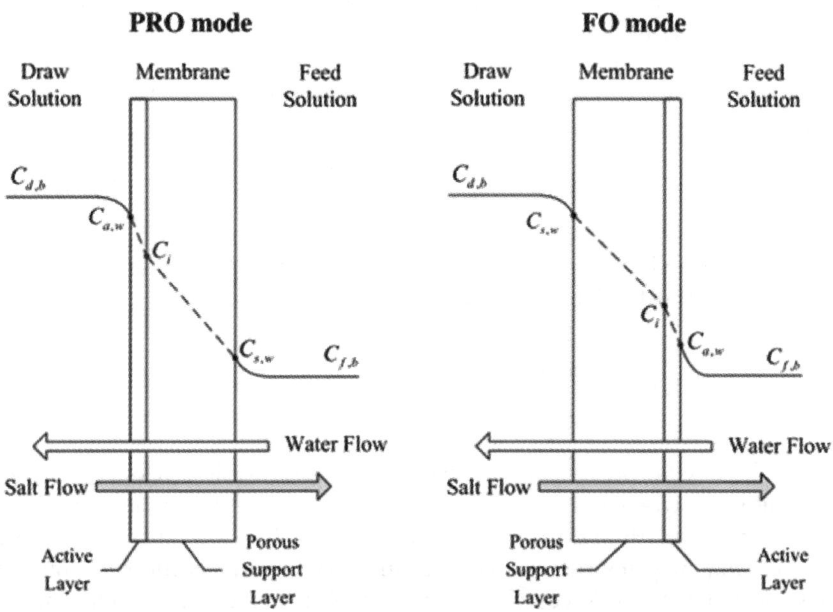

Figure 3.2. Schematics of osmotically driven membrane processes of PRO and FO modes. In actual PRO process applying PRO mode, the pressure is exerted on draw solution side

3.2 EXTERNAL CONCENTRATION POLARIZATION (ECP)

ECP has been extensively studied in the area of RO since membrane processes started to be applied in commercial (Mulder, 1996). ECP is simply elucidated by introducing mass transfer coefficient, k. With the calculation of mass transfer coefficient, the increased (usually in RO separation) or decreased wall concentration externally on the active layer surface can be obtained. Although ECP layer develops along the channel, it is normally assumed as uniform layer over the membrane surface. Hence, the changed concentration in the vicinity of active layer as a result of CP is calculated by the following equation derived from solute mass balance. The value of k is negative if it is dilutive CP and positive for concentrative CP.

$$\frac{c_{a,w} - c_p}{c_{\text{bulk}} - c_p} = \exp\left(\frac{v}{k}\right) \tag{3.1}$$

where $c_{a,w}$, c_p, and c_{bulk} are the solute concentration on the active layer surface, that of permeate, and that of bulk solution respectively; v is the channel averaged permeate velocity. k is the mass transfer coefficient, which is physically defined as the ratio of the diffusion coefficient of solute to the thickness of concentration boundary layer ($k = D/\delta$). The wall concentration $c_{a,w}$ can be estimated with the calculation of mass transfer coefficient; otherwise, it is difficult to measure the exact value experimentally.

Apart from the physical meaning of mass transfer coefficient, it can be related to Sherwood number (Sh) described as (Sherwood et al. (1967), Tan and Ng (2008)):

$$\text{Sh} = a\text{Re}^b\text{Sc}^c \tag{3.2}$$

where Sh, Re, and Sc are the Sherwood, Reynolds, and Schmidt number respectively and their definitions are as follows:

$$\text{Re} = \frac{u \cdot d_h \cdot \rho}{\mu} \qquad \text{Sc} = \frac{\mu}{\rho \cdot D} \qquad \text{Sh} = \frac{k \cdot d_h}{D} \tag{3.3}$$

where d_h refers to the hydraulic diameter, u is the cross flow velocity, μ is the dynamic viscosity, and ρ is the density. The coefficients—a, b, and c-can be obtained by regression analysis of experimental data. The ECP phenomenon can be reduced by enhancing mass transfer in stream channel. Mass transfer coefficient is mainly dependent on the diffusion coefficient, the flow velocity, and module configuration. During the process operation, k can be simply increased by increasing the flow velocity or promoting turbulence while other conditions such as diffusivity and module configuration are not easy to vary. It can be inferred that mass transfer coefficient represents the hydrodynamic characteristics of membrane channel. Similar to RO modeling, this Sherwood relation is most widely used in FO modeling.

3.3 INTERNAL CONCENTRATION POLARIZATION (ICP)

Mathematical model to predict ICP in FO separation is proposed by Lee et al. published in 1981. In the process of solving the convective and diffusion equation for FO separation, solute resistivity (K) is introduced. It is physically defined as

$$K = \frac{\tau t_s}{D\varepsilon} = \frac{S}{D} \tag{3.4}$$

where τ is the tortuosity of support layer, t_s is the thickness of support layer, ε is the porosity of support layer, and S is the membrane structure parameter ($S = \tau t_s/\varepsilon$). Solute resistivity, K, indicates the degree of how much the support layer is resistant against salt transport. Smaller value of K is preferred to reduce the effect of ICP. By using the value of solute resistivity, the concentration at the interface of active layer and support layer, c_i, can be determined. As depicted in Figure 3.2, concentrative ICP occurs in FO mode and dilutive ICP happens in PRO mode. ICP is the dominant factor for reducing effective driving force water permeation compared to ECP in FO separation. PRO mode is less affected by ICP than FO mode since draw solution of high concentration faces active layer. Under the same concentration of bulk feed and draw solutions, PRO mode has higher flux.

3.4 CONCENTRATION POLARIZATION INDEX

As we discussed in earlier, CP plays am imperative role for FO flux performance. However, it is challenging to quantify the effect of CP since dilutive and concentrative CP coexist as well as external and internal CP occur simultaneously.

To measure the degree of CP in FO processes, previous studies usually adopt CP modulus, the ratio of the osmotic pressures at interfaces. As depicted in Figure 3.3, the dilutive ECP and concentrative ICP modulus in RPO mode are as follows:

$$\frac{\pi_{a,w}}{\pi_{d,b}} = \exp\left(-\frac{J_W}{k}\right) \quad \text{DECP moduli}$$

$$\frac{\pi_i}{\pi_{f,b}} = \exp(J_W K) \quad \text{CICP moduli}$$

In the case of FO mode, overall CP consists of dilutive ICP and concentrative ECP. Each CP module is described as:

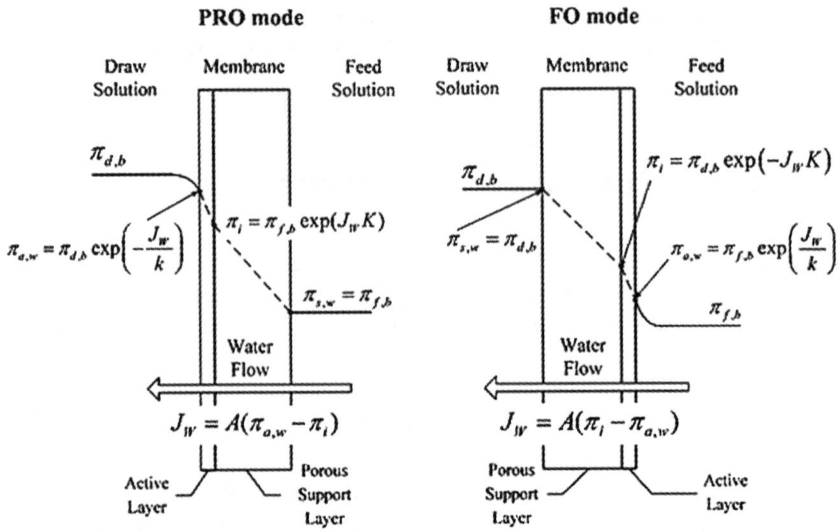

Figure 3.3. Schematic diagram of the CP modulus proposed by McCutcheon and Elimelech (2006) with the assumption of perfect rejection

$$\frac{\pi_i}{\pi_{d,b}} = \exp(-J_W K) \qquad \text{DICP moduli}$$

$$\frac{\pi_{a,w}}{\pi_{f,b}} = \exp\left(\frac{J_W}{k}\right) \qquad \text{CECP moduli}$$

where $\pi_{a,w}$ and π_i refers to the osmotic pressure on the membrane wall of active layer and at the active-support layer interface as a consequence of CP; $\pi_{d,b}$ and $\pi_{f,b}$ are the osmotic pressure of the feed and draw bulk solutions; J_W indicates water flux; and k is the mass transfer coefficient for ECP and K is the solute resistivity for ICP. The value of dilutive module is less than unity whereas that of concentrative module is higher than unity.

In an attempt to evaluate the FO performance, flux performance reflecting the influence of CP should be compared based on objective criteria. However, above CP modulus are apparently ambiguous for direct comparison due to its inconvenience. Concentrative and dilutive modulus may require different approach of interpretation according to operation condition. In addition, each module only represents the CP in the corresponding channel, and it cannot provide overall information for the whole FO operation. Therefore, it is highly required to develop a simple index that can analyse CP effect comprehensively.

Required criterion for the index is the ability to elucidate the CP pertaining to the system characteristics such as flux. In order to meet this criterion, the index needs to imply the degree of overall CP as well as the impact of CP on flux performance. The concentration polarization index (CPI) is suggested, which was derived from the flux equation (Park and Kim, 2013).

$$
\begin{aligned}
v_W &= A(\Delta P - \Delta\pi) && \text{RO} \\
v_W &= A\Delta\pi = A(\pi_{\text{draw}} - \pi_{\text{feed}}) = A(\pi_i - \pi_{a,w}) && \text{FO mode} \\
v_W &= A\Delta\pi = A(\pi_{\text{draw}} - \pi_{\text{feed}}) = A(\pi_{a,w} - \pi_i) && \text{PRO mode}
\end{aligned}
$$

To calculate water flux and solute flux in both RO and FO, we need to know $\Delta\pi$, which is the osmotic pressure difference between each side of the active layer. The flux equation for FO mode is manipulated to derive CPI as follows:

$$v_W = A\Delta\pi = A(\pi_i - \pi_{a,w}) \tag{3.5}$$

$$
\begin{aligned}
v_W &= A\left(\frac{\pi_i}{\pi_{s,w}}\frac{\pi_{s,w}}{\pi_{b,\text{draw}}}\pi_{b,\text{draw}} - \frac{\pi_{a,w}}{\pi_{b,\text{feed}}}\pi_{b,\text{feed}}\right) \\
&= A(\text{MOD}_{\text{DICP}}\,\text{MOD}_{\text{DECP}}\,\pi_{b,\text{draw}} - \text{MOD}_{\text{CECP}}\,\pi_{b,\text{feed}}) \\
&= A\,\pi_{b,\text{draw}}\,\text{MOD}_{\text{DICP}}\,\text{MOD}_{\text{DECP}}\left(1 - \frac{\text{MOD}_{\text{CECP}}}{\text{MOD}_{\text{DICP}}\,\text{MOD}_{\text{DECP}}}\frac{\pi_{b,\text{feed}}}{\pi_{b,\text{draw}}}\right) \\
&= A\,\pi_{b,\text{draw}}\,\text{MOD}_{\text{DCP}}\left(1 - \frac{\text{MOD}_{\text{CCP}}}{\text{MOD}_{\text{DCP}}}\frac{\pi_{b,\text{feed}}}{\pi_{b,\text{draw}}}\right)
\end{aligned} \tag{3.6}
$$

where $\pi_{b,\text{draw}}$ and $\pi_{b,\text{feed}}$ are the osmotic pressure of the feed and draw bulk solutions (refer to Figure 3.4 for the subscripts); MOD refers to the modulus; $\text{MOD}_{\text{DICP}} = \pi_i/\pi_{s,w}$; $\text{MOD}_{\text{DECP}} = \pi_{s,w}/\pi_{b,\text{draw}}$; $\text{MOD}_{\text{CECP}} = \pi_{a,w}/\pi_{b,\text{feed}}$; $\text{MOD}_{\text{DCP}} = \text{MOD}_{\text{DICP}}\text{MOD}_{\text{DECP}} = \pi_i/\pi_{b,\text{draw}}$; and $\text{MOD}_{\text{CCP}} = \text{MOD}_{\text{CECP}}$. PRO mode also has an equation identical to the final line of above derivation. Consequently, let us define

$$\text{CPI} = \frac{\text{MOD}_{\text{CCP}}}{\text{MOD}_{\text{DCP}}}\frac{\pi_{b,\text{feed}}}{\pi_{b,\text{draw}}} \tag{3.7}$$

The CPI is a dimensionless number and can be utilized to quantitatively compare FO systems as a quick diagnosis. The value of CPI is less than unity and higher value close to unity means severe CP. CPI, newly defined index, is substituted into above Eq. (3.6); then, an equation similar to the original flux equation is obtained.

$$v_W = A^*(1 - \text{CPI}) \tag{3.8}$$

or, it can be described in a different way as follows for FO and PRO mode individually:

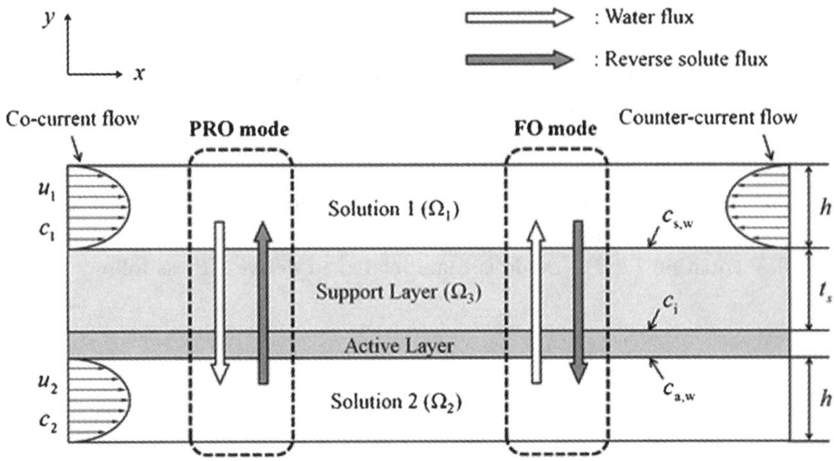

Figure 3.4. Schematic of simulation domain by FEM for FO modeling

$$v_W = A^*(1 - \text{CPI}) = A\pi_i(1 - \pi_{a,w} / \pi_i) \quad \text{FO mode}$$
$$= A\pi_{a,w}(1 - \pi_i / \pi_{a,w}) \quad \text{PRO mode} \qquad (3.9)$$

Note that although eq. (3.6) and (3.8) are possible to be applied to both FO and PRO modes, the value of MOD and CPI varies according to the membrane orientation as seen in eq. (3.9). Under FO mode, DECP and DICP can be integrated into DCP as a product of MOD_{DECP} and MOD_{DICP} and only CECP comprises CCP without CECP. On the contrary, in PRO mode, CECP and CICP are consolidated into CCP as a product of MOD_{CECP} and MOD_{CICP} while only DECP forms DCP.

For the application of CPI, numerical modelling was conducted based on finite element method (FEM). Figure 3.4 displays a simulation domain consisting of three parts (Ω_1, Ω_2, and Ω_3). Domains 1 and 2 are stream channels in which feed and draw solution flows and Domain 3 is s support layer region. ECP develops in the vicinity of membrane surfaces in Domain 1 and 2. Domain 3 where ICP layer grows is considered as porous medium.

The powerful advantage of FEM for FO moleling is that it is possible to predict FO performance under various conditions with a constant value of membrane structure parameter *S*, which is an intrinsic property of membrane itself. The inconsistency in *S* value in former experimental studies is resolved (Park et al. (2011)), Ma and Song (2006), Song and Ma (2005)).

FEM enables us to understand the momentum and mass transport entirely in the membrane channel by solving three governing equation: continuity, steady-state incompressible Navier-Stokes for the laminar regime, and convective diffusion equations. In domain 1 and 2, all three equations are simultaneously solved to compute velocity and concentration profile. For the support layer, ICP is computed with the assumption that support layer has a homogeneous structure (identical effective diffusivity within the support layer defined as $D_{eff} = D\varepsilon/\tau$) and the velocity in the support layer varies only in the horizontal direction meaning that no axial velocity exists. Thus, the vertical velocity in the support layer is identical to that of active-support interface at the same axial position. Lastly, the solution-diffusion model was employed as a boundary condition for active layer; then, water and solute flux is determined.

Simulation results for the slit channel with no spacers in Figure 3.4 are shown in Figure 3.5. NaCl solution is used for feed and draw solution. It proves that the developed model by FEM has considerably high predictability under various conditions. Based on the reliability of this model, the practicality of the CPI as a CP indicator is verified further.

The CPI is inversely proportional to the permeate flux and directly proportional to the degree of overall CP. Its association is illustrated in Figure 3.6 by comparing permeate flux and channel averaged CPI value. The concentration of feed and draw solutions is varied while keeping a concentration difference of 0.5 M. Similar value of chemical potential differences at various conditions makes the assessment of CP impact on flux even. As the feed concentration increases with 0.5 M difference, permeate velocity declines while averaged CPI value increases in both FO and PRO mode. This reciprocal relationship between the CPI and permeate flux indicates that it is feasible to use the CPI as a CP indicator.

In terms of direct comparison of FO mode and PRO mode, CPI cannot be applied although having the same bulk concentration of feed and draw solutions. In Figure 3.6, PRO mode has slightly higher CPI value at the range of lower concentration even though it shows higher flux. There is a limitation of CPI application, which is attributed to the distinctively different value of MOD_{DCP} in both orientations. The degree of DCP in FO mode is much higher than that of

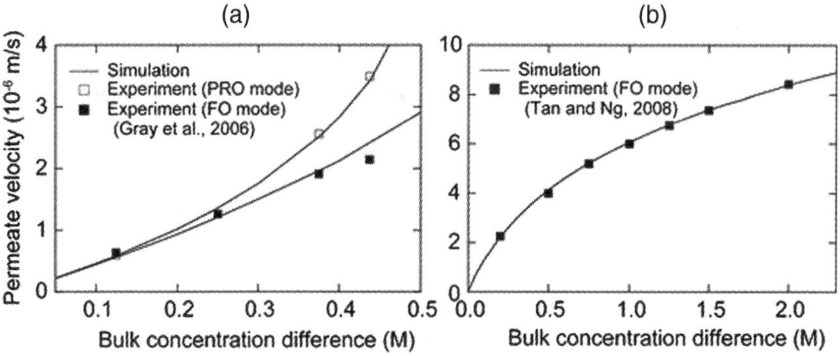

Figure 3.5. Comparison of simulation result and experimental data from literatures. Simulation conditions: (a) draw solution concentration of 0.5 M; feed solution concentrations of 0.0625, 0.125, 0.25, and 0.375 M; $S = 369$ μm; $D = 1.61 \times 10^{-9}$ m²/s; and (b) feed solution concentration of 0 M; draw solution concentrations of 0.2, 0.5, 0.75, 1.0, 1.25, 1.5, and 2 M; $S = 190$ μm; $D = 1.71 \times 10^{-9}$ m²/s

SOURCE: Adapted from Park and Kim 2013; reproduced with permission from Elsevier

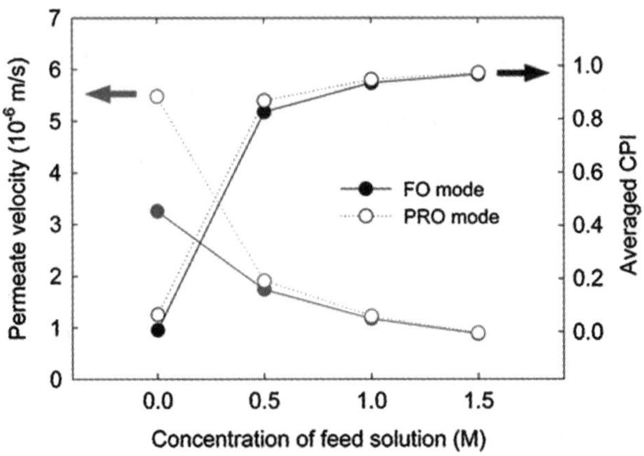

Figure 3.6. Permeate velocity and CPI values in the slit channels for FO and PRO modes

SOURCE: Adapted from Park and Kim 2013; reproduced with permission from Elsevier

PRO mode because DCP in PRO mode only includes DECP whereas it is comprised of DECP and DICP in FO mode. Hence, MOD_{DCP} is a dominant factor to determine water permeation Eq. (3.8) because it is a function of A^* and CPI that are immediately dependent on MOD_{DCP}.

Consequently, it is found that the CPI is a useful index to evaluate the extent of overall CP in FO separation with same orientation. Also, it can provide information pertaining to water flux due to its inverse proportionality with CPI value. However, its application to compare between membrane orientations will not be recommendable. In a subsequent section, CPI will be utilized to estimate the effect of spacers on the CP separately for each membrane orientation.

3.5 SPACER IMPACTS ON CONCENTRATION POLARIZATION

3.5.1 CPI as a CP indicator

Spacer is a common turbulence promoter to disturb the ECP layer in spiral wound modules (SWMs) of RO. Spacers enhance mass transfer in membrane channel and keep the membrane leaves apart (Schwinge et al., 2002a). The positive effect of hydrodynamic mixing by spacers has been widely investigated in literatures (Fimbres-Weihs and Wiley (2008), Geraldes et al. (2003), Li et al. (2005)). However, few research about FO spacer has been carried out so far. The effect of spacers on FO membrane needs to be investigated since the aspects of CP in FO separation are complicated.

Figure 3.7(a) shows the schematic of simulation domain and spacer configurations. In this simulation domain, only spacers are added to slit channel of Figure 3.4. For the spacer configurations, three types have typically been used;

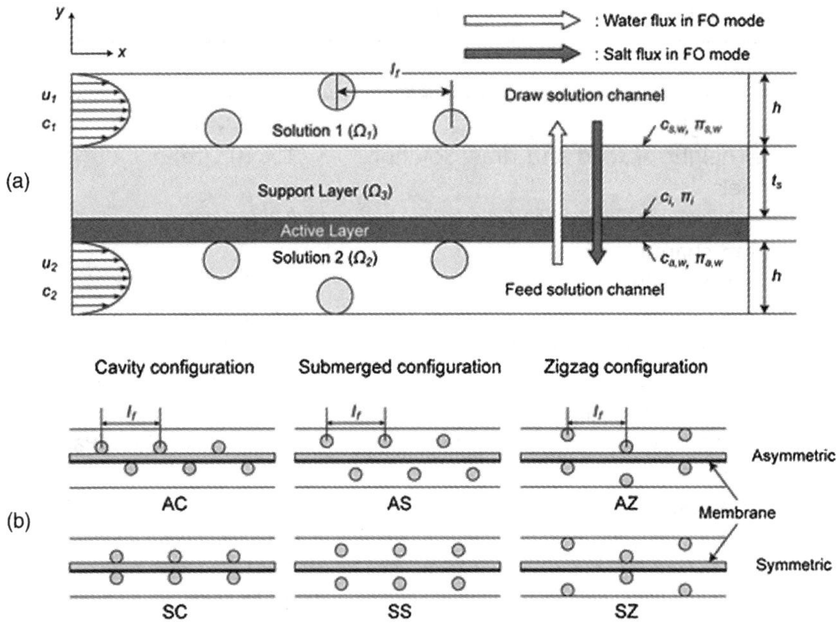

Figure 3.7. Schematic of (a) simulation domain of FO mode (b) spacer configurations. AC, AS, and AZ refers to asymmetric cavity, submerged, and zigzag configurations; SC, SS, SZ refers to symmetric cavity, submerged, and zigzag configurations
SOURCE: Adapted from Park and Kim 2013; reproduced with permission from Elsevier

cavity, submerged, and zigzag configurations (Schwinge et al., 2002a, 2004). Only feed channel is subjected to spacer study in RO separation; however, both feed and draw channel should be considered. Considering the symmetricity of spacer disposition at both feed and draw channels, total six configurations is fixed as in Figure 3.7(b). Their name is in abbreviated form according to geometric characteristics. If cavity spacers are arranged symmetrically based on membrane, it is named as symmetric cavity (SC) configurations. The cavity spacers with asymmetric alignment are referred as asymmetric cavity (AC) configurations. Other abbreviation is stated in Figure 3.7(b).

Table 3.1 summarizes the simulation conditions. Those employed values are referenced form other FO studies (Geraldes et al. (2001), Gray et al. (2006), Guillen and Hoek (2009), McCutcheon and Elimelech (2006), Phillip et al. (2010)). Channel configuration is analogous to lab-scale device and other parameters such as water and solute permeability are similar with those in other studies.

Figure 3.8 compares the water flux of the six configurations of spacer. The difference between the feed and draw solution concentrations is maintained at 0.5 M. Permeate velocities display no significant differences due to short channel length. Simulation results of permeate velocities of each configuration are normalized to those of slit channel for corresponding concentration. As the feed

Table 3.1. Simulation conditions in FEM model

Parameter	*Value*
Channel length	0.077 (m)
Channel height of feed and draw solution channels	1×10^3 (m)
Average inlet velocity of feed and draw solution channels	0.1 ($m\,s^{-1}$)
Water permeability, *A*	1.2 ($L m^{-2} h^{-1} bar^{-1}$)
Solute permeability, *B*	0.5 ($L m^{-2} h^{-1}$)
Membrane structure parameter, *S*	300 (μm)
Diffusion coefficient of NaCl, *D*	1.5×10^{-9} ($m^2 s^{-1}$)
Density of feed and draw solutions*, ρ	$997 + 40.6c$ ($kg\,m^{-3}$)
Dynamic viscosity of feed and draw solutions*, μ	$0.89\,(1 + 0.183c)$ (Pa s)
Osmotic pressure*	$4.5032 \times c^2 + 43.6426 \times c$ (bar)
Spacer diameter	5×10^{-4} (m)
The number of spacers	20 in total
Distance between transverse filaments, l_f	4.5×10^{-3} (m)

*The unit of c is M

SOURCE: Adapted from Park and Kim (2013); reproduced with permission from Elsevier

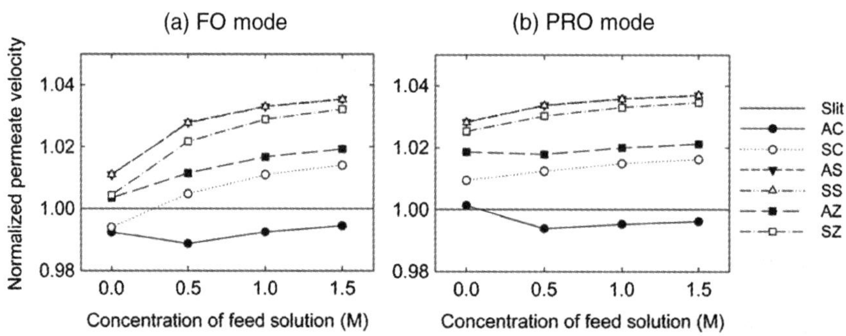

Figure 3.8. Comparison of normalized fluxes of six spacer configurations. The difference between the feed and draw solution concentrations is maintained at 0.5 M. Other conditions are identical to those in Table 3.1

SOURCE: Adapted from Park and Kim 2013; reproduced with permission from Elsevier

and draw solution concentrations increase, the normalized fluxes of most configurations are gradually enhanced regardless of spacer type and membrane orientation. The spacers in FO mode are more efficient to improve water permeation particular for higher concentration range. Among the spacer types, the submerged configuration displays the highest performance whereas the cavity shows the lowest performance. In the case of submerged type, similar trend of performance improvement is observed, irrespective of symmetricity. Interestingly, the AC configuration even worsens permeate flux compared to the slit channel without spacers. It seems that spacers attached on the membrane surface induce undesirable effects to aggravate CP at the exterior of membrane.

At the feed concentration of 0 M, each spacer type results in similar value of normalized permeate velocity in FO mode regardless of symmetricity, but in PRO mode spacer configuration excluding submerged type shows different value depending on symmetricity. As no ECP exists in feed side, it can be inferred that the dissimilarity in permeate flux is primarily attributed to the type of draw spacers. However, in PRO mode, the permeate fluxes varies with regard to spacer symmetricity at the zero feed concentration except for submerged type. This difference in PRO mode from FO mode indicates that feed spacers only in PRO mode impart an additional impact on CP, or, that feed spacers might have extra effect on both orientations.

To investigate the additional effect of feed spacers, another simulation is conducted with spacers being employed only feed channel. The concentration of feed and draw solutions is 0 M and 0.5 M respectively. Other conditions are identical to Table 3.1. Then, longitudinal CPI values along the membrane channel are calculated as presented in Figure 3.9. The CPI values near the region to which spacers are attached (pertinent to cavity and zigzag type) are abruptly increased although the feed concentration is zero. On the other hand, the channel with no spacers attached such as submerged type displays CPI values analogous with the slit channel. This result reveals that spacers attached on membrane surface magnifies the CP. The reason of adverse effect of spacers on CP is that the solutes penetrating through active layer by reverse diffusion of draw solute are captured near the spacers attached on membrane. Accordingly, the solutes trapped by spacers exacerbate the local CP. It is conceptually illustrated in Figure 3.10. The reversely diffused solutes are possibly swiped by shear force if there is no obstacle such as

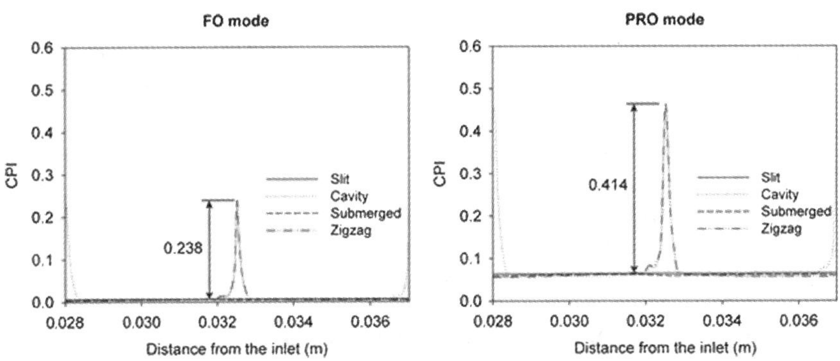

Figure 3.9. CPI valued obtained for spacer types employed only in feed channels. The concentration of feed and draw solutions is 0 M and 0.5 M respectively
SOURCE: Adapted from Park and Kim 2013; reproduced with permission from Elsevier

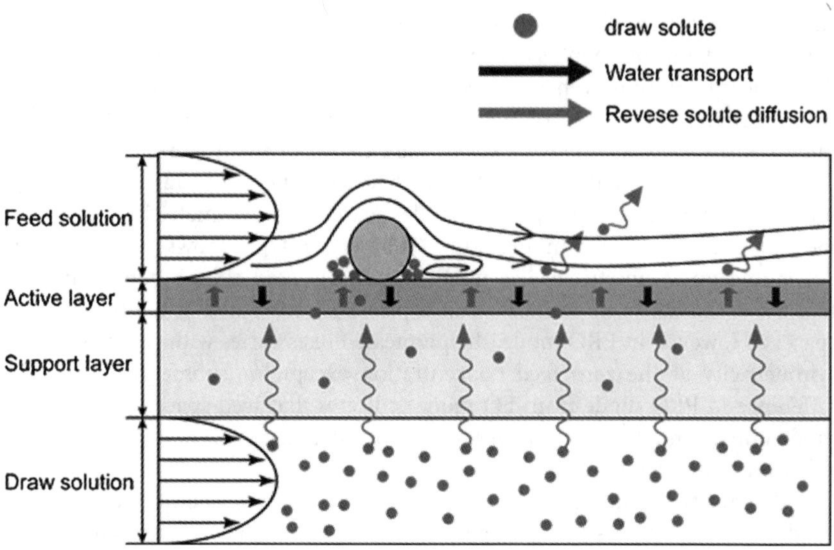

Figure 3.10. A conceptual illustration for the adverse impacts of spacers adjacent to the membrane in FO mode. The solute passing through active layer by reverse diffusion cannot be swiped by shear force thereby severely aggravating local permeate flux
SOURCE: Adapted from Park and Kim 2013; reproduced with permission from Elsevier

spacer. To verify this hypothesis, same simulation is conducted again with the assumption of complete solute rejection ($B = 0$) for cavity type. As a result, no sudden peak in CPI value is observed near spacers and the profile is almost identical to the channel without spacers (data not shown). It confirms that the solutes captured by spacers adjacent to the membrane deteriorate CP.

In addition, the increase in CPI value is compared in detail in Figure 3.9. For the slit channel, the CPI value is relatively higher in PRO mode than in FO mode. In PRO mode, the CPI value is locally elevated at the maximum by 0.414 compared to that at the same horizontal position in the slit channel. FO mode shows relatively lower increase by 0.23. Consequently, the local effect of feed spacer adjacent to the membrane surface on intensifying CP is more influential in PRO mode than in FO.

Based on these observations, the negative effect of spacers in FO separation is identified to play an important role. The design of FO spacer should be developed in the direction of avoiding the negative effect. However, it needs to be noted that positive effect of spacers usually surpasses negative one. The efficient application of spacers in FO processes will promote the commercial feasibility of its technology.

3.5.2 Influence of Spacer Type Based on CPI

In the preceding subsection, we found that a small amount of solutes leakage by reverse diffusion of draw solute could have negative impact on increasing CP near the spacer attachment when the feed concentration is zero. In this section, additional simulation is carried out for different condition when the feed concentration is not zero. If the feed solution has a certain amount of concentration, the contribution of reverse diffusion becomes minor. We investigate the spatial profiles of shear stress and CPI for the six spacer types in Figure 3.11. This Figure illustrates the longitudinal values of CPI and the corresponding wall shear stresses on the surface of active and support layers. The feed and draw solution concentrations are 0.5 M and 1 M, respectively. Since PRO mode has very similar results with FO mode, we only demonstrate the results of FO mode. In the case of cavity type of Figure 3.11(a), the CPI is rapidly increased at the region where spacers are attached on the surface of either active layer or support layer. The increased value is up to almost unity at peak point, which makes the dimensionless driving force in Eq. (3.8) close to zero. Although the maximum value of CPI in SC configuration is slightly higher than in AC configuration, there exists obvious gap in normalized permeate velocity between AC and SC configurations as identified in Figure 3.8. The SC arrangement enhances water permeation compared to the case without spacers whereas AC arrangement even decreases the flux performance. More frequent peaks observed in AC arrangement finally results lower flux.

Wall shear stress exerting on membrane surface boosts hydrodynamic mixing for alleviating ECP. The absolute value of the wall shear stress indicates that the magnitude of stress tangentially exerted on the wall, or surface, of the membrane. It disturbs flow streams thereby dispersing reversely diffused solutes by convection. Sharp peak abruptly appearing at the spacer attachment in the wall shear stress profile are due to the existence of infinitesimal gap between spacers and membrane during the numerical simulation, which is negligible. The higher value of wall shear stress implies the ability to swipe concentration boundary layer. It accordingly enhances mass transfer.

The wall shear stress in channel with cavity type is generally lower than in slit channel. Nevertheless, CPI value is slightly lower than that of slit channel except

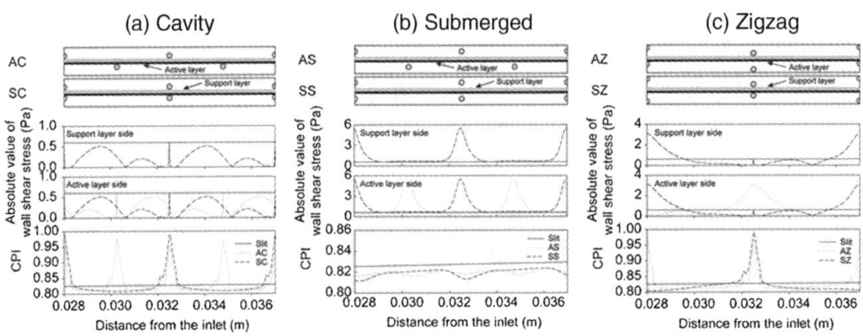

Figure 3.11. CPI and wall shear stress of FO mode with six spacer configurations. The feed and draw solution concentration are 0.5 M and 1 M. Other simulation conditions are identical to Table 1

SOURCE: Adapted from Park and Kim 2013; reproduced with permission from Elsevier

for the region of spacer attachment. Boundary layer disruption induced by spacers helps the concentration boundary layer not to be fully developed yet its beneficial influence is attenuated by the adverse effect mentioned earlier. In the channel with submerged spacers (Figure 3.11(b)), they are not attached on the membrane surface. The overall CPI values for both AS and SS are lower than that of slit channel without any sudden increase. The wall shear stress is also promoted showing the highest magnitude among three spacer types. At the position where spacers are placed, dramatic increase in shear stress swipes the concentration boundary layer hence reducing CPI values.

Zigzag-type spacers (Figure 3.11(c)) show intermediate characteristics between cavity and submerged type. Similarly in cavity spacers, CPI values at the attachment of spacers irrespective of whether on active or support layer increase drastically. Interestingly, CPI value at both horizontal ends of profile in AZ arrangement is significantly increases compared to the SZ configuration although shear stress is relatively higher due to spacers attached on the top and bottom walls. The negative effect of spacers attached on the membrane surface surpasses the positive effect of moderately high value of shear stress. Therefore, boundary layer compression induced by submerged spacer is likely to be more efficient to moderate CP than boundary layer disruption in FO separation, since boundary layer disruption inherently accompanies solute accumulation near the spacer attachment, which makes the CPI almost up to unity, i.e., zero dimensionless driving force.

3.5.3 Discussion of Difference between RO and FO Spacers

Unlike in RO, FO spacer can have adverse effect on flux performance under a specific situation because the aspect of CP development is quite different. As discussed in preceding section, submerged type shows the highest efficiency and cavity type shows the lowest efficiency. Schwinge et al. (2004) and Ma and Song (2006) carried out numerical simulation of RO membrane channels with spacers. The results revealed higher permeate flux in the order of zigzag, cavity, and submerged types. Zigzag-type spacers easily disturb the CP layer by recirculation arising adjacent to the membrane. The periodic disruption and redevelopment of the boundary layer due to consecutive spacers are effective for CP control (Focke, 1983). Therefore, boundary layer disruption is more effective than the boundary layer compression in RO membrane. Proper spacing is required for effective mixing between the successive spacers as well as for preventing the boundary layer from being fully developed.

Figure 3.12 illustrates the concentration contour of AS and AC configurations respectively for PRO mode. Boundary layer compression can be seen in the channel with submerged spacers as in Figure 3.12(a). The applied condition of feed and draw solutions is 1.5 M and 2 M. Refer to Figure 3.7 for the schematic of simulation domain. Submerged-type spacers do not disrupt the boundary layer but tenuously compress the concentration boundary layer. Also, the concentration contours within support layer are almost parallel. Although this compression is less effective than boundary layer disruption, it has no negative effect on the permeate flux. The concentration profile of AC configuration in Figure 3.12(b) is

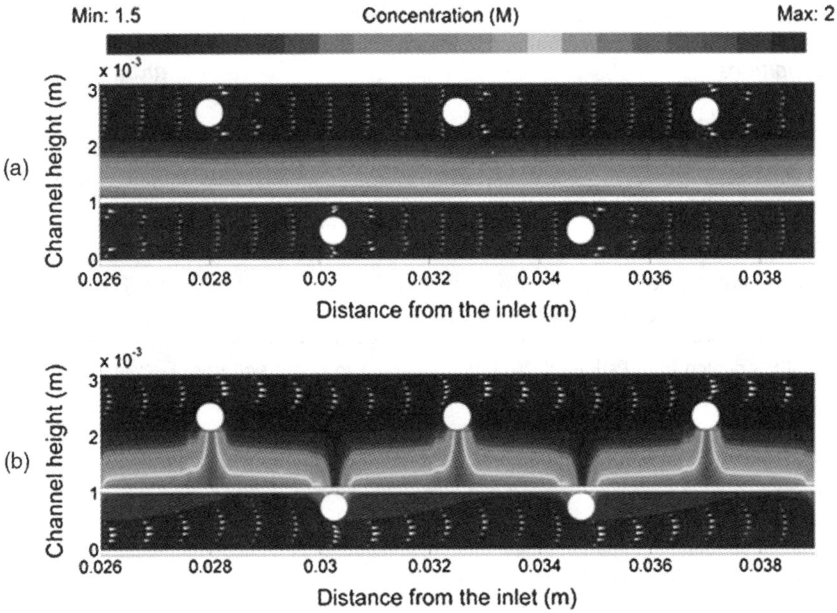

Figure 3.12. Concentration contours of (a) AS and (b) AC configuration in PRO mode. The concentration of feed and draw solution is 1.5 M and 2 M. Other simulation conditions are identical to those in Table 3.1
SOURCE: Adapted from Park and Kim 2013; reproduced with permission from Elsevier

largely fluctuated within support layer. The spacers attached on membrane surface eventually change the ICP profile inside the porous support layer. Steep concentration gradient is observed particularly in the vicinity of the attachment regions of feed and draw spacers. This abrupt gradient in concentration significantly declines local permeate flux.

However, the purpose of spacer is to maintain the space between membrane sheets, the attachment of spacers on the membrane surface is unavoidable. Thus, optimizing proper spacing between consecutive spacers will be imperative role for enhancing mass transfer in the vicinity of membrane.

3.6 SENSITIVITY ANALYSIS: CHARACTERISTICS OF FORWARD OSMOSIS

In order to define how much the process parameters are sensitive to flux performance, sensitivity analysis can be applied to FO modeling. The Latin-hypercube one-at-a-time (LH-OAT) method is implemented (Holvoet et al. (2005), van Griensven et al. (2006)). As a result of sensitivity analysis, the effects of each modeling parameter such as membrane properties and stream velocities on the objective function (i.e. permeate flux) are quantitatively provided in terms of index

Table 3.2. Simulation conditions for the developed model

Conditions	*Parameter*
Current direction	Counter-current
Channel height of feed and draw solution (h)	3×10^{-3} m
Channel length (L)	7.7×10^{-2} m
Channel width (W)	2.6×10^{-2} m
Temperature	20°C
Feed solution concentration (NaCl)	0–1 M
Draw solution concentration (NaCl)	0.2–2 M (fixed)

SOURCE: Adapted from Park et al. 2011; reproduced with permission from Elsevier

*Table 3.3. Six modeling parameters used in the sensitivity analysis**

Parameter	*Min. value*	*Max. value*
Membrane structure parameter, S (m)	272	333
Diffusion coefficient, D ($m^2 s^{-1}$)	1.45×10^{-9}	1.77×10^{-9}
Water permeability, A ($Lm^{-2} h^{-1} bar^{-1}$)	1.00	1.22
Solute permeability, B ($Lm^{-2} h^{-1}$)	0.42	0.50
Average inlet velocity of feed solution, u_f ($m s^{-1}$)	0.412	0.504
Average inlet velocity of draw solution, u_d ($m s^{-1}$)	0.412	0.504

*A sensitivity index in percent (%) was obtained by varying the modeling parameters within $\pm 10\%$ of the value employed in the simulation
SOURCE: Adapted from Park et al. 2011; reproduced with permission from Elsevier

in percentage. We can determine the rank of parameters in the order of higher sensitivity to flux performance. The FEM model developed in preceding subsection is utilized and the simulation conditions are summarized in Table 3.2 (McCutcheon and Elimelech (2006), Gray et al. (2006)). Schematic diagram for simulation domain is identical to Figure 3.4 and no spacers are applied. We determine six parameters for sensitivity analysis as selected in Table 3.3. The range of values is decided by varying their values $\pm 10\%$ from the initially developed simulation.

Figure 3.13 illustrates the results of sensitivity analysis in descending order downward. Several analyses are carried out for various concentration conditions. During the analysis, membrane properties (water permeability A, solute permeability B, structure parameter S) and operation parameters (average inlet velocity of feed and draw solutions, diffusion coefficient D) are selected as model parameters; then they are varied within the range as in Table 3.3 to measure their sensitivity index. The higher the value is, the more influential the parameter is on the permeate flux.

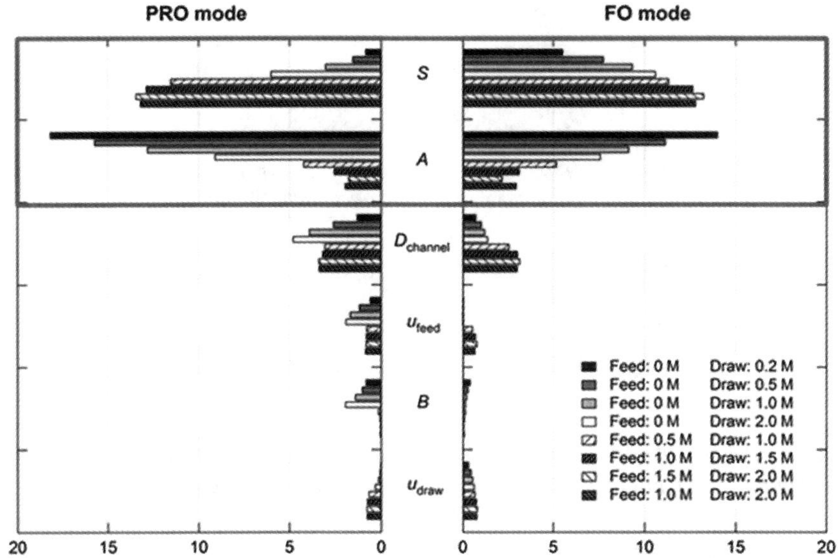

Figure 3.13. Results of sensitivity analysis using the LH-OAT method
SOURCE: Adapted from Park et al. 2011; reproduced with permission from Elsevier

In Figure 3.13, membrane structure parameter and water permeability is the most influential parameters. This result implies that membrane intrinsic property needs to be improved for better performance in the direction of increasing water permeability *A* (higher water permeability) and decreasing structure parameter *S* (less resistance in the porous support layer against solutes) because they are the most sensitive parameters that can easily increase flux. The relatively low sensitivity index of feed and draw solution velocities suggests that the development of FO processes should be focused on membrane improvement. In addition, it can be inferred that minimizing ICP is more important rather than minimizing ECP.

Another graph, Figure 3.14, compares sensitivity indices to provide informative insights into FO membrane development. This Figure summarizes sensitivity indices with respect to feed solution concentration while maintaining draw solution concentration fixed at 1.5 M. As the feed solution concentration increases, the sensitivity of *S* increases while that of *A* is decreased. The criteria for membrane development can be dependent on the concentration range of subjected process. Under the range of brackish water concentration, both *S* and *A* are influential although the index of *S* is almost twice that of *A*. However, the *S* parameter becomes dominant in seawater desalination region because ICP becomes increasingly significant. Thus, sensitivity analysis concludes that decrease in *S* parameter would be the most efficient solution to improve FO performance.

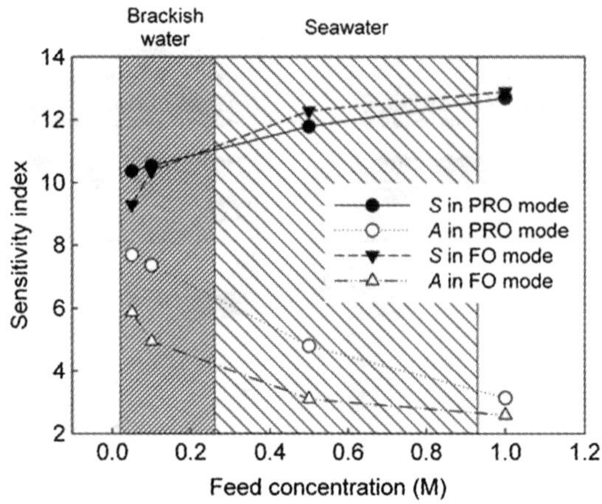

Figure 3.14. Sensitivity indices of the membrane structure parameter S and water permeability A in PRO and FO modes. Sensitivity indices were calculated by varying the feed solution concentration from 0.05 M to 1.0 M while keeping the draw solution concentration fixed at 1.5 M

SOURCE: Adapted from M. Park et al., 2011; reproduced with permission from Elsevier

3.7 CONCLUDING REMARKS

The effects of spacers on CP development in FO separation were investigated based on newly introduced index, CPI (Concentration Polarization Index). CPI is proportional to the degree of CP and reversely proportional to the permeate flux; therefore, it can be used as a CP indicator. Among the spacer types—cavity, submerged, and zigzag, the submerged-type spacer has the highest efficiency compared to the slit channel without spacers. The boundary disruption by zigzag-type spacer shows the highest performance in RO separation; on the contrary, boundary layer compression by submerged spacer in FO separation gives the best efficiency. Other type of cavity and zigzag spacers can have negative effect to augment CP since reversely diffused draw solutes are accumulated near the spacer attachment, kind of dead-zone. However, in membrane processes, spacers are inherently attached on the membrane surface to keep membrane leaves apart. Hence, further studies on FO spacers will be the optimization of spacer configuration to minimize the adverse effect on flux performance.

Apart from the studies on spacers, sensitivity analysis was also conducted to identify sensitive parameters in FO separation. It is found that water permeability and structure parameter are the most influential parameters on flux performance. Hence, a membrane with high water permeability and low solute resistivity might be the most efficient way to maximize FO performance, rather than optimizing operating conditions such as feed and draw solution velocities. The strategy for the development of FO membrane needs to consider the concentration range of applied water source because the sensitivity of parameters are changed with regard to water concentration.

3.8 NOMENCLATURE

A	water permeability ($Lm^{-2}h^{-1}bar^{-1}$)
B	solute permeability ($Lm^{-2}h^{-1}$)
c	concentration (M)
$c_{a,w}$	concentration on active layer wall (active layer surface)
c_{bulk}	concentration of bulk solution
c_i	concentration at active layer-support layer interface (AS interface)
c_p	concentration of permeate
D	diffusion coefficient ($m^2 s^{-1}$)
δ	thickness of concentration boundary layer
d_h	hydraulic diameter
ε	porosity of support layer
k	mass transfer coefficient (m/s)
J_W	flux ($m^3 m^{-2} s^{-1}$)
K	solute resistivity ($s m^{-1}$)
ρ	density ($kg m^{-3}$)
μ	dynamic viscosity (Pa s)
$\pi_{a,w}$	osmotic pressure on active layer wall (active layer surface) (bar)
$\pi_{d,b}$ or $\pi_{b,draw}$	osmotic pressure of bulk draw solution (bar)
$\pi_{f,b}$ or $\pi_{b,feed}$	osmotic pressure of bulk feed solution (bar)
π_i	osmotic pressure at active-support layer interface (bar)
$\pi_{s,w}$	osmotic pressure on support layer wall (support layer surface) (bar)
S	membrane structure parameter (m)
τ	tortuosity of support layer
t_s	support layer thickness
u	horizontal velocity ($m s^{-1}$)
v_W	permeate velocity ($m s^{-1}$)

References

AWWA Research Foundation, Lyonnaise des eaux-Dumez (Firm), and South Africa. Water Research Commission. (1996). *Water Treatment Membrane Processes*, New York: McGraw-Hill.

Chou, S., Shi, L., Wang, R., Tang, C. Y., Qiu, C., and Fane, A. G. (2010). "Characteristics and potential applications of a novel forward osmosis hollow fiber membrane." *Desalination*, 261(3), 365–372.

Fimbres-Weihs, G. A. and Wiley, D. E. (2008). "Numerical study of two-dimensional multilayer spacer designs for minimum drag and maximum mass transfer." *Journal of Membrane Science*, 325(2), 809–822.

Focke, W. W. (1983). "On the mechanism of transfer enhancement by eddy promoters." *Electrochimica Acta*, 28(8), 1137–1146.

Geraldes, V., Semião, V., and Norberta Pinho, M. (2003). "Hydrodynamics and concentration polarization in NF/RO spiral-wound modules with ladder-type spacers." *Desalination*, 157(1–3), 395–402.

Geraldes, V. t., Semião, V., and de Pinho, M. N. (2001). "Flow and mass transfer modelling of nanofiltration." *Journal of Membrane Science*, 191(1–2), 109–128.

Gray, G. T., McCutcheon, J. R., and Elimelech, M. (2006). "Internal concentration polarization in forward osmosis: role of membrane orientation." *Desalination*, 197(1–3), 1–8.

van Griensven, A., Meixner, T., Grunwald, S., Bishop, T., Diluzio, M., and Srinivasan, R. (2006). "A global sensitivity analysis tool for the parameters of multi-variable catchment models." *Journal of Hydrology*, 324(1–4), 10–23.

Guillen, G. and Hoek, E. M. V. (2009). "Modeling the impacts of feed spacer geometry on reverse osmosis and nanofiltration processes." *Chemical Engineering Journal*, 149(1–3), 221–231.

Holvoet, K., van Griensven, A., Seuntjens, P., and Vanrolleghem, P. A. (2005). "Sensitivity analysis for hydrology and pesticide supply towards the river in SWAT." *Physics and Chemistry of the Earth, Parts A/B/C*, 30(8–10), 518–526.

Kimura, S. and Sourirajan, S. (1968). "Concentration Polarization Effects in Reverse Osmosis Using Porous Cellulose Acetate Membranes." *Industrial and Engineering Chemistry Process Design and Development*, 7(1), 41–48.

Lee, K. L., Baker, R. W., and Lonsdale, H. K. (1981). "Membranes for power generation by pressure-retarded osmosis." *Journal of Membrane Science*, 8(2), 141.

Li, F., Meindersma, W., de Haan, A. B., and Reith, T. (2005). "Novel spacers for mass transfer enhancement in membrane separations." *Journal of Membrane Science*, 253(1–2), 1–12.

Ma, S. and Song, L. (2006). "Numerical study on permeate flux enhancement by spacers in a crossflow reverse osmosis channel." *Journal of Membrane Science*, 284(1–2), 102–109.

McCutcheon, J. R. and Elimelech, M. (2006). "Influence of concentrative and dilutive internal concentration polarization on flux behavior in forward osmosis." *Journal of Membrane Science*, 284(1–2), 237–247.

Mi, B. and Elimelech, M. (2008). "Chemical and physical aspects of organic fouling of forward osmosis membranes." *Journal of Membrane Science*, 320(1–2), 292–302.

Mulder, M. (1996). *Basic principles of membrane technology*, (2nd ed.), Dordrecht; Boston: Kluwer Academic.

Park, M. and Kim, J. H. (2013). "Numerical analysis of spacer impacts on forward osmosis membrane process using concentration polarization index." *Journal of Membrane Science*, 427(0), 10–20.

Park, M., Lee, J. J., Lee, S., and Kim, J. H. (2011). "Determination of a constant membrane structure parameter in forward osmosis processes." *Journal of Membrane Science*, 375(1–2), 241–248.

Phillip, W. A., Yong, J. S., and Elimelech, M. (2010). "Reverse draw solute permeation in forward osmosis: Modeling and experiments." *Environmental Science and Technology*, 44(13), 5170–5176.

Schwinge, J., Wiley, D. E., and Fane, A. G. (2004). "Novel spacer design improves observed flux." *Journal of Membrane Science*, 229(1–2), 53–61.

Schwinge, J., Wiley, D. E., and Fletcher, D. F. (2002a). "Simulation of the flow around spacer filaments between channel walls. 2. Mass-transfer enhancement." *Industrial and Engineering Chemistry Research*, 41(19), 4879–4888.

Schwinge, J., Wiley, D. E., and Fletcher, D. F. (2002b). "Simulation of the flow around spacer filaments between narrow channel walls. 1. Hydrodynamics." *Industrial and Engineering Chemistry Research*, 41(12), 2977–2987.

Sherwood, T. K., Brian, P. L. T., and Fisher, R. E. (1967). "Desalination by Reverse Osmosis." *Industrial & Engineering Chemistry Fundamentals*, 6(1), 2.

Song, L. and Ma, S. (2005). "Numerical studies of the impact of spacer geometry on concentration polarization in spiral wound membrane modules." *Industrial and Engineering Chemistry Research*, 44(20), 7638–7645.

Song, L. and Yu, S. (1999). "Concentration polarization in cross-flow reverse osmosis." *AIChE Journal*, 45(5), 921–928.

Tan, C. H. and Ng, H. Y. (2008). "Modified models to predict flux behavior in forward osmosis in consideration of external and internal concentration polarizations." *Journal of Membrane Science*, 324(1–2), 209–219.

Winzeler, H. B. and Belfort, G. (1993). "Enhanced performance for pressure-driven membrane processes: The argument for fluid instabilities." *Journal of Membrane Science*, 80, 35–48.

CHAPTER 4

Forward Osmosis Element Design

Jack Herron

4.1 INTRODUCTION

The invention of asymmetric cellulosic membranes by Loeb and Sourirajan (Loeb, 1966) led to rapid development of membrane module designs. By 1970 DuPont was marketing hollow fiber reverse osmosis modules and companies such as Desal and UOP were selling spiral wound membrane modules. Almost all semipermeable membrane development was focused on reverse osmosis (RO). However, there was academic interest in forward osmosis (FO) by people such as Popper (Popper 1969).

That the industry focuses on RO instead of FO was because RO can produce clean fresh water while FO can only concentrate one fluid by diluting another. Commercial applications of FO were not envisioned until the early 1990's when Osmotek (now Hydration Technology Innovations or HTI) began to investigate applications which take advantage of the high fouling resistance of the FO process. Side- by-side tests of FO versus disc and tubular RO systems in the concentration of landfill leachate showed FO to be much more stable and able to achieve higher concentrations (York et al. 1999). In this application the salt brine used as osmotic agent or draw solution was reconcentrated by seawater RO. FO was also used to dewater high-value shear-sensitive fluids such as fruit juices and nutraceuticals.

Much of the design of commercial FO equipment has been focused on designs such as plate and frame stacks and spiral modules with very open feed spacers in order to handle very fouling streams. Compared to RO desalination, the membrane packing densities in these FO systems are low and the pumping volumes are high. Therefore, they are only used in applications where RO is unable to concentrate the fluids. Recently, applications for FO are arising which are not entirely driven by the fouling resistance of the FO process. Examples include water recovery by spent brine in the chlor-alkalai industry and the use of ammonium bicarbonate as a draw solution to concentrate produced water from oil and gas wells to very high salinities.

This chapter will discuss the modifications which have been made to standard pressure filtration element designs to transform them into FO element designs. Primarily this entails redesigning the permeate channels to accept a pumped draw solution.

4.2 PLATE AND FRAME

Conceptually, the simplest FO element to visualize is the plate and frame design since it is merely a stack of feed and draw solution channels separated by membranes. In practice, however, it tends to be an expensive and bulky design due to the need to maintain stable and uniform membrane spacing.

Typically, a membrane plate is a stiff sheet with membranes on both sides. Draw solution is pumped through an access manifold into the interior of the plate where it is directed into gaps between the plate and the membranes. This is the location in the device where osmosis occurs. The gap between the membrane and plate is usually around 1 mm, and the velocity of the draw solution is kept to a few cm/sec. Membranes are oriented with their active sides toward the feed (away from the plate) so the draw solution is subject to all the internal concentration polarization (ICP) issues discussed in previous chapters. Low draw solution velocity is adequate since the ICP causes a much larger resistance to flux than concentration polarization in the draw solution boundary layer, or in other words, increasing the draw velocity has little effect on flux. The draw solution flow rate needs to only be enough that the bulk concentration remains adequate for osmosis. In most plate designs the gap between the membrane and the plate is maintained by texturing the plate surface or by inserting a fabric permeate spacer. During operation, a higher pressure is maintained on the feed solution than on the draw solution, and draw solution seeps laterally through the permeate spacer or surface texturing.

The velocity of the draw solution only weakly affects flux, but it is highly important that the contact points between the membrane and the permeate spacer or the textured surface do not prevent draw solutes from diffusing to the membrane surface. In other words, the contact points should be small, preferably not more than twice the thickness of the membrane. Draw solutes can diffuse laterally to provide flux if the contact area is small, but lateral diffusion distances over 1 mm essentially stop flux. Figure 4.1 shows a picture of the diffusion paths around wide and narrow contact points.

It is tempting to design a plate with an unsupported membrane but in practice unsupported designs have little improvement in flux, and they have large problems with stretching, flapping (leading to membrane fatigue) and non-uniform flow distribution.

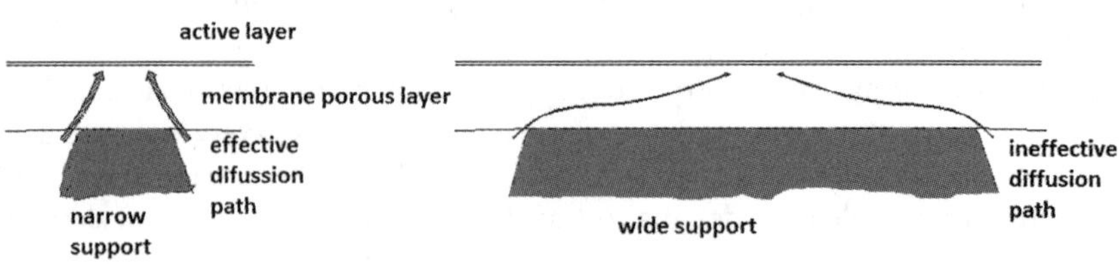

Figure 4.1. Proper membrane support for FO processes showing the diffusion paths around wide and narrow contact points

The big advantage to plate and frame designs which often makes them worth building in spite of their high cost is that the feed flow channels don't require spacers. This is critically important in situations where the feed solution is highly fouling, and since one of the major selling points of FO is its ability to concentrate fouling fluids, plate and frame designs are often attractive.

4.2.1 Pumped Plate and Frame Designs

Pumped plate and frame stacks usually have the feed and draw solution manifolds formed by penetrations through the membrane plates themselves which seal when the membrane stack is compressed by structural plates on both ends. A picture of a plate for a small stack is shown in Figure 4.2(a), and a picture of a plate and frame system for landfill leachate is shown in Figure 4.2(b).

4.2.2 Submerged Plate Designs

The submerged plate design is primarily used in membrane bioreactors (MBRs) which have become a popular technology for wastewater treatment. In MBRs, waste which is being aerobically digested in a tank is simultaneously dewatered by microfiltration plates immersed in the tank. Removing water from the waste provides faster digestion and allows much smaller treatment plants. MBR membrane plates are fixed a set distance from each other in a frame and air bubbles are released beneath the frames so that they rise in the gaps between the plates. The bubbles have a dual purpose, they induce high fluid shear at the membrane surface which increases flux and reduces fouling, and they oxygenate the water which promotes biological digestion of the waste. A drawback to microfiltration MBRs is the water passing through the membrane, although

Figure 4.2. Pumped plate and frame design (a) small test plate for tomato concentration and (b) plate and frame system for landfill leachate treatment (Courtesy: HTI)

free of suspended solids, carries virtually all of the dissolved solids present in the wastewater. Further treatment is required if the water is to be reused. In addition, microfiltration membranes foul quickly and must be frequently cleaned.

Recently there has been interest in applying FO to MBRs. Instead of pulling water through a MF membrane by suction, a draw solution is pumped through the plates and the waste is dewatered by osmosis.

4.3. SPIRAL WOUND ELEMENTS

4.3.1 Spiral Elements for Pressure Processes

The most common UF or RO elements use the spiral wound design in which a number of membrane leaves are formed by wrapping membrane and spacers around a perforated pipe. The basic components of the element are the perforated tube, membrane sheets, net spacers that provide feed fluid flow channels, and a second set of net spacers which provide flow channels for the permeate. The membranes are glued so that the permeate channels are in communication with the center tube but sealed elsewhere. The feed spacers are unglued and allow feed to flow axially through the element. A drawing of an unwound element is shown in Figure 4.3.

Spiral elements are often the cheapest membrane configuration due to the channel spacing being set by simple fabrics or nets. This avoids the expensive plates and manifolds required in plate designs.

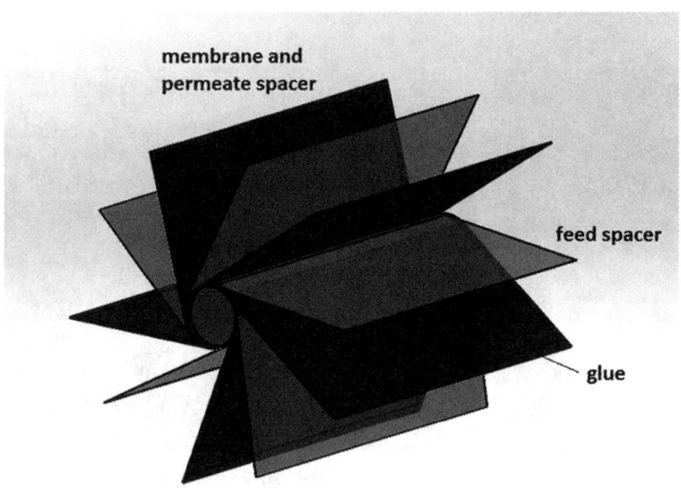

Figure 4.3. Unfurled spiral RO element

The typical method for assembly is a number of permeate spacers are fixed to a perforated tube so that when they are wound around the tube they are equally spaced. Before rolling, membrane packs are placed between each of the permeate spacers. Packs are formed by folding a flat sheet membrane over a net feed spacer so that the active side of the membrane is in contact with both sides of the feed spacer. The folded end of the pack is placed adjacent to the perforated tube and the edges perpendicular to the tube are glued to the adjacent packs through the permeate spacer. The entire stack of membrane and spacer is then wound around the pipe and the permeate channels are closed by gluing the outermost seams. Lastly, a shell is formed around the element to provide mechanical strength.

The length of elements is typically 1 meter and the individual membrane envelopes in the spiral are often about 1 meter square. In RO desalination elements both the feed and permeate spacers are less than 1 mm in thickness, so more than 30 m^2 of membrane can be fitted into a 20-cm diameter element.

In operation, a number of spiral elements are installed in a long tubular housing and the fluid to be filtered is pumped the length of the tube through the feed spacers of the elements. A portion of the feed fluid permeates the membranes and spirals inward in the permeate channels to the perforated center tube where it flows axially from the elements through a permeate exit in the tubular housing. A high feed cross-flow velocity is applied to maintain flux and only a portion of the feed passes through the membranes during a transit of the tube.

4.3.2 FO Spiral Elements

FO spiral elements are somewhat more complex because instead of the permeate flowing through the permeate spacer and out through the perforated tube, a draw solution must be pumped across the permeate side of the membrane. A common method for accomplishing this is shown in Figure 4.4. The design employs a plug in the center of the perforated tube and a glue line down a portion of the permeate channel. Draw solution pumped into the tube is forced to flow through the tube perforations, between the membranes, to the far end of the permeate channel then around the end of the center glue line and back to the center tube where it exits the element. During its transit of the permeate channel the draw solution is diluted by osmotically pulling water through the membranes from the feed.

4.3.3 Element Design for Multi-Element Housings FO

Multi element housings are by far the most economical design for spiral systems but in FO they present a difficulty not seen in pressure filtration applications. In pressure filtration applications, the center tubes of the elements are fitted with o-ring seals so that permeate can pass from one element to the next to the permeate exit at the end of the housing. In contrast, if the FO element design

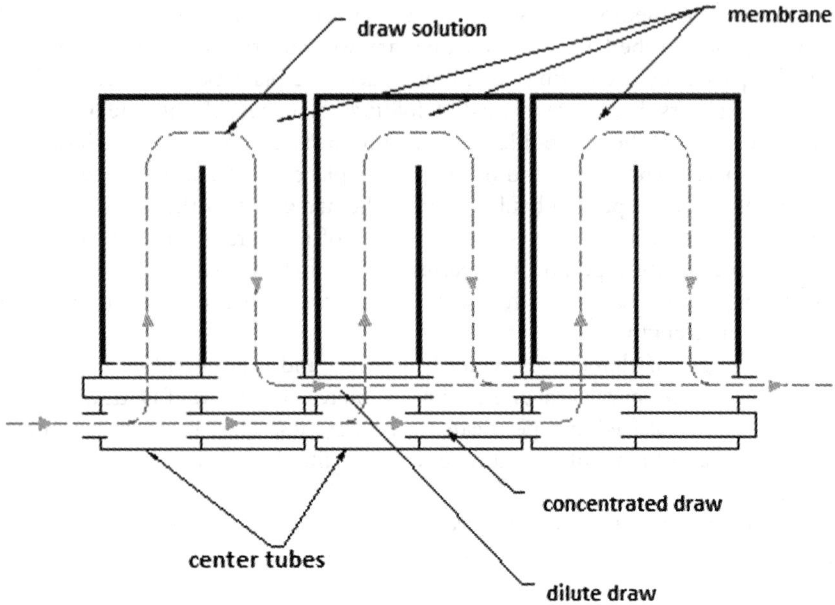

Figure 4.4. Draw solution flow through Spiral FO leaf

shown in Figure 4.4 were to be used in a multiple element housing, the draw solution would pass through the permeate spacer channel in each element in series. If there were 6 elements in a housing, all of the draw solution would be forced to spiral out and in through the permeate channels of each of the 6 elements in series, picking up water in every element. The volume flow rates required to provide adequate draw solution strength to every element would be completely impractical.

A design that overcomes this difficulty requires the center tube to be redesigned so there are two flow channels in the center tube. One channel transmits the concentrated draw solution to every element and the other conducts the diluted draw solution out of the housing. The only path from the concentrate channel to the dilute is to pass through the permeate channel of an element. Figure 4.5 shows a schematic of this. Cut-away drawings of an early design used by HTI for the center tube are shown in Figure 4.6.

Two problems with the early design pictured in Figure 4.6 were that, it required a centertube with a large diameter (8.75 cm) to accommodate the draw solution flow volumes in multi-element housings, and elements were difficult to install due to it having two off-center o-ring interconnectors. HTI now uses a smaller diameter (7 cm) centertube with concentric flow channels for the concentrated and dilute draw solutions (Figure 4.7). The centertubes contain flow inverters at the midpoint which make the inner and outer flows switch places thus maintaining the flow regime depicted in Figure 4.5.

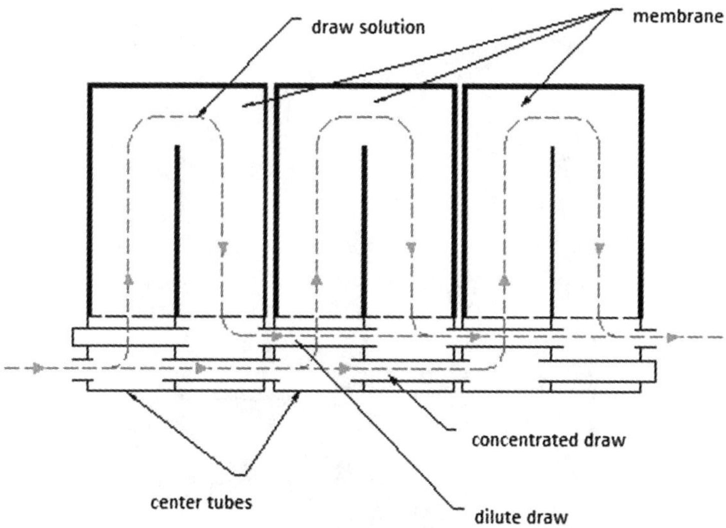

Figure 4.5. Parallel flow of draw solution through multiple elements

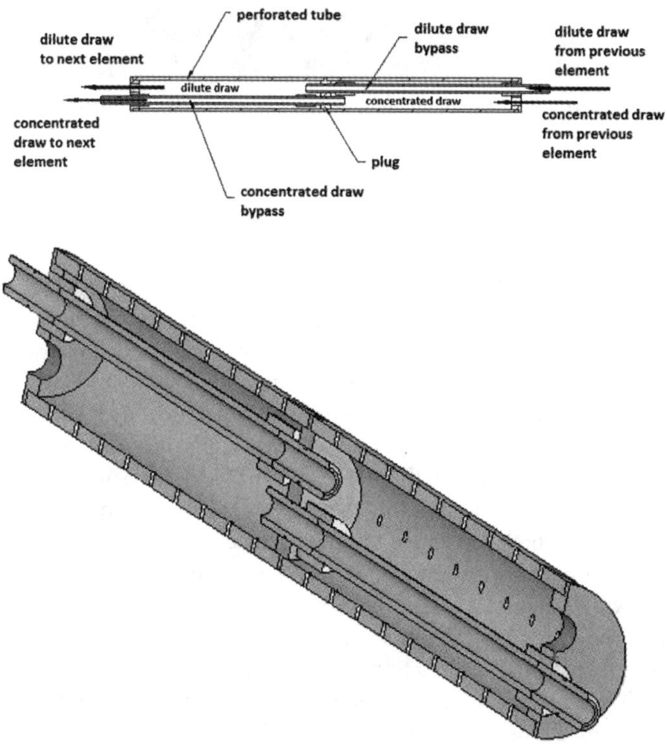

Figure 4.6. Cut-away drawing of multi-element center core showing (a) longitudinal section and (b) isometric view of the longitudinal section

Figure 4.7. New generation of multi-element center core (Courtesy: HTI)

4.4 PERMEATE SPACER DESIGN

Permeate spacers need to be open enough to allow adequate draw solution flow (typically more than 500 ml/min/spacer), while being thin in order to allow a large membrane area per element. They need to be stiff to not collapse over years of use, have numerous enough supports to prevent the membrane sagging into the flow channels, and they should also promote uniform flow over their entire area. Figure 4.8 shows a knitted polyester epoxy-coated permeate spacer used in RO, which is attractive in that it is thin, stiff, and has low pressure drops along the channels formed by the knitted ribs. The major drawback to this type of spacer is they have relatively high resistance to flow in the direction perpendicular to their ribs. If they are used as a permeate spacer for FO the flow will distribute in a fashion shown in Figure 4.9. Uneven flow distribution is a problem in that the areas which see little flow velocity have the draw solution become diluted to the point where flux drops to a fraction of the expected value.

The RO style permeate spacers, however, can be modified to deliver uniform flow by attaching a cross flap at the far end of the spacer as shown in the exploded view in Figure 4.10. The flow in the permeate spacer then is more uniform as shown in Figure 4.11.

Figure 4.8. Ribbed permeate spacer

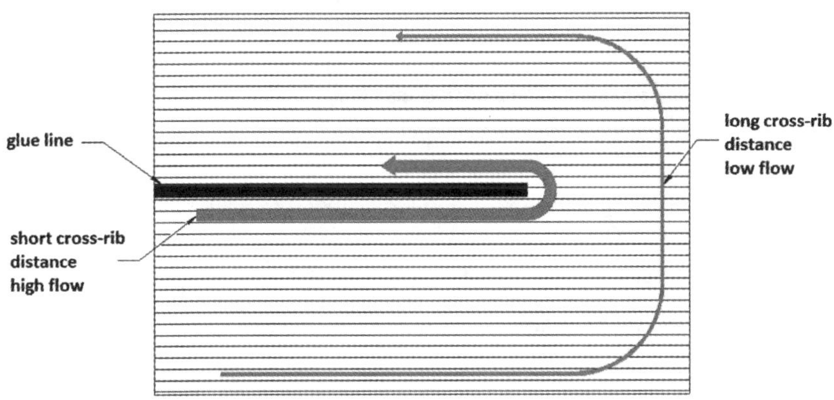

Figure 4.9. FO flow if a simple RO permeate spacer is used

4.5 FUTURE DEVELOPMENTS OF FO ELEMENTS

Currently, commercial FO applications use first generation CTA membranes and have been primarily in highly fouling steams. Feed spacers in spiral elements for these fluids are often 2.5-mm thick, which allows high sweeping velocities across the membrane, but consumes large amounts of power for circulation pumps. Thick spacers and high pumping costs for fluids such as landfill leachate or oil well drilling mud will continue to be viable, however there are emerging applications such as dewatering high salinity oil and gas produced water where much thinner feed spacers will be attractive.

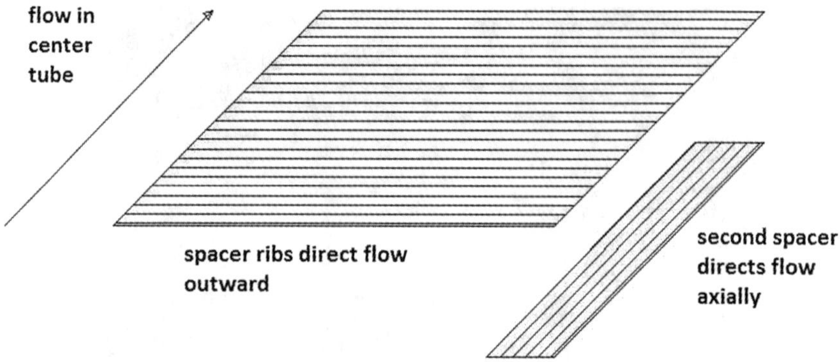

Figure 4.10. Permeate spacer to promote good draw distribution

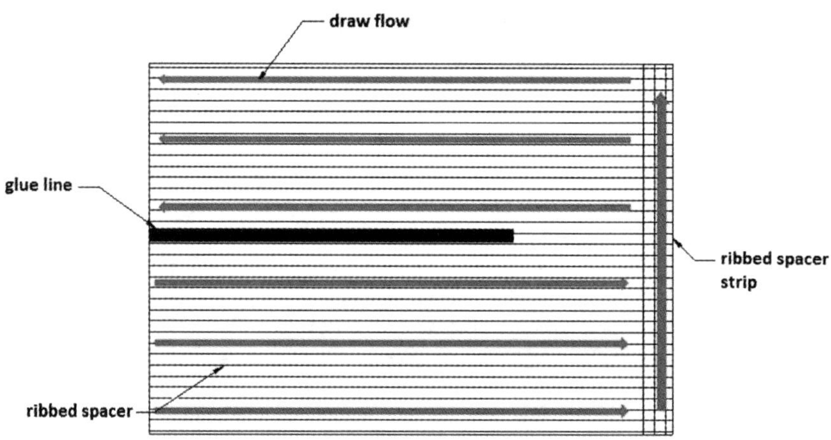

Figure 4.11. Draw solution flow in FO permeate spacer

FO technology has progressed to the point where commercial elements and systems are available. Developments which would currently be of the greatest advantage are higher flux membranes with greater pH and temperature tolerance and economical methods of draw solution recovery which can produce solutions of higher osmotic potential than can be achieved with RO.

4.6 ABBREVIATIONS/NOMENCLATURE

CP concentration polarization
CTA cellulose triacetate
DO dissolved oxygen
DOC dissolved organic carbon

References

Loeb, S. (1996). "Method of forming porous membranes." United States patent 3283042, 1966.

Popper, K. (1969). "Osmotic processes and apparatus." United States Patent 3423310, 1969.

York, R., Thiel, R., and Beaudry, E. (1999). "Full scale experience of direct osmosis concentration applied to leachate management." *Proceedings of the Seventh International Waste Management and Landfill Symposium (Sardinia '99)*, S. Margherita di Pula, Cagliari, Sardinia, Italy, 1999.

CHAPTER 5

Draw Solutes in Forward Osmosis Processes

Ho Kyong Shon
Laura Chekli
Sherub Phuntsho
Jungeun Kim
Jaeweon Cho

This chapter aims to provide valuable insight for the selection of suitable draw solution (DS) and review different DS characteristics affecting the performance of forward osmosis (FO) processes. It also presents a comprehensive classification of the different DS that have been tested so far. Although there are some commercial applications of FO technology, the development of an effective large-scale process is currently limited due to the lack of both suitable DS and membrane. The main challenge regarding the development of appropriate DS is their separation and recovery from the product water, especially for drinking water applications. The recovery process should not be expensive and energy intensive otherwise the FO process cannot be advantageous compared with other membrane technologies such as RO.

5.1 GENERAL CRITERIA FOR SELECTION OF SUITABLE DS

The selection of a suitable DS is crucial for the development of the FO technology and will not only improve the performance of the process but also save significant costs regarding the separation and recovery process. The selection of appropriate DS include an initial screening based on the intrinsic properties of the DS, laboratory-scale performance, full-scale performance and finally commercialisation as depicted in Figure 5.1. The information provided remains general and applicable to any applications. However, it is obvious that further specific criteria, depending on the targeted application, also need to be evaluated.

Initial screening can be carried out using thermodynamic modelling softwares such as OLI Stream Analyzer. These softwares can be very useful to

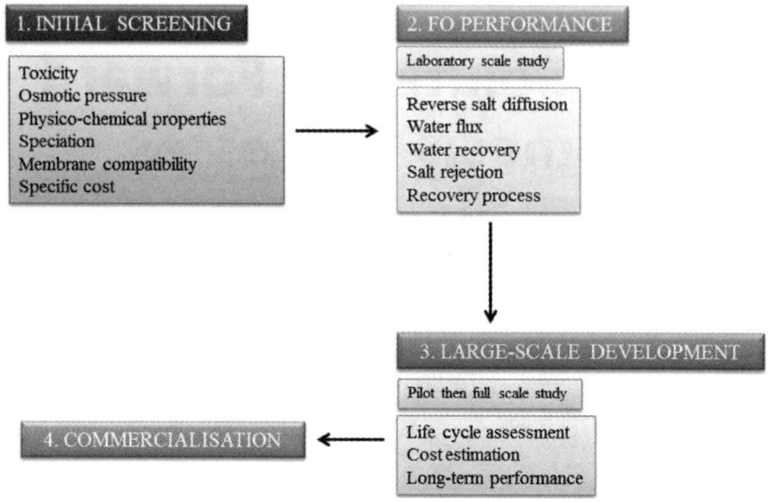

Figure 5.1. General criteria for selection of suitable DS

calculate some basic intrinsic properties including water solubility, pH, speciation and osmotic pressure, which are important properties that affect the performance of the FO process (see section 2 for more details). It is also essential to ensure that the DS is inert, of near neutral pH, stable and have minimal toxicity, especially if the targeted application is the production of drinking water. In that specific case, the DS must be safe for human consumption. International norms on water quality and human health can be found on the WHO guidelines for drinking-water quality (WHO 2011). Another important criterion to evaluate is the compatibility between the DS and the FO membrane. In fact, the selected DS should not modify the properties and performance of the membrane through fouling, reaction, adsorption or dissolution. Finally, the DS should be produced at low cost to ensure the economic viability of the process.

Once the preliminary selection has been done, laboratory experiments can be conducted to evaluate the performance of the pre-selected DS in terms of reverse salt diffusion (RSF), water flux and water recovery, which are the main parameters used to assess the performance of the FO process. In the case of desalination applications, salt rejection is also an essential parameter to consider. Another crucial parameter to evaluate in most applications is the recovery process of the DS. This process should not be complex and be able to reconcentrate and recover the DS at low-energy cost.

Finally, the remaining DS should be tested at full-scale, and life cycle assessment should be conducted to assess the environmental impacts of each stage of the process (i.e. from the production of raw materials to the treatment and disposal of waste).

5.2 DS CHARACTERISTICS AFFECTING FO PERFORMANCE

There are several factors affecting the FO process performance, and these factors are generally associated to the membrane properties, DS characteristics and operating conditions. Some recent studies investigated the effect of these different parameters on the process performance (Xu et al. 2010; Zhao and Zou 2011; Zhao et al. 2011; Lay et al. 2012; Mamisaheby et al. 2012). This section will focus on the DS characteristics.

5.2.1 General DS Characteristics

As the osmotic pressure across the FO membrane is the driving force of the process, the DS must exhibit high solubility in water and high osmotic pressure, much higher than the osmotic pressure of the feed solution in order to induce high water flux across the membrane. Based on the theory proposed by Van't Hoff (1887), the osmotic pressure (π) of an ideal dilute solution is expressed as follow:

$$\pi = n\left(\frac{c}{M}\right)RT \tag{5.1}$$

where n is the number of moles of species formed by the dissociation of solutes in the solution; c is the concentration of the solute in g/L; M is the molecular weight of the solute in g/mol; R is the gas constant ($R = 0.0821$); and T is the absolute temperature of the solution in K. Based on equation (5.1), it is clear that a DS combining a low molecular weight and high water solubility will produce a high osmotic pressure (on equal mass basis) and generate high water flux across the membrane (McCutcheon et al. 2005).

The diffusion coefficient (D_s) of the DS will also affect the performance of the process and can be described as (Van't Hoff 1887):

$$K = \frac{t\tau}{\varepsilon D_s} \tag{5.2}$$

where K is the solute resistance to diffusion within the membrane support layer, t, τ and ε represent the thickness, tortuosity and porosity of the membrane support layer respectively. This equation clearly shows that a DS with a high diffusion coefficient will readily diffuse through the membrane support layer and thus decrease the ICP effects. In general, solutes with lower molecular weight exhibit higher diffusion coefficient and can thus theoretically promote higher water fluxes across the membrane. However, low molecular weight solutes also show high RSF (Ng and Tang 2006; Achilli et al. 2010; Wang et al. 2010; Yen et al. 2010; Phuntsho et al. 2011) and several studies demonstrated that this phenomenon can jeopardise the process (Hancock and Cath 2009; Lee et al. 2010). The reverse salt diffusion (J_s) of a DS can be calculated from the increase of the DS concentration in the feed solution:

$$J_s = \frac{(C_t V_t) - (C_0 V_0)}{\Delta t}\frac{1}{A_m} \tag{5.3}$$

where C_t (mol/L) and V_t (L) are the salt concentration and feed volume at a time t, C_0 (mol/L) and V_0 (L) are the salt concentration and feed volume at t_0, $\Delta t = t - t_0$ (h) is the duration of the experiment and A_m (m^2) is the effective membrane area. When DS exhibit electrical conductivities in aqueous solution, their concentration in the feed solution can be easily monitored when pure water is used as feed solution. RSF of the DS can not only decrease the performance of the process by decreasing the net osmotic pressure across the membrane, but can also complicate concentrate management. The accumulation of DS in the feed solution can also have adverse effect on sensitive receiving environment or alter adjacent treatment processes (Hancock and Cath 2009).

As for any pressure-based membrane process, temperature is also affecting the performance of the FO process; and more specifically, the DS temperature may impact significantly on the efficiency of the process. Many studies have demonstrated that increasing the DS temperature enhanced the water flux due to reduced water viscosity and thus improved mass transfer across the membrane (Petrotos et al. 1998; McCutcheon et al. 2006; Garcia-Castello et al. 2009; Zhao and Zou 2011). Increasing the DS solution temperature will also increase its diffusion coefficient in solution and therefore decrease the value of K (i.e. solute resistance to diffusion within the membrane support layer) based on equation (2) which will ultimately improve the water flux (Petrotos et al. 1998; McCutcheon et al. 2005). However, it was also demonstrated recently that increasing temperature improved water flux only to a certain critical point; after which membrane scaling occurred, causing flux decline and in overall decreasing the process performance (Garcia-Castello et al. 2009; Zhao and Zou 2011).

Finally, the DS concentration has also been shown to influence the process performance (McCutcheon et al. 2006; Achilli et al. 2009; Choi et al. 2009; Xu et al. 2010). However, the relationship between DS concentration and water flux is not linear, especially at high DS concentration where the relation has been observed to become logarithmic. This has been attributed to ICP effects in the membrane support layer which become more important at higher permeate flux resulting in less effective water flux improvement (Tan and Ng 2010).

5.2.2 Other DS Characteristics

Other specific DS characteristics can impact on the process performance depending on the targeted application. For instance, Achilli et al. (2010) demonstrated that the use of DS containing scale precursor ions (e.g. Mg^{2+}, Ca^{2+}, Ba^{2+}, SO_4^{2-} and CO_3^{2-}) has to be limited to application involving the use of pure water as feed solution such as the food industry. In fact, the presence of scale precursor ions will promote mineral scaling on the membrane surface when the feed solution concentration is above the solubility limit. Novel synthetic DS have been recently developed and their unique properties may also impact on the process efficiency. For instance, magnetic nanoparticles (MNPs) as DS, have been developed by Chung and co-workers (Ling et al. 2010;

Ling and Chung 2011b; Ling and Chung 2011a; Ling et al. 2011) and demonstrated a strong tendency to aggregate which could enhance membrane fouling and decrease the performance of the process.

5.3 CLASSIFICATION OF OSMOTIC DS

In the past few decades, there has been a rapid progress in the development of suitable draw solutions for FO process; however, only few reviews on draw solutions have been published to date (Chekli et al. 2012; Ge et al. 2013). Draw solutions can be generally classified as volatile compounds, organic compounds, inorganic compounds and novel synthetic compounds including MNPs and polymer hydrogel. Sub-classification may include electrolyte and non-electrolyte solutions depending on the surface charge of the DS. An overview of the different DS that has been tested so far, including some of their properties and performance is summarised in Table 5.1.

5.3.1 Volatile Compounds

The development of volatile compounds as draw solute started in the early 60's. Neff (1964) was the first to propose a mixture of ammonia and carbon dioxide gases for seawater desalination. In the patent, the experimental conditions to prepare ammonium bicarbonate (NH_4HCO_3) solution and its regeneration process (i.e. a gas-liquid separating column) are explained. However, due to its low solubility in water, NH_4HCO_3 produced a low osmotic pressure, limited its application for FO process. To overcome this issue, McGinnis and co-workers (McCutcheon et al. 2006; McGinnis et al. 2007) developed a novel thermolyte solution using a mixture of ammonium hydroxide (NH_4OH) and NH_4HCO_3 in specific proportion or by fluctuating the ratio of ammonia and carbon dioxide for the preparation of NH_4HCO_3 to increase its solubility. Higher osmotic pressure was achieved (i.e. up to 238 bars) as well as high water fluxes, close to 25 LMH. They also demonstrated that the mixture can be easily recovered by distillation process using low heat energy as this DS can decompose into NH_3 and CO_2 by heating up to only 60°C (McCutcheon et al. 2006). However, the proximity of low-grade heat (e.g. from thermal power plants) is required to ensure that the recovery process is economically viable. The use of NH_4HCO_3 as DS was also investigated in other recent studies for desalination application (Ng and Tang 2006; Hancock and Cath 2009; Achilli et al. 2010). The solubility of NH_4HCO_3 was found to be the lowest in comparison to other tested DS and its RSF was among the highest due its low molecular weight (Achilli et al. 2010). Ng and Tang (2006) also found that this DS can suffer from early decomposition if the operated temperature is higher than 30°C.

In order to develop a continuous FO process including the DS recovery process, McGinnis (2002) introduced the use of potassium nitrate (KNO_3) and

Table 5.1. Overview of the different draw solutions used in FO process

Draw solution types	*Draw solution*	*Osmotic pressure (concentration)*	*Water flux (Feed solution – membrane)*	*Reverse salt flux*	*Recovery process(es)*	*Drawbacks*	*Example(s) of application*	*Commercial status*
Volatile compounds	Ammonium Bicarbonate (NH_4HCO_3)	66.40 atm (2.0 M)	2.04 μm/s (DI water – CTA flat sheet)	18.2 $g/m^2 \cdot h$ at 2.8 MPa	Heating – decomposition into NH_3 and CO_2	Low solubility in water; High reverse salt diffusion; High replenishment cost; Not thermally stable	Seawater desalination (McCutcheon et al. 2005; McCutcheon et al. 2006; Ng et al. 2006; Hancock and Cath 2009; Achilli et al. 2010)	Commercial (Oasys Water) for brine concentration
	Sulfur dioxide (SO_2)	Not reported	Not reported	Not reported	Heating, air stripping or distillation	Volatile; Corrosive; Unstable in solution	Seawater desalination (Batchelder 1965; Glew 1965)	—
Organic compounds	Glucose ($C_6H_{12}O_6$)	55.03 atm (2.0 M)	0.37 LMH (tomato juice – thin film composite aromatic polyamide)	Not reported	Direct application, Low pressure RO	High molecular size causes high internal concentration polarization (ICP) effect	Emergency drinking solutions (Kravath and Davis 1975; Wallace et al. 2008); Seawater desalination (Yaeli 1992; Ng and Tang 2006); Tomato juice concentration (Petrotos et al. 1998)	—
	Fructose ($C_6H_{12}O_6$)	55.02 atm (2.0 M)	2.5 LMH (0.5 M NaCl –thin layer of cotton-derived cellulose-ester plastics embedded on top of a microfiltration membrane)	Not reported	Direct application		Nutritious drink production (Stache 1989); Seawater desalination (Ng et al. 2006); Brine control (Ng and Tang 2006)	—

	Sucrose ($C_{12}H_{22}O_{11}$)	26.7 atm (1.0 M)	12.9 LMH (DI water – CA hollow fibre)	Not reported	Nanofiltration	Low water flux	Wastewater treatment (Su et al. 2012)	—
	Ethanol (C_2H_6O)	49.93 atm (2.0 M)	Not reported	Not reported	Pervaporation-based separation	High reverse salt diffusion and low water flux	Recovery of water from highly impaired sources (McCormick et al. 2008)	—
	Sodium formate (HCOONa)	31.4 MPa (at saturation)	2.6 μm/s at 2.8 MPa (DI water – CTA flat sheet)	6.04 $g/m^2 \cdot h$ at 2.8 MPa	RO process	High reverse salt flux; High replenishment cost	Wastewater treatment (Bowden et al. 2012)	—
	Sodium acetate ($C_2H_3NaO_2$)	27.0 MPa (at saturation)	2.5 μm/s at 2.8 MPa (DI water – CTA flat sheet)	2.73 $g/m^2 \cdot h$ at 2.8 MPa		Relatively high replenish-ment cost compared to inorganic salts		—
	Sodium propionate ($C_3H_5NaO_2$)	N.A.	2.41 μm/s at 2.8 MPa (DI water – CTA flat sheet)	1.47 $g/m^2 \cdot h$ at 2.8 MPa				
	Magnesium acetate ($Mg(CH_3COO)_2$)	10.3 MPa (at saturation)	2.25 μm/s at 2.8 MPa (DI water – CTA flat sheet)	1.07 $g/m^2 \cdot h$ at 2.8 MPa				
	Switchable polarity solvents (SPS)	~325 atm (7.6 mol/kg)	10 LMH at 7.6 mol/kg (2 mol/kg NaCl – CTA flat sheet)	1.8 wt.% of the concentration	Polar to non-polar phase shift induced by CO_2	Degradation of FO membrane	Industrial purification of water (Stone et al. 2013a)	Under develop-ment (Idaho National Laboratory)

(Continued)

Table 5-1. Overview of the different draw solutions used in FO process (Continued)

Draw solution types	*Draw solution*	*Osmotic pressure (concentration)*	*Water flux (Feed solution – membrane)*	*Reverse salt flux*	*Recovery process(es)*	*Drawbacks*	*Example(s) of application*	*Commercial status*
Inorganic salts	Sodium Chloride (NaCl)	100.40 atm (2.0 M)	2.68 μm/s (DI water – CTA flat sheet)	7.2 g/m^2· h at 2.8 MPa	RO process, Distillation/RO, Direct application	High reverse salt diffusion; Protein deterioration	Algal biodiesel production (Zou et al. 2011); Seawater desalination (Achilli et al. 2010); Sucrose concentration (Garcia-Castello et al. 2009); Reuse of domestic wastewater (Achilli et al. 2009); Juice concentration (Petrotos et al. 1998); Emergency drinking solutions (Wallace et al. 2008)	Pre-commercial (IDE Technologies, Porifera)
	Potassium Chloride (KCl)	89.30 atm (2.0 M)	6.337 μm/s (DI water – CA embedded in polyester woven mesh)	0.222 mmoles/m^2· s	Direct application	High reverse salt diffusion	Direct fertigation (Phuntsho et al. 2011); Desalination (Achilli et al. 2010)	—
	Ammonium Chloride (NH_4Cl)	87.70 atm (2.0 M)	5.348 μm/s (DI water – CA embedded in polyester woven mesh)	0.333 mmoles/m^2· s	Direct application	High reverse salt diffusion	Direct fertigation (Phuntsho et al. 2011); Desalination (Achilli et al. 2010)	—
	Ammonium Nitrate (NH_4NO_3)	64.90 atm (2.0 M)	4.177 μm/s (DI water – CA embedded in polyester woven mesh)	0.791 mmoles/m^2· s	Direct application	High reverse salt diffusion	Direct fertigation (Phuntsho et al. 2011)	—

	Potassium Bromide (KBr)	89.70 atm (2.0 M)	2.84 μm/s (DI water – CTA flat sheet)	22 g/m^2· h at 2.8 MPa	RO process	Very high reverse salt diffusion; High replenishment cost	Desalination (Achilli et al. 2010)	—
	Sodium Bicarbonate ($NaHCO_3$)	46.70 atm (2.0 M)	2.47 μm/s (DI water – CTA flat sheet)	1.7 g/m^2· h at 2.8 MPa	RO process	Low water solubility; Contain scale precursor ions;	Desalination (Achilli et al. 2010)	—
	Potassium Bicarbonate ($KHCO_3$)	79.30 atm (2.0 M)	2.25 μm/s (DI water – CTA flat sheet)	1.4 g/m^2· h at 2.8 MPa	RO process	Contain scale precursor ions; Not easily recovered by RO	Desalination (Achilli et al. 2010)	—
	Magnesium Chloride ($MgCl_2$)	256.50 atm (2.0 M)	2.33 μm/s (DI water – CTA flat sheet)	4.8 g/m^2· h at 2.8 MPa	Nanofiltration or direct application	High viscosity; Low diffusion coefficient; Mg^{2+} may affect membrane fouling by complexing with some functional group	Seawater desalination (Hancock and Cath 2009; Achilli et al. 2010); Emergency drinking solutions (Wallace et al. 2008); Algal biodiesel production (Zou et al. 2011)	—
	Calcium Chloride ($CaCl_2$)	217.60 atm (2.0 M)	2.64 μm/s (DI water – CTA flat sheet)	7.9 g/m^2· h at 2.8 MPa	RO process	Contain scale precursor ions	Tomato juice concentration (Petrotos et al. 1998); Emergency drinking solutions (Wallace et al. 2008); Seawater desalination (Achilli et al. 2010)	—

(Continued)

Table 5-1. Overview of the different draw solutions used in FO process (Continued)

Draw solution types	*Draw solution*	*Osmotic pressure (concentration)*	*Water flux (Feed solution – membrane)*	*Reverse salt flux*	*Recovery process(es)*	*Drawbacks*	*Example(s) of application*	*Commercial status*
	Ammonium Sulfate ($(NH_4)_2SO_4$)	92.10 atm (2.0 M)	5.391 μm/s (DI water–CA embedded in polyester woven mesh)	0.006 mmoles/$m^2 \cdot s$	Direct application	High replenishment cost	Direct fertigation (Phuntsho et al. 2011); Desalination (Achilli et al. 2010)	—
	Sodium Sulfate (Na_2SO_4)	95.2 atm (2.0 M)	2.14 μm/s (DI water – CTA flat sheet)	2.7 g/$m^2 \cdot h$ at 2.8 MPa	Nanofiltration	Contain scale precursor ions	Desalination (Achilli et al. 2010)	—
	Potassium Sulfate (K_2SO_4)	32.40 atm (2.0 M)	2.52 μm/s (DI water – CTA flat sheet)	3.7 g/$m^2 \cdot h$ at 2.8 MPa	RO process	Low water solubility; High replenishment cost	Desalination (Achilli et al. 2010)	—
	Magnesium Sulfate ($MgSO_4$)	54.80 atm (2.0 M)	1.54 μm/s (DI water – CTA flat sheet)	1.2 g/$m^2 \cdot h$ at 2.8 MPa	Nanofiltration	High viscosity; Low water solubility; Contain scale precursor ions	Desalination (Achilli et al. 2010)	—
	Copper Sulfate ($CuSO_4$)	29.94 bar (220,000 ppm)	3.57 LMH (5,050 ppm NaCl – CTA flat sheet)	Not reported	Metathesis precipitation with barium hydroxide then sulphuric acid	Low water flux; FO process severely affected by concentration polarization effect	Brackish water desalination (Alnaizy et al. 2013)	—
	Sodium Nitrate ($NaNO_3$)	81.10 atm (2.0 M)	5.706 μm/s (DI water – CA embedded in polyester woven mesh)	0.278 mmoles/$m^2 \cdot s$	Direct application	High reverse salt diffusion	Direct fertigation (Phuntsho et al. 2011)	—

	Potassium Nitrate (KNO_3)	64.90 atm (2.0 M)	4.429 μm/s (DI water – CA embedded in polyester woven mesh)	0.486 mmoles/$m^2 \cdot s$	Direct application	High reverse salt diffusion; toxic; energy intensive	Direct fertigation (Phuntsho et al. 2011)	—
	Calcium Nitrate ($Ca(NO_3)_2$)	108.50 atm (2.0 M)	50.22 μm/s (DI water – CA embedded in polyester woven mesh)	0.009 mmoles/$m^2 \cdot s$	Direct application	High replenishment cost; Poor water extraction capacity	Direct fertigation (Phuntsho et al. 2011); Desalination (Achilli et al. 2010)	—
	Ammonium Phospate ($NH_4H_2PO_4$)	86.30 atm (2.0 M)	4.349 μm/s (DI water – CA embedded in polyester woven mesh)	0.069 mmoles/$m^2 \cdot s$	Direct application	Low water flux	Direct fertigation (Phuntsho et al. 2011)	—
	Diammonium Phosphate ($(NH_4)_2HPO_4$)	95.00 atm (2.0 M)	3.892 μm/s (DI water – CA embedded in polyester woven mesh)	0.01 mmoles/$m^2 \cdot s$	Direct application	Low water flux	Direct fertigation (Phuntsho et al. 2011)	—
Novel synthetic materials	Polyacrylic acid MNPs (PAA MNPs)	Up to 70 atm (0.08 mol/L)	10 to 17 LMH (DI water – HTI membrane)	Not reported	Magnetic field separator, FO process using RO brine as DS, Ultrafiltration	Slightly drop of water flux due to agglomeration of the MNPs	Protein enrichment (Ling and Chung 2011b); Water reuse (Ling et al. 2010); Desalination (Ling and Chung 2011a)	—
	2-pyrolidone based MNPs (2-Pyrol MNPs)	Not reported	0.5 to 5 LMH (DI water – HTI membrane)	Not reported				
	Triethylene glycol MNPs (TREG MNPs)	Not reported	0.5 to 5 LMH (DI water – HTI membrane)	Not reported				

(Continued)

Table 5-1. Overview of the different draw solutions used in FO process (Continued)

Draw solution types	*Draw solution*	*Osmotic pressure (concentration)*	*Water flux (Feed solution – membrane)*	*Reverse salt flux*	*Recovery process(es)*	*Drawbacks*	*Example(s) of application*	*Commercial status*
	Polyethylene glycol diacid MNPs (PEG-$(COOH)_2$ MNPs)	55 to 73 atm (0.065 mol/L)	5.3 to 9.1 LMH (DI water – CTA flat sheet)	Not reported			No specific application (Ge et al. 2011)	—
	Nano size dextran coated Ferric oxide MNPs (Fe_3O_4)	Not reported	3.25 to 4 LMH (DI water – HTI membrane)	Not reported	External magnet	Slightly drop of water flux due to agglomeration of the MNPs	Brackish water desalination (Bai et al. 2011)	—
	2-methylimidazole based compounds with monovalent and divalent charges	50 to 150 bar (2.0 M)	0.1 to 20 LMH (DI water – CTA flat sheet)	5 to 80 $g/m^2 \cdot h$	FO-MD integrated process	High ICP effect when using compound with divalent charge; High reverse solute flux; High cost	Seawater desalination (Yen et al. 2010)	—
	Polyelectrolytes of polyacrylic acid sodium (PAA-Na)	25-45 atm (0.72 mg/L)	13 to 21 LMH (DI water – CA hollow fiber)	0.15 $g/m^2 \cdot h$	FO-MD integrated process, Ultrafiltration	High viscosity	Wastewater treatment (Ge et al. 2012b)	—
	Thermo-sensitive polyelectrolytes	Up to 88 atm (14.28 wt% poly-electrolytes with different sodium acrylate content)	0.05 to 0.75 LMH (DI water – HTI membrane)	Not reported	Hot ultrafiltration	Poor water flux	Desalination (Ou et al. 2013)	—

	Polymer hydrogels	2.7 MPa	0.55 to 1.1 LMH (2000 ppm NaCl – HTI membrane)	Not reported	Direct application, Heating, Pressure Stimuli	Energy intensive; Poor water flux	Seawater desalination (Li et al. 2011b; Li et al. 2013)	—
	Acyl-TAEA	N.A.	N.A.	N.A.	High temperatures	Poor water flux	Seawater desalination (Noh et al. 2012)	—
	Micelles close to the Kraft point	95.1 Pa	4.73 to 16.14 LMH (n.a.)	Not reported	Temperature swing with low grade heat and crystallisation	Low diffusivity	Seawater desalination (**Gadelha and Hankins 2011**)	—
	Dendrimers	330 psi (22.5 atm) (20 wt.%)	Not tested	Not tested	Wide range of pH value and ultrafiltration	Not feasible	Dewatering Reverse Osmosis concentrate (Adham et al. 2009)	—
	Albumin	7 psi (30 wt.%)	Not tested	Not tested	Denatured and solidified upon heating			—
	Concentrated RO Brines	Not measured	8.8 to 11 LMH	Not tested	RO process	Precipitation of organic salts on membrane surface	NPs regeneration (Ling and Chung 2011a); Impaired water concentration (Bamaga et al. 2011); Seawater desalination (Martinetti et al. 2009)	—
	Hexavalent phosphazene	Not measured	6 (Na salt) to 7 (Li salt) LMH (DI water – HTI membrane)	Not tested	Direct application	Hydrolysis the CTA membrane	No specific application (Stone et al. 2013b)	—
	Polymer-graphene composite hydrogels	Not measured	6.8 to 8.2 LMH (DI water or 2000 ppm NaCl – HTI membrane)	Not reported	Heating	Poor water flux	Seawater desalination (Zeng et al. 2013)	—

Note: The osmotic pressure, water flux and reverse salt flux data were obtained from studies highlighted in bold.

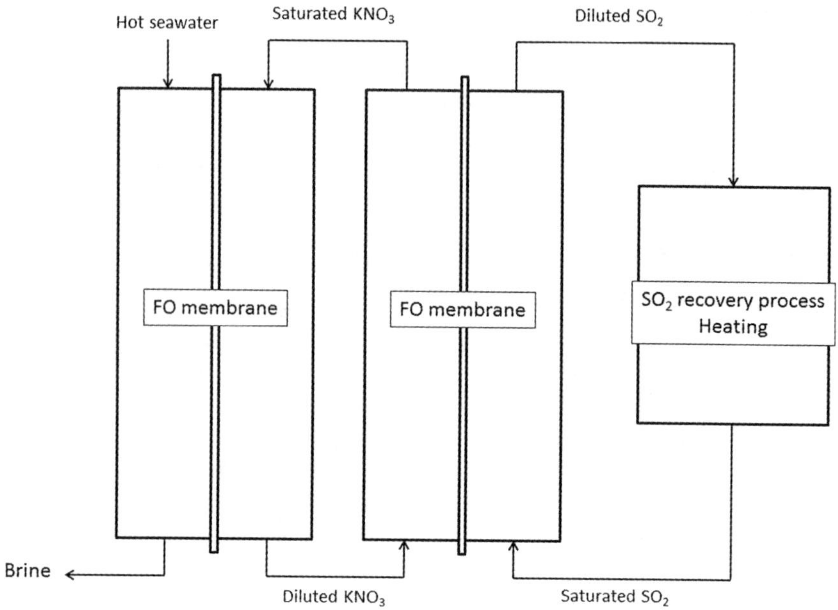

Figure 5.2. Schematic representation of the two-stage FO process for desalination

SO_2 as DS for desalination application. This concept consisted of a two-stage FO process (Figure 5.2): Hot seawater and saturated KNO_3 solution were used as feed and draw solution respectively in the first unit while the diluted KNO_3 from the first FO unit is used as feed solution in the second unit and reconcentrated by a saturated SO_2 solution, used as DS. The SO_2 is removed by standard means such as heating. However, performance data are not available till now.

SO_2 was also suggested as potential DS in the early 1960's. Batchelder (1965) was the first to test SO_2 for desalination application. In the patent, the separation of SO_2 from the product water was made by heating or air stripping. However, although a positive water flux was observed in the experiments; neither the osmotic pressure, nor the water flux and salt rejection were reported in the patent. In the same year, Glew (1965) proposed a mixture of water and SO_2 or aliphatic alcohol as DS. The purpose of adding SO_2 or aliphatic alcohol to water is to increase the osmotic pressure of the resulting solution. In this study, the recovery of SO_2 was proposed by means of distillation.

5.3.2 Organic Compounds

Over the past decades, organic compounds have been tested as DS for various applications, including mainly food production (Beaudry and Lampi 1990; Wrolstad et al. 1993; Herron et al. 1994; Petrotos et al. 1998) and seawater desalination (Kravath and Davis 1975; Kessler and Moody 1976; Stache 1989; Ng and Tang 2006). Although organic compounds are generally non-charged compounds, they display high solubility in water and can generate high osmotic pressure (Ng and Tang 2006).

5.3.2.1 Nutrient Compounds

The study of nutrient compounds as DS originally focused on desalination applications for drinking water production, where the diluted DS can be used directly, excluded the need for its recovery. Kravath and Davis (1975) were the first to explore glucose as a DS for emergency water supply in lifeboats. The diluted glucose solution, present in the final product, was edible and did not require further treatment. Following the work of Kravath and Davis, Kessler and Moody (1976) investigated the combination of glucose and fructose as DS. Better performance was achieved in comparison to pure glucose solution. Glycine was also suggested as DS for the same application as it can produce a higher osmotic pressure than fructose; however, it can be toxic for human consumption at high concentration (Stache 1989). In the work, Stache (1989) demonstrated that, to be efficient, the fructose content must be at least 90% (w/w) to deliver an appropriate water flux. He also showed that below 74% (w/w), the water flux may be reversed from the draw solution to the saline feed solution. Later, Yaeli (1992) developed a hybrid system combining FO process and low pressure RO process for the recovery of glucose.

The use of nutrient compounds as DS is still attracting attention recently (Wallace et al. 2008; Su et al. 2012; Yong et al. 2012) and HTI has recently developed and manufactured the first commercial hydration bags using FO technology where sugar or beverage powder is used as DS (HTI 2010). In their recent work, Su et al. (2012) found that sucrose was more effective than $MgCl_2$ as DS. This was explained by the large molecular size of sucrose resulting in a very low reverse draw solution flux; producing an almost constant osmotic gradient across the membrane during the time of the experiments.

5.3.2.2 Organic Salts

Organic salts are defined as the mixture of an organic anion and an organic or inorganic cation. In a recent study, Bowden et al. (2012) tested a series of organic salts as potential DS in an osmotic membrane bioreactor (OMBR) for wastewater treatment as shown in Figure 5.3. After an initial screening based on ten organic salts, they selected four potential DS for further tests in OMBR, RO reconcentration and biodegradation. The advantage of using biodegradable organic salts is to moderate the accumulation of DS inside the bioreactor. They found that the organic salts with shorter carbon chain displayed better performance that the ones with longer carbon chains when they have the same cation species at the same osmotic pressure. This is because the organic salts with shorter carbon chain have higher diffusion coefficient and thus generate less ICP effects. Compared to pure inorganic salts (i.e. NaCl and $MgCl_2$), organic salts were proved to display higher salt rejection in the RO reconcentration process but the produced water fluxes were much lower under the same conditions.

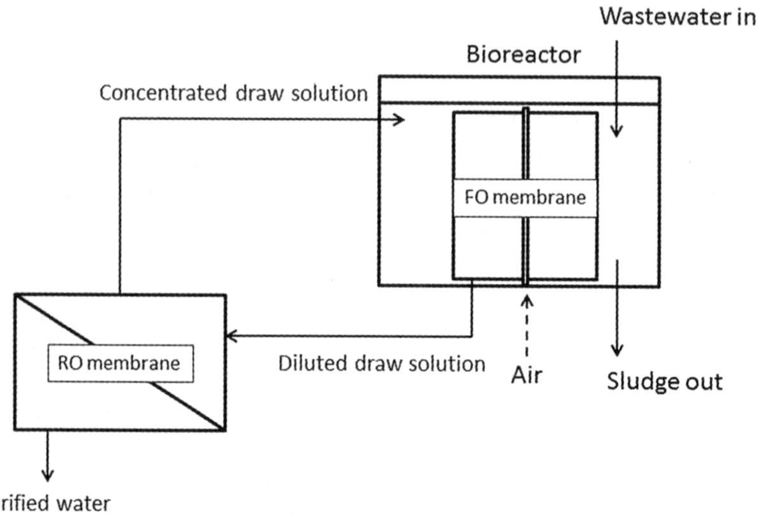

Figure 5.3. Schematic of the OMBR process
SOURCE: Adapted from Bowden et al. 2012; reproduced with permission from Elsevier

5.3.2.3 Switchable Polarity Solvents (SPS)

In a recent work carried out by the Idaho National Laboratory (Stone et al. 2013a), switchable polarity solvents (SPS) (Jessop et al. 2005, 2011, 2012) have been tested as potential novel DS for the industrial purification of water. SPS are mixtures of CO_2, water and tertiary amines. Their principal characteristic is their ability to switch from polar to non-polar form in the presence or not of CO_2, as shown in Figure 5.4.

This unique characteristic makes their recovery very straightforward, simply by adding 1 atm of air or N_2 with mild heating to form hydrophobic SPS. Then, the hydrophobic SPS can be regenerated and recycled to hydrophilic SPS by adding 1 atm of CO_2. Further treatment can be done to remove the trace amount of hydrophobic SPS from the produced water with RO process; as depicted in Figure 5.5.

Regarding FO performance, these DS were able to generate a very high osmotic pressure, up to 325 atm, and produce a reasonable water flux, up to 35 LMH, depending on the concentration of NaCl in the feed solution. Poor RSF was observed at NaCl concentration below 0.75 mol/kg. At higher concentration, RSF was also relatively low (e.g., up to 1.8 wt% of the initial DS concentration). However, under the tested experimental conditions, authors found that SPS degraded the FO membrane.

$$NR_3 + CO_2 + H_2O \leftrightarrow HNR_3^+ + HCO_3^-$$

Non-polar Polar

Figure 5.4. General example of an SPS from its non-polar to its polar form

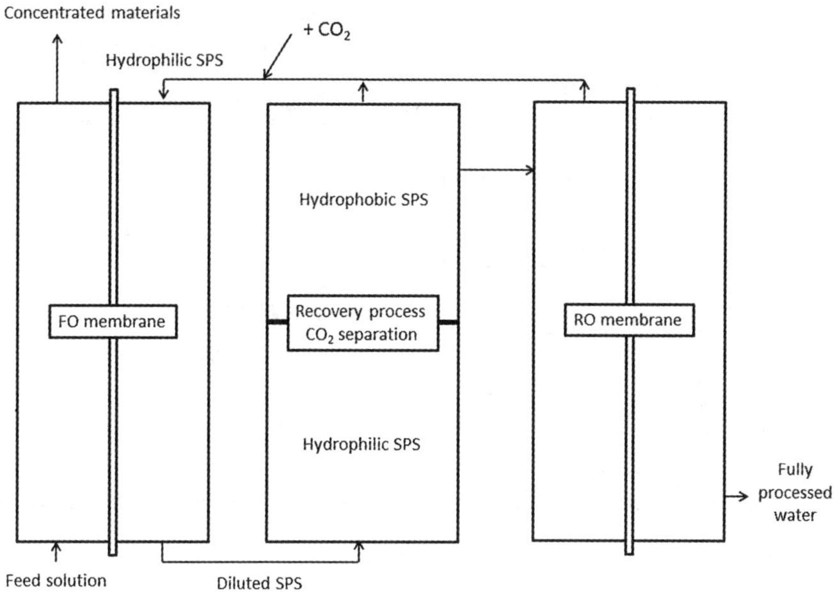

Figure 5.5. Schematic representation of the proposed SPS FO process

5.3.3 Inorganic Compounds

The vast majority of FO studies have tested inorganic-based compounds as DS, and they are still widely used nowadays. Inorganic-based compounds are mainly composed of electrolyte solutions although some non-electrolyte solutions could be also encountered. The use of inorganic compounds as DS has attracted many researchers since they can generally produce reasonably high water fluxes and can be easily recovered by RO (Achilli et al. 2010) or NF (Tan and Ng 2010) or used directly without further treatment (e.g. for fertigation) (Phuntsho et al. 2011).

The study of inorganic compounds as DS started in 1970 when Hough tested a variety of acids such as manganic acid, valeric acid and glucuronic acid and a range of inorganic salts such as calcium, potassium, sodium, barium and cesium salts as DS for desalination application. In the patent, their temperature-sensitive solubility or reactions with sulphuric acid or carbon dioxide to extract water by precipitating these DS were utilised. Later, Frank (1972) investigated the use of aluminium sulphate as DS for seawater desalination and their recovery by precipitation with calcium sulphate. The major drawback of these DS was the excessive use of chemicals for their recovery.

The most recent and comprehensive studies of inorganic compounds as DS were proformed by Phuntsho et al. (2011), Achilli et al. (2010) and Tan and Ng (2010). In their study, Achilli and co-workers tested and compared 14 inorganic compounds as DS for potential application in desalination. These DS were selected from among 500 inorganic compounds due to their higher water solubility and

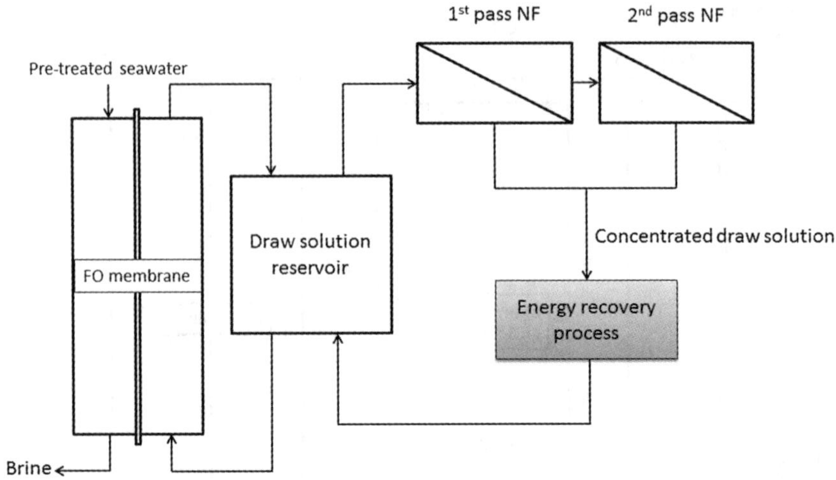

Figure 5.6. Schematic representation of the proposed FO-NF process for seawater desalination

SOURCE: Adapted from Tan and Ng, 2010; reproduced with permission from Taylor and Francis

osmotic pressure and lower cost and toxicity. The performance of the selected DS was evaluated in terms of water fluxes and RSF through the FO process and their recovery were tested using a RO process design software. DS containing larger hydrated anions (i.e. $MgSO_4$, $KHCO_3$, $NaHCO_3$, Na_2SO_4, $(NH_4)_2SO_4$ and K_2SO_4) showed the lowest RSF. However, based on the overall performance (i.e. FO performance and recovery test), none of the 14 tested DS outperformed the others. Tan and Ng (2010) investigated a novel hybrid system combining FO and NF for seawater desalination (Figure 5.6); seven potential DS were tested at laboratory scale, including six inorganic compounds (i.e. NaCl, KCl, $MgCl_2$, $CaCl_2$, $MgSO_4$, Na_2SO_4). Their results showed that KCl and NaCl exhibit the highest water flux of more than 25 LMH while $MgCl_2$ and $CaCl_2$ produced the highest osmotic pressure due to their high solubility in water. Experiments also showed that all DS displayed high solute rejection (i.e. above 99 %). Only $MgCl_2$, $MgSO_4$ and Na_2SO_4 were tested on the hybrid FO/NF system and experiments showed that the system was able to produce high quality water when using $MgSO_4$ and Na_2SO_4 as DS. Finally, in a recent study by Phuntsho et al. (2011), fertilisers were proposed as DS to extract water from brackish sources for direct irrigation. The diluted fertilisers can be applied directly without any further treatment as a source of nutrient for irrigation. In their study, They investigated nine commonly used fertilisers as DS and assessed their performance in terms of pure water flux and reverse solute flux. KCl, $NaNO_3$ and KNO_3 exhibited the highest water fluxes while $NH_4H_2PO_4$ and $(NH_4)_2HPO_4$ showed the lowest. The ammonium compounds of phosphate and sulphate (i.e. $NH_4H_2PO_4$, $(NH_4)_2HPO_4$, $(NH_4)_2SO_4$ and $Ca(NO_3)_2$) had the lowest reverse solute fluxes since they possess ionic species with large hydrated diameter. Based on the results, they concluded that the choice of DS for fertigation mainly depends on the compatibility with the FO membrane as well as the plant nutrient requirements. In a more recent study (Phuntsho et al. 2012), they tested blended fertilisers as DS but the obtained osmotic pressures and water fluxes were lower than the sum of the two individual DS.

Sodium chloride (i.e. NaCl) has been widely used as DS in a variety of applications, including food production (Petrotos et al. 1998; Garcia-Castello et al. 2009) and water and wastewater treatment (York et al. 1999; Holloway 2006; Achilli et al. 2009; Martinetti et al. 2009). The advantages of NaCl are that it is abundant on earth, making seawater a natural and inexpensive source of DS, it is relatively easy to reconcentrate with RO process and it has high water solubility and osmotic pressure. Besides, the thermodynamic properties of NaCl have been widely investigated, making it easier to study.

Finally, a recent study by Alnaizy et al. (2013) investigated the use of copper sulfate ($CuSO_4$) as draw solution for desalination applications. They found that $CuSO_4$ cannot produce enough osmotic pressure to desalinate seawater but could be a viable candidate for brackish water desalination. The proposed recovery method consisted in metathesis precipitation reaction with barium hydroxide followed by reaction with sulphuric acid.

5.3.4 Novel Synthetic Compounds

5.3.4.1 Hydrophilic Magnetic Nanoparticles (MNPs)

Hydrophilic magnetic nanoparticles (MNPs) have gained growing interest in the past few years as promising DS for FO process (Adham et al. 2009; Ling et al. 2010; Warne 2010; Bai et al. 2011; Ge et al. 2011; Ling and Chung 2011b; Ling and Chung 2011a; Ling et al. 2011; Ling and Chung 2012). Warne et al. (2010) were the first to introduce the concept of using MNPs as DS. In the patent, they synthesised MNPs via multi-step reactions and their regeneration process consisted of an external magnetic field. However, very few data were provided in their study to demonstrate the feasibility of this novel class of DS. The same year, Chung and co-workers successfully applied MNPs as DS for FO processes (Ling et al. 2010; Ge et al. 2011; Ling and Chung 2011b; Ling and Chung 2011a; Ling et al. 2011). In their different studies, they synthesised various MNPs via one-pot reactions and capped them with different surface coatings. Four different MNPs were investigated as potential DS, including: the polyacrylic acid magnetic nanoparticles (PAA MNPs), the 2-Pyrrolidonemagnetic nanoparticles (2-Pyrol MNPs), the triethylene glycol magnetic nanoparticles (TREG MNPs) and the polyethylene glycol diacid (PEG-$(COOH)_2$ MNPs). Moderate performances were achieved in terms of water fluxes (i.e. up to 18 LMH) (Ling et al. 2010; Ge et al. 2011). However, performance of MNPs in terms of reverse fluxes was very high compared to inorganic compounds such as NaCl and $MgCl_2$ due to their large particle sizes.

The recovery of MNPs from the produced water was proposed by various methods including magnetic field (Ling et al. 2010; Bai et al. 2011; Ge et al. 2011) and low pressure membrane processes such as ultrafiltration (UF) (Ling and Chung 2011a). However, it was observed that under high-strength magnetic field, MNPs have a strong tendency to agglomerate which decreases significantly the performance of the FO process. Ultrasonication was used to overcome this issue but was not able to effectively restore MNPs without changing their magnetic properties (Ling and Chung 2011a). Therefore, UF was found to be more effective in recycling the MNPs. Thermo-responsive MNPs were also synthesised (Ling et al. 2011) and their recovery process only consisted in temperature stimulus. In fact, by heating the diluted MNPs solution above the critical temperature of the thermo-responsive MNPs, the MNPs will agglomerate spontaneously and can then be separated from the produced water by UF or low-strength magnetic field.

The dextran-coated Fe_3O_4 MNPs as a new draw solute was successfully synthesized and used for the FO desalination process (Bai et al. 2011). Fe_3O_4 MNPs can be separated from water by an external magnet field simply and it has good water solubility because it is coated with highly soluble organic substances. Dextran, a natural polysaccharide with a variety of molecular weight, is highly water soluble, which can generate as high osmotic pressure as inorganic salts (Cath et al. 2006). Therefore, Fe_3O_4 MNPs can be a candidature for the FO process as DS due to its strong magnetism resulting in easy separation with high osmotic pressure. The performance of the dextran coated Fe_3O_4 MNPs were investigated in terms of water flux in a lab-scale FO system for water reuse and desalination. Thus, two kinds of the dextran coated Fe_3O_4 MNPs containing different concentration of the dextran were demonstrated as draw solutes. Experimental results showed that the water flux of the dextran coated Fe_3O_4 MNPs containing higher dextran as DS was higher than that of the dextran coated Fe_3O_4 MNPs containing lower dextran as DS. This result indicated that the dextran coated Fe_3O_4 MNPs containing higher dextran can generate higher osmotic pressure leading to the higher driving force through the membrane. It was proved that the dextran coated Fe_3O_4 MNPs DS can draw clean water from the saline water and can be easily separated from the diluted DS by an external magnet depending on its super-paramagnetism.

Most recently, surface-ionized MNPs have been investigated as potential DS in FO (Ling and Chung 2012). These MNPs showed improved FO performance after the surface ionization and could be recovered via an integrated process comprising an electric field and a NF system. More details on MNPs are given in another chapter.

5.3.4.2 2-methylimidazole Based Compounds

Yen et al. (2010) synthesised novel organic compounds, both neutral and charged 2-methylimidazole based compounds and investigated their application for the FO process. The synthesised compounds were able to extract water from seawater and recovered by an integrated FO-MD (i.e. membrane distillation) process. However, this novel class of DS exhibited high reverse solute flux and the cost of synthesis remains very high.

5.3.4.3 Polyelectrolytes of Polyacrylic Acid Sodium (PAA-Na)

Recently, Ge and co-workers developed polyelectrolytes of polyacrylic acid sodium (PAA-Na) and tested its potential as DS (Ge et al. 2012a; Ge et al. 2012b). Their high solubility, flexible configuration and easy recovery via UF process make them suitable DS candidate for FO process. Experimental results showed that these new DS can achieve high water fluxes with very low reverse solute fluxes. A high salt rejection was also obtained when recycling PAA-Na with UF process. The recycled DS showed repeatable performance and indicated no aggregation problems in comparison with MNPs. However, at high concentration, PAA-Na solution showed a very high viscosity, limited its application at ambient conditions. To overcome this issue, a combined FO/MD process has been applied to increase the performance of this DS since MD works at high temperature (Ge et al. 2012b). In this recent study, PAA-Na salts were used as DS in an FO-MD hybrid system to recover water from a model wastewater containing acid dye as shown in Figure 5.7.

More recently, new thermo-sensitive polyelectrolytes were evaluated as DS for the FO process (Ou et al. 2013). Polyelectrolytes were synthesised by copolymerization of N-isopropylacrylamide with different amounts of sodium acrylate (SA) and thus nine different polyelectrolytes were prepared and tested in a

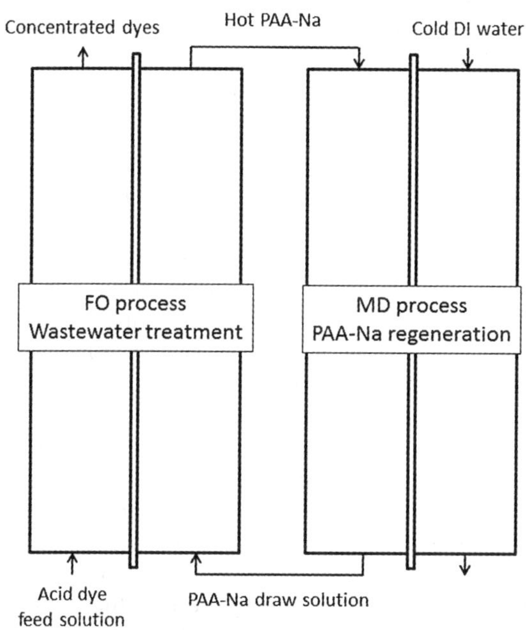

Figure 5.7. Schematic representation of the proposed FO-MD process for dye wastewater treatment

SOURCE: Adapted from Ge et al. 2012b; reproduced with permission from American Chemical Society lab-scale FO process as DS; these prepared draw solutes were thermo-sensitive and water soluble. In this study, hot ultrafiltration (HUF) process was used as a low-energy technique to separate the water from the polyelectrolyte draw solutions which was operated at 45°C and at 2 bar. Among nine prepared draw solutes, 4% poly N-isopropylacrylamide (PNIPAM)-SA solution showed the best performance in terms of the water flux (i.e. 0.347 LMH) in the FO process and water recovery fraction was around 65% in the HUF process. Although the produced water flux was relatively low, the ease of water recovery and reused would prove that new thermo-sensitive polyelectrolytes used in this study would have a potential of a good candidate for the FO process.

5.3.4.4 Stimuli-Responsive Polymer Hydrogels

Recently, Li et al. (2011a) have developed a new class of DS, the stimuli-responsive polymer hydrogels for desalination application. Polymer hydrogels are three-dimensional networks of polymer chains linked by either physical or chemical bonds. They are able to extract and release water in response to environmental stimuli (e.g. temperature, pressure or light). One interesting response to these stimuli is the change from hydrophilic to hydrophobic which make the polymer hydrogels to release water. This unique property is also advantageous for their recovery which becomes very easy and at low energy cost (compared to membrane or thermal processes). In a more recent study (Li et al. 2011c), polymer hydrogels were combined with light-absorbing carbon particles to enhance their heating and dewatering. In both studies, the polymer hydrogels showed high osmotic pressure of about 2.7 MPa at 27°C. However, FO performances were very poor at room temperature, with water fluxes ranging from 0.55 to 1.1 LMH only. Increasing the temperature to 50°C did improve the performance but required a high hydraulic pressure of 30 bars during the dewatering process.

In order to enhance the swelling ratio of the polymer hydrogel used as DS for the FO process, carbon filler particles were incorporated in the polymer hydrogel (Li et al. 2013). The poly (sodium acrylate) hydrogel was selected in this study due to its good performances in terms of high swelling ratio and water flux as compared with the non-ionic poly (N-isopropylacrylamide) or the copolymers of sodium acrylate and N-isopropylacrylamide. Although the addition of carbon particles increased the swelling pressure of hydrogels and led to enhanced water flux, excessive carbon fillers also contributed to a dramatic decrease in the swelling ratio of the composite hydrogels resulting in a lower flux in the FO process. In addition, the effect of hydrogel particle size on the FO process was examined. Smaller composite polymer hydrogel particles (ranging from 100 μm to 200 μm as draw agents) induced higher water fluxes in FO process as compared with those with larger particle sizes (500–700 μm as draw agents). Furthermore, increased osmotic pressure of the feed solution attributed to the decreasing water flux in the FO process. More details on these DS are given in another chapter.

5.3.4.5 Acyl-TAEA

Lee and co-workers (Noh et al. 2012) recently developed thermo-sensitive DS based on N,N',N"- triacylated tris (2-aminoethyl) amine (acyl-TAEA) derivatives. These new types of DS were able to extract water from saline water when temperatures were lower than the phase separation temperature while the recovery of DS required higher temperature. However, the FO performance of this novel DS is still poor and need further investigation.

5.3.4.6 Micellar Solutions

Gadêlha and Hankins (2011) recently introduced the concept of using micellar solutions as DS in FO process. Micellar solutions have interesting characteristics; they exhibit almost constant osmotic pressure above the critical micelle concentration which enables many dilutions prior to regeneration. Their solubility is highly temperature sensitive around the Krafft point (i.e. critical minimum temperature under which micelles cannot form). This property makes their regeneration very straightforward by simply applying small temperature fluctuations by using low-grade heating and cooling. They can also act as monomeric concentration buffers which attenuate the ICP effects within the membrane.

5.3.4.7 Dendrimers

Dendrimers are symmetrical globular or spheroid nanostructures, engineered to carry molecules. They consisted of a highly branched tree-like structure linked to a central core via covalent bonds. They were suggested and tested as DS for dewatering RO concentrate in a recent study by Adham et al. (2009). These macromolecules were able to produce a very high osmotic pressure, up to 330 psi (i.e. 22.5 atm), which is much higher than the osmotic pressure of RO concentrate. Another advantage is that they can be easily regenerated by low-pressure membrane processes such as UF. However, FO performances, in terms of water fluxes and reverse fluxes, were not assessed in this study.

5.3.4.8 Concentrated RO Brines

Some recent studies have investigated the use of RO brines as DS in order to solve the issue of its disposal. Ling and Chung (2011b) proposed a new FO system for protein enrichment in which MNPs are used as DS in the first FO process to concentrate proteins and RO brine is used as DS in the second FO process to reconcentrate the MNPs. Bamaga et al. (2011) also investigated the use of RO brine as DS in a hybrid FO/RO/FO system (Figure 5.8) where the first FO process is used as a pre-treatment for RO desalination to minimise scaling risk. The second FO process, using RO brine as DS, is used to concentrate the impaired water to minimise its volume for post-treatments. The combined FO/RO system has recently been successfully designed and commercialised by Modern Water (Thompson and Nicoll 2011) and the first commercialised FO desalination plant has been opened in Oman on the Arabian Sea.

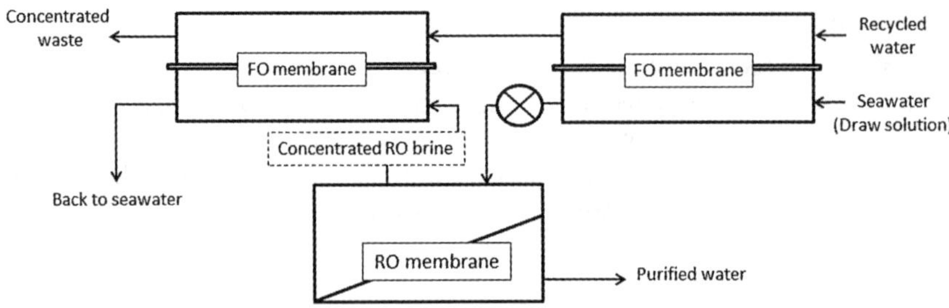

Figure 5.8. Schematic of the hybrid FO/RO system
SOURCE: Adapted from Bamaga et al. 2011; reproduced with permission from Elsevier

5.3.4.9 Hexavalent Phosphazene Salts

Phosphazenes are hybrid organic-inorganic materials whose structure is centred on phosphorous and nitrogen and the phosphazene system has some advantages such as hydrolytic stability and versatile chemistry. Therefore, a novel draw solute based on the phosphazene molecule has been recently proposed and initially investigated (Stone et al. 2013b). In their study, trimeric phosphazene carboxylic acids were treated with either NaOH or LiOH to yield stable and highly water soluble sodium and lithium phasphazene salts. As a result, the Li salt showed higher initial fluxes (~7 LMH) as compared to the Na salt (~6 LMH) at identical 0.067 M DS concentrations at 30°C. In addition, when the water flux was measured at 50°C, the initial water flux increased (~8.5 LMH). However, hydrolysis of the CTA membrane in some degree was observed, probably due to higher pH of the phosphazene DS (pH= ~8).

5.3.4.10 Polymer-Graphene Composite

Grapene is a two-dimensional carbon material with outstanding properties thus reduced graphene oxide (rGO)-composite hydrogels were used as draw agents to enhance water flux in FO desalination (Zeng et al. 2013). The DS was prepared by combining rGO and two different hydrogels. The amounts of rGO ranged from 0.3 wt% to 3 wt%, and poly (sodium acrylate) (PSA) and poly (sodium acrylate)-poly (N-isopropylacrylamide) (PSA-NIPAM) hydrogel draw agents were used in this study. The water fluxes for the pure polymer hydrogels (3.1 LMH for PSA and 1.6 LMH for PSA-NIPAM) were significantly smaller than a water flux of 6.8 LMH (1.2 wt%rGO + PSA – NIPAM as DS) and 8.2 LMH (1.2 wt%rGO + PSA as DS) in the FO process. In addition, although these water fluxes of the two composite hydrogels with 1.2 wt% rGO were significantly higher than other emerging draw agents such as MNP, the results were relatively lower than more traditional draw agents such as 1.5 M NH_4HCO_3 DS The water recovery rate in the solar dewatering process for composites with 1.2 wt% rGO was found to be twice as fast as that for pure hydrogels (i.e. 21% for pure PSA-NIPAM, 45% for PSA-NIPAM with 1.2 wt% rGO).

5.4 CONCLUDING REMARKS

This chapter discusses an overview of the presently available DS. It also provides some general criteria for the selection of suitable DS. Based on the current literature, there is no particular DS that can be regarded as "ideal" and the selection of a suitable DS mainly depends on the targeted application. The success of most FO applications also relies on how the DS can be recovered from the produced water. Therefore, the commercialised FO processes such as FO followed by RO seawater desalination, emergency drinks and osmotic dilution are utilised without a DS recovery system, which leads to its simplicity and significant savings of energy. Research is still needed in this area and further study need to be carried out in the development of more suitable DS to achieve full-scale commercialisation of the FO process.

5.5 ABBREVIATIONS

FO	forward osmosis
DS	draw solution
RSF	reverse salt flux
RO	reverse osmosis
MNPs	magnetic nanoparticles
SPS	switchable polarity solvents
UF	ultrafiltration
NF	nanofiltration

References

Achilli, A., Cath, T. Y., and Childress, A. E. (2010). "Selection of inorganic-based draw solutions for forward osmosis applications." *Journal of Membrane Science*, 364(1–2), 233–241.

Achilli, A., Cath, T. Y., Marchand, E. A., and Childress, A. E. (2009). "The forward osmosis membrane bioreactor: A low fouling alternative to MBR processes." *Desalination*, 239(1–3), 10–21.

Adham, S., Oppenheimer, J., Liu, L., and Kumar, M. (2009). *Dewatering Reverse Osmosis concentrate from water reuse applications using Forward Osmosis*, WaterReuse Foundation.

Alnaizy, R., Aidan, A., and Qasim, M. (2013). "Copper sulfate as draw solute in forward osmosis desalination." *Journal of Environmental Chemical Engineering*, 1(3), 424–430.

Bai, H., Liu, Z., and Sun, D. D. (2011). "Highly water soluble and recovered dextran coated Fe_3O_4 magnetic nanoparticles for brackish water desalination." *Separation and Purification Technology*, 81(3), 392–399.

Bamaga, O. A., Yokochi, A., Zabara, B., and Babaqi, A. S. (2011). "Hybrid FO/RO desalination system: Preliminary assessment of osmotic energy recovery and designs of new FO membrane module configurations." *Desalination*, 268, 163–169.

Batchelder, G. (1965). "Process for the demineralization of water." US Patent 317179903.

Beaudry, E. G., and Lampi, K. A. (1990). "Membrane technology for direct osmosis concentration of fruit juices." *Food Technology*, 44, 121.

Bowden, K. S., Achilli, A., and Childress, A. E. (2012). "Organic ionic salt draw solutions for osmotic membrane bioreactors." *Bioresource Technology*, 122, 207–216.

Cath, T., Childress, A., and Elimelech, M. (2006). "Forward osmosis: Principles, applications, and recent developments." *Journal of Membrane Science*, 281(1–2), 70–87.

Chekli, L., Phuntsho, S., Shon, H. K., Vigneswaran, S., Kandasamy, J., and Chanan, A. (2012). "A review of draw solutes in forward osmosis process and their use in modern applications." *Desalination and Water Treatment*, 43(1–3), 167–184.

Choi, Y.-J., Choi, J.-S., Oh, H.-J., Lee, S., Yang, D. R., and Kim, J. H. (2009). "Toward a combined system of forward osmosis and reverse osmosis for seawater desalination." *Desalination*, 247(1–3), 239–246.

Frank, B. S. (1972). "Desalination of seawater." US Patent 367089720.

Gadelha, G. A. F., and Hankins, N. P. (2011). *Assessment of draw-solution performance in combination with assymetric membranes for forward osmosis*, IDAWC, Perth, Australia.

Garcia-Castello, E. M., McCutcheon, J. R., and Elimelech, M. (2009). "Performance evaluation of sucrose concentration using forward osmosis." *Journal of Membrane Science*, 338(1–2), 61–66.

Ge, Q., Ling, M., and Chung, T.-S. (2013). "Draw Solutions for Forward Osmosis Processes: Developments, Challenges, and Prospects for the Future." *Journal of Membrane Science*, 447, 1–11.

Ge, Q., Su, J., Amy, G. L., and Chung, T.-S. (2012a). "Exploration of polyelectrolytes as draw solutes in forward osmosis processes." *Water research*, 46(4), 1318–1326.

Ge, Q., Su, J., Chung, T.-S., and Amy, G. (2011). "Hydrophilic Superparamagnetic Nanoparticles: Synthesis, Characterization, and Performance in Forward Osmosis Processes." *Industrial & Engineering Chemistry Research*, 50(1), 382–388.

Ge, Q., Wang, P., Wan, C., and Chung, T.-S. (2012b). "Polyelectrolyte-Promoted Forward Osmosis–Membrane Distillation (FO–MD) Hybrid Process for Dye Wastewater Treatment." *Environmental science & technology*, 46(11), 6236–6243.

Glew, D. N. (1965). "Process for liquid recovery and solution concentration." US Patent 9, 930.

Hancock, N. T., and Cath, T. Y. (2009). "Solute coupled diffusion in osmotically driven membrane processes." *Environmental Science & Technology*, 43(17), 6769–6775.

Herron, J. R., Beaudry, E. G., Jochums, C. E., and Medina, L. E. (1994). "Osmotic concentration apparatus and method for direct osmosis concentration of fruit juices." US Patent 5, 430.

Holloway, R. W. (2006). *Forward osmosis for concentration of anaerobic digester centrate*, M.S., University of Nevada, Reno.

HTI. (2010). "Hydration bags-technology overview." ⟨www.hydrationtech.com.⟩ (accessed Oct. 2013).

Jessop, P. G., Heldebrant, D. J., Li, X., Eckert, C. A., and Liotta, C. L. (2005). "Green chemistry: Reversible nonpolar-to-polar solvent." *Nature*, 436(7054), 1102–1102.

Jessop, P. G., Kozycz, L., Rahami, Z. G., Schoenmakers, D., Boyd, A. R., Wechsler, D., et al. (2011). "Tertiary amine solvents having switchable hydrophilicity." *Green Chemistry*, 13(3), 619–623.

Jessop, P. G., Mercer, S. M., and Heldebrant, D. J. (2012). "CO_2-triggered switchable solvents, surfactants, and other materials." *Energy & Environmental Science*, 5(6), 7240–7253.

Jessop, P. G., Phan, L., Carrier, A., Robinson, S., Dürr, C. J., and Harjani, J. R. (2010). "A solvent having switchable hydrophilicity." *Green Chemistry*, 12(5), 809–814.

Kessler, J. O., and Moody, C. D. (1976). "Drinking water from sea water by forward osmosis." *Desalination*, 18(3), 297–306.

Kravath, R. E., and Davis, J. A. (1975). "Desalination of seawater by direct osmosis." *Desalination*, 16(1975), 151–155.

Lay, W. C., Zhang, J., Tang, C., Wang, R., Liu, Y., and Fane, A. G. (2012). "Factors affecting flux performance of forward osmosis systems." *Journal of Membrane Science*, 394, 151–168.

Lee, S., Boo, C., Elimelech, M., and Hong, S. (2010). "Comparison of fouling behavior in forward osmosis (FO) and reverse osmosis (RO)." *Journal of Membrane Science*, 365(1–2), 34–39.

Li, D., Zhang, X., Simon, G. P., and Wang, H. (2013). "Forward osmosis desalination using polymer hydrogels as a draw agent: Influence of draw agent, feed solution and membrane on process performance." *Water Research*, 47(1), 209–215.

Li, D., Zhang, X., Yao, J., Simon, G. P., and Wang, H. (2011a). "Stimuli-responsive polymer hydrogels as a new class of draw agent for forward osmosis desalination." *Chemical Communications*, 47, 1710–1712.

Li, D., Zhang, X., Yao, J., Simon, G. P., and Wang, H. (2011b). "Stimuli-responsive polymer hydrogels as a new class of draw agent for forward osmosis desalination." *Chem. Commun.*, 47(6), 1710–1712.

Li, D., Zhang, X., Yao, J., Zeng, Y., Simon, G. P., and Wang, H. (2011c). "Composite polymer hydrogels as draw agents in forward osmosis and solar dewatering." *Soft Matter*, 7(21), 10048–10056.

Ling, M. M., and Chung, T.-S. (2011a). "Desalination process using super hydrophilic nanoparticles via forward osmosis integrated with ultrafiltration regeneration." *Desalination*, 278, 194–202.

Ling, M. M., and Chung, T.-S. (2011b). "Novel dual-stage FO system for sustainable protein enrichment using nanoparticles as intermediate draw solutes." Journal of *Membrane Science*, 372(1–2), 201–209.

Ling, M. M., Chung, T.-S., and Lu, X. (2011). "Facile synthesis of thermosensitive magnetic nanoparticles as "smart" draw solutes in forward osmosis." *Chemical Communications*, 47(38), 10788–10790.

Ling, M. M., and Chung, T. S. (2012). "Surface-ionized nanoparticles draw solutes and their regeneration in an integrated electric field and nanofiltration system." *Industrial & Engineering Chemistry Research*, 51, 15463–15471.

Ling, M. M., Wang, K. Y., and Chung, T.-S. (2010). "Highly Water-Soluble Magnetic Nanoparticles as Novel Draw Solutes in Forward Osmosis for Water Reuse." *Industrial & Engineering Chemistry Research*, 49(12), 5869–5876.

Mamisaheby, S., Phuntsho, S., Shon, H., Lotfi, F., and Kim, J. (2012). "Factors Affecting the Performances of Forward Osmosis Desalination Process." *Procedia Engineering*, 44, 1449–1451.

Martinetti, C. R., Childress, A. E., and Cath, T. Y. (2009). "High recovery of concentrated RO brines using forward osmosis and membrane distillation." *Journal of Membrane Science*, 331, 31–39.

McCormick, P., Pellegrino, J., Mantovani, F., and Sarti, G. (2008). "Water, salt, and ethanol diffusion through membranes for water recovery by forward (direct) osmosis processes." *Journal of Membrane Science*, 325(1), 467–478.

McCutcheon, J. R., McGinnis, R. L., and Elimelech, M. (2005). "A novel ammonia–carbon dioxide forward (direct) osmosis desalination process." *Desalination*, 174, 1–11.

McCutcheon, J. R., McGinnis, R. L., and Elimelech, M. (2006). "Desalination by ammonia-carbon dioxide forward osmosis: Influence of draw and feed solution concentrations on process performance." *Journal of Membrane Science*, 278, 114–123.

McGinnis, R. (2002). *Osmotic desalination process*, US patent 7, 560, 029. 1 Feb. 2002.

McGinnis, R. L., McCutcheon, J. R., and Elimelech, M. (2007). "A novel ammonia-carbon dioxide osmotic heat engine for power generation." *Journal of Membrane Science*, 305(1–2), 13–19.

Neff, R. A. (1964). "Solvent extractor." *Google Patents*.

Ng, H. Y., and Tang, W. (2006). "Forward (Direct) Osmosis: A novel and prospective process for brine control." *Water Environment Foundation*, 4345–4352.

Ng, H. Y., Tang, W., and Wong, W. S. (2006). "Performance of Forward (Direct) Osmosis Process: Membrane Structure and Transport Phenomenon." *Environ. Sci. Technol.*, 40(7), 2408–2413.

Noh, M., Mok, Y., Lee, S., Kim, H., Lee, S. H., Jin, G.-w., et al. (2012). "Novel lower critical solution temperature phase transition materials effectively control osmosis by mild temperature changes." *Chemical Communications*, 48(32), 3845–3847.

Ou, R., Wang, Y., Wang, H., and Xu, T. (2013). "Thermo-sensitive polyelectrolytes as draw solutions in forward osmosis process." *Desalination*, 318(0), 48–55.

Petrotos, K. B., Quantick, P., and Petropakis, H. (1998). "A study of the direct osmotic concentration of tomato juice in tubular membrane" module configuration. I. The effect of certain basic process parameters on the process performance." *Journal of Membrane Science*, 150(1), 99–110.

Phuntsho, S., Shon, H. K., Hong, S. K., Lee, S. Y., and Vigneswaran, S. (2011). "A novel low energy fertilizer driven forward osmosis desalination for direct fertigation: evaluating the performance of fertilizer draw solutions." *J. Membr. Sci.*, 375(2011), 172–181.

Phuntsho, S., Shon, H. K., Majeed, T., El Saliby, I., Vigneswaran, S., Kandasamy, J., et al. (2012). "Blended fertilizers as draw solutions for fertilizer-drawn forward osmosis desalination." *Environmental Science & Technology*, 46(8), 4567–4575.

Stache, K. (1989). "Apparatus for transforming sea water, brackish water, polluted water or the like into a nutritious drink by means of osmosis." US Patent 4, 030.

Stone, M. L., Rae, C., Stewart, F. F., and Wilson, A. D. (2013a). "Switchable polarity solvents as draw solutes for forward osmosis." *Desalination*, 312, 124–129.

Stone, M. L., Wilson, A. D., Harrup, M. K., and Stewart, F. F. (2013b). "An initial study of hexavalent phosphazene salts as draw solutes in forward osmosis." *Desalination*, 312, 130–136.

Su, J., Chung, T.-S., Helmer, B. J., and de Wit, J. S. (2012). "Enhanced double-skinned FO membranes with inner dense layer for wastewater treatment and macromolecule recycle using Sucrose as draw solute." *Journal of Membrane Science*, 396, 92–100.

Tan, C. H., and Ng, H. Y. (2010). "A novel hybrid forward osmosis–nanofiltration (FO-NF) process for seawater desalination: Draw solution selection and system configuration." *Desalination and Water Treatment*, 13, 356–361.

Thompson, N. A., and Nicoll, P. G. (2011). "Forward osmosis desalination: A commercial reality." *International Desalination Association World Congress*, Perth, Australia.

Van't Hoff, J. H. (1887). "Die Rolle der osmotischen Druckes in der Analogie zwischen Lösungen und Gasen." *Z. Phys. Chem.*, 1, 481–508.

Wallace, M., Cui, Z., and Hankins, N. P. (2008). "A thermodynamic benchmark for assessing an emergency drinking water device based on forward osmosis." *Desalination*, 227(1–3), 34–45.

Wang, K. Y., Ong, R. C., and Chung, T.-S. (2010). "Double-Skinned Forward Osmosis Membranes for Reducing Internal Concentration Polarization within the Porous Sublayer." *Industrial & Engineering Chemistry Research*, 49(10), 4824–4831.

Warne, B. B. R., Mayes, E., Oriard, T., and Norris, I. (2010). Water purification method using a field separable osmotic agent. GB2464956.

WHO. (2011). "Guidelines for drinking-water quality." 4th ed., ⟨http://www.who.int/water_sanitation_health/dwq/guidelines/en/⟩ (accessed Oct. 2013).

Wrolstad, R. E., McDaniel, M. R., Durst, R. W., Micheals, N., Lampi, K. A., and Beaudry, E. G. (1993). "Composition and sensory characterization of red raspberry juice concentrated by direct-osmosis or evaporation." *Journal of Food Science*, 58, 633–637.

Xu, Y., Peng, X., Tang, C. Y., Fu, Q. S., and Nie, S. (2010). "Effect of draw solution concentration and operating conditions on forward osmosis and pressure retarded osmosis performance in a spiral wound module." *Journal of Membrane Science*, 348 (1–2), 298–309.

Yaeli, J. (1992). "Method and apparatus for processing liquid solutions of suspensions particularly useful in the desalination of saline water." US Patent 509857524.

Yen, S. K., Mehnas Haja, N. F., Su, M., Wang, K. Y., and Chung, T.-S. (2010). "Study of draw solutes using 2-methylimidazole-based compounds in forward osmosis." *Journal of Membrane Science*, 364(1–2), 242–252.

Yong, J. S., Phillip, W. A., and Elimelech, M. (2012). "Coupled reverse draw solute permeation and water flux in forward osmosis with neutral draw solutes." *Journal of Membrane Science*, 392, 9–17.

York, R. J., Thiel, R. S., and Beaudry, E. G. (1999). *Full-scale experience of direct osmosis concentration applied to leachate management*, Seventh International Waste Management and Landfill Symposium.

Zeng, Y., Qiu, L., Wang, K., Yao, J., Li, D., Simon, G. P., et al. (2013). "Significantly enhanced water flux in forward osmosis desalination with polymer-graphene composite hydrogels as a draw agent." *RSC Advances*, 3(3), 887–894.

Zhao, S., and Zou, L. (2011). "Effects of working temperature on separation performance, membrane scaling and cleaning in forward osmosis desalination." *Desalination*, 278(1–3), 157–164.

Zhao, S., Zou, L., and Mulcahy, D. (2011). "Effects of membrane orientation on process performance in forward osmosis applications." *Journal of Membrane Science*, 382(1), 308–315.

Zou, S., Gu, Y., Xiao, D., and Tang, C. Y. (2011). "The role of physical and chemical parameters on forward osmosis membrane fouling during algae separation." *Journal of Membrane Science*, 366(1–2), 356–362.

CHAPTER 6

Nanoparticles as Draw Solution for Forward Osmosis Processes

Z. Y. Liu
H. W. Bai
D. D. Sun

In this chapter, a new kind of forward osmosis draw solute, dextran coated Fe_3O_4 magnetic nanoparticles (DF-MNPs), is introduced, and its performance as draw solutes for forward osmosis (FO) process evaluated. This DF-MNP combines the advantages of high osmotic pressure of dextran and strong magnetism of Fe_3O_4 MNPs. This new draw solute shows good performance in brackish water desalination with the merits of high water flux and easy recovery by an external magnet, which is favorable for significantly reducing the operation cost in forward osmosis processes.

6.1 INTRODUCTION

The increasing drinking water shortage has stimulated intensive research on finding alternative solutions to provide sufficient water sources (Ward et al. 2008). The forward osmosis (FO) process has shown great potential in wastewater reuse and seawater/brackish desalination with low cost energy consumption compared with traditional processes, such as, microfiltration (MF)/ultrafiltration (UF) and nanofiltration (NF)/reverse osmosis (RO) (McGinnis et al. 2008). FO is an osmotically driven membrane process that takes advantage of osmotic pressure gradient to drive water across the semi-permeable membrane from the feed solution (low osmotic pressure) side to the draw solution (high osmotic pressure) side. The novelty of the FO process is the utilization of natural osmotic pressure rather than energy-intensive hydraulic pressure as in the conventional pressure driven membrane processes such as RO (Chung et al. 2012).

There are two key components in the FO process: one is FO membrane and the other is draw solute. This chapter will focus on draw solute. An ideal draw solute for the FO process should have the characteristics of (1) high osmotic pressure that may generate high water flux; (2) easy recovery with low energy consumption; and (3) minimal reverse diffusion through FO membranes (McCutcheon et al. 2006). Many efforts have been devoted to develop suitable draw solutes, for example, sulfur dioxide or ammonium bicarbonate could be removed by a heated gas stripping operation (McGinnis et al. 2007); glucose or fructose could be recovered by RO (Stache 1989); aluminum sulfate could be removed with lime via precipitation (Frank 1972). However, these small molecules are not economical and practical as draw solutes because of significant reverse diffusion and high cost in draw solute recycling process. Therefore, finding an ideal FO draw solute is still a pressing demand.

The recent advance of nanotechnology has opened a new door for the development of FO draw solutes. In a previous study (Liu et al. 2011), a low-energy forward osmosis process to produce drinking water was designed by Prof. Sun and Dr. Liu's group, in which a novel draw solute separation system mimics a natural "destabilization" phenomenon with the help of superparamagnetic nanoparticles (Figure 6.1) to separate drinking water from draw solution without any intensive energy, such as, hydraulic pressure or heat, as shown in Figure 6.2.

In the FO part, an $Al_2(SO_4)_3$ draw solution is used to draw fresh water from wastewater across a semipermeable FO membrane. Then, in the draw solution reuse part, the diluted $Al_2(SO_4)_3$ draw solution with fresh water is sent to a $Al_2(SO_4)_3$ separation system, which will separate $Al_2(SO_4)_3$ draw solute, meanwhile producing drinking water. A special kind of negatively-charged and magnetic nanoparticles are introduced to the $Al_2(SO_4)_3$ separation system, which

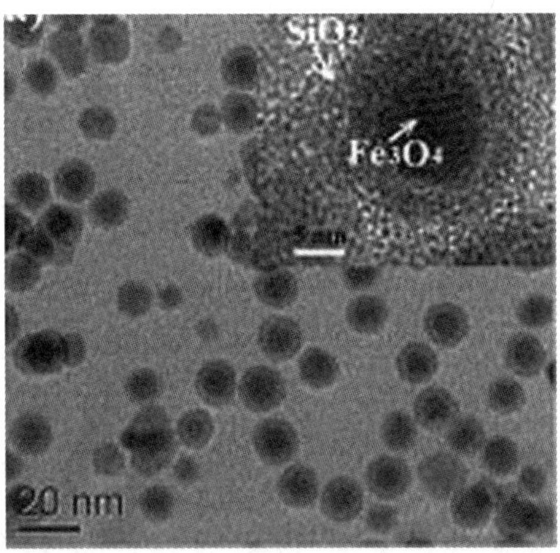

Figure 6.1. TEM of the Fe_3O_4@SiO_2 nanoparticles with Fe_3O_4 core and SiO_2 shell. Insert is the enlarged TEM of a Fe_3O_4@SiO_2 nanoparticle, showing different contrast of the SiO_2 shell and the Fe_3O_4 core

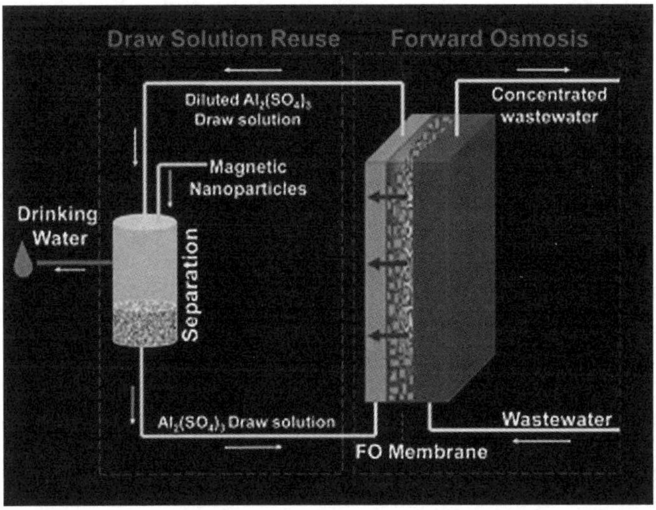

Figure 6.2. The scheme of the low-energy FO process to produce drinking water

mimics a natural "destabilization" phenomenon to economically separate $Al_2(SO_4)_3$ draw solute with low energy consumption. This novel draw solution separation process does not require any intensive energy input, such as hydraulic pressure or heat.

Although the above process is a low-energy process, it still consumes certain chemicals to recover the draw solutes. Recently, Prof. Sun and Dr. Liu's group further developed a novel draw solute, DF-MNPs, which do not consume chemical and energy in its recovery process (adapted from Bai et al. 2011).

A highly hydrophilic nano-sized magnetic particle was designed as draw solute, which has high osmotic pressure and can be easily recovered using an external magnetic field. Fe_3O_4 MNPs were chosen as a typical magnetic material because they can be easily separated from water by an external magnet field (Li et al. 2004; Lu et al. 2007; Hong et al. 2008). When Fe_3O_4 MNPs are coated with highly soluble organic substances, the requirements for draw solutes with high osmotic pressure and easy separation can be met concurrently. Both surface hydrophobicity and size of magnetic nanoparticle is crucial for these nanoparticles as draw solutes. In principle, the polymers with low molecular weight are preferred due to its high solubility and osmotic pressure in water. The surfaces of these magnetic nanoparticles were functionalized by hydrophilic dextran. Dextran is such a polymer, which is highly water-soluble and has high osmotic pressure as inorganic salts (Cath et al. 2006). Some researches have fabricated DF-MNPs for the application in biomedicine (Chen et al. 2011; Cole et al. 2011; Zhang et al. 2011). However, the study of DF-MNPs as draw solute in the FO process has rarely been reported. The current study is to integrate the merits of high osmotic pressure of dextran and easy recovery of Fe_3O_4 MNPs as draw solute, as shown in Figure 6.3.

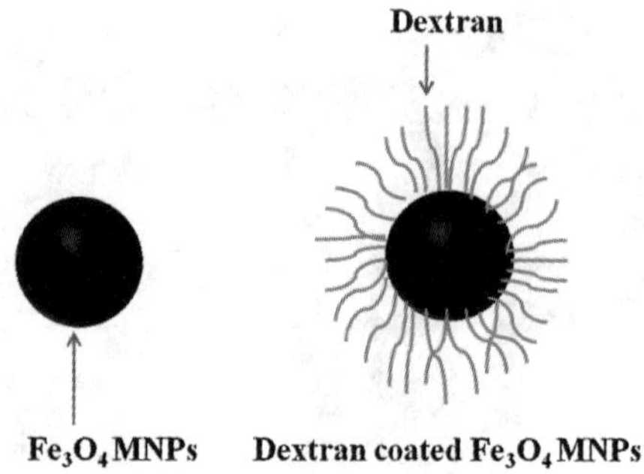

Figure 6.3. The scheme of Fe_3O_4 MNPs and DF-MNPs
SOURCE: Adapted from Bai et al. 2011; reproduced with permission from Elsevier

6.2 FABRICATION OF NANOPARTICLE DRAW SOLUTION

The DF-MNPs were synthesized by a wet chemical method (Hong et al. 2008). A three-neck flask located with nitrogen gas protection was used to synthesize DF-MNPs. Firstly, 50 mL of dextran and iron (III) chloride hexahydrate were filled in the three-neck flask with mechanical stirring, followed by the addition of 1 mL hydrazine hydrate and some sulfate heptahydrate under nitrogen gas. After constantly stirring for 30 min, 10 mL ammonia solution was added drop-wise by a syringe into the mixture under vigorous stirring until the pH was adjusted around 10. After stirring for 4 h at 60 °C, a black suspension was formed in the flask. Dextran was coated on the surface of Fe_3O_4 MNPs via covalent binding interaction between the functional group of dextran and the surface hydroxyl groups of Fe_3O_4 (Banerjee et al. 2007), was verified by FTIR spectra. The black suspension was extracted by centrifugation at a speed of 10,000 rpm for 10 min. The liquid at the top of the centrifugal tube was dialyzed using a membrane bag (Thermo Scientific, SnakeSkin pleated dialysis tubing, 3500 molecular weight cut-off (MWCO)) for 12 h, in order to remove the excess ammonia, hydrazine, iron cation and dextran macromolecules.

The weight ratio of dextran in DF-MNPs was measured. The thermal degradation of Fe_3O_4 MNPs, dextran, and DF-MNPs were investigated under atmospheric conditions using Perkin Elmer TGA7 thermogravimetric analyzer, from room temperature to 900°C, at a heating rate of 10°C/ min. The ratio of dextran in DF-MNPs at different weight ratio of dextran to Fe_3O_4 was obviously revealed in TGA curves as 64.44 wt.% (curve B of Figure 6.4) and 78.03 wt.% (curve C of Figure 6.4), respectively. The weight loss (15%) of Fe_3O_4 MNPs after 100°C was attributed to the loss of physically bonded water. For DF-MNPs, a great weight loss occurred after 300°C due to the decomposition of dextran. After dextran was decomposed completely, the residual substance is mainly Fe_3O_4 MNPs.

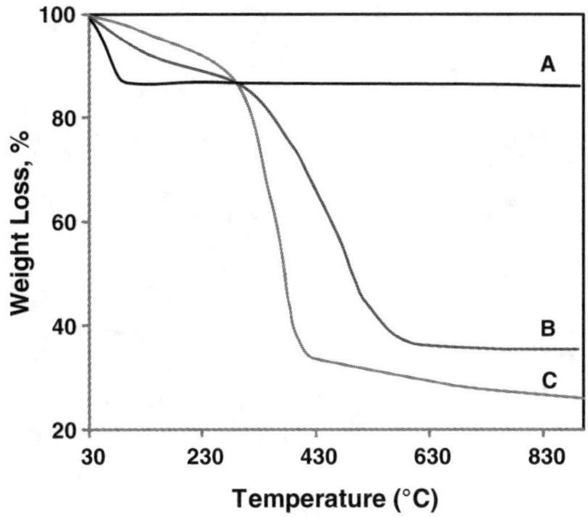

Figure 6.4. TGA curves of (A) Fe_3O_4 MNPs, (B) DF-MNPs containing dextran 64.44 wt.% and (C) DF-MNPs containing dextran 78.03 wt.%
SOURCE: Adapted from Bai et al. 2011; reproduced with permission from Elsevier

6.3 CHARACTERIZATIONS OF NANOPARTICLE DRAW SOLUTION

6.3.1 Morphology

The morphology of DF-MNPs was observed by field-emission scanning electron microscopy (FESEM, Jeol JSM-6340F) and transmission electronic microscopy (Jeol TEM-2010). For TEM sample preparation, DF-MNPs were firstly dispersed in ethanol, and then one droplet of the diluted solution was dipped on the copper grid for TEM measurement. As shown in Figure 6.5, the TEM image of DF-MNPs appeared to be spherical, with a uniform diameter size around 30 nm. The TEM at different magnifications also shows that DF-MNPs have good dispersive ability in water, therefore they will generate high osmotic pressure in the FO process. The size of DF-MNPs is related to the dextran ratio, that is, the higher ratio of dextran coated on the surface of Fe_3O_4 MNPs, the smaller the diameter size of DF-MNPs. This is confirmed in Fig. 6.5. The size of DF-MNPs containing dextran 64.44 wt.% in Figure 6.5(a) is around 30 nm, while that containing dextran 78.03 wt.% in Figure 6.5(b) is around 10 nm. In principle, the higher the dextran content ratio, the higher the osmotic pressure generated; and the smaller the size of DF-MNPs, the better the solubility of DF-MNPs. Therefore, the merits of small size and good solubility indicate DF-MNPs an ideal draw solute for FO applications.

FESEM samples were prepared by dipping one droplet of DF-MNPs solution on sample holders before Pt coating. As shown in Figure 6.6, DF-MNPs are spherical, with a uniform diameter size around 30 nm. During the drying process, these DF-MNPs are easy to aggregate, due to their high specific surface area and surface energy.

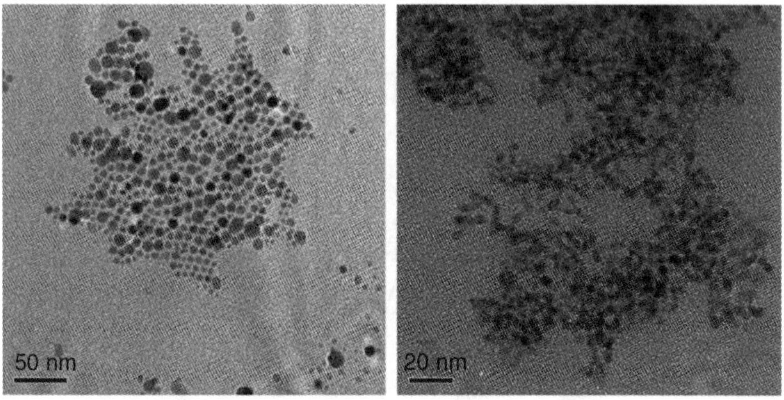

(a) (b)

Figure 6.5. TEM image of DF-MNPs with dextran (a) 64.44 wt.%, and (b) 78.03 wt.%
SOURCE: Adapted from Bai et al. 2011; reproduced with permission from Elsevier

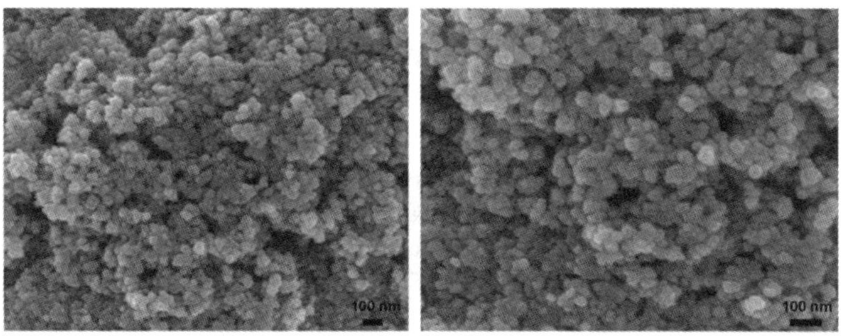

(a) (b)

Figure 6.6. FESEM images of DF-MNPs containing dextran 64.44 wt.% at different magnifications
Source: adapted from Bai et al. 2011; reproduced with permission from Elsevier

6.3.2 X-ray Diffractometer

The structure and crystal phase of DF-MNPs were measured by a Shimadzu XRD-6000 X-ray diffractometer with monochromated high-intensity Cu Ka radiation ($k = 1.5418$ Å) operated at 40 kV and 30 mA. The scanning rate was 2°/ min. As shown in Figure 6.7, the positions and the relative intensities of all these characteristic diffraction peaks: 2.96(220), 2.52(311), 2.09(400), 1.71 (422), 1.61(511), 1.47(440) and 1.27(533) match well with those of Fe_3O_4, which indicates the successful fabrication of Fe_3O_4 MNPs (Qi et al. 2009; Liu et al. 2011).

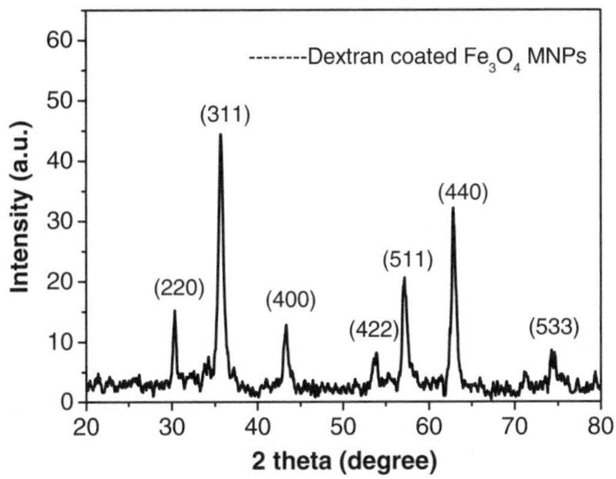

Figure 6.7. XRD pattern of DF-MNPs
SOURCE: Adapted from Bai et al. 2011; reproduced with permission from Elsevier

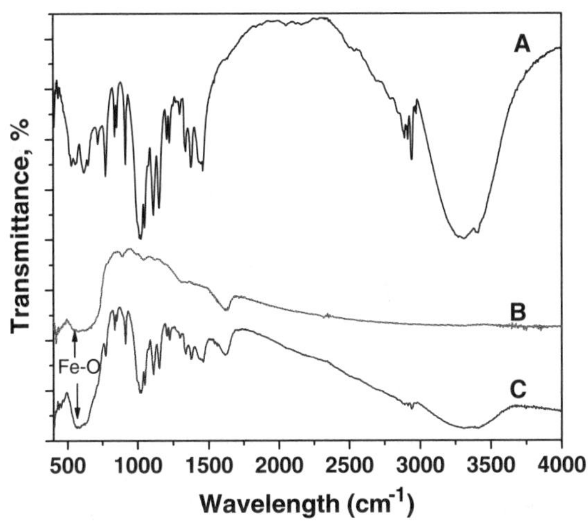

Figure 6.8. FTIR spectra of (A) pure dextran, (B) Fe_3O_4 MNPs and (C) DF-MNPs containing dextran 64.44 wt.%
SOURCE: Adapted from Bai et al. 2011; reproduced with permission from Elsevier

6.3.3 Fourier Transform Infrared

Fourier transform infrared (FTIR) spectra of Fe_3O_4 MNPs, dextran, and DF-MNPs were recorded on a Perkin Elmer spectrum GX FTIR spectrometer based on the KBr pellet technique. The resolution was kept at 5 cm^{-1}. As shown in Figure 6.8, the peak at 580 cm^{-1} is attributed to the characteristic absorption of Fe—O bond. In contrast to pure dextran, some new peaks appeared in the spectra of DF-MNPs, such as, the peak at 2939 cm^{-1} is attributed to the stretching vibration of $-CH_2-$ groups, the peak at about 1633 cm^{-1} is the characteristic peaks of C═O stretching

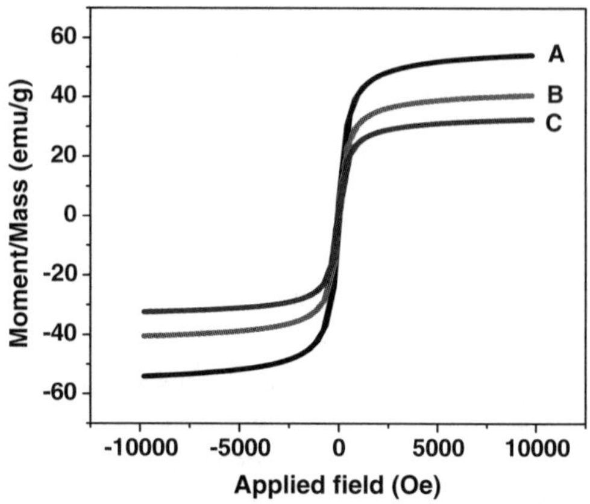

Figure 6.9. Hysteresis loops of (A) Fe_3O_4 MNPs and DF-MNPs containing dextran (B) 64.44 wt.% and (C) 78.03 wt.%

SOURCE: Adapted from Bai et al. 2011; reproduced with permission from Elsevier

vibration, and the peak at 1458 cm^{-1} is attributed to the bending vibration of C—H bond. The peaks at about 1024, 1111, and 1151 cm^{-1} are attributed to the stretching vibration of the alcoholic hydroxyl (C—OH), and the peak at 3410 cm^{-1} indicates the existence of hydrolxyls on pure dextran and DF-MNPs. All of these characteristic peaks obviously show that dextran was successfully coated on the surface of Fe_3O_4 MNPs by covalent binding between the functional group of dextran and the surface hydroxyl groups of Fe_3O_4, and Vander Waals force, hydrogen bond and electrostatic interactions also played significant roles.

6.3.4 Magnetic Properties

The magnetic properties of Fe_3O_4 MNPs and DF-MNPs were recorded in vibrating sample magnetometer (VSM, LakeShore 7300) (1.5 Tesla) with an applied field between 5000 and 5000 Oe. Christ freeze dryer (Alpha 1–2) was used for drying samples. As shown in Figure 6.9, the room-temperature magnetization of Fe_3O_4 MNPs and DF-MNPs indicates both of them have superparamagnetic characteristics. And the saturation magnetizations of Fe_3O_4 MNPs, DF-MNPs containing dextran 64.44 wt.% and DF-MNPs containing dextran 78.03 wt.%, are 54.04, 40.54, and 32.43 emu/g, respectively. The saturation magnetization of DF-MNPs is higher than that those of Fe_3O_4 MNPs functioned with 2-pyrrolidone, due to the surface spin canting effects of strong polar 2-pyrrolidone (Palma et al. 2007). The reduction of saturation magnetization of DF-MNPs in comparison with Fe_3O_4 MNPs is because the dextran on the surfaces of Fe_3O_4 MNPs blocks the magnetic properties Fe_3O_4 MNPs to some extent, and therefore the magnetization of DF-MNPs is reduced with the increase of dextran content.

The easy separation of DF-MNPs from water by an external magnet was demonstrated in Fig. 6.10. The homogeneously DF-MNPs draw solution is shown

Figure 6.10. Demonstration of the recovery of DF-MNPs with a magnet. (a) A photo of DF-MNPs contains dextran 64.44 wt.%. A photo shows that DF-MNPs containing dextran (b) 64.44 wt.% and (c) 78.03 wt.% was recovered by a magnet after 10 min for (b) and 15 min for (c)

SOURCE: Adapted from Bai et al. 2011; reproduced with permission from Elsevier

in Figure 6.10(a). With a magnet placed outside of the bottles in Figure 6.10(b) and (c), DF-MNPs were moved and collected at the walls close to the magnet after a certain time. This phenomenon clearly shows that DF-MNPs can be easily recovered by an external magnet without any chemical and energy input. It was observed that the solution of DF-MNPs containing dextran 64.44 wt.% takes 10 min to become clear in Figure 6.10(b), while the solution of DF-MNPs containing dextran 78.03 wt.% takes 15 min to become clear in Figure 6.10(c). This is because of the difference in surface magnetization of the Fe_3O_4 MNPs coated with different dextran content. It is obvious that DF-MNPs meet the stringent requirement as an ideal FO draw solute, with the merit of easy recovery with the help of an external magnet.

6.4 APPLICATION OF NANOPARTICLE DRAW SOLUTION IN FO PROCESSES

The performance of DF-MNPs as draw solute was investigated in a lab-scale setup (Figure 6.11). The commercial available cellulose triacetate (CTA) membrane from HTI (Hydration Technologies Inc.) was used for FO tests, and the effective membrane area was 48 cm^2 in a rectangular membrane module. The synthetic

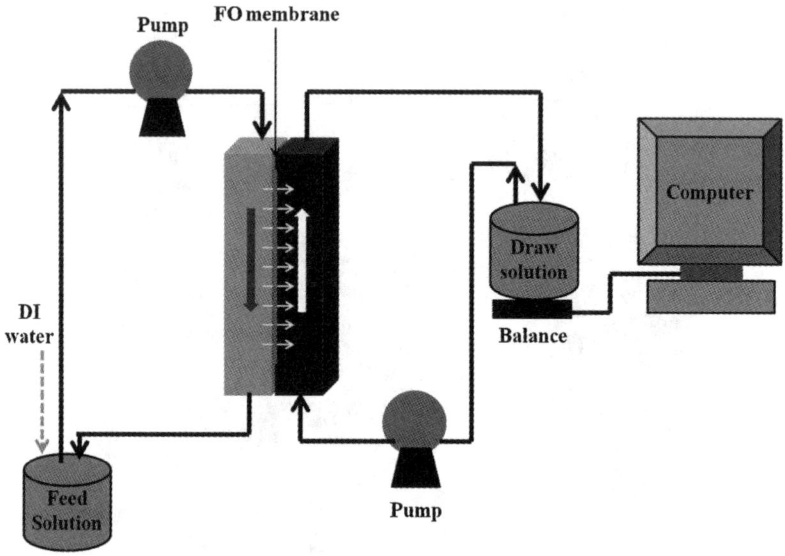

Figure 6.11. Schematic diagram of the lab-scale FO setup
SOURCE: Adapted from Bai et al. 2011; reproduced with permission from Elsevier

brackish water containing $MgSO_4$ (2 g/L) was used as the feed solution (Ling et al. 2010).

The velocities of feed solution and draw solution were controlled to be the same with two peristaltic pumps (Masterflex, Console Drive, Model 7518-10), and the draw solution and feed solution counter currently passed through the permeation cell channel (Cartinella et al. 2006). The temperature of the whole FO system was maintained at 22 ± 0.5°C. As the size of DF-MNPs is larger than the pore size of HTI FO membrane, DF-MNPs can be 100% rejected by HTI FO membrane. The weight increase of draw solution is attributed to the water permeated from feed solution. The DS concentration changes due to dilution by the drawing water from feed solution. That feed solution becomes more concentrated after water is drawn away has a significant role in the permeation results during the FO process. In order to control and minimize this influence, at a given interval (every 2 h), identical DI water was replenished into the feed solution to balance the concentration of feed solution. The water flux (J_v, $\mathrm{L m^{-2} h^{-1}}$, abbreviated as LMH) was calculated by dividing the produced water volume (ΔV, L) over the interval of time (Δt, h) and the effective surface area of membrane (A, m^2). Assume the density of water in the experimental condition is 1000 g/L, so the weight increase (ΔW) of the draw solution automatically recorded by computer is equal to the produced water volume (ΔV, L) (Yang et al. 2009). The draw solution faces the selective layer of the membrane, while the feed solution faces the support layer of the membranes.

$$J_v = \Delta V / (A \Delta t) \tag{6.1}$$

The concentration of dissolved ions in water is related to the ionic strength of solution, so the ionic strength (I) of the draw solution was measured by a conductivity meter (Cyberscan, CON 6000, Euteoh Instrument) during the FO process in order to monitor whether the Mg^{2+} and SO^{2-} have passed through the HTI FO membrane. The rejection rate (R) of $MgSO_4$ was calculated as follows:

$$R = 1 - C_t / C_0 = 1 - I_t / I_0 \tag{6.2}$$

where C_t and I_t are the concentration of $MgSO_4$, and its resulted ionic strength of draw solution at the moment of time t of the FO process, and C_0 and I_0 are the concentration of $MgSO_4$, and its resulted ionic strength of the original prepared $MgSO_4$ feed solution before the FO process. Because all the solutions are prepared in DI water, and the influence of the well synthesized dextran coated Fe_3O_4 MNPs

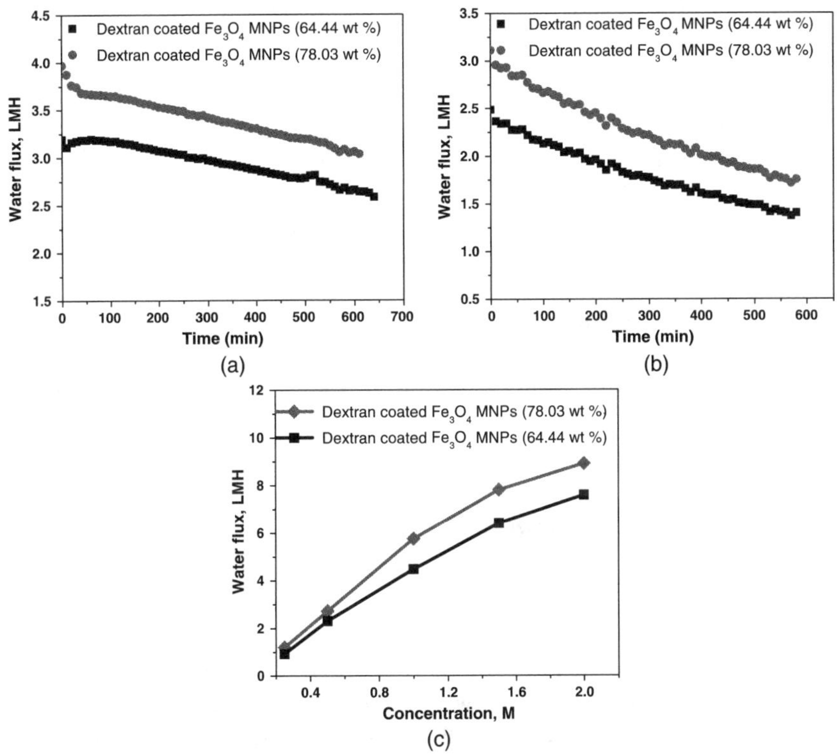

Figure 6.12. Water flux of DF-MNPs as draw solute in FO application. (a) DI water as feed solution, and DF-MNPs at an initial concentration of 0.5 M as draw solution, (b) $MgSO_4$ *solution of 2 g/L as feed solution, and DF-MNPs at an initial concentration of 0.5 M as draw solution, and (c) water fluxes at an time internal of 10 min as the function of DF-MNPs concentrations*

SOURCE: Adapted from Bai et al. 2011; reproduced with permission from Elsevier on ionic strength is negligible, hence the ionic strength of the draw solution is attributed to $MgSO_4$.

Figure 6.12(a) shows the water flux when DI water was used as feed solution, and two DF-MNPs of different size and different dextran content at an initial concentration of 0.5 M (mol/L) was used as draw solution. The water flux gradually reduced along with time, due to the reduced osmotic pressure across the membrane when the draw solution was diluted by the permeated water from feed solution. It should be noted that the water flux of DF-MNPs containing dextran 78.03 wt.% is higher than that of DF-MNPs containing dextran 64.44 wt.% as a result of the relative higher osmotic pressure generated from DF-MNPs containing dextran 78.03 wt.% that has smaller size and higher dextran content. Concentrated NaCl solution is a conventional inorganic salt as FO draw solute for clean water production. As shown in Figure 6.12, DF-MNPs presented water flux similar as NaCl draw solutes under the same conditions. However, DF-MNPs is easily to be separated from draw solution for recycle, while the separation of NaCl has to adopt energy-intensive RO membrane process.

Due to the good solubility of dextran, DF-MNPs are also highly soluble in water, generating high osmotic pressure to draw fresh water from brackish water. In this study, synthetic brackish water (2 g/L $MgSO_4$ in DI water) was used as the feed solution, and two DF-MNPs of different dextran contents and particle sizes were used as the draw solutions at an initial concentration of 0.5 M. As shown in Figure 6.12(b), two DF-MNPs with different dextran contents demonstrated to be able to draw fresh water across the HTI FO membrane from the brackish water as feed solution, and the water flux reduction trend is similar as that of DI water as feed solution in Figure 6.12(a). But the water flux reduction seems to be faster than that in Figure 6.12(a), due to the relatively high osmotic pressure of brackish water as feed solution. The water fluxes with brackish water as feed solution was investigated, as shown in Figure 6.12(c). It is clear that the water flux increases with the increase in the concentration of DF-MNPs draw solutes.

6.5 CONCLUDING REMARKS

A new kind of draw solute, nano-sized DF-MNPs, has been successfully synthesized and tested in the FO process for brackish water desalination. DF-MNPs were systematically characterized by FESEM, TEM, XRD, TGA and VSM. The increase of dextran content in Fe_3O_4 MNPs will reduce the particle size of DF-MNPs, but will increase the solubility of DF-MNPs in water. DF-MNPs are highly soluble in water, generating high osmotic pressure to draw fresh water from brackish water. Moreover, DF-MNPs can be easily recovered from the diluted draw solution by an external magnet. The intriguing advantages of high osmotic pressure and easy recovery of DF-MNPs meet the stringent criteria for an ideal draw solute. This study further promotes the FO process as an environmentally friendly process for clean water production. In future research, the properties of DF-MNPs should be further optimized with the aim to further improve osmotic pressure for possible commercial applications.

6.6 ABBREVIATIONS/NOMENCLATURE

CTA	cellulose triacetate
DF-MNP	dextran coated Fe_3O_4 magnetic nanoparticles
DS	draw solution
FESEM	field-emission scanning electron microscopy
EMU	electromagnetic unit
FO	forward osmosis
FTIR	fourier transform infrared
HTI	Hydration Technologies Inc
LMH	$L\,m^{-2}\,h^{-1}$
MF	microfiltration
MWCO	molecular weight cut-off
NF	nanofiltration
RO	reverse osmosis
TEM	transmission electron microscopy
TGA	thermal gravimetric analysis
UF	ultrafiltration
VSM	vibrating sample magnetometer
XRD	X-ray diffractometer

References

Bai, H. W., Liu, Z. Y., and Sun, D. D. (2011). "Highly water soluble and recovered dextran coated Fe_3O_4 magnetic nanoparticles for brackish water desalination." *Separation and Purification Technology*, 81, 392–399.

Banerjee, S. S. and Chen, D. H. (2007). "Fast removal of copper ions by gum arabic modified magnetic nano-adsorbent." *Journal of Hazardous Materials*, 147, 792–799.

Cartinella, J. L., Cath, T. Y., Flynn, M. T., Miller, G. C., Hunter, K. W., and Childress, A. E., (2006). "Removal of natural steroid hormones from wastewater using membrane contactor processes." *Environmental Science and Technology*, 40, 7381–7386.

Cath, T. Y., Childress, A. E., and Elimelech, M. (2006). "Forward osmosis: principles, applications, and recent developments." *Journal of Membrane Science*, 281, 70–87.

Chen, J. P., Yang, P. C., Ma, Y. H., and Wu, T. (2011). "Characterization of chitosan magnetic nanoparticles for in situ delivery of tissue plasminogen activator." *Carbohydrate Polymers*, 84, 364–372.

Chung, T. S., Zhang, S., Wang, K. Y., Su, J. C., and Ling, M. M. (2012). "Forward osmosis processes: Yesterday, today and tomorrow." *Desalination*, 287, 78–81.

Cole, A. J., David, A. E., Wang, J. X., Craig, J. G., Hannah, L. H., and Yang, V. C. (2011). "Polyethylene glycol modified, cross-linked starch-coated iron oxide nanoparticles for enhanced magnetic tumor targeting." *Biomaterials*, 32, 2183–2193.

Hong, R. Y., Feng, B., Chen, L. L., Liu, G. H., Li, H. Z., Zheng, Y., and Wei, D. G. (2008). "Synthesis, characterization and MRI application of dextran-coated Fe_3O_4 magnetic nanoparticles." *Biochemical Engineering Journal*, 42, 290–300.

Frank, B.S. (1972). "Desalination of Sea Water." US Patent 3, 670, p. 897.

Li, Z., Chen, H., Bao, H., and Gao, M. (2004). "One-pot reaction to synthesize water-soluble magnetite nanocrystals." *Chemistry of Materials*, 16, 1391–1393.

Ling, M. M., Wang, K. Y., and Chung, T. S. (2010). "Highly water-soluble magnetic nanoparticles as novel draw solutes in forward osmosis for water reuse." *Industrial and Engineering Chemistry Research*, 49, 5869–5876.

Liu, Z. Y., Bai, H. W., Lee, J., and Sun, D. D. (2011a). "A low-energy forward osmosis process to produce drinking water." *Energy & Environmental Science*, 4, 2582–2585.

Liu, Z. Y., Bai, H. W., and Sun, D. D. (2011). "Facile fabrication of porous chitosan/TiO_2/Fe_3O_4 microspheres with multifunction for water purifications." *New Journal of Chemistry*, 35, 137–140.

Lu, A. H., Salabas, E. L., and Schüth, F. (2007). "Magnetic nanoparticles: synthesis, protection, functionalization, and application." *Angewandte Chemie- International Edition*, 46, 1222–1244.

McCutcheon, J. R., McGinnis, R. L., and Elimelech, M. (2006). "Desalination by ammonia – carbon dioxide forward osmosis: influence of draw and feed solution concentrations on process performance." *Journal of Membrane Science*, 278, 114–123.

McGinnis, R. L. and Elimelech, M. (2007). "Energy requirements of ammonia – carbon dioxide forward osmosis desalination." *Desalination*, 207, 370–382.

McGinnis, R. L. and Elimelech, M. (2008). "Global challenges in energy and water supply: the promise of engineered osmosis." *Environmental Science & Technology*, 42, 8625–8629.

Palma, R. D., Peeters, S., Van Bael, M. J., Van Den Rul, H., Bonroy, K., Laureyn, W., Mullens, J., Borghs, G., and Maes, G. (2007). "Silane ligand exchange to make hydrophobic superparamagnetic nanoparticles water-dispersible." *Chemistry of Materials*, 19, 1821–1831.

Qi, D., Lu, J., Deng, C., and Zhang, X. (2009). "Magnetically responsive Fe_3O_4@C@SnO_2 core-shell microspheres: synthesis, characterization and application in phosphoproteomics." *Journal of Physical Chemistry C*, 113, 15854–15861.

Stache, K. (1989). "Apparatus for Transforming Seawater, Brackish Water, Polluted Water or the Like into a Nitrous Drink by Means of Osmosis." *US Patent* p. 4, 879, 030.

Ward, F. A. and Pulido-Velazquez, M. (2008). "Water conservation in irrigation can increase water use." *Proceedings of the National Academy of Sciences*, 105, 18215–18220.

Yang, Q., Wang, K. Y., and Chung, T. S. (2009). "Dual-layer hollow fibers with enhanced flux as novel forward osmosis membranes for water production" *Environmental Science and Technology*, 43, 2800–2805.

Zhang, M., He, X. W., Chen, L. X., and Zhang, Y. K. (2011). "Preparation and characterization of iminodiacetic acid-functionalized magnetic nanoparticles and its selective removal of bovine hemoglobin." *Nanotechnology*, 22, 065705.

CHAPTER 7

Polymer Hydrogels as Smart Draw Agents in Forward Osmosis Processes

Amir Razmjou
George P. Simon
Huanting Wang

In this chapter, thermo-sensitive polymer hydrogels are introduced as a new class of draw agent for forward osmosis (FO) desalination. The effect of hydrogel chemistry is reviewed, including the effect of incorporation of ionic chemical moieties and light-absorbing carbon particles and reduced graphene oxide nanosheets on FO desalination performance. Other key experimental and processing parameters for the FO and water recovery step are explored, including hydrogel particle size, membrane type, feed solution, particle-particle contact areas, particle-membrane contact zones and hydrogel thickness.

7.1 INTRODUCTION TO HYDROGELS AS DRAW AGENTS

In recent years, stimuli responsive hydrogels have been used for numerous applications such as in the biochemical and biomedical fields (Gil and Hudson 2004), sensors and size selective separation agents as well as controlling growth agent in inorganic material synthesis (Qiu and Park 2001; Wang et al. 2003; Han et al. 2008; Yao et al. 2008). The highly concentrated hydrophilic groups in the three-dimensional networks of polymer chains allow a dehydrated or deswollen hydrogels to take up large volumes of water by extending polymer chain and generating a swelling pressure. For instance, a swelling pressure ranging from 0.20–4.23 MPa can be produced by weakly crosslinking of poly(acrylic acid)/poly(sodium acrylate) copolymers with polymer volume fractions between 0.03 and 0.30 (Wack and Ulbricht 2009).

Theoretically, it is possible to drastically raise the swelling pressure by strategically engineering polymer structures. As a result, a sufficient driving force can be created to draw water from high salinity seawater through the membrane. This primary characteristic of polymer hydrogels makes them a suitable option to

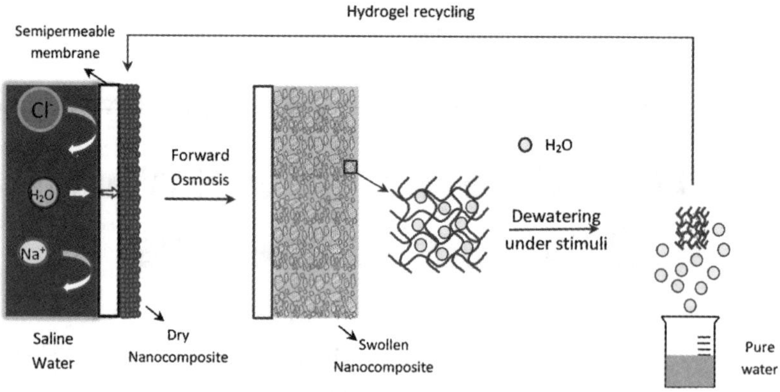

Figure 7.1. Schematic of the polymer hydrogel-driven FO desalination process

be used as draw agent in the FO desalination process. What has made the polymer hydrogels an attractive option as draw agent is their ability to be readily manipulated by appropriate and desirable environmental stimuli, enabling reversible volume change or solution–gel phase transitions. A variety of stimuli such as temperature, electric fields, solvent composition, light, pressure, sound, magnetic field, pH, ions and specific molecular recognition events can all be used to induce various responses, in particular to change the hydrophilic-hydrophobic balance, thus also stimulating the release of trapped water (Tanaka et al. 1982; Serizawa et al. 2001; Lietor-Santos et al. 2009; Qiu and Park 2012).

Utility of fast stimuli-responsive polymer hydrogels as draw media has the advantage of recovering high purity water without the concern of degradation of products (as in conventional forward osmosis draw agents), and potentially allowing the regeneration of the draw agent at a low energy cost. The concept of using polymer hydrogels as draw agent in FO desalination is presented in Figure 7.1. As can be seen from Figure 7.1, a high swelling pressure hydrogel encourages water molecules from the feed side to move through a selective membrane, whilst salt is rejected. The hydrogel particles can subsequently be stimulated to release the absorbed water by hydraulic pressure and/or low temperature thermal processing, following which they can readily be reused in the FO desalination process.

7.2 FLUX ENHANCEMENT

7.2.1 Ionic Hydrogels

The presence of ionic groups in the hydrogel networks provides unique capacity for hydrogels to attract more water. These covalently-incorporated ionic groups on the co-monomer unit are balanced by mobile counter ions to provide a positive osmotic pressure inside the hydrogel and consequently drive greater water sorption.

Synthesis of Ionic Hydrogels

The most common way by which the hydrogels are synthesized is radical polymerization of different monomers and cross-linker in an aqueous solution using initiator. Li and coworkers synthesized four different types of polymer hydrogels including two ionic polymer hydrogels: poly(sodium acrylate) (PSA), and poly (sodium acrylate)-co-poly(N-isopropylacrylamide) (PSA-NIPAM), and two non-ionic hydrogels: poly(acrylamide) (PAM), and poly(N-isopropylacrylamide) (PNIPAM) (Li et al. 2011a). In their work, N,N'-methylenebisacrylamide and ammonium persulfate were used as cross-linker and initiator, respectively. The reaction was maintained at around 70°C for a few hours to ensure complete polymerization. Once the hydrogel was prepared, it was kept in fresh water for 2 to 3 days in order to remove unreacted monomers and cross-linker, followed by drying at 80°C overnight. After being ground into fine powder, they were sieved into the desired particle size ranges.

Chemistry and Structural Analysis

The FTIR spectra of four different polymer hydrogels, including PAM, PSA, PNIPAN and PSA-NIPAM are shown in Figure 7.2. The broad absorption band at 3430 cm^{-1} for PAM is due to the N—H stretching of amide groups of acrylamide units. The units also have two peaks from amide-I and amide-II at around 1663 and 1608 cm^{-1}. The spectrum of PSA revealed the $-CH_2$ stretching vibration at 2940 cm^{-1} and also the $-CH_2$ bending vibration at 1455 cm^{-1}. The strong absorption peaks at around 1570 and 1410 cm^{-1} are attributed to the C=O stretching vibration of the ionic carboxyl group (—COO—). The main peak assignments of PNIPAM are 3292 cm^{-1}, 2970 cm^{-1}, 1650 cm^{-1} and 1550 cm^{-1}

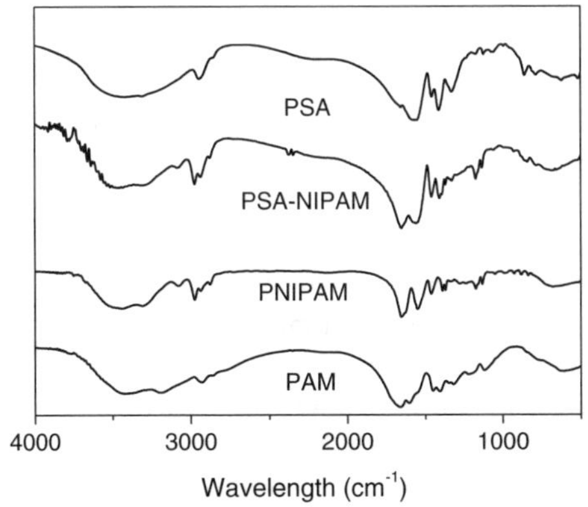

Figure 7.2. FTIR spectra of PAM, PSA, PNIPAM and PSA-NIPAM
SOURCE: Adapted from Li et al. 2011a; reproduced with permission from Royal Society of Chemistry

Figure 7.3. SEM images of polymer hydrogel particles: (a) at low magnification and (b) at high magnification

SOURCE: Adapted from Li et al. 2011a; reproduced with permission from Royal Society of Chemistry

which are assigned to the secondary amide N-H stretching, $-CH_3$ asymmetric stretching, secondary amide C=O stretching (amide I bond) and secondary amide C=O stretching (amide II bond), respectively. From Figure 7.2, the presence of NIPAM and SA units in the PSA-NIPAM sample can also be observed from its spectrum (Li et al. 2011a).

The structures of different polymer hydrogels are presented in Figure 7.3. The samples sizes range from 50 μm to 150 μm, with some smaller, finer particles apparent. The SEM images show that PNIPAM and PSA-NIPAM have rough surfaces, while PAM and PSA particles possess smoother surfaces.

Effect of Ionic Group Addition on the FO Flux Desalination

The effect of addition of ionic group into polymer hydrogels on the water flux is shown in Figure 7.4. The PSA hydrogel with the highest charge density resulted in the highest water flux, whereas PSA-NIPAM hydrogel with a 50 mol% reduced charge showed a significantly lower flux of 0.55 $L/m^{2\text{-}h}$. The flux was even lower when the non-ionic PAM and PNIPAM hydrogels were used. It can be seen that the fluxes decrease with time for both ionic and non-ionic hydrogels. The average flux was 0.96, 0.55, 0.36 and 0.30 L/m^{2h} for PSA, PSA-NIPAM, PAM, and PNIPAM, respectively, in the first hour, and reached $73.1 \pm 1.5\%$, $62.4 \pm 1.2\%$, $53.4 \pm 4.5\%$, and $42.2 \pm 3.4\%$ for PSA, PNIPAM-SA, PAM, and PNIPAM, respectively after 24 h (Li et al. 2011a).The swelling pressure (osmotic pressure) of polymer hydrogels plays a key role (driving force) in the water absorption. The inclusion of ionic groups into hydrogel backbone thus generates higher degrees of

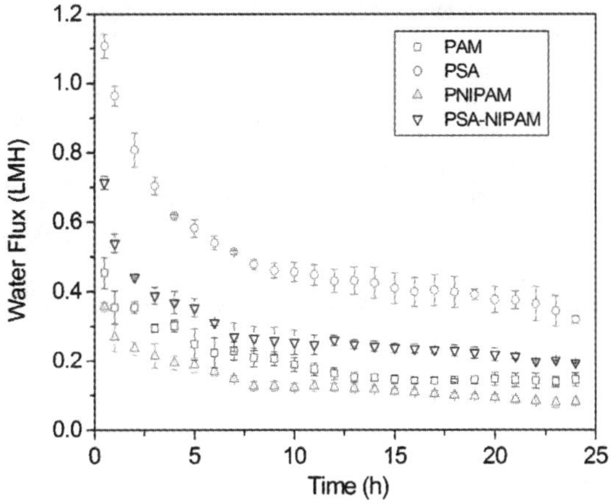

Figure 7.4. Water fluxes in the 24 h FO process using polymer hydrogels and 2000 ppm NaCl solution as feed
SOURCE: Adapted from Li et al. 2011a; reproduced with permission from Royal Society of Chemistry

swelling than neutral hydrogels (Kayaman et al. 1998), and thus a higher driving force in FO process. As the FO process progresses, the degree of swelling increases and leads to water flux decline.

7.2.2 Composite Hydrogels

In recent times significant attention has been devoted into organic-inorganic composite superabsorbents. The inclusion of inorganic particles such as clay (Haraguchi et al. 2005; Can et al. 2007; Haraguchi et al. 2007), zeolites (Ziesmer and Stock 2008) and metals (Ménager et al. 2004; Wang et al. 2004) into polymer networks could be achieved via crosslinking to the hydrogel, adsorbing or attaching to the polymer chains, or simply entrapping them within the hydrogel networks (Schexnailder and Schmidt 2009). The incorporation of the inorganic fillers provides superior properties to the polymer hydrogels such as higher water absorbency, absorption rate, gel strength and responsiveness to some stimuli (e.g., mechanical, optical, thermal, magnetic, electric stimulation, *etc.*). These new and unique properties broaden their range of applications in drug delivery systems, electronics, sensors, actuators and microfluidics sectors, separation devices, water absorbents, catalysts, and optics (Haraguchi et al. 2007; Schexnailder and Schmidt 2009).

Li et al. (2011b) reported the development of composite hydrogels with the inclusion of black solar-absorbing carbon particles as draw agents in FO desalination. In another attempt, Zeng et al. (2013) introduced reduced graphene oxide (rGO) nanosheets into polymer hydrogels as draw agents in FO desalination via the addition of chemically-reduced graphene oxide into the hydrogel network.

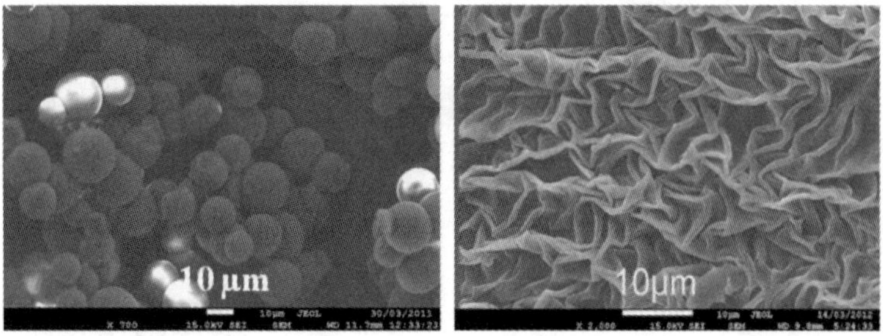

Figure 7.5. SEM images of (a) carbon particles and (b) polymer-graphene composite hydrogels

SOURCE: (a) Li et al. 2011b, (b) adapted from Zeng et al. 2013; both reproduced with permission from Royal Society of Chemistry

Synthesis of Composite Polymer Hydrogels

The composite polymer hydrogels were synthesized using a similar procedure to that described above except that the dried carbon or rGO nanosheets were dispersed into monomer solution before polymerization was initiated. A vigorous stirring and/or sonication throughout polymerization is necessary to avoid agglomeration of the particles (Li et al. 2011b; Zeng et al. 2013). A low temperature hydrothermal carbonization of sucrose was used to produce micron-sized carbon particles ranging from 10 µm to 25 µm with a smooth, dense surface as shown in Figure 7.5(a). The as-synthesized particles were then added directly to the monomer solution (Li et al. 2011b). In contrast, rGO nanosheets were created *in-situ* by the reduction of graphene oxide into graphene using hydrazine inside the monomer solution with the weight ratio of 7:10 for hydrazine to graphene oxide, see Figure 7.5(b) (Zeng et al. 2013).

Chemistry and Structural Analysis

The comparison between the PSA and PSA-C hydrogels (Figure 7.6(a)) shows broad bands in the range of 3000–3700 cm^{-1}, in which a wide band is attributed to O—H stretching vibrations of hydroxyl groups on the surface of carbon particles. However, for other composite hydrogels, no significant characteristic peaks of the hydrophilic carbon particles can be detected due to the obscuring of strong transmittance peaks arising from polymer structures (Li et al. 2011b). For the graphene-polymer composite hydrogel, the two peaks at 1348 cm^{-1} (D band) and 1580 cm^{-1} (G band) for rGO appeared on the composite hydrogels curves in Figure 7.6(b). The 15 cm^{-1} shift of G band toward higher wavenumbers is ascribed to the strong interaction between the hydrogel network and graphene (Wang et al. 2010).

In the SEM image (Figure 7.7(i)), neat non-ionic NIPAM polymer hydrogels demonstrate a rougher surface than those with PSA. The addition of carbon particles into PSA leads to a smooth surface with no obvious interfacial voids

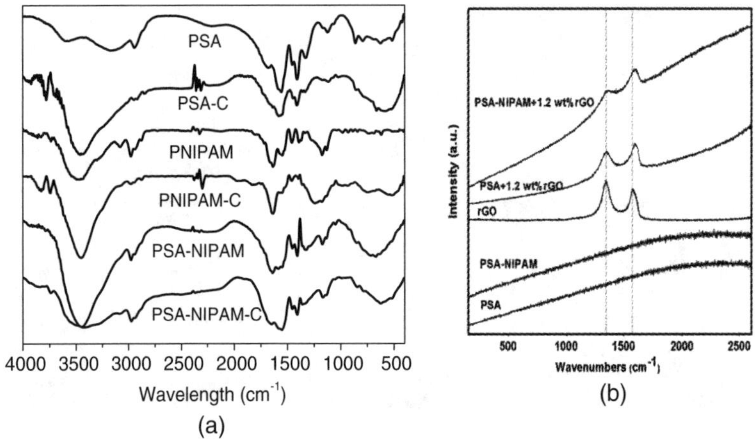

Figure 7.6. (a) FT-IR spectra of pure and composite polymer hydrogels with the incorporation of C carbon particles and (b) Raman spectra of pure polymers, reduced graphene oxide (rGO) and composite polymer hydrogels
SOURCE: (a) Li et al. 2011b, (b) adapted from Zeng et al. 2013; both reproduced with permission from Royal Society of Chemistry

between the inorganic and organic phases, indicating a high degree of compatibility. On the other hand, the incorporation of carbon particles into hydrogel networks with PNIPAM portions resulted in a macroporous structure. This is due to the presence of the hydrophobic isopropyl group in the backbone of PNIPAM based hydrogels that can reduce the interaction between the polymer network and hydrophilic carbon particles, forming large pore sizes around the carbon particles during polymerization at above lower solution critical temperature (LSCT). Similar behavior was also observed when graphene was added to the polymer hydrogels. As can be seen in Figure 7.7(ii) the pure PSA and NIPAM hydrogels are smooth while their composites show slightly rougher surface for PSA-graphene and wrinkled appearance of the PSA-NIPAM-graphene hydrogels.

Effect of Adding Solar Absorbent Particles on the FO Flux

As presented in Figure 7.8, the pure ionic and composite ionic hydrogels showed higher flux than that of non-ionic hydrogels. As mentioned before, ionic hydrogel (e.g., PSA) with mobile counter-ions can create higher osmotic pressure (or higher degrees of swelling) than neutral hydrogel (PNIPAM) (Kayaman et al. 1998) and as a result the draw agent with the higher charged density provides a higher driving force in FO process. This is also verified by comparing the swelling ratios of ionic and non-ionic hydrogels in Figure 7.9.

From Figure 7.8, the incorporation of solar absorbent carbon particles led to a higher flux regardless of the ionic or non-ionic nature of the hydrogels. During the first half an hour, the flux for PSA-C (1.32 LMH) is 18% higher than that of PSA (1.12 LMH). A similar behavior was also observed in 0.5 h FO test for

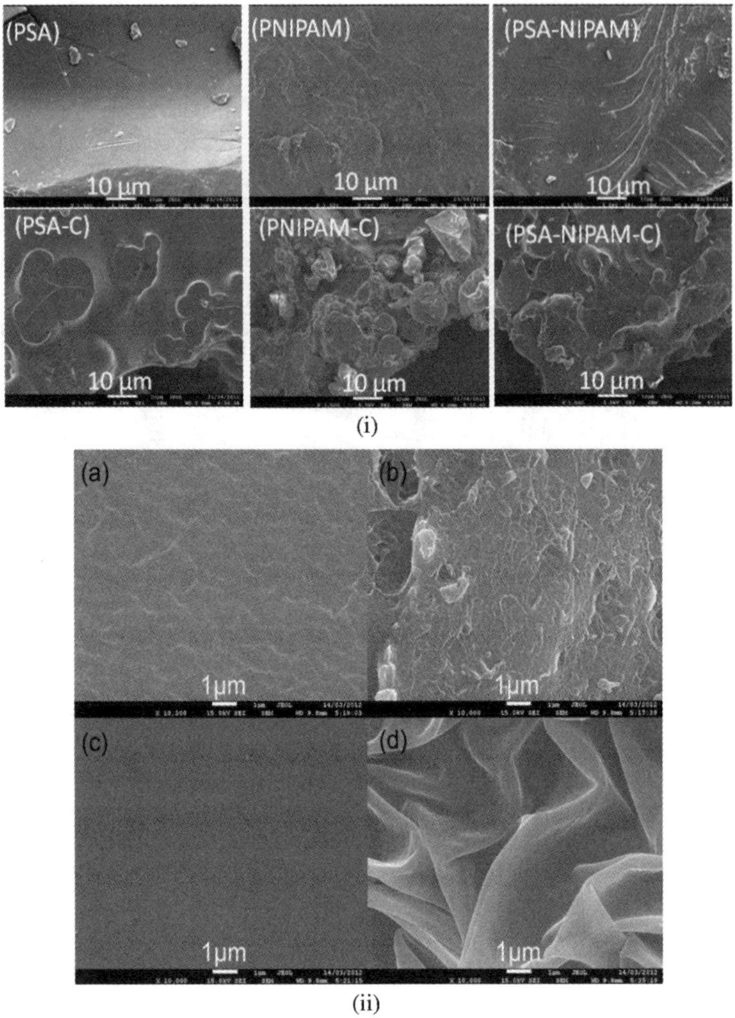

Figure 7.7. SEM images of pure and compositepolymer hydrogels: (i) carbon-polymer composite hydrogels and (ii) graphene-polymer composite hydrogels: (a) pure PSA, (b) PSA+1.2 wt% rGO, (c) pure PSA-NIPAM and (d) PSANIPAM+1.2 wt % rGO

SOURCE: (i) Li et al. 2011b, (ii) adapted from Zeng et al. 2013; both reproduced with permission from Royal Society of Chemistry

NIPAM-Cand PSA-NIPAM-C hydrogels. At the end of 24 h FO experiment, the pure polymers showed a lower water content compared with those of composite hydrogels: the water content of swollen PSA, PNIPAM, and PSA-NIPAM were 74.6 ± 3.1%, 39.8 ± 4.2%, and 63.8 ± 2.7%, respectively, whilst the water loadings of composite polymers after 24 h FO test are 81.3 ± 4.2%, 53.2 ± 3.1%, and 71.3 ± 1.0% for PSA-C, PNIPAM-C, and PSA-NIPAM-C(Li et al. 2011b). This

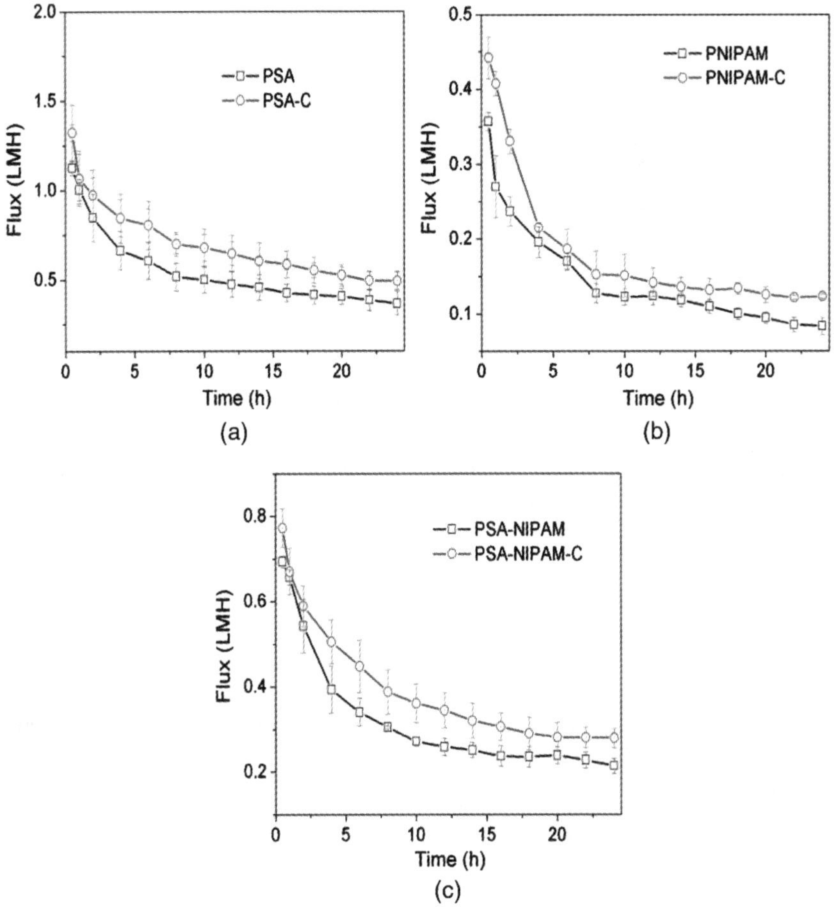

Figure 7.8. Effect of the addition of carbon particles into polymer hydrogels as draw agents on the water fluxes in 24 h FO process. Pure hydrogels were denoted PSA, PNIPAM and PSA-NIPAM and composite polymer hydrogels were denoted PSA-C, PNIPAM-C and PSA-NIPAM-C. 2000 ppm NaCl solution was used as feed solution

SOURCE: Adapted from Li et al. 2011b; reproduced with permission from Royal Society of Chemistry

improved performance of the composite hydrogels is attributed to their higher swelling ratios. The swelling ratios of PSA-C and PSA-NIPAM-C are around 9.7% and 8.0% greater than those of PSA and PSA-NIPAM-C, respectively (Figure 7.9). According to Li et al.(Li et al. 2011b), the addition of carbon particles can enhance the water uptake through different mechanisms:

1. The polar functional groups on the carbon particle surfaces such as hydroxyl or carboxyl groups increase the hydrophilicity of the composite and encourage water uptake, thus enhancing the swelling behavior of hydrogel (Gua and Yea 2009).

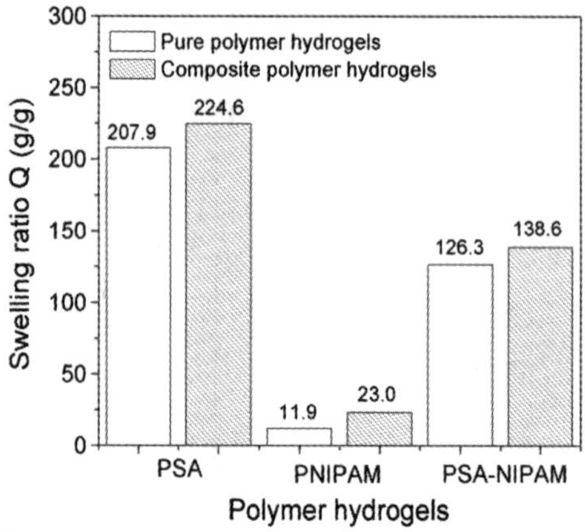

Figure 7.9. Swelling ratios of ionic and non-ionic pure (PSA, PNIPAM, and PSA-NIPAM) and composite (PSA-C, PNIPAM-C and PSA- NIPAM-C) polymer hydrogels
SOURCE: Adapted from Li et al. 2011b; reproduced with permission from Royal Society of Chemistry

2. The inclusion of inorganic particles with polar surfaces into polymer networks facilitates the accessibility of the charged groups inside the hydrogels, and subsequently the osmotic pressure. This enhanced accessibility alongside with the higher hydrophilicity of polymer network results in a greater water uptake capacity and swelling ratio of hydrogels (Santiago et al. 2007; Zheng et al. 2007).
3. The addition of inorganic particles (with the aim of increasing the porosity of bulk polymers and accessible surface area) has been proven as an effective way to raise the swelling rates of hydrogels (Chen et al. 1999; Kabiri and Zohuriaan-Mehr 2004; Wu et al. 1992). The addition of carbon particles into hydrogels with NIPAM portions resulted in a more porous structure (Figure 7.7(i)).

The effect of inclusion of graphene into polymer hydrogel on the water flux is presented in Figure 7.10(a) and (b). As can be seen, the composite hydrogels show a significantly higher initial flux than that of pure hydrogels particularly at 1.2 wt.% rGO concentration. In addition, the flux increases by the addition of rGO content up to 1.2 wt.% and then it reduces for further addition of rGO content (3 wt.%). These results are in correspondence with the variations in swelling ratios presented in Figure 7.10(c), where the swelling ratio increases as the rGo concentration in the composite increases and then it levels off before is reduced. This is an indication of the significant effect of a small amount of rGO (e.g., 1.2 wt.% or lower) on improving the swelling properties of polymer hydrogels and their resultant osmotic pressure. To understand why the hydrogels with the small amount of rGO performs much better than the neat one, the swelling process of pure and composite hydrogels with the optimum concentration of rGO in the first 70 s is shown in Figure 7.10(d). From the calculated projected area in the first 10 s,

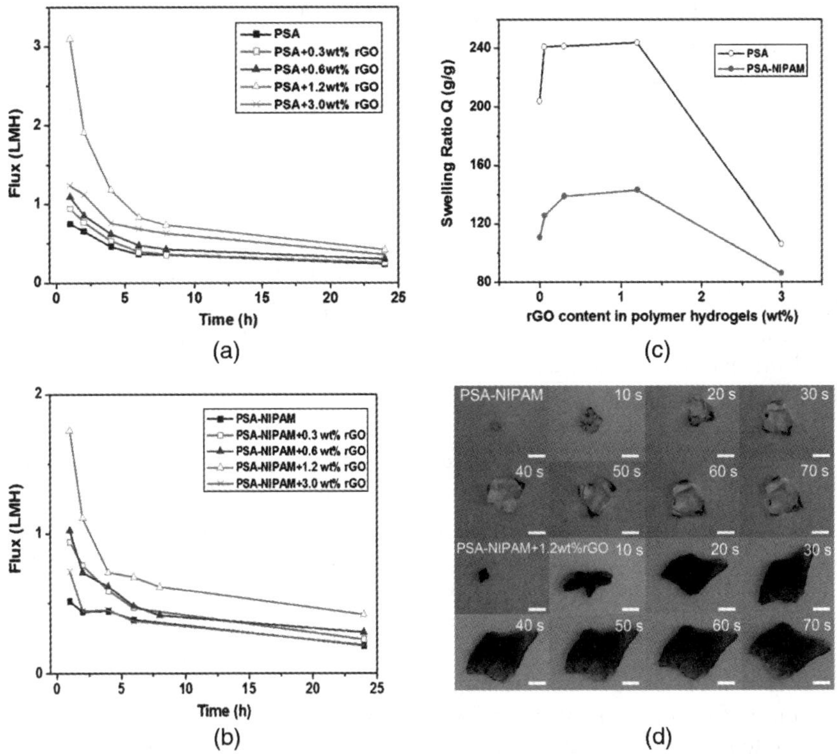

Figure 7.10. Water flux of pure and composite graphene polymer hydrogel with 2000 ppm NaCl as feed solution (a-b), swelling ratio as a function of graphene concentration (c) and the swelling process of pure and composite hydrogels in the first 70 s (scale bar = 1 mm)

SOURCE: Adapted from Zeng et al. 2013; reproduced with permission from Royal Society of Chemistry

the pure PSA expanded to 2.2 times of its original size (0.564 mm^2), while its composites expanded 4.78 times of its original size. After 70 s, the swelling degree of composite hydrogel was almost 2 times greater than that of the pure PSA. The addition of graphene into polymer hydrogel makes it much softer than the neat one (Alzari et al. 2011). This is the reason for the uneven swelling of composite hydrogels incorporated with rGO, whereas the pure hydrogels swell uniformly in all directions and retain their original particle shape.

According to the proposed mechanism by Zeng et al.(Zeng et al. 2013), the composite hydrogels with small amount of rGO (e.g., 1.2 wt% or less) have excellent shape adaptability and softness and each hydrogel composite particle can achieve maximal contact with other particles, as well as with the FO membrane, which leads to a faster water transport and neat hydrogels are more rigid nature and retain their original shapes during swelling, resulting in limited contact between the particles and FO membranes and among the other particles, and consequently low water flux. A further addition of rGO nanosheets (e.g., 3 wt. %) into the hydrogel mechanically reinforce the composite and acts as rigid, mechanically-percolating scaffolds to restrict the swelling of hydrogel, resulting in a low water absorption capacity and poor FO desalination performance.

7.3 DEWATERING

In any FO desalination process the separation of permeated water from the draw agent is also very important. The dewatering process of hydrogel needs the deswelling of the swollen hydrogels to release the water. Li et al. (Li et al. 2011a) used the hydraulic pressure and heating stimuli to release the water from PSA, PNIPAM, PSA-NIPAM and PAM hydrogels. They found that using only the hydraulic pressure at ambient temperature is not efficient, and a maximum of 3–5% total recovery can be achieved for NIPAM hydrogels when water content is more than 80%. However, significant water recovery of 54–75% was achieved when the dewatering of NIPAM carried out at 50°C. This is because NIPAM-containing polymer hydrogels undergo significant physical structural change above the LSCT, which is effective in assisting water release. PNIPAM hydrogel absorbs water and swells at temperature below LCST (34°C) and undergoes a dramatic shrinkage when environmental temperature exceeds the LSCT (Hirotsu et al. 1987). It is thought that the amphiphilic nature of the monomer unit itself is responsible for this onset of coil-globule transition (Cho and Kim 2010; Costa and Freitas 2002; Dhara and Chatterji 2000; Feil et al. 1993; Ilmain et al. 1991). Below the LSCT, the polymer–water interactions overcome the polymer–polymer interactions, where the H-bonding ability of —C=O and —N—H groups stabilizes the chain conformation in the favor of the PNIPAM hydrophilic behavior (Dhara and Chatterji 2000). Above the LSCT, however, the hydrophobic isopropyl groups of PNIPAM hydrogels destabilize the hydrogen bonds in the support of the polymer–polymer interactions (Cho and Kim 2010; Costa and Freitas 2002; Ilmain et al. 1991).

In another study, Li et al. (2011b) used solar energy as a sustainable method to release water from hydrogels with incorporated light-absorbing carbon particles. As shown in Figure 7.11, the water recovery rates of carbon-incorporated hydrogels are significantly higher than those of the neat hydrogels because the light absorbent particles in the bulk polymer absorb more solar energy and increase the bulk temperature faster than the pure hydrogels. Figure 7.11 demonstrates a water recovery rate for pure PSA-NIPAM hydrogels after 60 min exposure to sunlight with 0.4 kW/m^2 intensity of around 30%. In comparsion, carbon-incorporated PSA-NIPAM-C shows a recovery rate of 50% which increases to 90% if the light intensity is raised to 0.7 kW/m^2.

Solar energy was also used to release water from graphene-incorporated hydrogel (Zeng et al. 2013). As illustrated in Figure 7.12, water recovery rates rise by increasing the rGO content for both PSA and PSA-NIPAM hydrogels. However, PSA-NIPAM hydrogels also show a higher recovery rate at higher

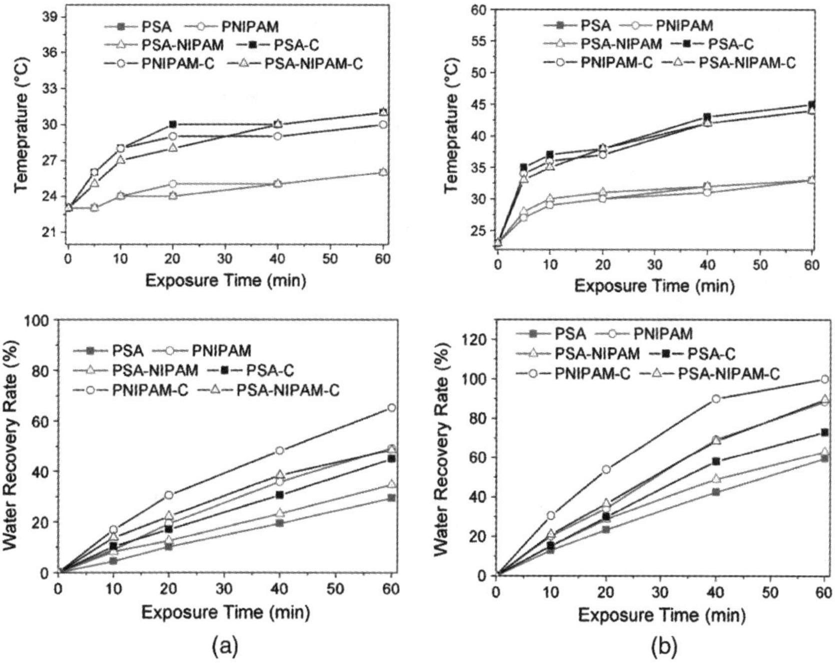

Figure 7.11. Temperature changes and water recovery rates of pure and composite (carbon incorporated) hydrogels after exposure to the simulated sunlight with, (a) 0.4 kW/m^2 and (b) 0.7 kW/m^2 intensity

SOURCE: Adapted from Li et al. 2011b; reproduced with permission from Royal Society of Chemistry

rGO content due to the thermoresponsive property of PNIPAM. The temperature profiles reveal that the inclusion of rGO can effectively absorb solar energy and convert it into heat to raise bulk temperatures and facilitate water release. It is worth mentioning that in the comparison between carbon loaded hydrogels and graphene loaded hydrogels, the latter composite shows similar performance at much lower concentrations of particle (graphene). For example, the temperature of 1.2 wt.% rGO-loaded hydrogel composite after 20 min of sunlight with 1 kWh^{-1}intensity was higher than that of 50 mol% carbon loaded hydrogels (37°C for PSA-NIPAM-C and 50°C for PSA-NIPAM-graphene). This better performance at low concentration of rGO is believed mostly due to the fast water transport through the two-dimensional spaces between polymer network and graphene nanosheets (Zeng et al. 2013).

7.4 PROCESS OPTIMIZATION

In the previous sections it was shown that the modification of the structure of the hydrogel network via the inclusion of ionic groups and light-absorbing particles can significantly improve the performance of hydrogels used as draw agents in FO

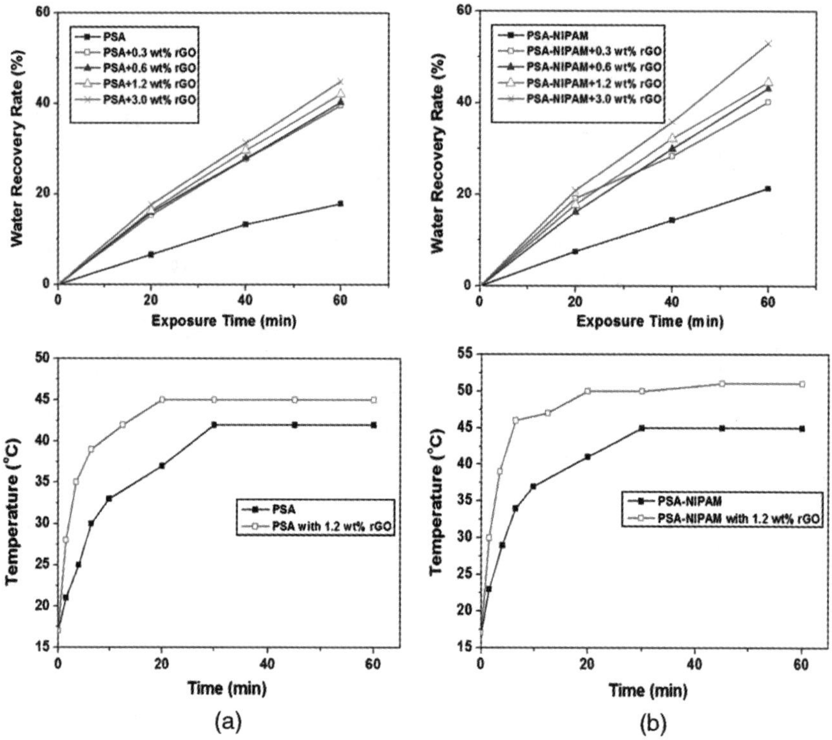

Figure 7.12. Water recovery rates and temperature plots of pure and composite hydrogels under sunlight with 1 kW/m^2 intensity

SOURCE: Adapted from Zeng et al. 2013; reproduced with permission from Royal Society of Chemistry

desalination. Apart from the chemical alteration of polymer hydrogels, other key parameters in determining the process performance are also important. Such parameters include particle size, particle-membrane contact zones, particle-particle contact areas, feed concentration and membrane type.

7.4.1 Effect of Hydrogels Particle Size on the Swelling kinetics

Effect of Hydrogels Particle Size on FO Flux desalination

Razmjou et al. (2013) prepared four different ranges of hydrogel particle size including 500–1000 μm, 350–500 μm, 190–350 μm and 2–25 μm. As can be seen in Figure 7.13(a), the water flux for the larger hydrogel particles (500–1000 μm) of PNIPAM-PSA is smaller than that of small particles (2–25 μm), which is related to the coverage of membrane surface. Small hydrogel particles can provide greater surface-to-volume ratio compared to the large particles. The effect of particle size on the flux of PSA and carbon incorporated PSA (PSA-C) with two different size of 150 and 500 μm was investigated in Figure 7.13(b). From the figure, it is

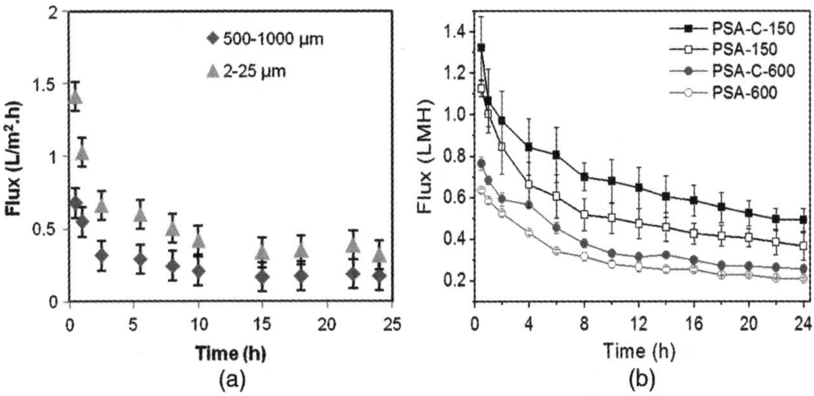

Figure 7.13. Effect of particle size on the water flux of (a) NIPAM-PSA and (b) pure PSA and carbon incorporated (PSA-C) hydrogels with two different particle sizes (150 and 600 μm) as the draw agent in a 24 h FO process using 2000 ppm NaCl solution as feed

SOURCE: (a) adapted from Razmjou et al. 2013, reproduced with permission from Elsevier, and (b) adapted from Li et al. 2013, reproduced with permission from Royal Society of Chemistry

clear that the hydrogel size significantly affects the water flux for both pure and composite hydrogels.

Understanding of the swelling process can be used to predict the effect of particle size on the driving force during the FO desalination. The swelling process has two main steps of water diffusion and polymer chains relaxation, forming a swollen (rubbery) polymer region on the outer shell advancing towards the non-swollen (glassy) rigid inner part of the hydrogel particles. The swelling is limited by the rigid inner core to the normal direction of the outer swollen shell. Once the advancing water front in the outer shell meets the inner core, the acceleration of the swelling kinetics takes place and the hydrogel particle begins to swell in all directions (Gerlach 2010). The point of acceleration depends on particle size and network thickness. The smaller the hydrogel particle size is, the faster it swells in all direction and the faster it absorbs water. Razmjou et al. (2013) fitted the initial swelling data ($M_t/M_\infty \leq 0.6$) of different hydrogel particle sizes to an exponential heuristic equation ($M_t/M_\infty = k.t^n$) to determine the diffusion mechanism. The corresponded mechanism is identified based on the value of n. Fickian diffusion ($n = 0.5$) or case I diffusion occurs whenthe rate of polymer chains relaxation is greater than that of the swelling agent (such as water) diffusion while Case II diffusion or non-Fickian diffusion ($n = 1$) occurs when water diffusion is faster than the polymer chains relaxation. Figure 7.14 shows that reducing the polymer hydrogel size shifts the swelling kinetics from non-Fickian to Fickian diffusion.

Effect of Hydrogels Particle Size on Water Recovery

In order to understand the contribution of hydrogel particle size to the water recovery, two different ranges of hydrogels with particle size of 2–25 μm and

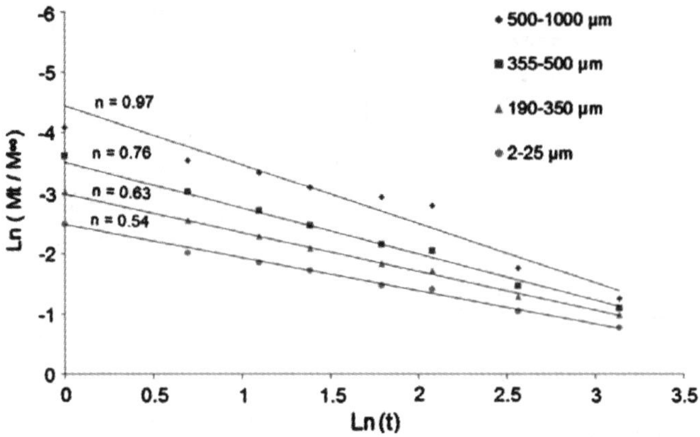

Figure 7.14. Swelling kinetics of NIPAM-PSA hydrogels with different particle sizes ("n" is characteristic exponent of the transport mode, "t" is time (h), Mt and M_∞ are the amount of absorbed water at time t and equilibrium hydration level, respectively)
SOURCE: Adapted from Razmjou et al. 2013; reproduced with permission from Elsevier

500–1000 μm, and swelling ratios of 10 and 200 were selected and subjected to gas pressure and temperature-stimuli (see Figure 7.15). From Figure 7.13(a), when gas pressure stimuli was used at low water content ($Q \sim 10$), there is no significant difference between water recoveries of small and large particles. However, when the swelling ratio increased to 200 the large hydrogel particles show more water recovery because the number of interstitial volumes and capillary spaces in the hydrogel bulk is significantly higher than that of small particles, and thus the water is more drainable under gas pressure stimuli. In addition, large hydrogel particles contain size a higher degree of unbound water which can readily be squeezed out of the particles as there is a very weak interaction between the unbound water molecules and the polymer network. Compared to large particles, hydrogels of small particle size contain more bound water and need a high energy input to overcome the strong interactions between water molecules and hydrogel network (Razmjou et al. 2013).

When a temperature stimulus is used, a reverse behavior was observed such that at high swelling ratio small particles show a higher level of water recovery after 50 min of exposure of the hydrogels to the environment with 60°C temperature (see Figure 7.15(b) and (c)). It is believed that a phenomenon known as "case hardening" in dehydration process (Bar et al. 2002; Khalloufi and Bongers 2012) is responsible for such low water recovery rates of large particles under temperature stimuli. When case hardening occurs, a glassy rigid skin forms at the surface of the hydrogels because the hydrogel surface moisture content reduces rapidly in atmosphere with low moisture content. This rigid skin significantly limits further macroscopic contraction and liquid water release consequently shows reduced water recovery. It is clear that the impact of case hardening on the dewatering rate of small particles is less than the large particles.

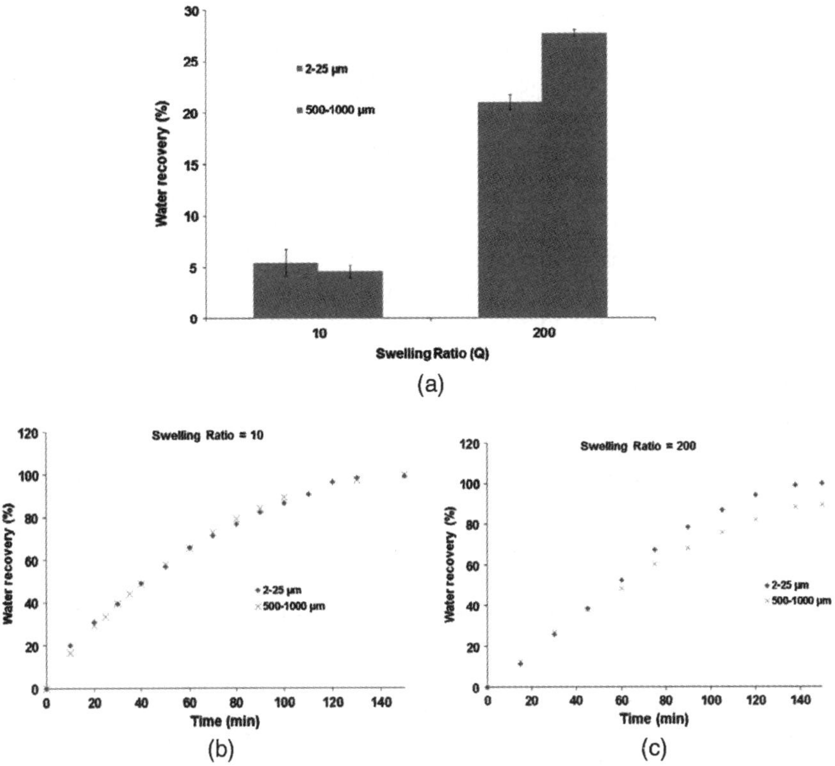

Figure 7.15. Water recovery of PNIPAM-PSA hydrogel with different particle sizes and swelling ratios under pressure (a) and temperature (b and c) stimuli
SOURCE: Adapted from Razmjou et al. 2013; reproduced with permission from Elsevier

7.4.2 Effect of Thickness, Contact Areas and Contact Zones

The effect of hydrogels thickness (added mass on the membrane surface) as draw agent on the membrane surface was investigated by Razmjou et al. (2013). Three different hydrogel mass of 0.05, 0.2 and 0.6 g on the membrane surface in a 24 h FO desalination process were examined. They found that for large particles (500–1000 μm), increasing the mass of the draw agent provided better membrane surface coverage and resulted in a significantly higher water flux, whereas using small hydrogel particles (2–25 μm) led to only a very modest increase. The effect of interstitial volume (the gap between individual particles and also the degree of contact between particles and membrane surface on the water flux in the FO process) was also investigated. The hydrogel powders (0.2 g) were pressed into a 0.2 g disc under hydraulic pressure to reduce the interstitial volumes and to minimize the particle-membrane surface contacting area. A significant reduction in water flux when the disc was used as draw agent was observed. However, when a small amount of powder (0.02 g) was added underneath of the disc, the flux increased sharply, implying the fact that the contribution of the contact areas between the hydrogel particles and the membrane surface on the water flux is more influential than the interstitial volumes.

7.4.3 Membrane Type and Feed solution

As shown in Figure 7.16(a), different saline solutions with NaCl concentrations of 2000, 5000 and 8000 ppm were used as feed solution in a FO desalination process. A significant reduction in water flux was experienced when the NaCl concentration was increased from 2000 ppm to 8000 ppm. This flux decline in the FO process is related to the increasing of the feed osmotic pressure, and thus a reduction in driving force and pressure difference across the FO membrane. In order to investigate the effect of membrane material, Li et al. tested two different membranes of AG-RO from GE Osmonics and CTA-FO from Hydration Technologies Inc., the first one being designed for reverse osmosis desalination, whilst the latter was prepared for forward osmosis desalination. From Figure 7.16(b), a substantial flux difference between the two membranes was seen. As expected, the AG-RO membrane is a thin film composite membrane which constitutes of a thin polyamide selective layer, a 50 μm microporous polysulfone membrane supported by a porous fabric layer (~90 μm), whereas the CTA-FO membranes is a 50 μm asymmetric cellulose acetate membrane with an embedded polyester mesh support. In addition, the AG-RO membrane has poor wettability because of its hydrophobic polysulfone support. Therefore, the CTA-FO membrane has a lower internal concentration polarization and thus higher fluxes in the FO process. This suggests that the structures and intrinsic properties of FO membranes can significantly affect the performance of hydrogel-drawn FO process.

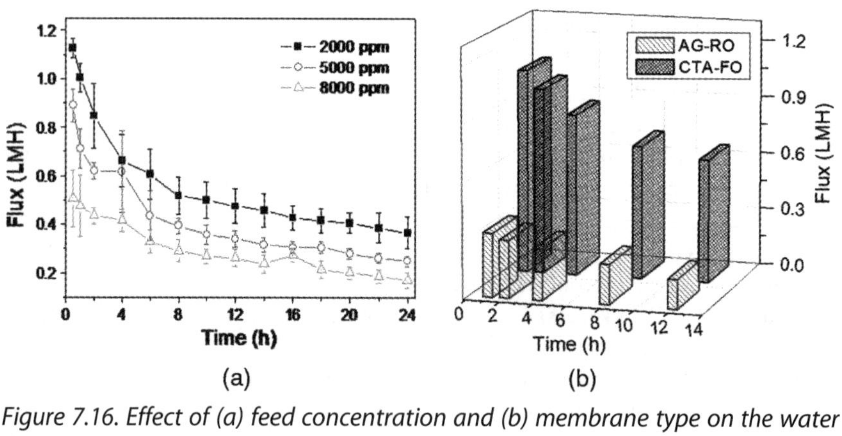

Figure 7.16. Effect of (a) feed concentration and (b) membrane type on the water flux

SOURCE: Adapted from Li et al. 2013; reproduced with permission from Royal Society of Chemistry

7.5 CONCLUDING REMARKS

The new strategy of using intelligent thermo-responsive polymer hydrogels as a draw agent in the FO desalination process has been introduced and demonstrated. Taking into account both the FO process and dewatering stage, the inclusion of ionic groups (e.g. PSA-NIPAM) into the hydrogel network is preferable for this innovative FO desalination application. Further engineering of the hydrogel structure via the addition of light absorbing materials such as carbon particles or graphene nanosheets into polymer hydrogel has shown a great influence on the FO desalination performance in terms of increasing both the water flux and dewatering rate. A significant, high level of carbon loading (50 mol %) is necessary to convert solar energy into heat during dewatering via a temperature stimulus. However, a small concentration of graphene nanosheets (1.2 wt.%) resulted in a similar or even greater performance. While this technology is in its infancy and has limitations, the manipulation of key parameters such as hydrogel particle size, particle-particle and particle-membrane surface contact area, feed concentration and membrane type could significantly improve the FO desalination performance. Although water flux in other types of draw agents reported in the literature are higher than those produced by hydrogel-driven FO process, the main advantage of this technology lies in the fact that the low, "green energy" input (i.e. solar energy) can be used during water recovery. More importantly, with suitable hydrogel structure and process design, the draw agent can be placed *in-situ* on the membrane surface to continuously absorb and release water at a low energy cost. Future investigations should be focused on further optimization of hydrogel properties and structures of membranes in order to enhance the performance of the hydrogel-driven FO process.

7.6 ABBREVIATIONS/NOMENCLATURE

CP	concentration polarization
CT	Acellulose triacetate
DO	dissolved oxygen
DOC	dissolved organic carbon

References

Alzari, V., Nuvoli, D., Scognamillo, S., Piccinini, M., Gioffredi, E., Malucelli, G., Marceddu, S., Sechi, M., Sanna, V., and Mariani, A. (2011). "Graphene-containing thermoresponsive nanocomposite hydrogels of poly(N-isopropylacrylamide) prepared by frontal polymerization." *J. Mater. Chem.*, 21, 8727–8733.

Bar, A., Ramon, O., Cohen, Y., and Mizrahi, S. (2002). "Shrinkage behaviour of hydrophobic hydrogel during dehydration." *J. Food Eng.*, 55, 193–199.

Can, V., Abdurrahmanoglu, S., and Okay, O. (2007). "Unusual swelling behavior of polymer-clay nanocomposite hydrogels." *Polymer*, 48, 5016–5023.

Chen, J., Park, H., and Park, K. (1999). "Synthesis of superporous hydrogels: Hydrogels with fast swelling and superabsorbent properties." *J. Biomed. Mater. Res.*, 44, 53–62.

Cho, S.M. and Kim, B.K. (2010). "Thermo-sensitive hydrogels based on interpenetrating polymer networks made of poly(N-isopropylacrylamide) and polyurethane." *J. Biomater. Sci., Polym. Ed.*, 21, 1051–1068.

Costa, R.O.R. and Freitas, R.F.S. (2002). "Phase behavior of poly(N-isopropylacrylamide) in binary aqueous solutions." *Polymer*, 43, 5879–5885.

Dhara, D. and Chatterji, P.R. (2000). "Swelling and deswelling pathways in non-ionic poly(N-isopropylacrylamide) hydrogels in presence of additives." *Polymer*, 41, 6133–6143.

Feil, H., Bae, Y.H., Feijen, J., and Kim, S.W. (1993). "Effect of comonomer hydrophilicity and ionization on the lower critical solution temperature of N-isopropylacrylamide copolymers." *Macromolecules*, 26, 2496–2500.

Gerlach, G.A. and Arndt, K.F. (ed.) (2010). "Hydrogel Sensors and Actuators." *Springer Series on Chemical Sensors and Biosensors*, Vol. 6, Springer, 2010.

Gil, E.S. and Hudson, S.M. (2004). "Stimuli-reponsive polymers and their bioconjugates." *Prog. Polym. Sci.*, 29, 1173–1222.

Gua, Y. and Yea, L. (2009). "Study on the polyvinylalcohol/montmorillonite composite hydrogel." *Polym. Plast. Technol. Eng.*, 48, 595–601.

Han, L., Yao, J.F, Li, D., Ho, J., Zhang, X., Kong, C.-H., Zong, Z.-M., Wei, X.-Y., Wang, H.T. (2008). "Hollow zeolite structures formed by crystallization in crosslinked polyacrylamide hydrogels." *J. Mater. Chem.*, 18, 3337–3341.

Haraguchi, K. (2007). "Nanocomposite hydrogels." *Curr. Opin. Solid State Mater. Sci.*, 11, 47–54.

Haraguchi, K., Li, H.-J., Matsuda, K., Takehisa, T., and Elliott, E. (2005). "Mechanism of forming organic/inorganic network structures during in-situ free-radical polymerization in PNIPA-clay nanocomposite hydrogels." *Macromolecules*, 38, 3482–3490.

Haraguchi, K., Li, H.-J., Song, L., and Murata, K. (2007). "Tunable optical and swelling/deswelling properties associated with control of the coil-to-globule transition of poly(N-isopropylacrylamide) in polymer-clay nanocomposite gels." *Macromolecules*, 40, 6973–6980.

Hirotsu, S., Hirokawa, Y., and Tanaka, T. (1987). "Volume-phase transitions of ionized N-isopropylacrylamide gels." *J. Chem. Phys.*, 87, 1392–1395.

Ilmain, F., Tanaka, T., and Kokufuta, E. (1991). "Volume transition in a gel driven by hydrogen bonding." *Nature*, 349, 400–401.

Kabiri, K. and Zohuriaan-Mehr, M.J. (2004). "Porous superabsorbent hydrogel composites: Synthesis, morphology and swelling rate." *Macromol. Mater. Eng.*, 289, 653–661.

Kayaman, N., Okay, O., and Baysal, B.M. (1998). "Swelling of polyacrylamide gels in polyacrylamide solutions." *J. Polym. Sci. Part B: Polym. Phys.*, 36, 1313–1320.

Khalloufi, S. and Bongers, P. (2012). "Mathematical investigation of the case hardening phenomenon explained by shrinkage and collapse mechanisms occurring during drying processes," in: Ian David LockhartB., MichaelF. (Eds.), *Computer Aided Chemical Engineering*, Elsevier, pp. 1068–1072.

Li, D., Zhang, X.Y., Simon, G.P., and Wang, H.T. (2013). "Forward osmosis desalination using polymer hydrogels as a draw agent: Influence of draw agent, feed solution and membrane on process performance." *Water Res.*, 47, 209–215.

Li, D., Zhang, X., Yao, J.F., Simon, G.P., and Wang, H.T. (2011a). "Stimuli-responsive polymer hydrogels as a new class of draw agent for forward osmosis desalination." *Chem. Commun.*, 47, 1710–1712.

Li, D., Zhang, X.Y., Yao, J.Y., Zeng, Y., Simon, G.P., and Wang, H.T. (2011b). "Composite polymer hydrogels as draw agents in forward osmosis and solar dewatering." *Soft Matter*, 7, 10048–10056.

Lietor-Santos, J.-J., Sierra-Martin, B., Vavrin, R., Hu, Z., Gasser, U., and Fernandez-Nieves, A. (2009). "Deswelling microgel particles using hydrostatic pressure." *Macromolecules*, 42, 6225–6230.

Ménager, C., Sandre, O., Mangili, J., and Cabuil, V. (2004). "Preparation and swelling of hydrophilic magnetic microgels." *Polymer*, 45, 2475–2481.

Qiu, Y. and Park, K. (2001). "Environment-sensitive hydrogels for drug delivery." *Adv. Drug Delivery Rev.*, 53, 321–339.

Qiu, Y. and Park, K. (2012). "Environment-sensitive hydrogels for drug delivery." *Adv. Drug Delivery Rev.*, 64, Supplement, 49–60.

Razmjou, A., Simon, G.P., and Wang, H.T. (2013) "Effect of particle size on the performance of forward osmosis desalination by stimuli-responsive polymer hydrogels as a draw agent." *Chem. Eng. J.*, 215–216, 913–920.

Santiago, F., Mucientes, A.E., Osorio, M., and Rivera, C. (2007). "Preparation of composites and nanocomposites based on bentonite and poly(sodium acrylate). Effect of amount of bentonite on the swelling behaviour." *Eur. Polym. J.*, 43, 1–9.

Schexnailder, P., and Schmidt, G. (2009). "Nanocomposite polymer hydrogels." *Colloid. Polym. Sci.*, 287, 1–11.

Serizawa, T., Wakita, K., and Akashi, M. (2001). "Rapid deswelling of porous poly(N-isopropylacrylamide) hydrogels prepared by incorporation of silica particles." *Macromolecules*, 35, 10–12.

Tanaka, T., Nishio, I., Sun, S.-T., and Ueno-Nishio, S. (1982). "Collapse of gels in an electric field." *Science*, 218, 467–469.

Wack, H. and Ulbricht, M. (2009). "Effect of synthesis composition on the swelling pressure of polymeric hydrogels." *Polymer*, 50, 2075–2080.

Wang, C., Flynn, N.T., and Langer, R. (2004). "Controlled structure and properties of thermoresponsive nanoparticle–hydrogel composites." *Adv. Mater.*, 16, 1074–1079.

Wang, H.T., Holmberg, B.A., and Yan, Y.S. (2003). "Synthesis of template-free zeolite nanocrystals by using in situ thermoreversible polymer hydrogels." *J. Am. Chem. Soc.*, 125, 9928–9929.

Wang, X., Wang, J., Zhao, W., Zhang, L., Zhong, X., Li, R., and Ma, J. (2010). "Synthesis and characterization of thermotropic liquid crystalline polyester/multi-walled carbon nanotube nanocomposites." *Appl. Surf. Sci.*, 256, 1739–1743.

Wu, X.S., Hoffman, A.S., and Yager, P. (1992). "Synthesis and characterization of thermally reversible macroporous poly(N-isopropylacrylamide) hydrogels." *J. Polym. Sci., Part A: Polym. Chem.*, 30, 2121–2129.

Yao, J.F, Li, D., Zhang, X.Y., Kong, C.-H., Yue, W., Zhou, W., and Wang, H.T. (2008). "Cubes of zeolite A with an amorphous core." *Angew. Chem. Int. Ed.*, 47, 8397–8399.

Zeng, Y., Qiu, L., Wang, K., Yao, J.F., Li, D., Simon, G.P., Wang, R., Wang, H.T. (2013). "Significantly enhanced water flux in forward osmosis desalination with polymer-graphene composite hydrogels as a draw agent." *RSC Adv.*, 3, 887–894.

Zheng, Y., Li, P., Zhang, J., and Wang, A. (2007). "Study on superabsorbent composite XVI. Synthesis, characterization and swelling behaviors of poly(sodium acrylate)/vermiculite superabsorbent composites." *Eur. Polym. J.*, 43, 1691–1698.

Ziesmer, S. and Stock, N. (2008). "Synthesis of bifunctional core-shell particles with a porous zeolite core and a responsive polymeric shell." *Colloid. Polym. Sci.*, 286, 831–836.

CHAPTER 8

Forward Osmosis Membranes: Synthesis and Characterization

Sui Zhang
Rui Chin Ong
Honglei Wang
Panu Sukitpaneenit
Duong Hoang Hanh Phuoc
Tai-Shung Chung

In this chapter, we aim to (1) summarize the background and state-of-the-art technologies for forward osmosis (FO) membranes fabrication, (2) analyze the mass transport phenomena in FO membranes and (3) discuss strategies to properly design FO membranes. We will also highlight basic characterization methods for FO membranes. We will focus on asymmetric membranes made of both phase inversion and composite membranes. The phase inversion membranes are mainly those made from cellulose acetate (CA), polybenzimidazole (PBI) and polyamide-imide (PAI), while the composite membranes include thin film composite (TFC), layer-by-layer (LbL) and biomimetic membranes. Each type of membranes is discussed in terms of material, structure, preparation methods and FO performance.

8.1 MASS TRANSPORT IN FORWARD OSMOSIS

There are three major components involved in the forward osmosis (FO) process, namely, (1) a draw solution comprising a high osmotic pressure, (2) a feed solution possessing a lower osmotic pressure and (3) a semipermeable membrane working as a barrier between the two solutions. Due to the osmotic pressure difference between the draw and feed solutions, water transports across the membrane naturally from the feed to the draw solution, and the water flux can be described by the following equation:

$$J_w = A\Delta\pi_m \tag{8.1}$$

where J_w is the water flux; A is the water permeability of the membrane; $\Delta\pi_m$ is the effective osmotic pressure difference across the membrane ($\pi_{D,m}-\pi_{F,m}$). If the membrane is not perfectly semipermeable, there is a solute reverse flux J_s from the draw solution to the feed solution due to the concentration gradient across the membrane. J_s can be calculated using Eq. (8.2):

$$J_s = B(C_{D,m} - C_{F,m}) \tag{8.2}$$

where B is the salt permeability, $C_{D,m}$ and $C_{F,m}$ are the salt concentrations on the surfaces of the selective layer in the draw and feed sides, respectively. By assuming that the osmotic pressure of the solutions obeys Van't Hoff Equation, the solute reverse flux can then be expressed as a function of J_w using Van't Hoff factor i (Tang et al. 2010; Zhang et al. 2013):

$$J_s = \frac{BJ_w}{iRTA} \tag{8.3}$$

Due to the asymmetry of most membranes for water treatment, there exist two different operational modes in FO as shown in Figure 8.1. The pressure retarded osmosis (PRO) mode refers to the membrane orientation where the active layer is facing towards the draw solution and the porous support layer is facing against the feed, while the FO mode refers to the reversed orientation.

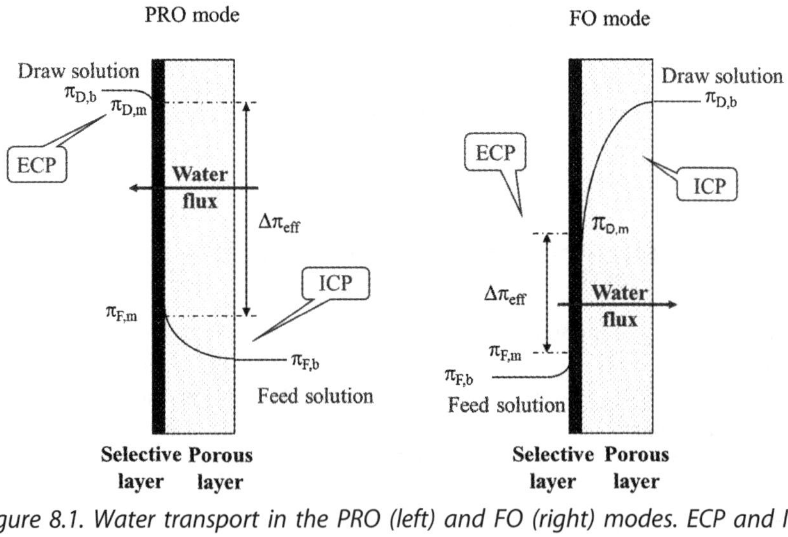

Figure 8.1. Water transport in the PRO (left) and FO (right) modes. ECP and ICP refer to external concentration polarization and internal concentration polarization, respectively

8.1.1 External Concentration Polarization

Similar to conventional pressure driven membrane processes such as RO and others, concentration polarization affects FO performance. Concentration polarization refers to the existence of a concentration gradient of solutes between the bulk solution and the membrane's active surface due to the permeate flow across the membrane towards the draw solution. As a result, feed solutes build up on the membranes' active surface and create additional osmotic pressure on the feed side, while draw solutes become diluted on the other membrane surface. The combined effects lower the overall effective osmotic driving force across the membrane and the water flux is therefore reduced. As illustrated in Figure 8.1, this phenomenon may occur outside the membrane (referred to as external concentration polarization (ECP)) and inside the membrane (referred to as internal concentration polarization (ICP)) facing the draw and feed solutions in osmosis-driven membrane systems.

Dilutive ECP occurs when the draw solution is placed against the selective layer in the PRO mode. At steady state, the permeate water flows from the feed side and dilutes the solute concentration on the surface of the active layer. The osmotic pressure at the active layer surface $\pi_{D,m}$ can be obtained with the following equation:

$$\pi_{D,m} = \pi_{D,b} \exp\left(-\frac{J_w}{k}\right) \tag{8.4}$$

The solution concentration on the membrane surface is not only related to the bulk osmotic pressure of the draw solution, $\pi_{D,b}$, the mass transfer coefficient, k, but also strongly dependent on the permeate water flux, J_w. An increase in permeate water flux would cause an exponential decrease in effective osmolality of the draw solution on the membrane surface.

On the contrary, when the feed solution is facing against the selective layer in the FO mode, the concentrative ECP would happen which results in an increase in osmotic pressure at the membrane surface $\pi_{F,m}$ as follows:

$$\pi_{F,m} = \pi_{F,b} \exp\left(\frac{J_w}{k}\right) \tag{8.5}$$

where $\pi_{F,b}$ is the bulk feed osmotic pressure. The concentrative ECP reduces the effective driving force across the membrane and decreases the water flux. The degree of ECP is characterized by the mass transfer coefficient k in the flow chamber of the membrane unit. Based on the film theory, k is determined by the physicochemical properties of the solute solution, the flow velocity and the chamber dimensions (McCutcheon and Elimelech 2007). Eq. (8.6) describes the mass transfer coefficient with relations to solute diffusion coefficient D, hydraulic diameter of the flow chamber d_h, and Sherwood number Sh:

$$k = \frac{ShD}{d_h} \tag{8.6}$$

Depending on whether the solution is a laminar or turbulent flow, the Sherwood number in a rectangular chamber could be calculated by either of the following equations:

$$Sh = 1.85\left(\mathrm{Re}SC\frac{d_h}{L}\right)^{0.33} \qquad \text{(Laminar flow)} \tag{8.7}$$

$$Sh = 0.04\mathrm{Re}^{0.75}Sc^{0.33} \qquad \text{(Turbulent flow)} \tag{8.8}$$

where, Re is the Reynolds number, *Sc* is the Schmidt number, and *L* is the length of the channel. As a result, the negative effects of ECP on water flux can be mitigated by altering the operating parameters to enhance mass transfer coefficient, e.g., increasing the input solution velocity and decreasing the hydraulic diameter of the chamber.

8.1.2 Internal Concentration Polarization

As shown in Figure 8.1, typical FO membranes consist of a thin semi-permeable layer to reject undesirable solutes and a thick porous layer to provide the mechanical support. Ideally, the semi-permeable layer determines the selectivity and water permeability of the membranes. Therefore, the effective force to drive the water across the membrane is $\pi_{D,m} - \pi_{F,m}$, which is the osmotic pressure difference between both surfaces of the active layer. However, the solute concentration within the support layer is different from the bulk solution. In the PRO mode where the feed solution is facing the supporting layer, the feed solute is dragged into the supporting layer and left beneath the dense layer when water is permeating through the membrane. As a result, the solute concentration beneath the dense layer is higher than that in the bulk feed. Meanwhile, the non-uniform solute concentration within the support layer also causes solute diffusion along its concentration gradient. In other words, the solute reverse flux within the support layer is contributed by two components: solute diffusion due to concentration gradient and convective flow arising from water flux (Lee et al. 1981; Zhang et al. 2010):

$$J_s = \frac{D\varepsilon}{\tau}\frac{dC(x)}{dx} - J_w C(x) \tag{8.9}$$

where ε is the porosity of the support layer and τ is the tortuosity. At steady state, the solute reverse flux across the imperfectly semi-permeable dense layer should be equal to the solute reverse flux within the supporting layer. In other words, eq. (8.3) and eq. (8.9) are equal as follows:

$$B(C_{D,m} - C_{F,m}) = \frac{D\varepsilon}{\tau}\frac{dC(x)}{dx} - J_w C(x) \tag{8.10}$$

In the FO mode, a similar phenomenon is observed but the solute concentration on the inner surface of the active layer is diluted because of water inflow from the feed. The boundary conditions of eq. (8.10) include the bulk solute concentration on the outer surface of the support layer ($x=0$) and the solute concentration on the inner surface of the active layer ($x=\chi$), as shown in eq. (8.11) and eq. (8.12), respectively. The symbols χ refers to the thickness of the support layer.

$$C(x)=C_{F,b} \quad \text{at}\, x=0 \tag{8.11}$$

$$C(x)=C_{F,b} \quad \text{at}\, x=\chi \tag{8.12}$$

Rearrangement of eq. (8.10) with integration leads to the following:

$$\int_{C_{F,b}}^{C_{F,m}} \frac{1}{B(C_{D,M}-C_{F,m})+J_w C_{(x)}} dC(x) = \int_0^{\chi} \frac{\tau}{D\varepsilon} dx \tag{8.13}$$

Eq. (8.13) becomes as follows if steady state is reached:

$$\frac{1}{J_w} \ln \frac{B(C_{D,m}-C_{F,m})+J_w C_{F,m}}{B(C_{D,m}-C_{F,m})+J_w C_{F,m}} = \frac{\tau\chi}{D\varepsilon} \tag{8.14}$$

Thus the unknown concentration $C_{F,m}$ can be expressed with the known bulk conditions:

$$C_{F,m} = \frac{BC_{D,m}\exp(J_w K) + J_w C_{F,b}\exp(J_w K) - BC_{D,m}}{J_w - B + B\exp(J_w K)} \tag{8.15}$$

where the capital K represents the resistance within the support layer for solute diffusion and can be written as:

$$K=\frac{S}{D} \tag{8.16}$$

where S is the structural parameter of the support layer which is related to the layer's tortuosity, thickness and porosity:

$$S=\frac{\tau\chi}{\varepsilon} \tag{8.17}$$

Then the concentration gradient across the dense membrane can be expressed by the following equation:

$$C_{D,m}-C_{F,m} = \frac{J_w C_{D,m} - J_w C_{F,b}\exp(J_w K)}{J_w - B + B\exp(J_w K)} \tag{8.18}$$

Assuming a linear relationship between solute concentration and osmotic pressure, the combination of eq. (8.18) and eq. (8.1) yields:

$$J_w = \frac{1}{K} \ln \frac{A\pi_{D,m} - J_w + B}{A\pi_{F,b} + B} \tag{8.19}$$

Eq. (8.19) explains how the water flux is affected by the retarded solute diffusion in the supporting layer. A large K, an indication of serious resistance of solute diffusion within the support layer, would cause a dramatic loss in water flux. This phenomenon is named as the internal concentration polarization (ICP). Since K is only related to the diffusion coefficients of the solutes and the porous structure within the supporting layer, it is difficult to reduce ICP by manipulating the operating conditions of FO processes. The design of FO membranes with a proper structure; namely, membranes with a fully porous, low tortuosity, thin but still mechanically stable supporting layer become especially important. Similarly, in the FO mode, the expression to model the flux under ICP can be derived as follows (Zhang et al. 2010):

$$J_w = \frac{1}{K} \ln \frac{A\pi_{D,b} + B}{A\pi_{F,m} + J_w + B} \tag{8.20}$$

8.1.3 Solute Reverse Flux

Due to the imperfection of FO membranes, solutes leak from the draw solution to the feed. Eqs. (8.2) and (8.3) provide the simplified equations to quantify the solute reverse flux. Eq. (8.2) reveals that the root for the reverse diffusion of solutes is the salt permeability across the active layer and eq. (8.3) indicates that the salt reverse flux is proportional to the water flux. Therefore, if water and salt permeability are known, the salt reverse flux can be simply calculated from the water flux.

The solute reverse flux contaminates the feed solution. Its accumulation in the feed solution would ultimately enhance the feed osmotic pressure and reduces water flux. In addition, its accumulation in the supporting layer reduces the effective osmotic driving force across the membranes. Rearrangement of eq. (8.19) gives a direct expression of the B effect on water flux in the PRO mode:

$$J_w = A\pi_{D,m} - A\pi_{F,b} \exp(J_w K) - (\exp(J_w K) - 1)B \tag{8.21}$$

For asymmetric membranes, $\exp(J_w K) > 1$, so the term $(\exp(J_w K) - 1)B$ works as a negative factor for water flux. The 2nd term of eq. (8.21) denotes the effect of the feed solute built up because of water outflow to the draw solution and the diffusion resistance within the support layer, while the 3rd term denotes the effect of reverse salt flux across the dense layer which enhances the osmotic pressure of the feed solution on the inner surface of the dense layer. Consequently, the water flux is decreased. Only if ICP is eliminated ($K = 0$), the solute permeability term becomes zero.

8.2 STRUCTURAL DESIGN AND CHARACTERIZATION OF FO MEMBRANES

Even though both RO and FO membranes consist of a thin dense selective layer and a porous support layer, their design strategies are different. In conventional pressure-driven processes, the feed solution flows against the dense layer and the product permeate is collected at the other side. While a thin dense layer is required to maximize productivity, a thick and mechanically strong supporting layer is preferred to stabilize the membrane structure under high pressure operations. However, numerous studies have shown that the traditional RO membranes produce very low water fluxes in FO processes (Mehta and Loeb 1979; Thorsen and Holt 2009). Comparing with RO, low or no hydraulic pressure is involved in FO. Therefore, there is no stringent demand on mechanical strength. The major concerns lie in the design strategy to maximize water flux and minimize solute reverse flux in the FO process. A comparison of the process and membrane structure for FO and PRO is illustrated in Figure 8.2.

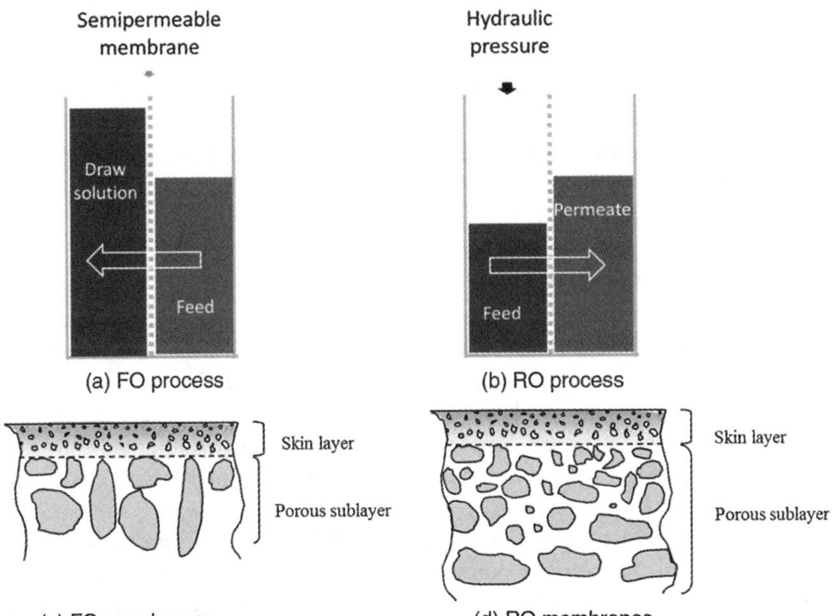

Figure 8.2. A comparison of the FO and RO processes. For FO, no hydraulic pressure is involved and two solutions run on both sides of the membranes; for RO, a hydraulic pressure is applied at the feed solution side and a clean permeate is collected on the other side of the membrane

8.2.1 Selective Layer

According to Eqs. (8.1) and (8.2), the ideal FO membranes ought to have a large A and a small B. Since both parameters are mainly determined by the thin dense selective layer, the thickness and perfection of this layer become especially important. In general, an ultrathin and non-defective layer is required to minimize transport resistance and increase selectivity. However, the selectivity is a relative concept which depends on the size and properties of the solutes involved.

Though the commercial applications of FO are still limited, FO has high potential to be applied in a broad range of fields, including wastewater treatment and water purification, food processing, pharmaceutical production, direct fertigation and seawater desalination (Cath et al. 2006; Chung et al. 2012a and b; Phuntsho et al. 2012; Zhao et al. 2012). The draw solutes can be salt, ammonium fertilizer, nutrient molecules, sugar, magnetic particles, electrolytes and many others (Ge and Chung 2013). Similarly, the feed solution may contain salts, heavy metals, organic matters, pharmaceutical materials, etc.

Depending on feed compositions and end applications, the requirements for membranes' selectivity are different. If sodium chloride is involved in the process, the selective layer should resemble the RO membranes with a very low pore size (close to the free volume of the corresponding polymers) and no defects. It allows the rapid passage of water but rejects most of the sodium and chloride ions. The materials suitable for such applications are limited. Most frequently, cellulose acetate (CA)/cellulose triacetate (CTA) and cross-linked polyamide formed by interfacial polymerization are used in the RO desalination processes. On the other hand, if larger molecules are used, a looser structure in the selective layer is tolerable as long as the solutes can be effectively rejected. In this case, more materials can be chosen such as polybenzimidazole (PBI) and layer-by-layer polyelectrolytes. Note that those materials are usually hydrophilic to ensure high water permeability.

8.2.2 Supporting Layer

Due to the ICP in the supporting layer, the supporting layer for FO membranes requires special designs to minimize the solute transport resistance K. Since K is proportional to the structural parameter S (eq. (8.16)) and S is a function of tortuosity, porosity and thickness of the supporting layer (eq. (8.17)), the supporting layer must have a low tortuosity, high porosity and small thickness to reduce S as long as the resultant membranes have sufficient mechanical strength to withstand the FO operation.

The structural parameters of conventional RO membranes are usually more than 1000 μm due to their low porosity and thick supporting layer (Thorsen and Holt 2009), while the FO membranes developed in the latest decade decreases to around 500 μm, or even less than 300 μm (McCutcheon and Elimelech 2006; Sukipaneenit and Chung 2012; Han et al. 2012a). The main approaches include employing a thin and fully porous woven/non-woven support, fixing a small membrane thickness, enhancing the porosity and the use of hydrophilic polymers.

Ways to increase porosity in the support layer are (1) reducing the polymer concentration, (2) adding pore-forming agents such as ethylene glycol (EG), calcium chloride and polyvinylpyrrolidone (PVP) into dopes, and (3) using weak non-solvents as the coagulant media (Yang et al. 2009; Li et al. 2012).

Hydrophilicity plays an important role in affecting ICP. When the substrate is hydrophobic, it is difficult to be wetted by aqueous solutions. The pores that are not wetted by water are not available for the transport of water and solute molecules. Therefore, the effective area for mass transport is reduced. More severe ICP is resulted and a lower water flux is observed. Studies have demonstrated that by increasing the hydrophilicity of the support layer, the performance of FO membranes could be largely enhanced (Widjojo et al. 2011; Arena et al. 2011; Han et al. 2012a, Wang et al. 2012; Widjojo et al. 2013).

8.2.3 Membrane Characterization

Usually, FO membranes are characterized by their water flux, solute reverse flux as well as structural parameter. Fundamental transport properties such as water permeability and salt permeability of the membrane materials are important to fully understand the transport properties and mechanisms.

The water permeability and salt permeability can be determined in a pressure-driven process (Zhang et al. 2010). At a given hydraulic pressure ΔP (usually lower than 10 bar to ensure that the membrane structure is similar to the one in the FO process), deionized (DI) water is used as the feed and the permeate is collected to get the water permeability using the following equation:

$$A = \frac{\Delta V}{S_m \Delta t \Delta p} \tag{8.22}$$

where ΔV is the volume of permeated water across the membrane area S_m in a certain time period Δt. Meanwhile, the salt rejection is tested using a low concentration salt solution (usually 100–5, 000 ppm) as the feed:

$$R = \left(1 - \frac{C_p}{C_f}\right) \times 100\% \tag{8.23}$$

The salt concentrations C_f and C_p in the feed and permeate, respectively, can be determined using a conductivity meter. Then the salt permeability B could be calculated based on eq. (8.24) where $\Delta\pi$ is the osmotic pressure difference between the feed and permeate in the pressure-driven process:

$$B = \frac{1 - R}{R}(\Delta P - \Delta\pi)A \tag{8.24}$$

In a lab-scale FO setup, membrane modules/cells are fixed in the middle of the setup, with feed and draw solutions flowing counter-currently at either side the membranes in the cross-flow mode. To standardize the membrane performance so that different experiments can be compared, 1M or 2M NaCl solution is commonly used as the draw solution, and DI water is the feed. The water permeation flux (J_w, L·m^{-2}·hr^{-1}, abbreviated as LMH) is calculated as below:

$$J_w = \frac{\Delta V}{S_m \Delta t} \tag{8.25}$$

The conductivity of the feed solution is recorded and salt reverse flux from the draw solution to the feed, J_s in g·m^{-2}·hr^{-1} (abbreviated as gMH), is determined from the increase in conductivity of the feed:

$$J_s = \frac{\Delta C_t V_t}{A \Delta t} \tag{8.26}$$

where ΔC_t is the concentration change of the feed during the testing period, and V_t is the volume of the feed at the end of the test. Data in both PRO and FO modes can be tested by simply changing the membrane orientation. Once water and salt permeability, and water flux in the FO mode are known, structural parameter can be calculated from eq. (8.27):

$$S = \frac{D}{J_w} \ln \frac{A\pi_{D,b} + B}{A\pi_{F,m} + J_w + B} \tag{8.27}$$

8.3 INTEGRALLY SKINNED ASYMMETRIC MEMBRANES BY PHASE INVERSION

Integrally skinned asymmetric membranes are formed by phase inversion where the dense skin layer and porous sublayer are made of the same material and formed in a one-step process. Figure 8.3(A) schematically depicts the cross-sectional views of two typical integrally skinned membranes.

The formation of integrally skinned asymmetric membranes via nonsolvent induced phase inversion was first demonstrated by Loeb and Sourirajan using CA

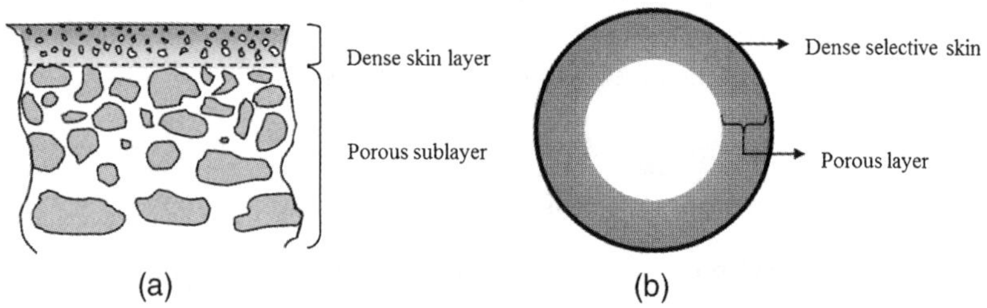

Figure 8.3. Schematic illustration of integrally skinned asymmetric (A) flat sheet membrane and (B) hollow fiber membrane structures

as the material (Loeb and Sourirajan 1964). They prepared a CA polymer solution containing acetone as the solvent and water and magnesium perchlorate as additives and cast evenly on a glass plate with a controlled thickness using a doctor blade. After subjecting to evaporation in air for a short duration, the as-cast membrane was then immersed in a water bath for nonsolvent induced phase inversion. The resultant membrane has the structure depicted in Figure 8.3(A). This type of membranes has been widely used in reverse osmosis (RO) applications. The study of integrally skinned FO membranes only took place about 10 years ago (Wang et al. 2007; Herron 2008].

By optimizing the chemistry of coagulants and casting plates as well as dope formulation, flat sheet membranes with various structures can be designed (Wang et al. 2010; Zhang et al. 2011; Ong et al. 2012). Similarly, integrally skinned hollow fiber membranes can be tailored made using the same concept (Wang et al. 2007; Su et al. 2010a & b) by choosing an appropriate external coagulant and a bore fluid as well as spinning conditions. Basically, a strong non-solvent coagulant such as water tends to induce a dense selective skin, while a weak non-solvent or solvent/non-solvent mixture tends to induce a sponge-like morphology as the support layer structure. A typical single layer hollow fiber membrane with a dense selective skin at the outer layer is depicted in Figure 8.3(B).

8.3.1 Cellulose Ester Membranes

Among all available polymers in the market, cellulose ester polymers, especially CA, have received the most attention as FO membrane materials due to vast availability, low costs, hydrophilicity, ease of fabrication and also chlorine resistance. Cellulose esters are green polymers as they are easily derived from esterification of cellulose with acetic acid, a commonly found natural organic compound. The Hydration Technology Innovations, HTI was the first to develop and manufacture CTA-based FO membranes with woven supports in the market (Herron 2008). The woven support is relatively porous and the membrane has an asymmetric structure formed via phase inversion as shown in Figure 8.4(A). This membrane achieves a superior FO performance than those standard RO membranes due to relatively thinner membrane thickness and more porous support structure.

Following the success of HTI CTA FO membranes, several CA FO membranes have been developed. The most widely explored cellulose ester is CA which has an acetyl content of less than 40 wt% and has high hydrophilicity. Its inherent free volumes in dry and wet states and water transport properties have been studied for in depth for both RO and FO (Zhang et al. 2012).

Among many CA FO membranes, the double-skinned CA membrane without support fabrics, as shown in Figure 8.4(B), has gained much attention due to its unique design which mitigates ICP and fouling phenomena. Contrary to the conventional Loeb-Sourirajan phase inversion membranes, the double-skinned CA membranes were invented by a slightly different method. Most conventional phase inversion membranes designed have dense selective skins on top of the asymmetric membranes via fast evaporation and outflow of solvent.

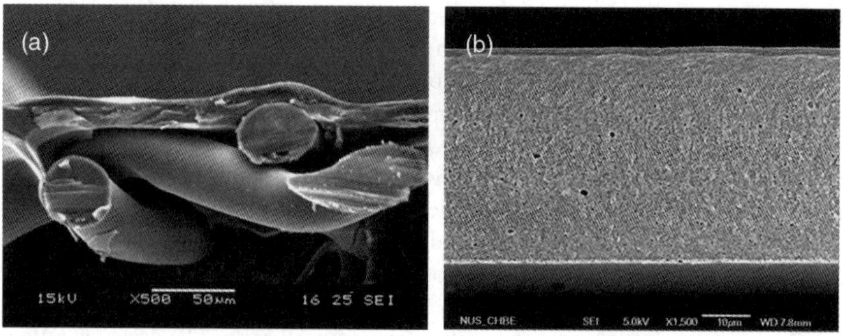

Figure 8.4. Cross-section images of (A) HTI CTA and (B) double-skinned CA membranes

However, the double-skinned CA membranes were fabricated without the solvent evaporation step (Wang K.Y. et al. 2010; Zhang et al. 2010; Ong et al. 2012). After casting the polymer solution on the glass plate, it was immersed into a water coagulation bath at room temperature immediately. The resultant double-skinned membranes has a distinctive three-layer structure consisting of a top loose dense skin with a larger mean effective pore size, a thick fully open cell porous middle layer and a thin dense bottom skin which has a small mean effective pore size to reject small solutes such as NaCl ions.

The top skin was formed due to (1) the rapid solvent evaporation during the casting as acetone was used as the solvent following by (2) the rapid solvent outflow to the coagulant when the nascent membrane was immersed into the coagulation bath. Thus the polymer concentration at the top skin increased rapidly and formed a relatively dense top skin. The fully porous middle layer was resulted from the delayed demixing in the bulk after the formation of the top dense skin. The bottom dense selective skin was formed via a hydrophilic-hydrophilic interaction between CA which has a high content of hydroxyl group and the hydrophilic glass casting substrate. During phase inversion, CA molecules near the bottom surface were in contact with the glass plate and due to the hydrophilicity of both CA and glass plate, the CA molecules would adhere on the glass surface and aggregate to form the bottom selective skin. With heat treatment in DI water as the post treatment, the double-skinned CA membranes achieved a high rejection towards NaCl ions.

CA FO membranes with a double-skinned structure have been tested and compared against single-skinned CA FO membranes (Zhang et al. 2010) using a colloidal feed solution containing aluminum oxide nanoparticles. The former has a significantly lower flux reduction due to less fouling than the latter. In addition, the backwashed membranes of the former have a higher recovered water flux than the latter. Clearly, the top loose dense layer of the double-skinned membranes can effectively prevent irreversible fouling where particles may trap easily within the membrane porous layer which is difficult to be removed by simple flushing. The advantages of the double-skinned structure for FO applications have been mathematically confirmed by Tang and his coworkers where they modeled the transport mechanism for double-skinned FO membranes by modifying the existing solution-diffusion model for single-skinned FO membranes (Tang et al. 2011).

Besides, integrally double-skinned hollow fiber membranes have been developed using CA (Su et al. 2012) to produce FO hollow fiber membranes with less susceptible to ICP and fouling. The dense inner selective skin was formed by using pure water as the bore fluid to induce instantaneous demixing. The outer skin was made with slightly looser pores by using a 50/50 wt % NMP/water coagulant to induce a slower incomplete demixing at first before the nascent fiber entered the external water bath. The inner selective skin was further heat treated by circulating 90°C DI water at the lumen side of the hollow fiber to improve the skin's selectivity as well as FO flux.

The post heat treatment is important to tailor CA FO membranes with proper pore sizes no matter the membranes are single- or double-layer and flat sheet or hollow fiber configurations. A two-step heat treatment, i.e., 60 min at 60°C and 20 min at 95°C using a DI water bath, was found effectively to shrink the membrane effective mean pore size of CA nanofiltration hollow fiber membranes for FO processes (Su et al. 2010a). Fundamental studies have revealed that heat treatment effectively increases FO water flux due to (1) the shrinkage of pore sizes and (2) the reduction of reverse salt flux to the feed solution which would otherwise lead to a drastic drop in the effective driving force (Su et al. 2010b). However, heat treatment at very high temperatures leads to a drop in water flux due to pore tightening that causes a high resistance for water transport.

Since CA has poor resistance towards biodegradation and a narrow pH operation window due to poor hydrolytic stability, other cellulose esters such as CTA, cellulose acetate propionate (CAP) and cellulose acetate butyrate (CAB) have also been explored as materials for FO membranes (Ong et al., unpublished manuscript 2013). CTA are cellulose esters containing very little or no content of hydroxyl functional groups while CAP and CAB have both hydroxyl and propionyl groups and hydroxyl and butyryl groups, respectively. Because of very little hydroxyl content, CTA is excellent in terms of hydrolytic stability and biodegradation resistance. FO membranes fabricated using CTA and CTA/CA blends have shown much higher salt resistance than pure CA FO membranes but at a tradeoff of lowering water flux. This is due to the higher CTA chain entanglement which results in the dense selective skin with a dense and highly packed structure (Ong and Chung 2012). Higher CTA content in CTA/CA blend FO membranes also lead to a higher rejection towards NaCl ions (Nguyen et al. 2013). Therefore, CTA is an excellent FO membrane material when a high selectivity is required. However, it is difficult to process CTA because of limited solvents available to dissolve it. CAP and CAB offer much more control over the desired FO membrane performance as the content of functional groups can be varied as depicted in Figure 8.5.

By tuning the content of various functional groups in the cellulose ester, the FO performance of the resultant membranes can significantly change (Ong et al. 2012 & unpublished manuscript 2013; Stamatialis et al. 2000). FO membranes

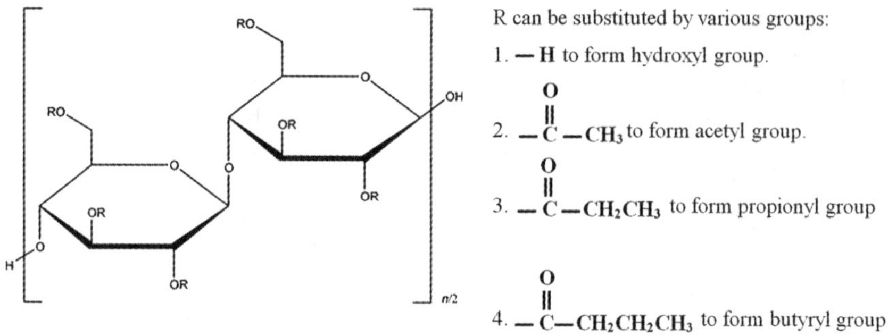

Figure 8.5 Repeating unit and various functional groups of cellulose esters

made from cellulose esters with very high hydroxyl content tend to have high water and solute permeability due to their relatively more hydrophilic. Since solutes alone cannot transport through the membranes without the aids from a solvent, the high water permeability across the membrane also facilitates a high solute transport. The bulkier and longer functional groups such as propionyl and butyryl groups are relatively more hydrophobic than hydroxyl and acetyl groups. Therefore, an increase in propionyl and butyryl content would result in cellulose ester membranes with higher hydrophobic and selectivity. On the other hand, high content of bulky propionyl and butyryl also causes an expansion in free volume and voids and reduces selectivity. To counter-balance these opposite effects, a moderate content of hydroxyl and propionyl or butyryl groups is preferred to achieve a reasonable FO water flux and solute rejection.

Though the CAP and CAB FO membranes exhibit better FO performance in terms of solute rejection, they still have the disadvantage of being relatively more hydrophobic. Therefore, dual layer hollow fiber membranes made of thin CAP or CAB selective skin layer on a thick and porous CA layer have been proposed (Su et al. 2013). Using CA as the material for the porous support layer offers a few advantages including high porosity and high hydrophicility which improve better water permeation whereas a selective skin layer made of CAP or CAB enhances the selectivity of the membrane.

8.3.2 Polybenzimidazole Membranes

Apart from cellulose esters, polybenzimidazole (PBI) is also a popular choice of materials for FO membranes. PBI offers some advantages over cellulose esters including excellent chemical resistance, high temperature stability and charged characteristics in aqueous environments. The chemical structure of PBI is shown in Figure 8.6.

PBI nanofiltration hollow fiber membranes were used for FO in the earlier research. However, this type of membranes did not have a good rejection towards monovalent ions such as NaCl even though the rejection towards divalent ions was high (Wang et al. 2007). Therefore, cross-linking of PBI selective skin using *p*-xylene dichloride was carried out to tailor the pore size and enhance the selectivity

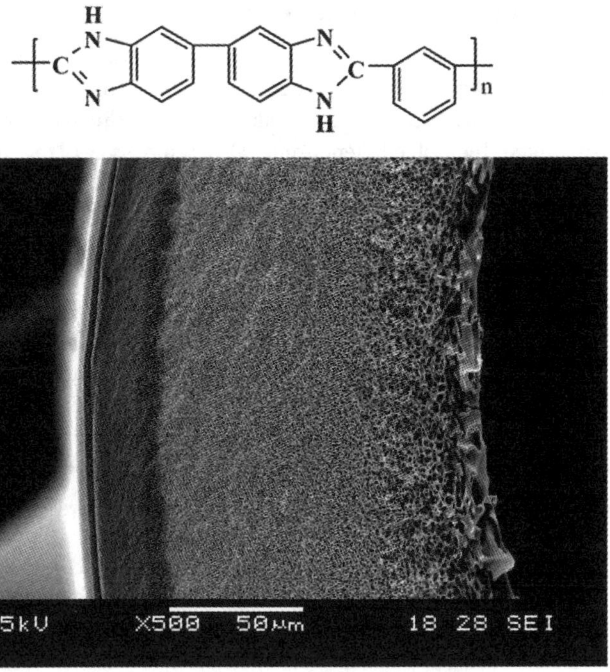

Figure 8.6. Chemical structure of polybenzimidazole (PBI) and cross-section morphology of the dual layer PBI hollow fiber membrane for FO

(Wang et al. 2009). Since pure PBI membranes are brittle and difficult to handle, dual layer hollow fiber membranes were then invented via coextrusion technique to solve this problem. A thin selective layer made of PBI was fabricated on a thick and porous polyethersulfone (PES) support in a one-step process (Yang et al. 2009). This not only eliminated the handling problem of PBI membranes, but also significantly reduced the material cost. Besides, the dual-layer structure can effectively mitigate ICP effects if one can design its sublayer with a fully porous structure. A cross sectional image of a dual layer hollow fiber membrane is shown in Figure 8.6.

Flat sheet FO membranes made from PBI were also evaluated. Surface modifications were conducted to increase hydrophilicity and surface charge of PBI membranes. The use of 4-(chloromethyl) benzoic acid (CMBA) as activator and *p*-phenylene diamine, ethylene diamine and taurine for the functionalization of the membrane surface resulted in increase in hydrophilicity and membrane surface charge (Flanagan et al. 2011; Hausman et al. 2010). Recently, poly (acrylamide-co-acrylic acid) (PAcA) was also used to functionalize PBI flat sheet membranes (Flanagan and Escobar 2013). From the zeta-potential tests, the functionalized PBI membranes had more charged surfaces than the virgin PBI membranes. The contact angles of the PBI membranes were also reduced significantly after functionalization indicating an increase in hydrophilicity which is favorable for FO applications.

8.3.3 Polyamide-imide Membranes

In recent years, Setiawan et al. (2011) employed polyamide-imide (PAI) as the base material to fabricate hollow fiber membranes for FO. The nascent PAI membrane has an ultrafiltration (UF) – like skin layer. Since PAI polymers can be easily crosslinked by polyethyleneimide (PEI) molecules, the membrane was modified to reduce its pore size of the selective layer. The degree of crosslinking was highly dependent on the molecular weight of PEI, PEI concentration, crosslinking time and temperature. The resultant membrane showed a poor rejection to NaCl, but reasonably good rejections to divalent ions. To avoid the excessive crosslinking in the supporting layer and to save the costly PAI material, PAI–PES dual layer hollow fiber membranes were developed (Setiawan et al. 2012a). PAI was the outer selective layer, while PES was the inner support layer. Similar to the case of PBI-PES dual hollow fiber membranes (Yang et al. 2009), the dual-layer PAI-PES membrane exhibited a better water flux than the single-layer one. However, this type of membranes is not efficient for NaCl rejection and hence can be mainly applied to the applications where no NaCl is involved.

Since the PEI-modified PAI hollow fiber membrane is positively charged which might induce severe fouling during applications. Therefore, a negatively charged polystyrene sulfonate sodium salt (PSS) was deposited on top of the PEI layer to alter its surface electrochemical properties (Setiawan et al. 2012a). Flat sheet membranes with a positively charged selective layer were also fabricated suing the same material on a woven fabric (Qiu et al. 2012).

8.4. COMPOSITE MEMBRANES

Composite membranes are usually prepared in two consecutive steps: the fabrication or selection of a proper supporting substrate and the coating of a thin selective layer onto the support by spin coating, on-site reaction, dip coating, etc. The interactions between the two layers might involve covalent bonding, electrostatic interaction or van der Waals force. The major advantage of composite membranes over integrally skinned membranes is the freedom of independent control of both layers to optimize the structure and maximize the performance for each specific application, i.e., the thickness, pore size, porosity and chemistry of each layer could be independently tailored. This is especially useful for FO membrane development since a fully porous support layer and a highly permeable and selective dense layer are required.

So far, three different types of composite membranes have been developed for FO, including thin film composite (TFC), layer-by-layer (LbL) and biomimetic membranes. The TFC membrane is the most important and common type of composite membranes for RO. Depending on the monomers used, the same interfacial polymerization technology could also be applied for NF. In other words, the selectivity of the thin dense selective layer can be designed for different applications. Therefore, this type of membrane holds a great potential for the desalination, wastewater treatment and many other separation purposes via FO. Another type of composite membranes for water separation is the LbL membrane. The selective layer is formed through the electrostatic interaction between different polyelectrolytes. The traditional LbL membranes are mainly designed for ultrafiltration (UF) and NF processes, and it is also possible to fabricate LbL membranes for some FO processes where large molecules are involved. Usually LbL membranes do not have sufficient rejection towards NaCl. The biomimetic membranes intend to mimic biological cell membranes to achieve high permeability and selectivity. Recent studies have shown their great potential for water reuse and desalination but their mechanical stability needs improvements for practical applications. Table 8.1 provides a list of the FO performance of the representative membranes developed in recent years.

8.4.1 Thin Film Composite Membranes

Thin film composite (TFC) membranes made of interfacial polymerization have gained considerable attention in the recent FO development and studies. Whilst the early concept of TFC membranes through interfacial polymerization (IP) was reported by Mogan in the 1960s (Morgan 1965), the major breakthrough on the fabrication of TFC membranes for various applications was discovered by Cadotte in the 1970s (Cadotte 1977; Cadotte et al. 1980; Cadotte 1981). Since then, TFC membranes have dominated the production of modern reverse osmosis (RO) and nanofiltration (NF) membranes (Petersen 1993; Verissimo 2005; Ghosh et al. 2008). Generally, TFC membranes consist of a top thin polyamide selective layer and a porous membrane support. The formation of a thin layer of polyamides can be attained via in-situ interfacial polymerization between two monomer solutions such as an aqueous polyfunctional amine solution, e.g. *m*-phenylenediamine (MPD) monomer, and a polyfunctional acyl chloride, e.g. trimesoyl chloride (TMC) monomer dissolved in an apolar organic solvent, e.g. hexane.

Because of solubility preference of monomers in two different phases, the aqueous and organic phrases are immiscible. The IP reaction takes place extremely fast (even less than one minute) at the organic phase near the interface, subsequently leading to form a dense polyamide thin film layer on top of the membrane support. Figure 8.7 shows a diagram of an interfacial polymerization between MPD and TMC monomers and the resultant TFC layer with a typical unique ridge and valley morphology.

In comparison with other integrally skinned membranes in terms of RO performance, the TFC membranes tend to exhibit a superior water permeation flux, greater solute rejection and better non-biodegradability (Roh and Khare 2002; Rao et al. 2003; Jin and Su 2009). Despite the fact that TFC membranes developed through IP are applicable for FO membranes, the fabrication schemes and optimizations for FO and RO membranes via IP are quite different. This is due to the fact that they have different essential requirements on their membrane support. RO membranes need a mechanically strong and robust supporting layer to withstand high transmembrane pressures (60–100 bars) during RO operation, whereas FO membranes need a thin and fully porous support layer to reduce ICP.

Table 8.1. A list of FO performance of representative phase inversion and composite membranes. DI water was used as the feed and tests were conducted at 20–25 °C.

Membrane	*Water flux $Lm^{-2} h^{-1}$ (FO/PRO)*	*Reverse salt Flux, $gm^{-2} h^{-1}$(FO/PRO)*	*Draw solution*	*Ref*
HTI flat sheet	13.0 (FO)	10.5	2.0 M NaCl	Phillip et al. 2010
CA double-skin flat sheet	22/32	16/25	2.0 M NaCl	Wang KY et al. 2010
CAP-CA dual layer hollow fiber	8.5/17.5	1/2	2.0 M NaCl	Su et al. 2013
PBI-PES dual layer hollow fiber	24.5/33.8	0.45/0.5	5.0 M $MgCl_2$	Yang et al. 2009
PAI-PES dual layer hollow fiber	15.5/27.5	5.5/83.7	0.5 M $MgCl_2$	Setiawan et al. 2012
TFC PES hollow fiber	29.5/68.0	2.6/5.8	2.0 M NaCl	Chou et al. 2010
TFC-PES hollow fiber	34.5/65.1	9.87/12.34	2.0 M NaCl	Sukitpaneenit et al. 2012
TFC-PESU E6020P flat sheet	62.8/78.8	14.9/13.5	2.0 M NaCl	Widjojo et al. 2013
LbL flat sheet	48/98	16/47	2.0 M $MgCl_2$	Qiu et al. 2011
LbL flat sheet	11.6/11.8	8.1/13.0	0.5 M NaCl	Duong et al. 2013
Biomimetic flat sheet	142 (PRO)	6.2	2.0 M NaCl	Wang H et al. 2012

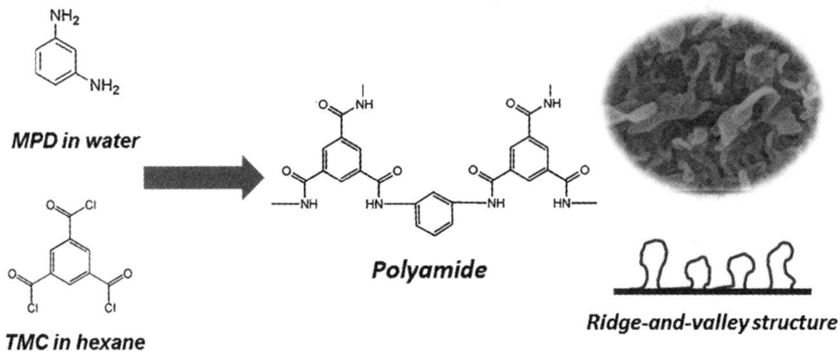

Figure 8.7. Schematic representation of chemical structure of polyamide layer via the interfacial polymerization between MPD and TMC monomers, and a typical ridge-and-valley surface structure of TFC membranes

As a result, typical TFC RO membranes consist of three layers, namely, (1) a thin polyamide layer as a selective layer, (2) a porous polysulfone layer as an intermediate layer for easy thin-film polymerization on top of it and better interactions with underneath non-woven fabrics, and (3) a thick layer of polyester (PET) nonwoven fabrics as the mechanical support layer. Except for the polyamide layer, other layers are usually thick, hydrophobic and have a relatively low porosity to tolerate high operating pressures. Therefore, the direct use of traditional TFC RO membranes in FO processes usually results in a relatively low water flux (McCutcheon and Elimelech 2006, 2008). Moreover, the thick nonwoven fabric may provide an additional resistance towards water transport in FO processes. As aforementioned, the desirable support layer for TFC FO membranes should be highly porous, hydrophilic and reasonably thin in order to minimize ICP and facilitate water transport across the membrane.

Since the early works on TFC FO membranes (Yip et al. 2010; Wang R et al. 2010), there have been many studies and new findings on the preparation, morphological design and performance of TFC membranes for FO applications. Generally, a non-solvent induced phase inversion process is used for the fabrication of porous supports, followed by interfacial polymerization for the formation of a thin polyamide active layer. The support layer structure with a thin sponge-like layer adhering on top of a highly porous macrovoids was preferred to form TFC FO membranes (Yip et al. 2010; Tiraferri et al. 2011).

Several studies have demonstrated the importance of hydrophilicity of porous supports on FO performance. Typically, TFC membranes that are interfacially polymerized on hydrophilic porous substrates not only show reduced ICP effects but also have a very high water flux. Wang KY et al. (2012) modified physico-chemical properties of TFC supports by blending a hydrophilic sulphonated polysulfone (SPSf) with a conventional polyethersulfone (PES), and then carried out the thin-film polymerization on top of the cast support. The resultant membranes exhibited a high flux up to 47.5 LMH in the PRO mode with a low salt leakage of 1.24 gMH using 2 M NaCl as a draw solution. A new approach to fabricate high performance TFC FO membranes using sulfonated materials in the preparation of membrane supports was demonstrated (Widjojo et al. 2011; Widjojo et al. 2013). Not only did the membranes show a fully sponge-like structure and likely anti-fouling characteristics but also attain superior water fluxes and low salt reverse fluxes. Based on their results, the best membranes could achieve a water flux up to 54 LMH with 8.8 gMH salt reverse flux against DI water, and 22 LMH against a 3.5 wt% NaCl model seawater using 2 M NaCl as the draw solution tested under the PRO operating mode, which is the highest value for seawater desalination ever reported. Moreover, the structural parameter was remarkably reduced with an increase in sulphonated material contents in the membrane support, thus minimizing the internal concentration polarization (ICP). A novel TFC FO membrane synthesized on top of a sulphonated poly (ether ketone) (SPEK) support also showed comparable FO performance (Han et al. 2012a).

The feasibility of using a new hydrophilic cellulose acetate propionate (CAP) as a porous support for the fabrication of TFC FO membranes was explored by Li X et al. (2012a). They found that CAP membrane supports with a hydrophilic macrovoid-free sponge-like structure could result in the formation of a good thin film layer for FO which is consistent with the previous studies mentioned above. Recently, other derivatives of cellulose-based polymers such as cellulose triacetate (Alsvik et al. 2013), acetylated cellulose ether (ACE) (Han et al. 2013) have been investigated as supporting materials for the TFC membrane fabrication.

Since the polydopamine (PDA) coating on the PSf support layer can alter its physicochemical and hydrophilic properties and remarkably enhance the water flux of TFC RO membrane by eight to fifteen folds in comparison with the uncoated membranes after removing their nonwoven fabric layer (Arena et al. 2011), a similar approach was conducted on the PSf support before interfacial polymerization for the fabrication of FO membranes. The resultant TFC FO membranes showed significant improvements in both water flux and salt rejection (Han et al. 2012).

Early works on FO membranes suggested that the desired support structure for FO membranes should have a thin sponge-like layer over a highly porous layer which is full of macrovoids. They also believed this type of structure could reduce transport resistance, ICP and structural parameter. However, recent works present a totally different view (Widjojo et al. 2011; Han et al. 2012a; Sukitpaneenit and Chung 2012; Widjojo et al. 2013). A fully sponge-like macrovoid-free and hydrophilic support may provide TFC FO membranes with better balance between FO performance and mechanical properties. For example, Sukitpaneenit and Chung (2012) proposed a new approach to fabricate high performance TFC FO hollow fiber membranes. Representative images of their membranes are given in Figure 8.8. The macrovoid-free and highly porous sponge-like PES hollow fiber substrate was fabricated using a dual-layer co-extrusion technology to design and effectively control the phase inversion during membrane formation, while a selective TFC layer was formed by interfacial polymerization. The developed

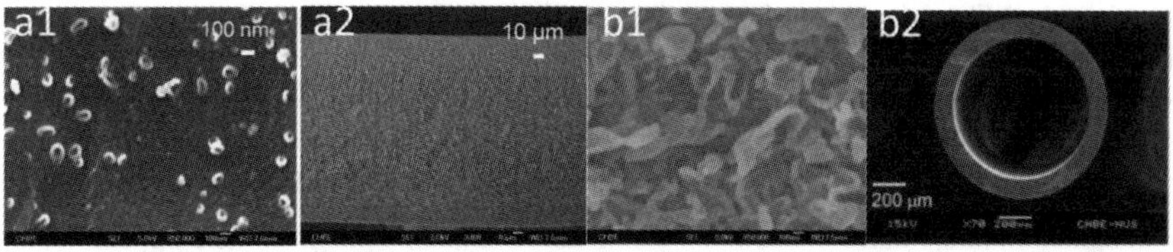

Figure 8.8. Representative SEM images of the TFC membranes for FO, where a and b are referred to the flat sheet and hollow fiber membranes, and 1 and 2 are referred to the active surface and cross section, respectively.

membranes displayed relatively high water fluxes of 32–34 LMH and up to 57–65 LMH against a pure water feed using 2 M NaCl as the draw solution tested under the FO and PRO modes, respectively, while consistently maintaining relatively low salt leakages below 13 gMH for all membranes. With model seawater solution (3.5 wt% NaCl) as the feed, the membranes could attain a water flux up to 15–18 LMH, which is reasonably comparable with the best value reported among investigated TFC FO membranes.

It is worth to note that so far most of TFC hollow fiber membranes for FO applications have been specifically designed with a thin film layer at the lumen side of the membranes. Even though much effort has been paid to develop TFC hollow fiber membranes with a selective thin film layer at the outer surface, there is no reports of these membranes available in the open literature so far. Clearly, there is still big room and more challenges to develop novel TFC hollow fiber membranes with desirable morphology and performance for FO applications.

In addition to the development of phase inversion integrally skinned and TFC membranes, an alternative concept of using nanofibers by means of an electrospinning technique has been applied to design a membrane substrate with high porosity and low tortuosity characteristics. As demonstrated through Bui and coworkers' works (Bui et al. 2011; Bui and McCutcheon 2013), TFC membranes with an electrospun nonwoven support exhibited relatively high water fluxes and reasonable low salt fluxes which are attributed to the superior porosity and pore interconnectivity of the eletrospun nanofiber supports. The best membrane produced water fluxes two to five times higher than the commercial HTI-CTA FO membrane. Song et al. (2011) prepared the same type of polymeric nanofiber supports using similar manners and found a very small gap between PRO and FO performance, indicating a remarkably reduced ICP. Such encouraging performance is mainly attributed to the unique scaffold-like porous structure and highly interconnected pores of the nanofiber supports.

8.4.2 Layer-by-layer Membranes

The LbL self-assembly of polymers onto a membrane is a simple, robust and flexible process based on the interactions between grafted layers via the functional groups of the polymers. The interactions between the nano-scale grafted layers can be a hydrophobic attraction (Lojou and Bianco 2004), a covalent bonding

Table 8.2. A list of common polyelectrolytes employed for LbL membrane fabrication.

Polycation	*Polyanion*
Poly(ethyleneimine) (PEI)	Poly(sodium styrene sulfonate)(PSS)
Poly(allylamine hydrochloride) (PAH)	Poly(acrylic acid) (PAA)
Chitosan (CS)	Hyaluronic acid (HA)
poly(methacryloyloxyethyl trimethylammonium chloride) (PDMC)	Carboxymethylcellulose (CMC)
Polydiallyldimethylammonium chloride (PDADMAC)	

(Johnson et al. 2012) or an electrostatic force (Decher 1997). Among them, the electrostatic force between highly charged polymer layers is the most commonly used method for LbL membranes due to its simplicity and non-toxicity.

In the typical LbL fabrication process, polymer multilayers are normally assembled onto a charged UF/NF membrane by the sequential adsorption of polyanions and polycations via dip-coating, spin-coating, or spraying methods with the aid of electrostatic interaction. Dip-coating is the traditional method but it is the most time-consuming technique among the three mentioned techniques. Therefore, spraying and spin-coating methods have been investigated recently to reduce the coating time from up to 30 min in dip-coating method to only few seconds in spraying/spin-coating methods. The commonly adopted polyanions and polycations are shown in Table 8.2. The permeability and solute rejection properties of LbL membranes can be controlled by varying the layer deposition time, the pH of the polyelectrolyte solutions or the ionic strength of the polyelectrolyte solution (Decher 1997). In addition, the number of the grafted polymer layers also affects the performance of LbL membranes due to the increase in the LbL thickness and the decrease in the defect as more layers are deposited.

Figure 8.9 displays the scheme of a LbL FO membrane prepared from the multiple deposition of polyelectrolyte. The first few generations of LbL FO membranes showed reasonably good rejections to $MgCl_2$ with high water fluxes up to 100 LMH using a 2 M $MgCl_2$ as the draw solution (Qiu et al. 2011; Saren et al. 2011; Li et al. 2012; Qi et al. 2012; Duong et al. 2013). However, a poor NaCl rejection was observed. Recently, Duong et al. (2013) employed additional crosslink of the polyelectrolyte layers and produced FO membranes with a relatively good rejection to NaCl.

8.4.3 Biomimetic Membranes

The biological membranes in the cells are characterized by their high permeability and exquisite selectivity towards water and different solutes, such as renal tubules, red cells and secretory glands. The high water permeability and selectivity of these

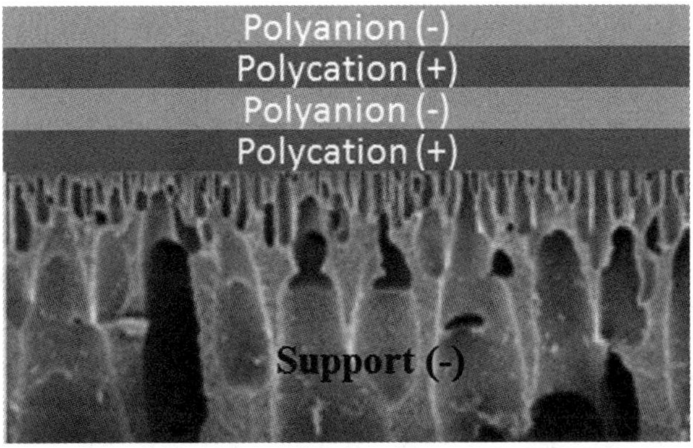

Figure 8.9. Scheme of the deposition of 2 bilayers of polyelectrolytes onto a porous support

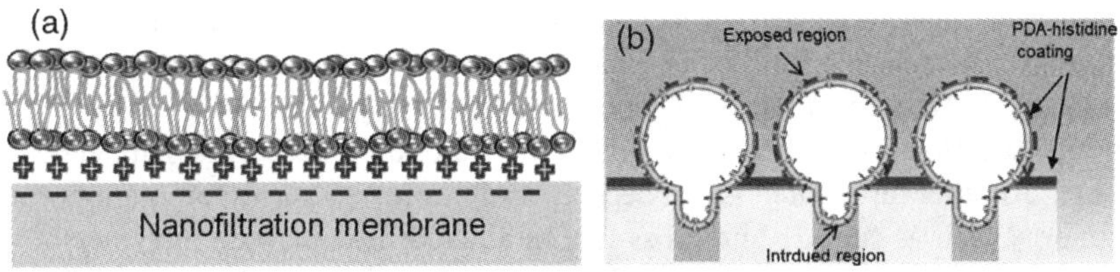

Figure 8.10. Schematic illustrations of (a) the biomimetic lipid membrane based on electrostatic charge interaction and (b) mechanically robust and highly permeable AquaporinZ-embeded biomimetic membranes for forward osmosis

membranes are resulted from the presence of the transmembrane proteins known as Aquaporins (AQPs).

Biomimetic membranes aim to achieve similar outstanding performance for practical use by mimicking the nature. Early efforts involved the design of black lipid membranes, vesicles and solid-supported lipid/block copolymer membranes. Figure 8.10 shows a design derived from solid-supported biomimetic membrane. In this design, a biomimetic lipid membrane has been coated on the surface of a commercial nanofiltration membrane and therefore the lipid bilayer becomes robust enough to withstand high pressures (Kaufman et al. 2010).

A big challenge for the fabrication of biomimetic membranes is the incorporation of aquaporin. The first essential step is to form Aquaporin-incorporated vesicles. Four different techniques have been used, including organic solvent-mediated incorporation, mechanical means, directly incorporation into preformed vesicles and detergent-mediated reconstitution (Rigaud et al. 1995). The most successful and frequently used method is the detergent-mediated method.

After Aquaporin-incorporated vesicles are prepared, the supported lipid/block copolymer layer can be formed by a spontaneous deposition of vesicles on a solid support. A general principle is that a sufficiently large deformation of vesicle will induce vesicle rupture and transform onto a flat bilayer disk. The bilayer disk may induce other adsorbed vesicle to rupture and to form bigger patches (Weirich et al. 2011). These rupture events are governed by the interaction between the membrane and the support as well as intra-membrane interactions. Specifically, these interactions are dependent on the nature of the vesicle (charge property, chemical composition and size), the surface characteristics of the solid support (surface chemical property and roughness) and the aqueous environment (ionic strength and pH). It is worth to mention that the block copolymers are more mechanically stable than lipids (Bermudez et al. 2002). Hence, it requires a stronger interfacial interaction for a polymer vesicle rupture to occur, such as electrostatic interaction and covalent bonding (Rakhmatullina et al. 2008; Dorn et al. 2011). The major concerns in the fabrication of biomimetic membrane are to mitigate the insufficient mechanical stability as well as eliminate defects between selective layer and substrate while avoiding the damage of the functionality of Aquaporins.

Several designs have been explored by researchers. For example, pore-suspending membranes have been established across multiple micron-sized apertures either as free-standing lipid or polymer membranes (Hansen et al. 2009a; Vogel et al. 2009; Hansen et al. 2009b; Pszon-Bartosz et al. 2011) or as hydrogel-encapsulated/ supported lipid or polymer membranes (Roerdink Lander et al. 2011). Another approach is to deposit and rupture vesicles on nanoporous supports, such as rupture and cross-link of aquaporin-containing polymersomes on methacrylate-functionalized cellulose acetate membranes (Zhong et al. 2010); charged lipid vesicle depositions on negatively charged nanofiltration membranes at pH 7 (Li et al. 2012b; Kaufman et al. 2010); rupture and cross-link of polymersome on the acrylate groups functionalized polycarbonate track-etched substrates (Wang et al. 2012). Most of these structures are not stable enough for industrial application. In recent years, a novel robust biomimetic membrane structure has been developed. The core technology of this design is to coat or to embed aquaporin-containing proteoliposomes or proteopolymersome in a cross-linked polyamide matrix (Zhao et al. 2012; Sun et al. 2013; Wang et al. 2013). An illustration is shown in Figure 8.10. Some of the membranes have been tested in FO for desalination and outstanding performance is observed as shown in Table 8.1.

8.5 ABBREVIATIONS/NOMENCLATURE

CP	concentration polarization
CTA	cellulose triacetate
DO	dissolved oxygen
DOC	dissolved organic carbon
FO	forward osmosis

References

Alsvik, I. L., Zodrow, K. R., Elimelech, M., and Hagg, M. B. (2013). "Polyamide formation on a cellulose triacetate support for osmotic membranes: Effect of linking molecules on membrane performance." *Desalination*, 312, 2–9.

Arena, J. T., McCloskey, B., Freeman, B. D., McCutcheon, J. R. (2011). "Surface modification of thin film composite membrane support layers with polydopamine: Enabling use of reverse osmosis membranes in pressure retarded osmosis." *J. Membr. Sci.*, 375, 55–62.

Bermudez, H., Brannan, A. K., Hammer, D. A., Bates, F. S., and Discher, D. E. (2002). "Molecular weight dependence of polymersome membrane structure, elasticity, and stability." *Macromolecules*, 35, 8203–8208.

Bui, N. N., Lind, M. L., Hoek, E. M. V., and McCutcheon, J. R. (2011). "Electrospun nanofiber supported thin film composite membranes for engineered osmosis." *J. Membr. Sci.*, 385-386, 10–19.

Bui, N. N., and McCutcheon, J. R. (2013). "Hydrophilic nanofibers as new supports for thin film composite membranes for engineered osmosis." *Environ. Sci. Technol.*, 47, 1761–1761.

Cadotte, J. E. (1977). "Reverse osmosis membrane." U.S. Patent 4, 039, 440, 2 August 1977.

Cadotte, J. E., Petersen, R. J., Larson, R. E., and Erickson, E. E. (1980). "A new thin-film composite seawater reverse osmosis membrane." *Desalination*, 32, 25–31.

Cadotte, J. E. (1981). "Interfacially synthesized reverse osmosis membrane." U.S. Patent 4, 277, 344, 7 July 1981.

Cath, T. Y., Childress, A. E., and Elimelech, M. (2006). "Forward osmosis: principles, applications, and recent developments." *J. Membr. Sci.*, 281, 70–87.

Chou, S., Shi, L., Wang, R., Tang, C. Y., Qiu, C., and Fane, A. G. (2010). "Characteristics and potential applications of a novel forward osmosis hollow fiber membrane." *Desalination*, 261, 365–372.

Chung, T. S., Li, X., Ong., R. C., Ge, Q. C., Wang, H. L., and Han, G. (2012). "Emerging forward osmosis (FO) technologies and challenges ahead for clean water and clean energy applications." *Current Opinion in Chemical Engineering*, 1, 246–257.

Chung, T. S., Zhang, S., Wang, K.Y., Su, J.C., and Ling, M.M. (2012). "Forward osmosis processes: yesterday, today and tomorrow." *Desalination*, 287, 78–81.

Decher, G. (1997). "Fuzzy nanoassemblies: Toward layered polymeric multicomposites." *Science*, 277, 1232–1237.

Dorn, J., Belegrinou, S., Kreiter, M., Sinner, E.K., and Meier, W. (2011). "Planar block copolymer membranes by vesicle spreading." *Macromol. Biosci.*, 11(4), 514–525.

Duong, P. H. H., Zuo, J., and Chung, T. -S. (2013). "Highly crosslinked layer-by-layer polyelectrolyte FO membranes: Understanding effects of salt concentration and deposition time on FO performance." *J. Membrane Sci.*, 427, 411–421.

Flanagan, M. F., and Escobar, I. C. (2013). "Novel charged and hydrophilized polybenzimidazole (PBI) membranes for forward osmosis." *J. Membr. Sci.*, 434, 85–92.

Flanagan, M., Hausman, R., Digman, B, Escobar, I. C., Coleman, M., and Chung, T. S. (2011). "Surface functionalization of polybenzimidazole membranes to increase hydrophilicity and charge." *Mod. Appl. Membr. Sci. Technol.*, 18, 303–321.

Ge, Q. C., and Chung, T.S. (2013). "Draw Solutions for Forward Osmosis Processes: Developments, Challenges, and Prospects for the Future." accepted by *J. of Membrane Science.*

Ghosh, A. K., Jeong, B. H., Huang, X. F., and Hoek, E. M. V. (2008). "Impacts of reaction and curing conditions on polyamide composite reverse osmosis membrane properties." *J. Membr. Sci.*, 311, 34–45.

Han, G., Chung, T.S., Toriida, M., and Tamai, S. (2012a). "Thin-film composite forward osmosis membranes with novel hydrophilicsupports for desalination." *J. Membrane Science*, 423–424, 543–555.

Han, G., Zhang, S., Li, X., Widjojo, N., and Chung, T. S. (2012b). "Thin film composite forward osmosis membranes based on polydopamine modified polysulfone substrates with enhancements in both water flux and salt rejection." *Chem. Eng. Sci.*, 80, 219–231.

Han, G., Zhang, S., Li, X., and Chung, T. S. (2013). "High performance thin film composite pressure retarded osmosis (PRO) membranes for renewable salinity-gradient energy generation." *J. Membr. Sci.*, 440, 108–121.

Han, J., Cho, Y. H., Kong, H., Han, S., and Park, H. B. (2013). "Preparation and characterization of novel acetylated cellulose ether (ACE) membranes for desalination applications." *J. Membr. Sci.*, 428, 533–545.

Hansen, J. S., Perry, M., Vogel, J., Groth, J., Vissing, T., Larsen, M., Geschke, O., Emneus, J., Bohr, H., and Nielsen, C.H. (2009a). "Large scale biomimetic membrane arrays." *Anal. Bioanal. Chem.*, 395(3), 719–727.

Hansen, J. S., Perry, M. E., Vogel, J., Vissing, T., Geschke, O., Emneus, J., and Nielsen, C. H. (2009b). "Development of an automation technique for the establishment of functional lipid bilayer arrays." *J. Micromech. Microeng.*, 19, 025014.

Hausman, R., Digman, B., Escobar, I. C., Coleman, M., and Chung, T. S. (2010). "Functionalization of polybenzimidazole membranes to impart negative charge and hydrophilicity." *J. Membr. Sci.*, 363, 195–203.

Herron, J. (2008). "Asymmetric forward osmosis membranes." *US Patent* 7,445,712.

Jin, Y., and Su, Z. (2009). "Effects of polymerization conditions on hydrophilic groups in aromatic polyamide thin films." *J. Membr. Sci.*, 330, 175–179.

Johnson, P. M., Yoon, J., Kelly, J. Y., Howarter, J. A., and Stafford, C. M. (2012). "Molecular layer-by-layer deposition of highly crosslinked polyamide films." *J. Polym. Sci. Pol. Phys.*, 50, 166–173.

Kaufman, Y., Berman, A., and Freger, V. (2010). "Supported lipid bilayer membranes for water purification by reverse osmosis." *Langmuir*, 26(10), 7388–7395.

Lee, K. L., Baker, R. W., and Lonsdale, H. K. (1981) "Membrane for power generation by pressure retarded osmosis." *J. Membr. Sci.*, 8, 141–171.

Li, X., Wang, K. Y., Helmer, B., and Chung, T. S. (2012a). "Thin-film composite membranes and formation mechanism of thin-film layers on hydrophilic cellulose acetate propionate substrates for forward osmosis processes." *Ind. Eng. Chem. Res.*, 51, 10039–10050.

Li, X., Wang, R., Tang, C. Y., Vararattanavech, A., Zhao, Y., Torres, J., and Fane, A. G. (2012b). "Preparation of supported lipid membranes for aquaporin Z incorporation." *Colloids Surf. B Biointerfaces*, 94, 333–340.

Li, Y., Wang, X., and Sun, J. (2012). "Layer-by-layer assembly for rapid fabrication of thick polymeric films." *Chem. Soc. Rev.*, 41, 5998–6009.

Loeb, S., and Sourirajan, S. (1964). " High flow porous membranes for separating water from saline solutions." *US Patent* 3,133,132.

Lojou, E., and Bianco, P. (2004). "Buildup of polyelectrolyte-protein multilayer assemblies on gold electrodes. Role of the hydrophobic effect." *Langmuir*, 20, 748–755.

McCutcheon, J. R., and Elimelech, M. (2006). "Influence of concentrative and dilutive internal concentration polarization on flux behavior in forward osmosis." *J. Membr. Sci.*, 284, 237–247.

McCutcheon, J.R., and Elimelech, M. (2007). "Modeling water flux in forward osmosis: implications for improved membrane design." *AIChE Journal*, 53, 1736–1744.

McCutcheon, J. R., and Elimelech, M. (2008). "Influence of membrane support layer hydrophobicity on water flux in osmotically driven membrane processes." *J. Membr. Sci.*, 318, 458.

Mehta, G. D., and Loeb, S. (1979). "Performance of permasep B-9 and B-10 membranes in various osmotic regions and at high osmotic pressures." *J. Membr. Sci.*, 4, 335–349.

Morgan, P. W. (1965) *Condensation polymers: By interfacial and solution methods. In Polymer Reviews*, Vol. 10. Wiley, New York.

Nguyen, T. P. N., Yun, E. T., Kim, I. C., and Kwon, Y. N. (2013). "Preparation of cellulose triacetate/cellulose acetate (CTA/CA)-based membranes for forward osmosis." *J. Membr. Sci.*, 433, 49–59.

Ong, R. C., and Chung, T. S. (2012). "Fabrication and positron annihilation spectroscopy (PAS) characterization of cellulose triacetate membranes for forward osmosis." *J. Membr. Sci.*, 394-395, 230–240.

Ong, R. C., Chung, T. S.Helmer, B. J., and de Wit, J. S. (2012). "Novel cellulose esters for forward osmosis membranes." *Ind. Eng. Chem. Res.*, 51, 16135–16145.

Petersen, R. J. (1993). "Composite reverse osmosis and nanofiltration membranes." *J. Membr. Sci.*, 83, 81–150.

Phillip, W. A., Yong, J. S., and Elimelech, M. (2010). "Reverse draw solute permeation in forward osmosis: Modeling and experiments." *Environ. Sci. Technol.*, 44, 5170–5176.

Phuntsho, S., Shon, H. K., Hong, S., Lee, S., and Vigneswaran, S. (2011). "A novel low energy fertilizer driven forward osmosis desalination for direct fertigation: evaluating the performance of fertilizer draw solutions." *J. Membr. Sci.*, 375, 172–181.

Phuntsho, S., et al. (2012). "Blended fertilizers as draw solutions for fertilizer-drawn forward osmosis desalination." *Environmental science & technology* 46(8): 4567–4575.

Pszon-Bartosz, K., Hansen, J. S., Stibius, K. B., Groth, J. S., Emneus, J., Geschke, O., and Helix-Nielsen, C. (2011). "Assessing the efficacy of vesicle fusion with planar membrane arrays using a mitochondrial porin as reporter." *Biochem. Biophys. Res. Commun.*, 406(1), 96–100.

Qi, S., Qiu, C. Q., Zhao, Y., and Tang, C. Y. (2012). "Double-skinned forward osmosis membranes based on layer-by-layer assembly–FO performance and fouling behavior." *J. Membr. Sci.*, 405–406, 20–29.

Qiu, C., Qi, S., and Tang, C. Y. (2011). "Synthesis of high flux forward osmosis membranes by chemically crosslinked layer-by-layer polyelectrolytes." *J. Membr. Sci.*, 381, 74–80.

Qiu, C., Setiawan, L., Wang, R., Tang, C. Y., and Fane, A. G. (2012). "High performance flat sheet forward osmosis membrane with an NF-like selective layer on a woven fabric embedded substrate." *Desalination*, 287, 266–270.

Rakhmatullina, E., and Meier, W. (2008). "Solid-supported block copolymer membranes through interfacial adsorption of charged block copolymer vesicles." *Langmuir*, 24, 6254–6261.

Rao, A. P., Joshi, S. V., Trivedi, J. J., Devmurari, C. V., and Shah, V. J. (2003). "Structure performance correlation of polyamide thin film composite membranes: Effect of coating conditions on film formation." *J. Membr. Sci.*, 211, 13–24.

Rigaud, J.L., Pitard, B., and Levy, D. (1995). "Reconstution of membrane-proteins into liposomes-application to energy transducing membrane-proteins." *Biochim. Biophys Acta-Bioenergetics*, 1231, 223–246.

Roerdink Lander, M., Ibragimova, S., Rein, C., Vogel, J., Stibius, K., Geschke, O., Perry, M., and Nielsen, C. H. (2011). "Biomimetic membrane arrays on cast hydrogel supports." *Langmuir*, 27, 7002–7007.

Roerdink Lander, M., et al. (2011). "Biomimetic Membrane Arrays on Cast Hydrogel Supports." *Langmuir* 27(11): 7002–7007.

Roh, I. J., and Khare, V. P. (2002). "Investigation of the specific role of chemical structure on the material and permeation properties of ultrathin aromatic polyamides." *J. Mater. Chem.*, 12, 2334–2338.

Saren, Q., Qiu, C. Q., and Tang, C. Y. (2011). "Synthesis and characterization of novel forward osmosis membranes based on layer-by-layer assembly." *Environ. Sci. Technol.*, 45, 5201–5208.

Setiawan, L., Wang, R., Li, K., and Fane, A. G. (2011). "Fabrication of novel poly(amide-imide) forward osmosis hollow fiber membranes with a positively charged nanofiltration-like selective layer." *J. Membr. Sci.*, 369, 196–205.

Setiawan, L., Wang, R., Li, K., and Fane, A. G. (2012a). "Fabrication and characterization of forward osmosis hollow fiber membranes with antifouling NF-like selective layer." *J. Membr. Sci.*, 394–395, 80–8.

Setiawan, L., Wang, R., Shi, L., Li, K., and Fane, A. G. (2012b). "Novel dual-layer hollow fiber membranes applied for forward osmosis process." *J. Membr. Sci.*, 421–422, 238–246.

Song, X., Liu, Z., and Sun, D. D. (2011). "Nono gives the answer: breaking the bottleneck of internal concentration polarization with a nanofiber composite forward osmosis membrane for a high water production rate." *Adv. Mat.*, 23, 3256–3260.

Stamatialis, D. F., Dias, C. R., and de Pinho, M. N. (2000). "Structure and permeation properties of cellulose esters asymmetric membranes." *Biomacromolecules*, 1, 564–570.

Su, J. C., Yang, Q., Teo, J. F., and Chung, T. S. (2010a). "Cellulose acetate nanofiltration hollow fiber membranes for forward osmosis processes." *J. Membr. Sci.*, 355, 36–44.

Su, J. C., Zhang, S., Chen, H., Chen, H., Jean, Y. C., and Chung, T. S. (2010b). "Effects of annealing microstructure and performance of cellulose acetate membranes for pressure-retarded osmosis processes." *J. Membr. Sci.*, 364, 344–353.

Su, J. C., Chung, T. S., Helmer, B. J., and de Wit, J. S. (2012). "Enhanced double-skinned FO membranes with inner dense layer for wastewater treatment and macromolecule recycle using Sucrose as draw solute." *J. Membr. Sci.*, 396, 92–100.

Su, J. C., Ong, R. C., Wang, P., and Chung, T. S. (2013). "Advanced FO membranes from newly synthesized CAP polymer for wastewater reclamation through an integrated FO-MD hybrid system." *AIChE J.*, 59, 1245–1254.

Sukitpaneenit, P., and Chung, T. S. (2012). "High performance thin-film composite forward osmosis hollow fiber membranes with macrovoid-free and highly porous structure for sustainable water production." *Environ. Sci. Technol.* 46, 7358–7365.

Sun, G. F., Chung, T. S., Jeyaseelan, K., and Armugam, A. (2013). "A layer-by-layer self-assembly approach to developing an aquaporin-embedded mixed matrix membrane." *RSC Adv.*, 3, 473–481.

Tang, C. Y., She, Q., Lay, W. C. K, Wang, R., and Fane, A. G. (2010). "Coupled effects of internal concentration polarization and fouling on flux behavior of forward osmosis membranes during humic acid filtration." *J. Membr. Sci.*, 354, 123–133.

Tang, C. Y., She, Q., Lay, W. C. K., Wang, R., Field, R., and Fane, A. G. (2011). "Modeling double-skinned FO membranes." *Desalination*, 283, 178–186.

Thorsen, T., and Holt, T. (2009). "The potential for power production from salinity gradients by pressure retarded osmosis." *J. Membr. Sci.*, 335, 103–110.

Tiraferri, A., Yip, N. Y., Phillip, W.A., Schiffman, J. D., and Elimelech, M. (2011). "Relating performance of thin-film composite forward osmosis membranes to support layer formation and structure." *J. Membr. Sci.*, 386, 340–352.

Verissimo, S., Peinemann, K. V., and Bordado, J. (2005). "Thin-film composite hollow fiber membranes: an optimized manufacturing method." *J. Membr. Sci.*, 264, 48–55.

Vogel, J., Perry, M.E., Hansen, J. S., Bollinger, P. -Y., Nielsen, C. H., Geschke, O. (2009). "Support structure for biomimetic applications." *J. Micromech. Microeng.*, 19, 025026.

Wang, H., Chung, T. S., Tong, Y. W., Jeyaseelan, K., Armugam, A., Chen, Z., Hong, M., and Meier, W. (2012). "Highly permeable and selective pore-spanning biomimetic membrane embedded with Aquaporin Z." *Small*, 8(8), 1185–1190.

Wang, H., Chung, T. S., Tong, Y. W., Jeyaseelan, K., Armugam, A., Duong, H. H. P., Fu, F., Seah, H., Yang, J., and Hong, M. (2013). "Mechanically robust and highly permeable AquaporinZ biomimetic membranes." *J. Membr. Sci.*, 10.1016/ j.memsci.2013.01.031.

Wang, K. Y., Chung, T. S., and Qin, J. J. (2007). "Polybenzimidazole (PBI) nanofiltration hollow fiber membranes applied in forward osmosis process." *J. Membr. Sci.*, 300, 6–12.

Wang, K. Y., Yang, Q., Chung, T. S., and Rajagolapan, R. (2009). "Enhanced forward osmosis from chemically modified polybenzimidazole (PBI) nanofiltration hollow fiber membranes with thin wall." *Chem. Eng. Sci.*, 64, 1577–1584.

Wang, K. Y., Ong, R. C., and Chung, T. S. (2010). "Double-skinned forward osmosis membranes for reducing internal concentration polarization within the porous sublayer." *Ind. Eng. Chem. Res.*, 49, 4824–4831.

Wang, K. Y., Chung, T. S., and Amy, G. (2012). "Developing thin-film-composite forward osmosis membranes on the PES/SPSf substrate through interfacial polymerization." *AIChE J.*, 58, 770–781.

Wang, R., Shi, L., Tang, C. Y., Chou, S., Qiu, C., and Fane, A. G. (2010). "Characterization of novel forward osmosis hollow fiber membranes." *J. Membr. Sci.*, 355, 158–167.

Weirich, K. L., Israelachvili, J. N., and Fygenson, D. K. (2010). "Bilayer edges catalyze supported lipid bilayer formation." *Biophys. J.*, 98(1), 85–92.

Widjojo, N., Chung, T. S., Weber, M., Maletzko, C., and Warzelhan, V. (2011). "The role of sulphonated polymer and macrovoid-free structure in the support layer for thin-film composite (TFC) forward osmosis (FO) membranes." *J. Membr. Sci.*, 383, 214–223.

Widjojo, N., Chung, T. S., Weber, M., Maletzko, C., and Warzelhan, V. (2013). "A sulfonated polyphenylenesulfone (sPPSU) as the supporting substrate in thin film composite (TFC) membranes with enhanced performance for forward osmosis (FO)." *Chem. Eng. J.*, 220, 15–23.

Xu, Y., Peng, X., Tang, C. Y., Fu, Q. S., and Nie, S. (2010). "Effect of draw solution concentration and operating conditions on forward osmosis and pressure retarded osmosis performance in a spiral wound module." *J. Membr. Sci.*, 348, 298–309.

Yang, Q., Wang, K. Y., and Chung, T. S. (2009). "Dual-layer hollow fibers with enhanced flux as novel forward osmosis membranes for water production." *Environ. Sci. Technol.*, 43, 2800–2805.

Yip, N. Y., Tiraferri, A., Phillip, W.A., Schiffman, J. D., and Elimelech, M. (2010). "High performance thin-film composite forward osmosis membrane." *Environ. Sci. Technol.*, 44, 3812–3818.

Zhang, S., Wang, K. Y., Chung, T. S., Chen, H., Jean, Y. C., and Amy, G. (2010). "Well-constructed cellulose acetate membranes for forward osmosis: Minimized internal concentration polarization with an ultra-thin selective layer." *J. Membr. Sci.*, 360, 522–535.

Zhang, S., Wang, K.Y., Chung, T.S., Jean, Y.C., and Chen, H.M. (2011). "Molecular design of the cellulose ester-based forward osmosis membranes for desalination." *Chem. Eng. Sci.*, 66, 2008–2018.

Zhang, S., Zhang, R. W., Jean, Y. C., Paul, D.R., and Chung, T. S. (2012). "Cellulose esters for forward osmosis: characterization of water and salt transport properties and free volume." *Polymer*, 53, 2664–2672.

Zhang, S., Fu, F. J., and Chung, T. S. (2013). "Substrate modifications and alcohol treatment on thin film composite membranes for osmotic power." *Chem. Eng. Sci.*, 87, 40–50.

Zhao, S., Zou, L., Tang, C. Y., and Mulcahy, D. (2012). "Recent developments in forward osmosis: opportunities and challenges." *J. Membr. Sci.*, 396, 1–21.

Zhao, Y., Qiu, C., Li, X., Vararattanavech, A., Shen, W., Torres, J., Hélix-Nielsen, C., Wang, R., Hu, X., Fane, A. G., and Tang, C. Y. (2012). "Synthesis of robust and high-performance aquaporin-based biomimeticmembranes by interfacial polymerization–membrane preparation and RO performance characterization." *J. Membr. Sci.*, 423-424, 422–428.

Zhong, P. S., Chung, T. -S., Jeyaseelan, K., and Armugam, A. (2012). "Aquaporin-embedded biomimetic membranes for nanofiltration." *J. Membr. Sci.*, 407-408, 27–33.

CHAPTER 9

Design of a Novel Nanofibre FO Membrane

Z. Y. Liu
X. X. Song
D. Chen
D. D. Sun

This chapter introduces a novel forward osmosis (FO) membrane, nanofiber FO membrane (NFM), which is made of an electrospun nanofiber support layer and a thin active layer. The electrospun nanofiber support layer has the special characters of low tortuosity, high porosity and thin thickness, which are critical for breaking the bottleneck of internal concentration polarization (ICP) that restrains conventional FO membranes to achieve high water flux. These NFM were systematically characterized in terms of structures and water permeability. This study paves the new way for the design and development of next-generation FO membranes with high water productivity.

9.1 INTRODUCTION

FO has obtained great interest as a low-energy process for various applications, including wastewater reuse (Holloway et al. 2007; Achilli et al. 2009), seawater desalination (McCutcheon et al. 2005; McCutcheon et al. 2006), and power generation (Lee et al. 1981; Seppälä 1999). In the FO process, driven by osmotic pressure difference, water diffuses naturally through a semipermeable FO membrane from low salinity solution side to high salinity solution side, while retaining the solutes on both sides of the FO membrane (Cath et al. 2006).

FO membrane is the key component of FO unit for high water production rate. Conventional FO membranes consist of a dense active layer and an asymmetric porous support layer that are made of phase inversion process. One challenge inevitably faced by this asymmetric porous support layer is the internal concentration polarization (ICP) phenomenon (McCutcheon 2006). In osmotically driven membranes, concentration polarization is caused by the concentration difference between the feed solution and the draw solution

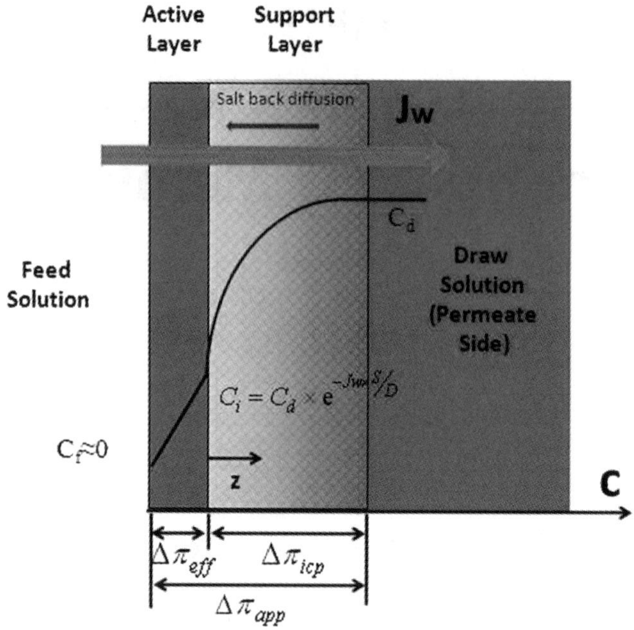

Figure 9.1. Draw solution gets diluted in the support layer of FO membranes, subsequently decreasing the effective driving force, which is determined by water flux J_w, membrane support structure parameter S and salt diffusion constant D

through an asymmetric FO membrane. ICP occurs within the porous support layers of FO membranes, which cannot be mitigated by altering external hydrodynamic conditions, such as, increasing the flow rate or turbulence (Zhao 2011). As shown in Figure 9.1, the ICP phenomenon is responsible for the water flux decline in FO process (Mehta 1978). Therefore, eliminating ICP problem in the support layer is critical to get a high-flux FO membrane (Sablani et al. 2001).

As shown in Figure 9.2, an ideal support layer for FO membrane should have the structure of low tortuosity, high porosity, and thin thickness (Loeb et al. 1997; McCutcheon 2008; Su 2011). However, conventional FO membranes made of phase inversion process (PI-FO) have tortuous sponge-like pore structure, which is formed with small segregated pores (Shi et al. 2011; Widjojo et al. 2011) and is not favorable for water and salt diffusion.

To tackle the ICP bottleneck in the support layers of FO membranes, Prof. Darren Sun's group designed and fabricated a novel nanocomposite FO (NC-FO) membrane with a nanofiber support layer (Song et al. 2011). This novel scaffold-like nanofiber support layer possesses unparalleled advantages, in terms of: i) low tortuosity, ii) high porosity, and iii) thin thickness, which are favorable for easy diffusion of salt and water.

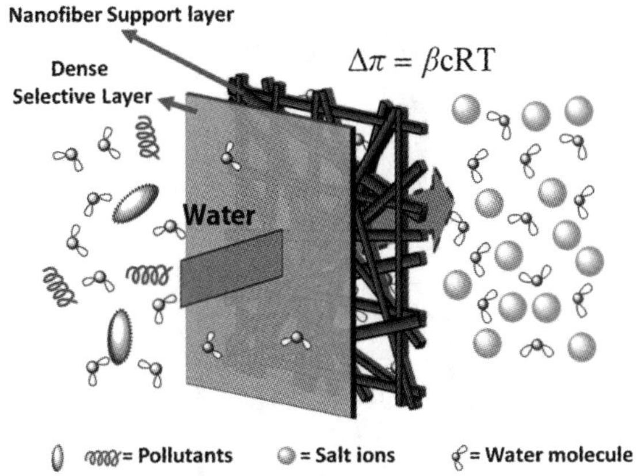

Figure 9.2. Schematic of the ideal support layer of FO membranes. The interconnected pores in between the scaffold-like nanofibers provide direct paths for water and salt diffusion

SOURCE: Adapted from Song et al. 2011; reproduced with permission from Wiley

9.2 FABRICATION OF FO MEMBRANES

9.2.1 Fabrication of Nanofiber Support Layers

As shown in Figure 9.3, a customer-designed electrospinning setup was used to fabricate the nanofiber support layer on a nonwoven fabric. The thickness of the nanofiber support layer can be adjusted by controlling the electrospinning time. The nanofibers were electrospun from 10% w/w polyethersulfone (PES) (molecular weight, Mw = 80,000, Solvay) solution in N-methyl pyrrolidone (NMP)/N,N-dimethyl formamide (DMF) mixed solvent (1:9 vol:vol). The stainless steel rotating drum was wrapped with a poly(ethylene terephthalate) (PET) nonwoven substrate (Ahlstrom Grade 3249), which was prewetted with NMP. A DC voltage (25 kV) was then applied between a spinneret (0.7 mm diameter) and the rotating drum (300 rpm), which was 15 cm below the spinneret orifice. The PES solution was fed to the spinneret at $10\,\mu L\ min^{-1}$. The spun nanofiber layer was collected on the surface of the rotating drum (300 rpm). This nanofiber support layer was then pressed at 220°C under 20 psi pressure (Li et al. 2003; Liu et al. 2007).

9.2.2 Fabrication of PI-FO Membrane Support Layer

PES polymer (12–17 wt%) was dissolved in a mixed solvent (DMF:NMP = 1 : 3) (Yip et al. 2010). A casting knife (Elcometer) at a gate height of 150 μm was used to cast the polymer solution on a PET nonwoven fabric, which was prewetted with NMP. The cast solution was immediately immersed in a water bath containing 3% NMP for 15 min. Then the PES support was transferred to DI water at 4°C for storage.

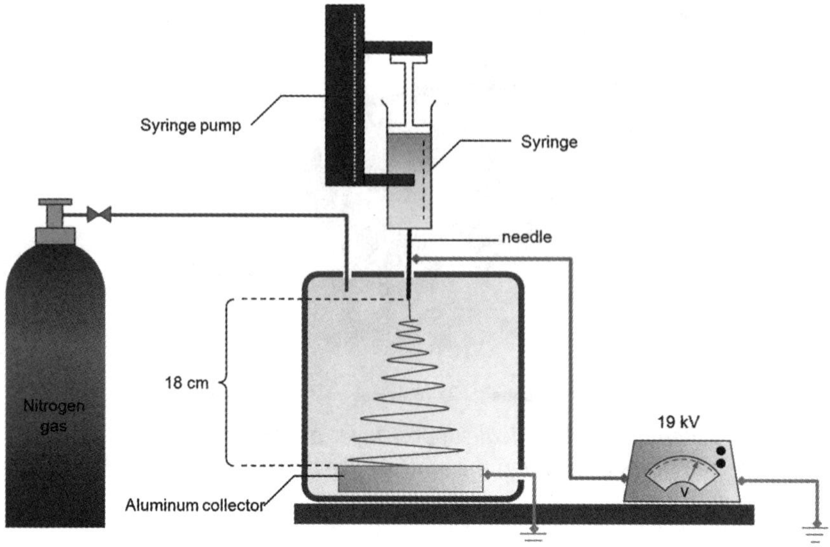

Figure 9.3. Electrospinning setup for the fabrication of nanofiber support layers

9.2.3 Interfacial Polymerization (IP) of Active Layers

The above support layer was dipped in 3% m-phenylenediamine (MPD) aqueous solution (containing 2% triethylamine) for 2 min. After withdrawn from the MPD solution, an air knife was used to remove excess MPD solution from the support layer until a dull color appeared. Then, 0.15% trimesoyl chloride (TMC) in isopar-G (Univar, China) solution was introduced to the surface of the support layer while all sides of the support layer were sealed with tape. After 1 min, the nascent composite membrane was cured at 90°C for 2 min and then rinsed with NaOCl (200 ppm) and $NaHSO_3$ (1000 ppm) aqueous solutions for 120 s and 30 s, respectively. Finally, the composite membrane was cured at 90°C for 5 min and then stored in DI water at 4°C (Ghosh et al. 2008).

9.3 CHARACTERIZATIONS OF FO MEMBRANES

9.3.1 Characterizations of the Support Layers

The pore structures of FO membranes were observed by a Zeiss Evo 50 Scanning Electron Microscopy (SEM) and a Joel JSM 7600F thermal FESEM. Cross-sections were acquired by fracturing the membrane samples immediately after flash-freezing in liquid nitrogen.

As shown in Figure 9.4(a), the cross-sectional SEM of the support layer made of phase inversion shows typical sponge-like structures. The sponge-like structure is formed with small segregated pores (Wang et al. 2010; Qiu et al. 2011),

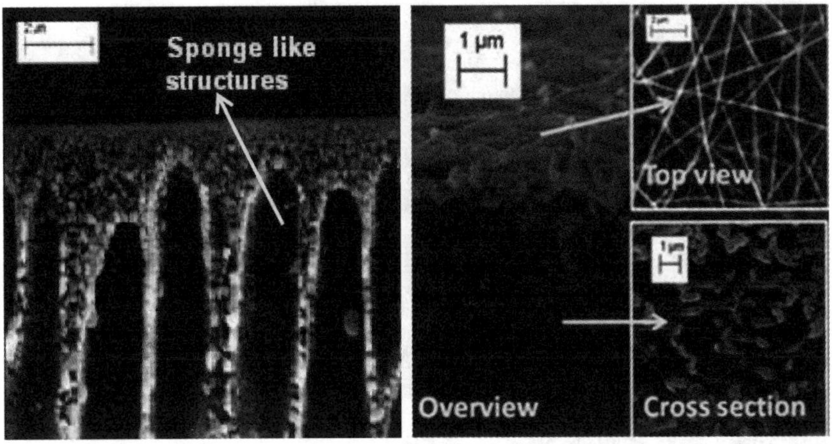

(a) (b)

Figure 9.4. (a) The cross-sectional SEM of conventional support layer made of phase inversion process, showing the tortuous sponge-like structure. (b) The overview SEM of nanofiber support layer made of electrospinning process. The top-view and cross-sectional SEM show that the pores are interconnected in between individual nanofibers

SOURCE: Adapted from Song et al. 2011; reproduced with permission from Wiley

which are the barriers for the free diffusion of water and salt. In contrast, as shown in Figure 9.4(b), the cross-sectional SEM of the nanofiber support layer shows scaffold-like porous structure, and the pores are interconnected between individual nanofibers. The diameters of nanofibers are in the range of 50–150 nm.

To easily make a comparison among different support layers, the nanofiber composite FO membrane (NC-FO), home-made phase inversion FO (PI-FO) membrane and commercial Hydration Technology Innovations FO (HTI-FO) membrane are shown in Figure 9.5. It is clear that the cross-sectional SEM of HTI-FO membrane shows a dense structure with nonwoven fabric imbedded in the polymer matrix.

9.3.2 Characterizations of the Active Layers

The active layers that are responsible for salt rejection were interfacially polymerized (IP) on top of the support layers. As shown in Figure 9.6, the typical ridge-and-valley morphology of polyamide active layers are formed on top of the support layers of both PI-FO and NC-FO membranes. Compared with the conventional phase-inversion support layer, the nanofiber support layer provides a different interface for the polymerization of polyamide. As shown in Figure 9.7, the high-resolution SEM shows the polyamide layer on NC-FO membrane is thicker than that on PI-FO membrane. When forming polyamide layers on highly porous NC-FO support layers, MPD monomers immediately come to contact with

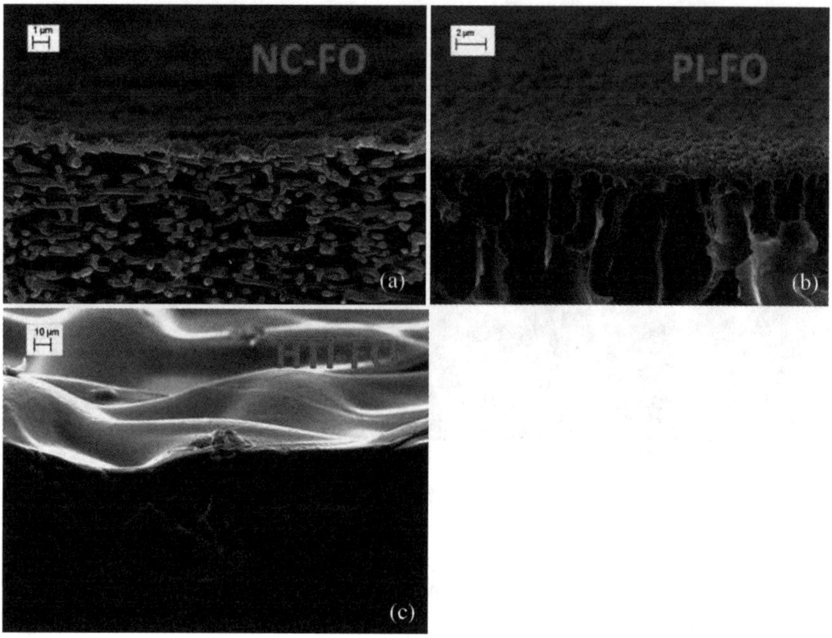

Figure 9.5. (a) The cross-section SEM of nanofiber support layer shows scaffold-like porous structure with inter-connected pores between individual nanofibers. (b) The cross-section SEM of the support layer of PI-FO membrane shows typical sponge-like structures. (c) The cross-section SEM of HTI-FO membrane shows a dense structure with non-woven fabric imbedded in polymer matrix. Abbreviations: NC-FO–nanofiber composite FO membrane; PI-FO–phase inversion FO membrane; HTI-FO–a commercial FO membrane

SOURCE: Adapted from Song et al. 2011; reproduced with permission from Wiley

TMC monomers in the initial stage, therefore the rate of reaction is accelerated and more polyamide is formed on top of NC-FO membrane. Hence, the polyamide layers grow thicker on nanofiber support layers, compared with conventional phase inversion support layers (Freger 2005).

The surface topology of NC-FO membrane was further investigated by a Park XE-100 Atomic Force Microscopy (AFM). As shown in Figure 9.8, the AFM image further confirms the ridge-and-valley morphology of the polyamide active layer.

9.4 PERFORMANCE TEST OF FO MEMBRANES

Performance parameters, including the intrinsic water permeability (A), the salt rejection rate (R), and the salt permeability (B) of FO membranes were determined using a cross-flow membrane cell and calculated based on Equations (9.1)–(9.3).

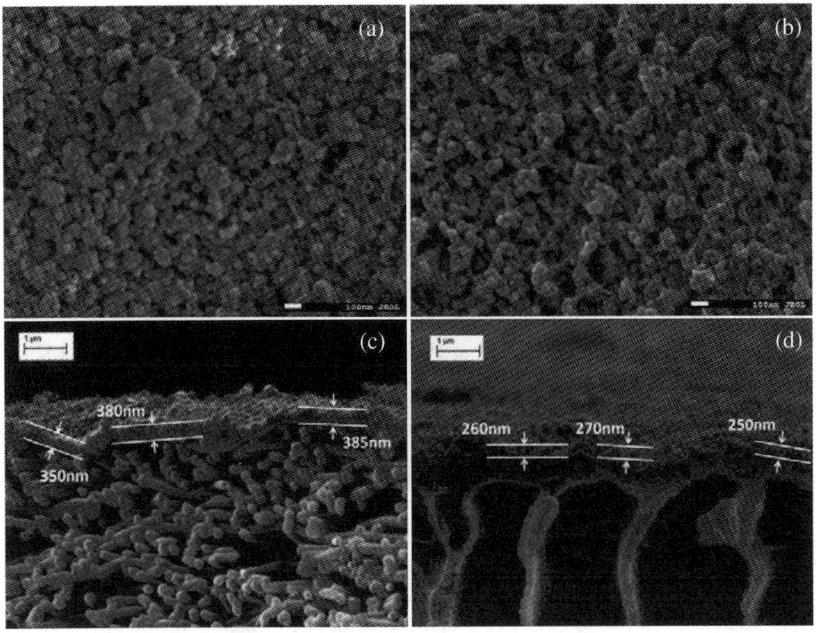

Figure 9.6. (a, b) The SEMs of NC-FO and PI-FO membrane surfaces show typical ridge-and-valley morphology made of interfacially polymerized polyamide. (c) The cross-section SEM of NC-FO membrane shows that the polyamide active layer is about 370 nm. (d) The cross-section SEM of PI-FO membrane shows that the polyamide active layer is about 260 nm

SOURCE: Adapted from Song et al. 2011; reproduced with permission from Wiley

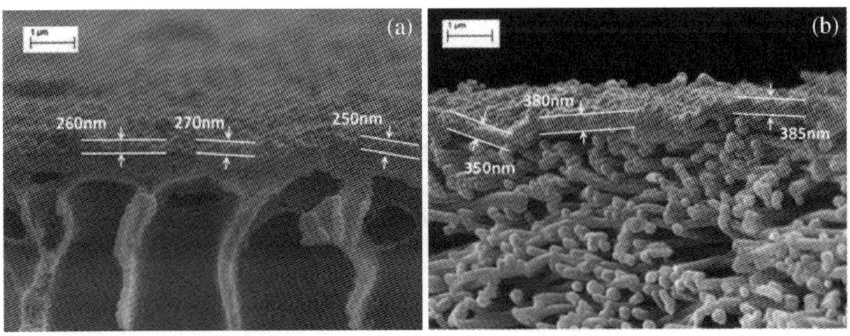

Figure 9.7. The cross-section SEM of (a) PI-FO membrane, showing the thickness of polyamide active layer is 260 nm and (b) NC-FO membrane, showing the thickness of polyamide active layer is 370 nm

SOURCE: Adapted from Song et al. 2011; reproduced with permission from Wiley

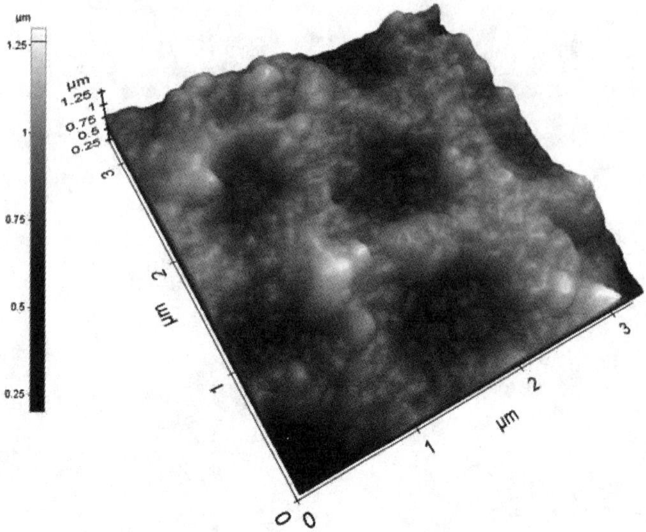

Figure 9.8. AFM image of NC-FO membrane surface, which further confirms the "ridge and valley" morphology of the polyamide active layer
SOURCE: Adapted from Song et al. 2011; reproduced with permission from Wiley

$$A = \frac{J_w}{\Delta P} \tag{9.1}$$

$$R = 1 - \frac{C_p}{C_f} \tag{9.2}$$

$$B = \frac{(1 - R)A(\Delta P - \Delta\pi)}{R} \tag{9.3}$$

where J_w is the volumetric water flux, ΔP is the applied pressure, C_p is the permeate salt concentration, C_f is the feed salt concentration, $\Delta\pi$ is the osmotic pressure difference across the membrane. FO performances were tested using a flat-sheet FO module (Mode C10-T, Nitto Denko, Japan) with an effective membrane area of 63 cm^2. Feed solution was DI water, and draw solutions were NaCl aqueous solutions with concentrations between 0.5 and 2 M. The feed and draw solutions were circulated with two peristaltic pumps. The test temperature was 23°C. The feed side was equipped with an electronic weight balance for automatic data logging.

RO performances, intrinsic water permeability (A) and salt rejection rate (R), were determined by a bench-scale crossflow RO test unit (SEPA CF II, GE, 14.5 cm × 9.7 cm effective area). The feed solution was 50 mM NaCl in DI water. The salt concentration was measured by a calibrated conductivity meter

(CON6000, Eutech, Netherlands). The cross flow rate was 15 $cm\,s^{-1}$ under the pressure of 24 ± 0.5 bar. The temperature was maintained at 25 ± 0.5°C. The S value is determined by fitting the reverse osmosis (RO) and FO performance parameters in Equation (9.4), where D is the salt diffusion coefficient and π_{draw} and π_{feed} are the osmotic pressures of the draw and feed solutions, respectively.

$$S = \frac{D}{J_w} \ln\left(\frac{Ax\pi_{draw} + B}{Ax\pi_{feed} + B + J_w}\right) \tag{9.4}$$

The performances and structure parameters of NC-FO, PI-FO, and commercial HTI-FO membranes are listed in Table 9.1. As shown in Table 9.1, the water permeability of the NC-FO membrane ($A = 1.70 \pm 0.1\ L\,m^{-2}\,h^{-1}\,atm^{-1}$) is about 3.5 times higher than that of commercial HTI-FO membranes ($A = 0.39 \pm 0.05\ L\,m^{-2}\,h^{-1}\,atm^{-1}$), and is about 0.36 times higher than that of PI-FO membranes ($A = 1.25 \pm 0.12\ L\,m^{-2}\,h^{-1}\,atm^{-1}$). As shown in Figure 9.9, nanofiber membranes have substantially higher water flux than conventional PI membranes. Thus, the higher A value for the NC-FO membrane may result from low water diffusion resistance in the nanofiber support layer. The salt rejection rate of NC-FO membrane is 97.0% ± 1.0%, similar to that of PI-FO membranes (96.5% ± 1.5%).

As shown in Figure 9.10, the NC-FO membrane, which has the smaller value of the structure parameter (S) and higher water permeability value (A), generates a water flux three times of that of phase inversion PI-FO membranes. At low salt concentration, the water permeability determines water flux, and this region is defined as the "A-governed region"; while at high salt concentration, the structure parameter (S) determines water flux, and this region is defined as the "S-governed region".

As shown in Figure 9.10, the NC-FO membranes #1 and #2 have water flux three times of the PI-FO membrane at all salt concentrations. The water permeability (A) of the NC-FO membrane is slightly higher than that of the PI-FO membrane and the salt rejection rates of the two membranes are similar, therefore the low water flux of the PI-FO membrane should be attributed to the reduced osmotic driving force that is resulted from ICP issue.

Herein, the structure parameter (S) is used to explain the low water flux performance. The *S* value is a direct indicator of ICP, and the higher *S* value results in the severer ICP problem. As shown in Table 9.1, the *S* value of the PI-FO membrane ($S = 450$ μm) is 5.6 times of #1 NC-FO ($S = 80$ μm) and 4.2 times of #2 NC-FO ($S = 106$ μm), which means that the structure of the NC-FO membrane is more favorable for generating higher water flux. Moreover, the S value is related with the tortuosity (positively), thickness (positively), and porosity (negatively) of membranes. The tortuosity (τ) of FO membranes can be calculated using Equation (9.5), wherein ϵ is the porosity and t is the membrane thickness.

$$\tau = Sx\frac{\varepsilon}{t} \tag{9.5}$$

Table 9.1. Performance and structure parameters of different FO membranes

Membrane[a]	*Thickness*[b] *[μm]*	*Porosity [%]*	*Water permeability* $[L\,m^{-2}\,h^{-1}\,bar^{-1}]$	*Salt rejection*[c] *[%]*	*Water flux In FO mode*[d] $[L\,m^{-2}\,h^{-1}]$	*Structure parameter S [μm]*	*Tortuosity*[e]
#1 NC-FO	50 ± 6	83 ± 1	1.70 ± 0.10	97.0 ± 1.0	37.8	80 ± 6	1.33
#2 NC-FO	70 ± 8	83 ± 1	1.65 ± 0.14	97.0 ± 1.2	33.9	106 ± 8	1.26
PI-FO	100 ± 12	81 ± 1	1.25 ± 0.12	96.5 ± 1.5	13.4	450 ± 50	3.65
HTI-FO	50 ± 12	40 ± 3	0.39 ± 0.05	92.0 ± 1.5	6.5	620 ± 80	4.96

[a]#1 NC-FO refers to the nanofiber composite FO membrane with a thickness of 50 μm; #2 NC-FO 70 μm; PI-FO refers to the phase inversion FO membrane; HTI-FO refers to a commercial FO membrane;
[b]non-woven substrate thickness;
[c]salt rejection was determined with 50 mM NaCl and an operation pressure of 24 bar;
[d]FO mode flux was determined with DI water as feed solution and 0.5 M NaCl as the draw solution; and
[e]tortuosity was calculated based on the average thickness, porosity, and S value.
SOURCE: Adapted from Song et al. 2011; reproduced with permission from Wiley.

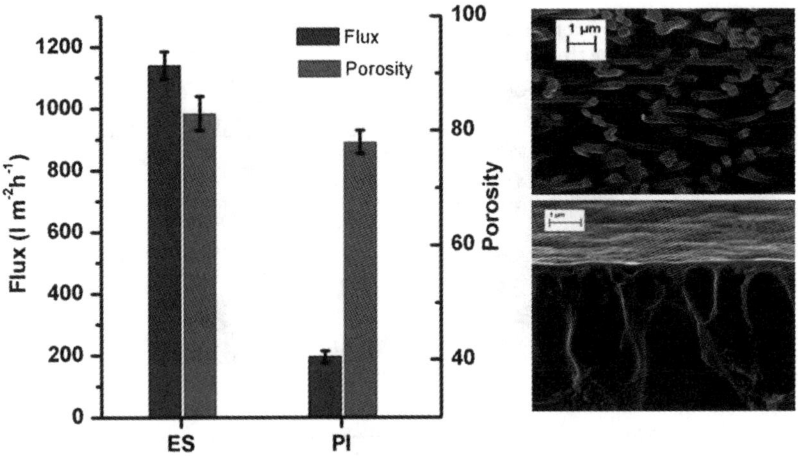

Figure 9.9. Comparison of Water fluxes and porosities between ES (electrospun) and PI (phase-inversion) support layers. The water flux of ES support layer is almost 6 times as high as that of PI support layer, resulting from the low-tortuosity of scaffold-like nanofiber structure

SOURCE: Adapted from Song et al. 2011; reproduced with permission from Wiley

As shown in Table 9.1, the porosity and thickness of the PI-FO membrane are similar as # 2NC-FO membrane, while the *S* value of the PI-FO membrane is 4.2 times of #2 NC-FO membrane. This means that the low tortuosity of NC-FO membranes is responsible for the small S value. According to Table 9.1 and Figure 9.5, it can be seen that the *S* value of HTI-FO membrane is the highest because the dense and sponge-like structure results in low porosity and high tortuosity. The sponge-like structure in the PI-FO membrane presents high porosity and high tortuosity, resulting in moderate *S* value. In contrast, the interconnected pore structure in NC-FO membrane presents not only high porosity but also low tortuosity, resulting in the lowest *S* value. For further investigation, PI-FO membranes were prepared with different porosities, thicknesses, and tortuosities, as shown in Figure 9.11. The performance and structure parameters of the above support layers are shown in Table 9.2. It can be seen that all PI-FO membranes have high tortuosity, which dramatically hinder salt diffusion in the support layers of PI-FO membranes and result in lower osmotic driving force. In contrast, the interconnected pore structure of NC-FO membranes is favorable for salt diffusion, resulting in high osmotic driving force. These also result in different scenarios of transverse salt concentration profiles in PI-FO membranes and NC-FO membranes, as shown in Figure 9.10(a).

As shown in Figure 9.12, conventional PI-FO membrane shows significant water flux decline in FO mode compared with PRO mode, which indicates

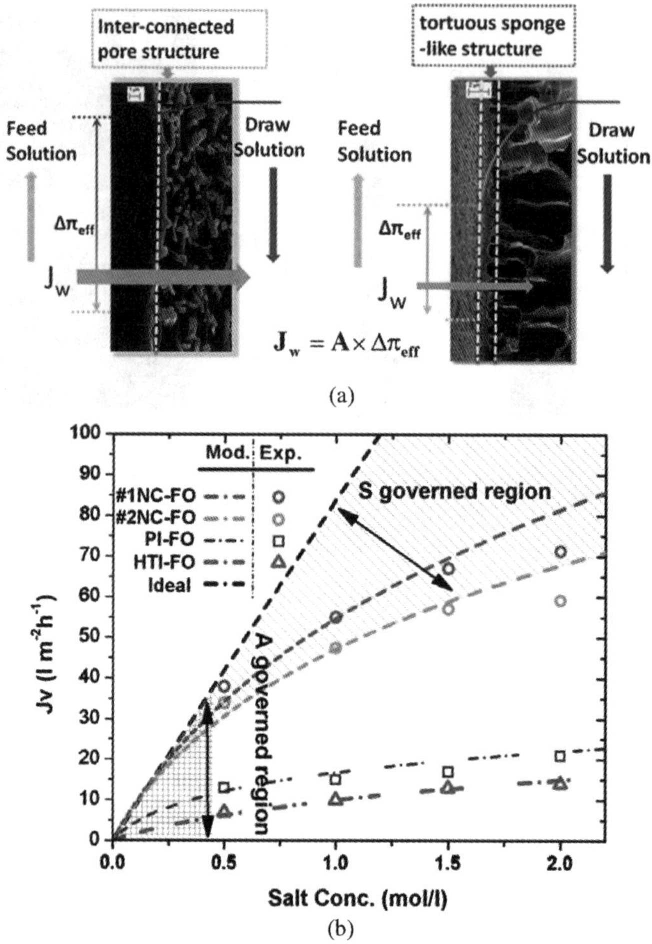

Figure 9.10. (a) Different salt concentration profiles for the NC-FO (left) and PI-FO (right) membranes. In NC-FO membrane, the interconnected pore structure is favorable for salt diffusion in the support layer and presents higher osmotic driving force. While in PI-FO membrane, the tortuous sponge-like structure dramatically hinders salt diffusion in the support layer, resulting in lower osmotic driving force. (b) The water fluxes of different FO membranes as a function of salt concentrations, wherein the feed solution is DI water and the draw solute is NaCl. Abbreviations: Mod.: Modeling; Exp.: Experiment. #1 NC-FO: 50-μm thick membrane; #2 NC-FO: 70-μm thick membrane

SOURCE: Adapted from Song et al. 2011; reproduced with permission from Wiley

a severe diffusive barrier in their tortuous support layers. In contrast, NC-FO membrane works equally well in both FO and PRO modes, which benefits from the greatly-reduced diffusive resistance in the low-tortuosity nanofiber support layers.

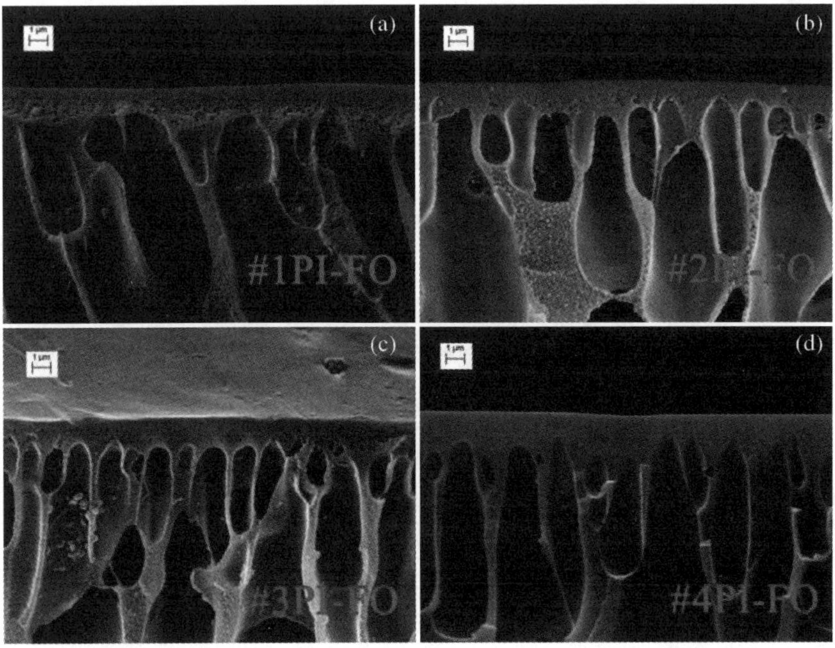

Figure 9.11. (a-d) cross-section SEMs of phase-inversion support layers with different PES polymer concentrations: (a) 12%; (b) 13%; (c) 15%; (d) 17%

SOURCE: Adapted from Song et al. 2011; reproduced with permission from Wiley

Table 9.2. Characteristics of PI-FO and NC-FO support layers

Support layer[a]	*Thickness [μm]*[b]	*Porosity [%]*	*Pure water flux [L m^{-2} h^{-1}]*[c]	*Structure parameter S [μm]*[d]	*Tortuosity e*
#1PI-FO	100 ± 12	81 ± 1	225 ± 18	450 ± 50	3.65
#2PI-FO	107 ± 8	80 ± 2	195 ± 16	467 ± 65	3.49
#3PI-FO	114 ± 10	78 ± 2	173 ± 15	524 ± 30	3.59
#4PI-FO	116 ± 10	77 ± 3	160 ± 12	575 ± 35	3.82
#1NC-FO	50 ± 6	83 ± 1	1139 ± 45	80 ± 6	1.33
#2NC-FO	70 ± 8	83 ± 1	1040 ± 50	106 ± 8	1.26

[a]The PES concentration for #1, 2, 3, 4 PI-FO are 12%, 13%, 15% and 17%, respectively. The thickness of #1 and #2 NC-FO support layers are 50 μm and 70 μm, respectively.

[b]Non-woven substrate thickness is included.

[c]Pure water flux is determined at operation pressure of 1 bar.

[d]Interfacial-polymerization processes are the same for all PI-FO membranes. S value is calculated based on the performances of FO and RO tests, using equation. Tortuosity is calculated based on averaged thickness, porosity and S value.

SOURCE: Adapted from Song et al. 2011; reproduced with permission from Wiley.

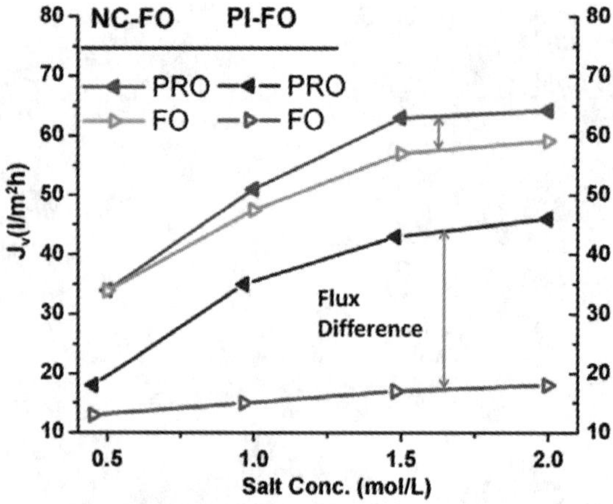

Figure 9.12. Water fluxes under FO and PRO membrane orientations. Conventional PI-FO membrane (red line) shows significant flux difference due to the tortuous support layer, while the novel NC-FO membrane (green line) exhibits insignificant flux difference due to the low-tortuosity nanofiber support layer, which indicates the successful breakdown of ICP bottleneck

SOURCE: Adapted from Song et al. 2011; reproduced with permission from Wiley

9.5 CONCLUDING REMARKS

In order to solve the ICP issue in conventional FO membranes, a novel nanocomposite FO (NC-FO) membrane with a nanofiber support layer was developed. This novel scaffold-like nanofiber support layer presents unique advantages of low tortuosity, high porosity, and thin thickness. This study paves a new way for the development of next-generation FO membranes with high water production rate. For future development of this novel FO membrane, the property of these nanofibers could be improved further in terms of hierarchical architecture (Li 2004), tunable hydrophilicity (Ma et al. 2007; Xue et al. 2009), adsorption ability (Sun et al. 2007), and self-cleaning abilities (Roy et al. 2010). The active layers could be modified with mesoporous materials to present ion selective and high flux performance.

9.6 ABBREVIATIONS

A	permeability
AFM	atomic forced microscope
B	salt permeability
C	concentration
CP	concentration polarization
CTA	cellulose triacetate

D	salt diffusion coefficient
DI	deionized
DMF	N,N-dimethyl formamide
DO	dissolved oxygen
DOC	dissolved organic carbon
ES	electrospun
FO	forward osmosis
ICP	internal concentration polarization
IP	interfacially polymerized
J	permeate flux
NC	nano composite
NMP	N-methyl pyrrolidone
PES	polyethersulfone
PET	poly(ethylene terephthalate)
PI	phase inversion
PRO	pressure retarded osmosis
RO	reverse osmosis
R	salt rejection rate or ideal gas constant
S	structure parameter
SEM	scanning electron microscope
T	temperature
t	membrane thickness.
β	dissociation number of an electrolyte
τ	tortuosity
ϵ	porosity
π	osmotic pressure
g	gravity
ΔP	applied pressure

References

Achilli, A., Cath, T. Y., Marchand, E. A., and Childress, A. E. (2009) "The forward osmosis membrane bioreactor: a low fouling alternative to MBR processes." *Desalination*, 239, 10–21.

Cath, T. Y., Childress, A. E., and Elimelech, M. (2006) "Forward osmosis: principles, applications, and recent developments." *Journal of Membrane Science*, 281, 70–87.

Freger, V. (2005) "Kinetics of film formation by interfacial polycondensation." *Langmuir*, 21, 1884–1894.

Ghosh, A. K., Jeong, B. H., Huang, X., and Hoek, E. M. V. (2008) "Impacts of reaction and curing conditions on polyamide composite reverse osmosis membrane properties." *Journal of Membrane Science*, 311, 34–45.

Holloway, R. W., Childress, A. E., Dennett, K. E., and Cath, T. Y. (2007) "Forward osmosis for concentration of anaerobic digester centrate." *Water Research*, 41, 4005–4014.

Lee, K. L., Baker, R. W., and Lonsdale, H. K. (1981) "Membranes for power generation by pressure-retarded osmosis." *Journal of Membrane Science*, 8, 141–171.

Lee, S. S., Bai, H. W., Liu, Z. Y., and Sun, D. D. (2013) "Optimization and an insightful properties-activity study of electrospun TiO2/CuO composite nanofibers for efficient photocatalytic H2 generation." *Applied Catalysis B: Environmental*, 140–141, 68–81.

Loeb, S., Titelman, L., Korngold, E., and Freiman, J. (1997) "Effect of porous support fabric on osmosis through a Loeb-Sourirajan type asymmetric membrane." *Journal of Membrane Science*, 129, 243–249.

Li, D., Wang, Y., and Xia, Y. (2003) "Electrospinning of polymeric and ceramic nanofibers as uniaxially aligned arrays." *Nano Letters*, 3, 1167–1171.

Li, D., and Xia, Y. (2004) "Direct fabrication of composite and ceramic hollow nanofibers by electrospinning." *Nano Letters*, 4, 933–938.

Liu, Z. Y., Sun, D. D., Guo, P., and Leckie, J. (2007) "An efficient bicomponent TiO2/SnO2 nanofiber photocatalyst fabricated by electrospinning with a side-by-side dual spinneret method." *Nano Letters*, 7, 1081–1085.

Ma, M., Gupta, M., Li, Z., Zhai, L., Gleason, K. K., Cohen, R. E., Rubner, M. F., and Rutledge, G. C. (2007) "Decorated electrospun fibers exhibiting superhydrophobicity" *Advanced Materials*, 19, 255–259.

McCutcheon, J. R. and Elimelech, M. (2008) "Influence of membrane support layer hydrophobicity on water flux in osmotically driven membrane processes." *Journal of Membrane Science*, 318, 458–466.

McCutcheon, J. R. and Elimelech, M. (2006) "Influence of concentrative and dilutive internal concentration polarization on flux behavior in forward osmosis." *Journal of Membrane Science*, 284, 237–247.

McCutcheon, J. R., McGinnis, R. L., and Elimelech, M. (2005) "A novel ammonia–carbon dioxide forward (direct) osmosis desalination process." *Desalination*, 174, 1–11.

McCutcheon, J. R., McGinnis, R. L., and Elimelech, M. (2006) "Desalination by ammonia-carbon dioxide forward osmosis: influence of draw and feed solution concentrations on process performance." *Journal of Membrane Science*, 278, 114–123.

Mehta, G. D. and Loeb, S. (1978). "Internal polarization in the porous substructure of a semipermeable membrane under pressure-retarded osmosis." *Journal of Membrane Science*, 4, 261–265.

Qiu, C., Qi, S., and Tang, C. Y. (2011). "Synthesis of high flux forward osmosis membranes by chemically crosslinked layer-by-layer polyelectrolytes." *Journal of Membrane Science*, 381, 74–80.

Roy, P., Dey, T., Lee, K., Kim, D., Fabry, B., and Schmuki, P. (2010). "Size-selective separation of macromolecules by nanochannel titania membrane with self-cleaning (Declogging) ability." *Journal of American Chemistry Society*, 132, 7893–7895.

Sablani, S. S., Goosen, M. F. A., Al-Belushi, R., and Wilf, M. (2001) "Concentration polarization in ultrafiltration and reverse osmosis: a critical review." *Desalination*, 141, 269–289.

Seppälä, A. and Lampinen, M. J. (1999). "Thermodynamic optimizing of pressure-retarded osmosis power generation systems." *Journal of Membrane Science*, 161, 115–138.

Shi, L., Chou, S. R., Wang, R., Fang, W. X., Tang, C. Y., and Fane, A. G. (2011). "Effect of substrate structure on the performance of thin-film composite forward osmosis hollow fiber membranes." *Journal of Membrane Science*, 382, 116–123.

Song, X. X., Liu, Z. Y., and Sun, D. D. (2011). "Nano gives the answer: breaking the bottleneck of internal concentration polarization with a nanofiber composite forward osmosis membrane for a high water production rate." *Advanced Materials*, 23, 3256–3260.

Su, J. and Chung, T. S. (2011). "Sublayer structure and reflection coefficient and their effects on concentration polarization and membrane performance in FO processes." *Journal of Membrane Science*, 376, 214–224.

Sun, X. Y., Shankar, R., Börner, H. G., Ghosh, T. K., and Spontak, R. J. (2007). "Field-driven biofunctionalization of polymer fiber surfaces during electrospinning." *Advanced Materials*, 19, 87–91.

Wang, R., Shi, L., Tang, C. Y., Chou, S., Qiu, C., and Fane, A. G. (2010). "Characterization of novel forward osmosis hollow fiber membranes." *Journal of Membrane Science*, 355, 158–167.

Widjojo, N., Chung, T.S., Weber, M., Maletzko, C., and Warzelhan, V. (2011). "The role of sulphonated polymer and macrovoid-free structure in the support layer for thin-film composite (TFC) forward osmosis (FO) membranes." *Journal of Membrane Science*, 383, 214–223.

Xue, Y., Wang, H., Yu, D., Feng, L., Dai, L., Wang, X., and Lin, T. (2009). "Superhydrophobic electrospun POSS-PMMA copolymer fibres with highly ordered nanofibrillar and surface structures." *Chemical Communications*, 6418–6420.

Yip, N.Y., Tiraferri, A., Phillip, W. A., Schiffman, J. D., and Elimelech, M. (2010). "High performance thin-film composite forward osmosis membrane." *Environmental Science & Technology*, 44, 3812–3818.

Zhao, S. and Zou, L. (2011). "Relating solution physicochemical properties to internal concentration polarization in forward osmosis." *Journal of Membrane Science*, 379, 459–467.

CHAPTER 10

Surface Modification by Zwitterionic Amino Acid L-DOPA to Increase Fouling Resistance of Forward Osmosis Membranes

L. Zou
A. Nguyen
S. Azari

10.1 INTRODUCTION

The global water crisis and the ever-increasing water demand have driven the extensive development of membrane technology for water treatment. Water and energy in modern society are intrinsically interlinked and both contribute to the high quality of human life. It is estimated that the overall energy consumption of modern seawater desalination plants is 3–4 times higher than the theoretical minimum energy demand for desalination, but the energy demand for the state of the art reverse osmosis (RO) process is only 25% higher than the thermal dynamic limit of energy for desalination, because the extensive pre-treatment and post-treatment steps demand a great deal of energy (Elimelech and Phillip 2011). It is argued that the future research to improve the energy efficiency of desalination should focus on the pre-treatment, so as to minimise the fouling of the membrane. The ideal solution for fresh water supply would be for a process to use very low energy input to produce drinking water from saline water sustainably. An engineered osmosis driven membrane process, forward osmosis (FO), is designed to fit these criteria (McGinnis and Elimelech 2008). Since the FO process utilises natural osmosis as the pressure source, the energy demand can be significantly reduced compared to other pressure-driven processes, such as RO.

10.1.1 Membranes for Forward Osmosis

Theoretically, any dense, non-porous, selectively permeable material can be used as the membrane for a FO process. Cath et al. (2006) and Miller and Evans (2006) summarized that an ideal membrane for FO should have several desired characteristics: high density of the active layer for high solute rejection; minimum porosity of the support layer for low internal concentration polarisation; high mechanical strength to support high hydraulic pressures; highly hydrophilic and thin; stable in contact with draw and feed solutions at the desired temperature and low fouling potential. These factors will help the membrane achieve minimal internal concentration polarisation (CP) and high performance such as high water flux and solute rejection in a FO process.

In recent years, various membranes have been used in FO processes by different research groups. In the 1990s, a novel FO membrane was developed by Osmotek Inc. (Albany, Oregon) (currently Hydration Technologies Inc. (HTI)). This proprietary membrane is made of cellulose triacetate (CTA). The thickness of the membrane is less than 50 μm and a polyester mesh is embedded within the polymer material for mechanical support. Compared to a RO membrane, the CTA FO membrane has a very thin active layer (less than 1 μm) and embedded polyester for mechanical support, without a thick support layer. Recent investigations (McCutcheon et al. 2006; Tang and Ng 2008) indicated that the CTA FO membrane made by HTI can achieve better performance than RO membranes in terms of water flux and salt rejection. During various investigations of FO membranes (McCutcheon et al. 2006; Ng et al. 2006; McGinnis and Elimelech 2007; Mi and Elimelech 2008; Tang and Ng 2008), the researchers indicated that as the hollow fibre membrane module is self-supportive, it may be very suitable for FO. Despite the relatively good performance of CTA membranes, significant improvements in physical structure and surface chemical properties (hydrophilicity) are necessary to achieve fresh water productivity that is commercially viable. Research innovation, in developing a very thin membrane with a highly permeable supporting layer is currently needed. A breakthrough membrane development will advance FO technology to the next level resulting in scale up to real applications in water desalination and recovery.

10.1.2 Membrane Fouling

Fouling is a significant problem associated with the use of membranes for water purification, include seawater desalination, wastewater treatment and water reuse systems. Membrane fouling, which is generally defined as the deposition of substances on the membrane surface or within the membrane pores, is a major hindrance to the widespread application of this technology in water purification processes (Baker 2004). Fouling deteriorates membrane performance by affecting the water quality of the permeate, increasing energy consumption and treatment costs and reducing membrane lifetime (Bartles et al. 2005). Organic compounds, such as proteins, polysaccharides, humic acids and extracellular polymeric substances, tend to be adsorbed on membrane surfaces to form a gel layer or biofilm, which not only significantly promotes the growth of bacteria but also causes severe reduction in water flux. Due to fouling, membrane-based processes require higher energy inputs, frequent chemical cleaning and premature membrane replacement, which lead to higher operational costs.

The types of membrane fouling include scaling and organic fouling (Tang et al. 2010) as well as bio-fouling (Pan et al. 2009). Dissolved organic matter and colloidal particles contribute to fouling by forming a fouling layer on the membrane surface. This fouling process, designated as organic fouling, mainly affects water permeability (Wilf and Alt 2000). It has been clearly shown that dissolved organic matter is the principal contributor to membrane fouling. A significant flux decline is expected if a membrane feed solution containing a high concentration of dissolved organic matter is used. Fouling mechanisms such as adsorption or pore blocking, which cannot be avoided by membrane back-washing, contribute largely to fouling and the corresponding flux decline (Saravia et al. 2006). Organic fouling, through its encouragement of bacterial growth on the membrane surface, is considered a precursor to bio-fouling. Moreover, it has been suggested that membranes with neutral, smooth and hydrophilic surfaces are less prone to fouling (Pasmore et al. 2001).

Although the fouling in FO membrane was reported as low, the membrane biofouling and inorganic fouling found in the FO process still can affect and limit its performance (McGinnis and Elimelech 2007; Achilli et al. 2009; Mi and Elimelech 2010). Membrane fouling in osmotic-driven membrane processes is different from that in pressure-driven membrane processes because low hydraulic pressure is employed in the former processes. With the onset of membrane fouling, a significant flux decline is expected. Fouling mechanisms such as adsorption or pore blocking, which cannot be avoided by membrane backwashing, contribute largely to fouling and the corresponding flux decline (Saravia et al. 2006). Organic fouling, through its encouragement of bacterial growth on the membrane surface, is considered a precursor to bio-fouling. Moreover, it has been suggested that membranes with neutral, smooth and hydrophilic surfaces are less prone to fouling (Pasmore et al. 2001).

10.1.3 Current Anti-fouling Coating Methods

Depending on the type of materials involved, current anti-fouling coating methods can be placed into three categories, i.e., anionic coating, hydrophilic polymer coating, and bioinspired zwitterionic coating. Each of these coating methods are described briefly as follows.

Anionic Coating

Since organic materials, such as proteins, extracellular polymeric substances and bacteria, have negatively charged surfaces, they can be efficiently repelled from membrane surfaces if the membrane surfaces are negatively charged. The strength of the electrostatic interactions can be measured in terms of the extent of the electric double layer surrounding the charged species. A drawback of this approach is that electrostatic repulsion between membranes and organic compounds is vulnerable to the ionic strength of the solution. Therefore, in a high ionic strength environment, such as seawater or wastewater, negatively charged surfaces do not have long-term, anti-fouling properties (Kim et al. 2002; Lee et al. 2006).

When negatively charged polymers such as poly(acrylic acid) are coated on the surfaces of membranes, particularly in the form of brushes anchored on the membrane surface, the polymer coating provides strong electrostatic repulsion as well as strong steric repulsion (collectively known as electrosteric repulsion) that prevents the adsorption of negatively charged organic matter (Kang et al. 2007a; Krishnan et al. 2008). However, the efficacy of the long-term anti-fouling effect of polymer chains anchored on membranes, when used for practical processes, is debatable, as the chains tend to detach from the membrane surfaces or degrade with time, especially during chemical cleaning of the membranes.

Hydrophilic Polymer Coating

Numerous methods for membrane surface modification to increase surface hydrophilicity have been examined (Kang et al. 2007b; Yu et al. 2008; Sagle et al. 2009). These methods include blending with hydrophilic polymers, surface modification by grafting hydrophilic monomers and surface coating. However, most methods require multiple complex steps, involve the use of chemicals and are time consuming, which restricts their use in practical applications.

When a hydrophilic polymer such as polyethylene glycol (PEG) is in contact with bulk water, a hydrogen bond network is formed. The highly hydrated layer is responsible for the polymer's anti-fouling properties (Chen et al. 2010). Success in the use of PEG for the provision of fouling resistance in biomedical applications had led to its use for membrane modification. PEG and PEG-based modifiers can be introduced to membranes via different methods such as blending, grafting and coating (Mansouri et al. 2010). For example, Belfer et al. (1998) studied redox-initiated radical grafting of vinyl monomers, including poly(ethylene glycol) methacrylate (PEGMA), to commercial polyamide RO membrane surfaces. Such modified membranes have lower contact angles and lower surface roughness than unmodified membranes. Sagle et al. (2009) reported improved fouling resistance for polyamide RO membranes coated with a PEG-based hydrogel. Su et al. (2009) synthesized a series of amphiphilic poly(ethylene glycol)-graft-polyacrylonitrile (PEG-g-PAN) copolymers. Subsequently, PEG-g-PAN ultrafiltration membranes were prepared by using a water phase-separation method; the prepared membranes were reported to have a lower bovine serum albumin (BSA) adsorption and a higher protein flux than the other tested membranes.

Bimodal PEG brushes have been reported as antifouling agent (Bosker et al. 2005). The antifouling properties of the PEG brushes depend on the polymer grafting density and orientations. The preparation of brushes requires multiple steps that involve many chemicals and solvents, is considered a rather chemically demanding procedure.

Although PEG shows excellent protein resistance ability, PEG-based polymers are susceptible to oxidation (Holmlin et al. 2001). In addition, they have been reported to be resistant to short-term bacterial adhesion (Roosjen et al. 2004); however, over the long-term, biofilm formation still occurs.

Bioinspired Zwitterionic Coating

It is well known that the plasma membranes of cells and bacteria can efficiently prevent adsorption of organics, proteins and even polysaccharides. The excellent anti-fouling effect of plasma membranes arises from the unique zwitterionic surface character of their phospholipid bilayers. Recently, bioinspired zwitterionic coatings have been developed to significantly reduce organic matter adsorption as they have both positively and negatively charged moieties with a uniformly mixed balanced charge (Zhu et al. 2009; Xi et al. 2009; Brault et al. 2010; Chen et al. 2010). A recent study by Lewis et al. (2000) focused on developing new ligands that bear a zwitterionic head, such as phosphorylcholine, sulphobetaine (bearing anionic sulphate and cationic quaternary ammonium groups), and carboxybetaine (bearing anionic carbonate and cationic quaternary ammonium groups). Kitano et al. (2005) reported more bound water molecules around the surface of zwitterionic materials than around other hydrophilic materials. In recent years, polysulphobetaine methacrylates have been grafted on polyvinylidene fluoride, polysulphone, polyethersulphone and microporous polypropylene membranes and surface resistance to proteins has been achieved (Yang et al. 2010). A critical factor determining the non-fouling property of polyzwitterionic materials is the need to control both uniformity of charge distribution and charge neutrality. Unfortunately, anchoring of these zwitterionic ligands on the surfaces of membranes involves multiple chemical reactions, and the methods cannot be easily adapted for water treatment applications.

10.1.4 Zwitterionic L-DOPA

L-DOPA and its derivatives, which were inspired by the adhesive proteins found in mussels, have been successfully incorporated into various synthetic polymers. The created polymers can be firmly attached to a surface in mild, aqueous environments (Lee et al. 2007); thus, compared to the other methods, the coating process is simple and can be easily carried out in large-scale industrial applications. Until recently, most studies on L-DOPA have focused on its use as a spacer for substrate functionalisation (Dalsin et al. 2003). There is little research on its application as the main membrane modifier agent (Xi et al. 2009). L-DOPA has both positively and negatively charged zwitterionic heads, which could lead to anti-fouling property and, at the same time, it is exceptionally adhesive to various surfaces, making it an ideal anti-fouling coating.

Many surface modification studies focus on pressure-driven membranes (e.g. UF, NF and RO). However, to our best knowledge, few surface modification studies have been conducted for FO membranes. Since the FO process utilises natural osmosis as the pressure source, the energy demand can be significantly reduced compared to other pressure-driven processes, such as RO. More recently, it has been demonstrated that bioinspired zwitterionic materials including phosphorylcholine, sulphobetaine, and carboxybetaine can significantly reduce organic

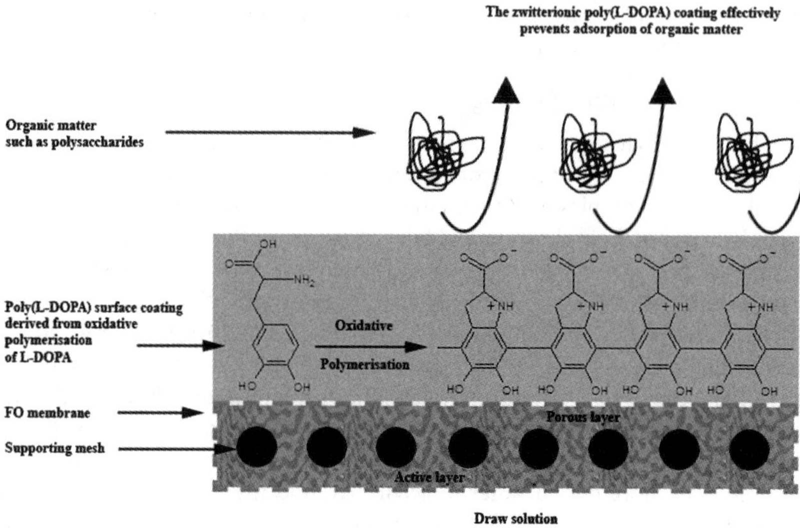

Fig. 10.1. Schematic of surface adsorption resistance for organic matter imparted by the zwitterionic poly(L-DOPA) coating
SOURCE: Adapted from Nguyen et al. 2013; reproduced with permission from Elsevier

matter adsorption (Zhao and Zou 2011). Azari and Zou (2012) have incorporated redox functional amino acid 3-(3,4-Dihydroxyphenyl)-L-alanine (L-DOPA) onto commercial reverse osmosis (RO) membranes to create a zwitterionic surface that resists membrane fouling. It was found in the accelerated fouling tests that the modified membranes exhibited less flux decline than the original untreated membrane and high water recovery.

In this study, the zwitterionic polymer L-DOPA was coated on membrane surfaces to further increase the fouling resistant of the membrane.The relationship between the zwitterionic coating and the improved anti-fouling effect by observing filtration flux and by quantifying the change of water contact angle on the membrane surfaces was investigated. Figure 10.1 is a schematic diagram of the structure of zwitterionic poly L-DOPA attached on the FO membrane via in-situ polymerisation. It is hypothesized that the coated membrane becomes strongly hydrated and the hydrated layer prevents the accumulation and adsorption of solutes on the membrane surface. Finally, the reversibility of FO membrane fouling by hydraulic cleaning was studied.

10.2 ZWITTERIONIC L-DOPA MEMBRANE SURFACE MODIFICATION

10.2.1 Materials

Flat sheet cellulose triacetate FO membranes were obtained from Hydration Technologies Innovation (USA). They are currently considered as the best available FO membranes which have asymmetric structure with an active layer and an embedded polyester mesh support layer to enhance mechanical stability (Zou et al. 2013).

The model organic foulant AAS, a protein extracted from brown algae, was purchased from Sigma-Aldrich (Australia and New Zealand). The feed solution used for FO fouling experiments was prepared by dissolving AAS in deionised water for over 16 hours, then adding 200 mg/L $CaCl_2$ to enhance the intermolecular adhesion between alginate molecules and promote synergistic fouling effects in FO process. This resulted in more severe flux decline during alginate fouling due to the formation of a cross-linked alginate gel layer on the membrane surface (Mi and Elimelech 2008). The foulant solution was finally kept in the fridge at 4°C to prevent bacteria growth.

Tris (hydroxymethyl) aminomethane (Tris) buffer and coating material, 3-(3,4-Dihydroxyphenyl)-l-alanine (L-DOPA) (formula weight (FW) 197 g) were also supplied by Sigma-Aldrich (Australia and New Zealand). A Tris-HCl buffer solution (10 mM, pH 8) was prepared in advanced as a solvent for L-DOPA (2 g/L).

10.2.2 Membrane Modification

The membranes were firstly soaked in DI water over-night to wash out protective agents. The clean membrane, which only the porous side of FO membranes was modified, was then placed into a flat sheet cell. A peristaltic pump, connecting with the cell, allowed the solution to flow and circulate L-DOPA solution from a beaker to over one side of membrane surface for a pre-designed coating time which varied from one hour to 12 hours. After coated, the membranes were taken out from the cell, carefully washed by deionized water at least three times and stored in the fridge at 4°C.

10.2.3 FO fouling and Cleaning Experiments

The FO setup is shown in Figure 10.2 and similar to that described in our previous study (Zhao and Zou 2011; Zhao et al. 2011; Azari and Zou 2012). It comprises a membrane holding cartridge (2 mm in depth, 80 mm in length, and 60 mm in width); a digital balance; two peristaltic pumps; a feed tank and a draw tank; and an in-stream conductivity sensor. As the surface modification was only taken place in porous side of membrane, all the experiments were conducted in PRO mode, where membrane porous layer faces the draw solution. The water flux was obtained with 25 M sodium chloride (NaCl) as the draw solution and the feed solution containing 1 g/L AAS and 200 mg/L $CaCl_2$. The two peristaltic pumps were used to generate the crossflow rates for the draw and feed solution which were optimised at 25.0 cm/s not only in our previous studies (Zhao and Zou 2011; Zhao et al. 2011) but also in other studies (Wang et al. 2010; Phillip et al. 2010). The feed solution tank was placed on a digital balance and weights of permeate were recorded every minute by a computer. The setup also includes two plastic mesh spacers that were used to reduce external concentration polarisation on both sides of the membrane surface. To keep the osmotic pressure constant and avoid

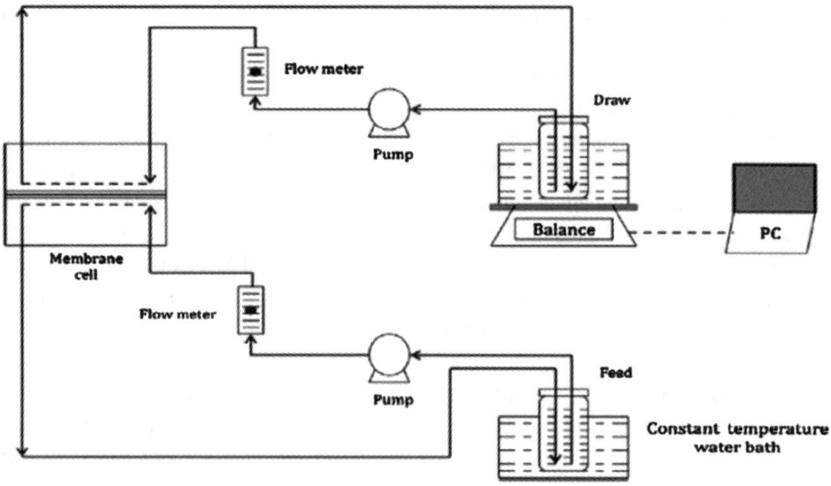

Fig. 10.2. Schematic diagram of the laboratory-scale FO system
SOURCE: Adapted from Zhao and Zou 2011; reproduced with permisson from Elsevier

the reduction of water flux due to the dilution of draw solution, during the experiments, the draw concentration was maintained at 2 M by using a dosing pump and an on-line conductivity sensor. Once the conductivity of the draw solution falls below the required value, the equipment will automatically pump the NaCl solution from a more concentrated reservoir to the draw solution reservoir.

Before the fouling generation was run, the FO membranes were stabilised for at least 30 minutes with deionised water as the feed and 2 M NaCl as the draw solution. The stable pure water flux was obtained right after this step. The fouling filtration was then conducted for three hours. The membrane fouling experiments were performed with a mixture of AAS and $CaCl_2$ solution as the feed and 2 M NaCl as the draw solution. Following the fouling tests was the cleaning process with the same method conducted in our previous studies (Zhao and Zou 2011). Experimental cleaning conditions include: deionised water in both the feed and draw solutions; cross-flow velocity of 30 cm/s; temperature of 25 ± 1°C; and 1 h rinsing. Finally, the pure water flux of the membrane was measured again with deionized water as the feed solution and 2M NaCl as the draw solution.

The water flux (Jw) was calculated by measuring the weight change of permeate during each experimental time period as in equation below (Zhao and Zou 2011):

$$Jw = \Delta\text{Weight}/(\text{Water density} \times \text{Effective area} \times \Delta\text{Time})$$

10.2.4 FO Membrane Characterisation

In order to determine the characteristics of the untreated and treated membranes, the contact angle analysis, UV-Vis absorption, FTIR spectra and Zeta potential measurement were carried on and described as below:

A Data-Physics OCA15 Contact Angle Analyser (Data-Physics Instruments GmbH, Filderstadt, Germany) was employed to determine the hydrophilicity of membrane surface. The contact angle measurement with the captive bubble method was applied, which suits for hydrophilic and porous cellulose acetate membrane surface (Grundke et al. 1996). The membrane samples were firstly fixed horizontally on a substrate, and then carefully immersed in a chamber with water. After that, a microsyringe was lowered into the water where a drop was formed on the syringe tip, snapped from the tip and allowed to rise to the membrane-water interface. The contact angles of pendant drop were finally calculated by OCA software. For each sample, an averaged data was obtained after at least five times of repetition.

UV-Vis adsorption analysis was carried out due to the dark brown color of the membrane surface after modified with L-DOPA. The Varian Cary 100 UV-Vis spectrophotometer with an ultraviolet to visible range of 200–800 nm, equipped with diffuse reflectance accessories, was employed to examine the adsorption of L-DOPA on modified membranes. The ATR-FTIR spectroscopy, using Spectrum100 spectrometer (Perkin Elmer) recording spectra over the range of 4000–400 cm^{-1}, was also conducted to characterise the functional groups of the membrane surface.

The Zeta potential values of uncoated and coated FO membranes were determined using an Anton Paar SurPASS ElectrokineticAnalyser (Anton Paar Austria). The result was obtained by using the adjustable gap cell apparatus, measuring the streaming current and applying the Helmholtz-Smoluchowski approximation to evaluate the data. The pH ranged from 8.5 to 3 with 10 mM KCl solution as the background solution for all experiments. The instrument, which automatically dosed 0.1 M HCl until the pH reduced to 3, recorded the data at 0.5 pH intervals. Four measurements were recorded and the average values with error bars in each pH value were reported in following section.

10.3 EFFECTS OF L-DOPA COATING ON FO MEMBRANE

10.3.1 FO Membrane Characterisations

Contact angle analysis was conducted to examine the hydrophilicity of membrane surfaces. As addressed in literature (Wavhal and Fisher 2002; Azari and Zou 2012), membrane surface hydrophilicity is highly correlated to membrane fouling resistance property. The hydrophilicity surfaces are able to absorb a layer of water molecules, establishing a buffer which helps to prevent adhesion of foulants to the membrane surface (Elimelech and Phillip 2011; Miller et al. 2012). The more hydrophilic surface is, the higher fouling resistance is expected. In this study, the modification and observation were taken place in the porous side of FO membrane which is rough and heterogeneous in the dry stage making it difficult to perform the contact angle measurement. According to Grundkeet et al. (1996), in this case, conventional contact angle measurements, such as the sessile drop method, do not

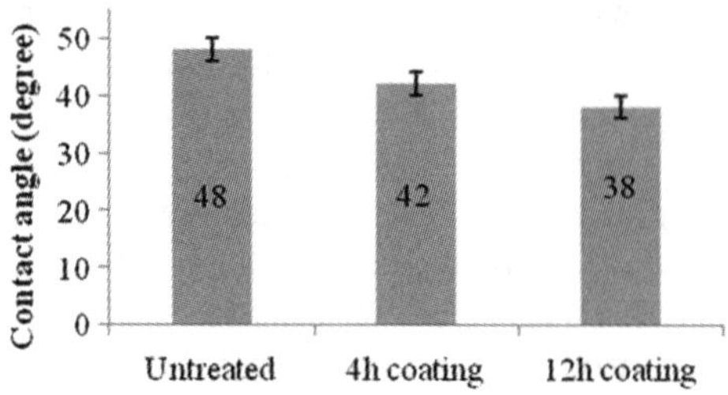

Fig. 10.3. Contact angles of untreated, 4 h and 12 h L-DOPA coated membranes
SOURCE: Adapted from Nguyen et al. 2013; reproduced with permission from Elsevier

provide reproducible contact angle values. The captive bubble method, on the other hand, was more suitable and provided more accurate contact angle data.

The results of contact angle measurement on untreated and coated membranes are illustrated in Figure 10.3. As shown in this figure, the contact angle of untreated membrane was higher than coated sample, initially 48° comparing to 44° for 4 h and to 38° for 12 h coated membranes. The decrease of contact angle as the time of L-DOPA modification increased was evident that the presence of L-DOPA coating layer enhanced the wettability of membrane surface. As mentioned our previous study (Azari and Zou 2012), poly L-DOPA comprises an equal number of positively and negatively charged moieties and strongly interacts with water via an ionic-dipole interaction which greatly increased its hydrophilicity. The form of hydrated surface also effectively resists the membrane fouling. The longer coating time was, the more interactions between L-DOPA moieties with membrane surfaces occurred, and the more improvement of membrane surface hydrophilicity was achieved.

Figure 10.4 illustrated the ATR-FTIR spectra of the unmodified and L-DOPA coating membranes. Since the CTA FO membrane surface and L-DOPA have some common functional groups, such as carbonyl —C=O, hydroxyl —O-H and ether —C-O-C— group, the ATR-FTIR spectra of untreated and L-DOPA coating membrane are not much different. The O—H group spectra appear at approximately $3,500\ cm^{-1}$; whereas, the band appearing at approximately $1{,}700\ cm^{-1}$ was corresponded to the Cdbond;O group. The peak at around $1,200\ cm^{-1}$ was identified as the —C-O-C spectra while the band appearing at approximately $1,000\ cm^{-1}$ was assigned to the —C-O-C bond in the pyranose ring (Ilharco and Brito de Barros 2000). As can be seen from Figure 10.4, there was a decrease in the intensities of the band of the O-H group of modified membranes. Figure 10.4 also shows that the intensities were more decreased as the coating time was longer. Amongst the L-DOPA coating samples, the 12 h L-DOPA coated membrane is the most decrease in intensity. This indicated that chemical interactions potentially happened between —O—H— group on FO CTA membrane surfaces and active catechol group of L-DOPA.

The UV-Vis absorbance analysis was carried on from ultraviolet to visible range of 200–800 nm. The UV-Vis absorbance results of the untreated membrane

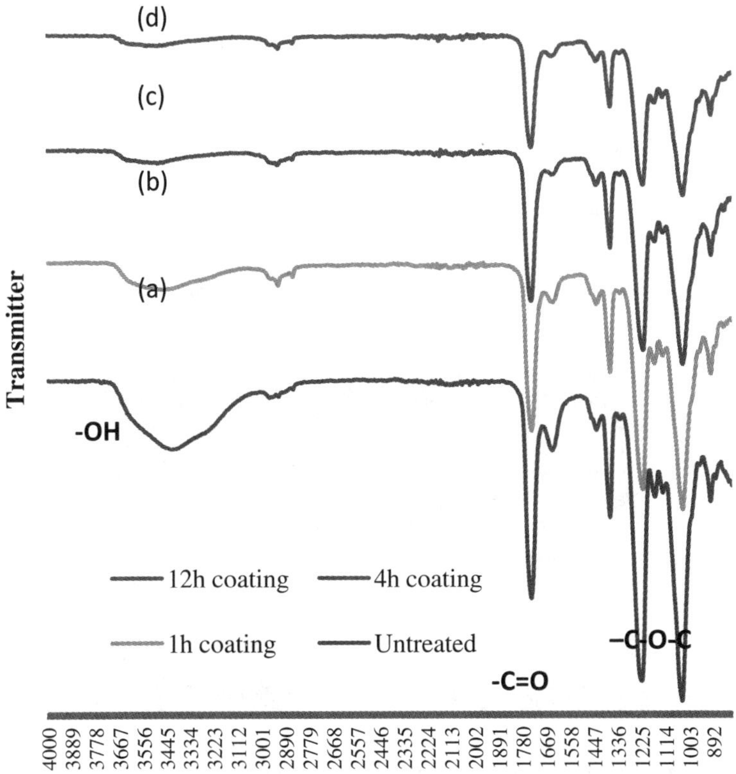

Fig. 10.4. ATR-FTIR spectra of (a) original membrane and membrane with (b) 1 h, (c) 4 h and (d) 12 h DOPA coating
SOURCE: Adapted from Nguyen et al. 2013; reproduced with permission from Elsevier

and L-DOPA coated membranes are shown in Figure 10.5(A), and Figure 10.5(B) shows the visual dark brown colour of coated membranes. The coated membrane surfaces had dark brown colour after modified with L-DOPA due to the interaction happening within L-DOPA molecule which a Dopa quinone was formed by the oxidation of the catechol part of the L-DOPA. The Dopa quinone further reacts to form poly L-DOPA oligomers which firmly attach to the surface. Having a smooth and white surface, the original FO membrane was almost zero absorption in the visible range. Meanwhile, the L-DOPA coated membranes which were darker in colour had higher absorbance than the original one. The absorbance increased with the longer coating time. Figure 5(A) shows that the 12 h coating sample has the highest absorbance than others.

The Zeta potential measurement confirmed the success of L-DOPA coating on membrane surfaces. Figure 10.6 shows Zeta potential values of uncoated and coated FO membranes. The untreated FO membrane was slightly positive charged at pH of three but mostly negative as pH was increased to 8.5. The Zeta potential of L-DOPA coated membranes, whereas, were fully negative charged. It is known

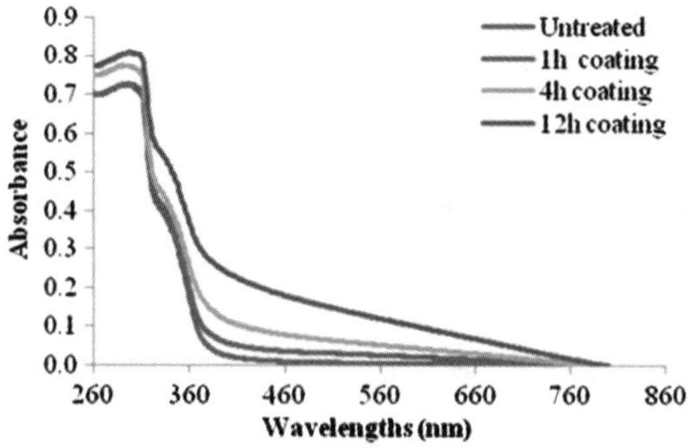

(a)

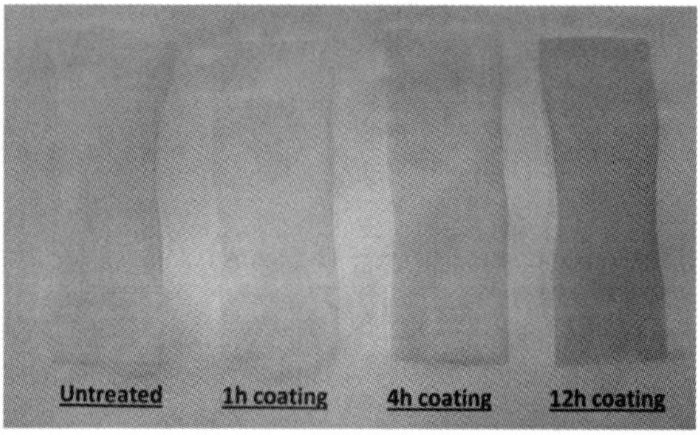

(b)

Fig. 10.5. (A) UV–Vis absorbance of untreated membrane and L-DOPA coated membranes. (B) Visible surface colour change of virgin and coated membranes
SOURCE: Adapted from Nguyen et al. 2013; reproduced with permission from Elsevier

that L-DOPA has both positive charged and negative charged, contributing by NH_{3+} groups and carboxylic acid (COO-) groups respectively. However, the overall charges of the coated membranes surfaces were mainly contributed by carboxylic acid groups which resulted in higher negative charged of modified membranes as shown in Figure 10.6.

10.3.2 Fouling Behaviour and Flux Recovery of FO Membranes

To evaluate the improvement of fouling behaviour of FO membranes after modified, the filtration experiments were carried on for 180 minutes with 2 M NaCl as the draw solution and mixture of 1 g/L AAS and 200 mg/L $CaCl_2$ as the feed. The membrane fouling fluxes and flux recovery ratios were normalised to demonstrate the fouling changes and reported in Figure 10.7. As can be seen from Figure 10.7, the coated membranes had less flux reduction than the original. A 70% flux reduction caused by fouling was obtained by the uncoated membrane while the coated samples showed less flux reduction, which was only a 40% reduction for the twelve-hour coating, a 50% reduction for the four-hour coating and about 60% for the one-hour of coating. With 30% less fouling than the

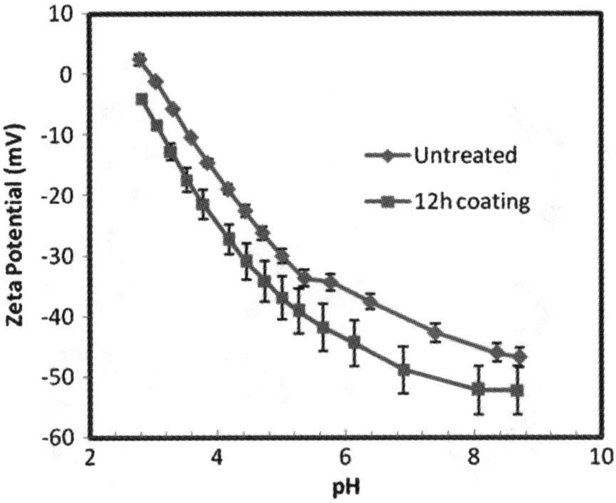

Fig. 10.6. Zeta potential of uncoated and L-DOPA coated FO membranes
SOURCE: Adapted from Nguyen et al. 2013; reproduced with permission from Elsevier

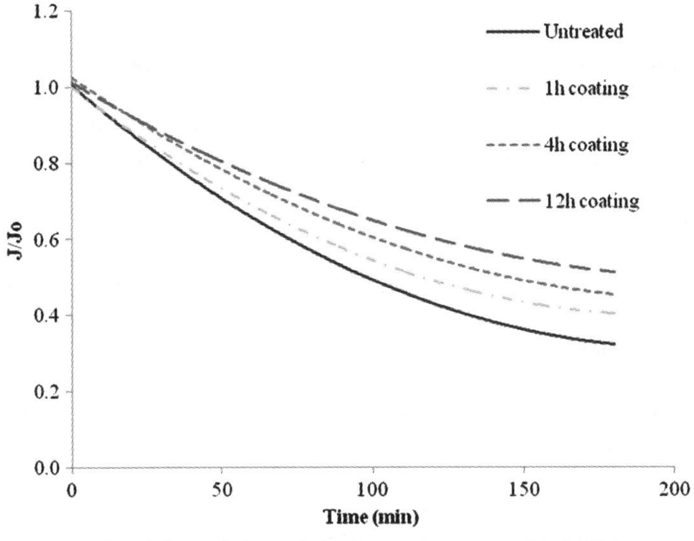

Fig. 10.7. Normalised flux of the original membrane and L-DOPA coated membranes
SOURCE: Adapted from Nguyen et al. 2013; reproduced with permission from Elsevier

uncoated sample, the membrane coated with L-DOPA for 12 h indicates that higher fouling resistant membranes are achieved after modified. This can be explained by the reduction of the adhesion force between membrane material and the foulants. Another factor, which also impacts to the fouling behaviour of FO membrane, is the internal concentration polarization (ICP). The ICP causes the

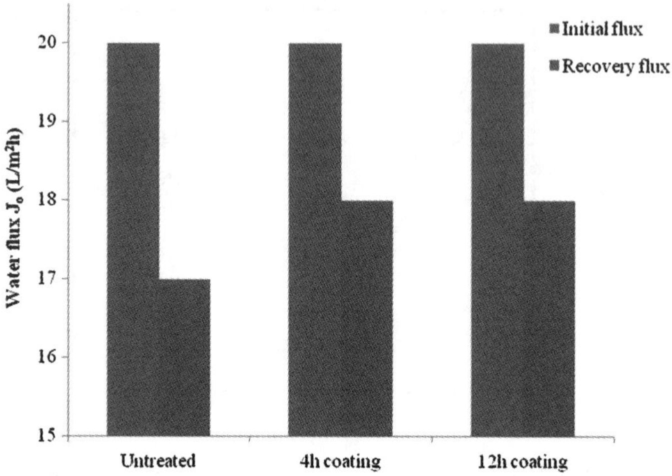

Fig. 10.8. Initial and recovered flux cleaned by water of virgin and L-DOPA coated membranes

SOURCE: Adapted from Nguyen et al. 2013; reproduced with permission from Elsevier

accumulation of solutes in the boundary layer adjacent to the membrane surface, promotes the attachment of foulants to the membrane surface and results in fouling. Thus, it is potential that the L-DOPA coating layer of modified membrane surface may reduce the ICP that helps to improve the fouling resistance of the membranes.

After 3 h of fouling generation, the membranes were cleaned by DI water for 1 hour with high cross flow velocity. The pure water flux was recorded again for both uncoated and coated membranes to evaluate the flux recovery. Figure 10.8 shows that the modified membranes obtained higher flux recovery than the original. The modified membrane was regained 90% of the initial flux while the pure water flux of unmodified membrane only recovered 85% after cleaned. It should be noted that in this study the fouling experiments were conducted in PRO mode, which the porous layer of FO membrane facing the feed and have lower flux recovery than the FO mode, where the dense active layer facing the feed. Thus, the membrane recovery results obtained from this study are reasonable and indicate that the L-DOPA coating layer successfully helped to prevent the tight attachment of organic foulants on membrane surface.

10.4 CONCLUSIONS

A zwitterionic L-DOPA coating method was applied to modify the commercial CTA FO membranes. The modification, which aimed to improve the anti-fouling performance of FO membrane, was taken place on the porous side of the membrane. The fouling experiments showed that the L-DOPA coated samples having less flux reduction caused by on-set of fouling than the untreated membranes. Achieving 30% less fouling than untreated membrane, the sample with a 12-h coating of L-DOPA proved that the membrane fouling resistance was enhanced after modified. A 90% flux recovery by hydraulic cleaning with DI water was also obtained by the L-DOPA coated membrane. In addition, the reduction of water contact angle indicated that the modified membrane surfaces were more hydrophilic than the uncoated membrane; meanwhile, the UV-vis, ATR-FTIR spectra and Zeta potential results confirmed the success of L-DOPA coating method. It can be concluded that L-DOPA coating modification is an effective method to enhance the fouling resistance of FO membranes. Moreover, this direct coating method is also simple and easy to operate without the requirement of high-energy input and the special equipment; thus, it has the potential to be scaled up for *in-situ* membrane modification in an industrial environment.

10.5 ABBREVIATIONS/NOMENCLATURE

CP	concentration polarization
CT	Acellulose triacetate
DO	dissolved oxygen
DOC	dissolved organic carbon

References

Achilli, A., Cath, T. Y., Marchand, E. A., and Childress, A. E. (2009). "The forward osmosis membrane bioreactor: A low fouling alternative to MBR processes." *Desalination*, 239, 10–21.

Azari, S., and Zou, L. (2012). "Using zwitterionic amino acid L-DOPA to modify the surface of thin film composite polyamide reverse osmosis membranes to increase their fouling resistance." *Journal of Membrane Science*, 401–402, 68–75.

Baker, R. W. (2004). *Membrane Technology and Applications*, 2nd ed., John Wiley & Sons Ltd.

Bartles, C. R., Wilf, M., Andes, K., and Iong, J. (2005). "Design considerations for wastewater treatment by reverse osmosis." *Water Science and Technology*, 51(6-7), 473–482.

Belfer, S., Purinson, Y., Fainstein, R., Radchenko, Y., and Kedem, O. (1998). "Surface modification of commercial composite polyamide reverse osmosis membranes." *Journal of Membrane Science*, 139, 175–181.

Bosker, W. T. E., Iakovlev, P. A., Norde, W., and Cohen Stuart, M.A. (2005). "BSA adsorption on bimodal PEO brushes." *Journal of Colloid and Interface Science*, 286, 496–503.

Brault, N. D., Gao, C., Xue, H., Piliarik, M., and Homola, J. (2010). "Ultra-low fouling and functionalizable zwitterionic coating grafted onto SiO_2 via a biomimetic adhesive group for sensing and detection in complex media." *Journal of Biosensors and Bioelectronics*, 25, 2276–2282.

Cath, T. Y., Childress, A. E., and Elimelech, M. (2006). "Forward osmosis: principles, applications, and recent developments." *Journal of Membrane Science*, 281, 70–87.

Chen, S., Li, L., Zhao, C., and Zheng, J. (2010). "Surface hydration: Principles and applications toward low-fouling/nonfouling biomaterials." *Polymer*, 51, 5283–5293.

Dalsin, J. L., Hu, B. H., Lee, B. P., and Messersmith, P. B. (2003). "Mussel adhesive protein mimetic polymers for the preparation of nonfouling surfaces." *Journal of the American Chemical Society*, 125, 4253–4258.

Elimelech, M., and Phillip, W. A. (2011). "The future of seawater desalination: energy, technology and the environment." *Science*, 333, 712–717.

Grundke, K., Bogumil, T., Werner, C., Janke, A., Poschel, K., and Jacobasch, H. -J. (1996). "Liquid-fluid contact angle measurements on hydrophilic cellulosic materials." *Colloids and Surfaces A: Physicochemical and Engineering Aspects*, 116, 79–91.

Holmlin, R. E., Chen, X., Chapman, R. G., Takayama, S., and Whitesides, G. M. (2001). "Zwitterionic SAMs that resist nonspecific adsorption of protein from aqueous buffer." *Langmuir*, 17 (9), 2841–2850.

Ilharco, L. M., and Brito de Barros, R. (2000). "Aggregation of pseudoisocyanine iodide in cellulose acetate films: Structural characterization by FTIR." *Langmuir*, 16, 9331–9337

Kang, S., Asatekin, A., Mayes, A. M., and Elimelech, M. (2007a). "Protein antifouling mechanisms of PAN UF membranesincorporating PAN-g-PEO additive." *Journal of Membrane Science*, 296, 42–50

Kang, G., Liu, M., Lin, B., Cao, Y., and Yuan, Q. (2007b). "A novel method of surface modification on thin-film composite reverse osmosis membrane by grafting poly(ethylene glycol)." *Polymer*, 48, 1165–1170.

Kim, K. S., Lee, K. H., Cho, K., and Park, C. E. (2002). "Surface modification of polysulfone ultrafiltration membrane by oxygen plasma treatment." *Journal of Membrane Science*, 199(1–2), 135–145

Kitano, H., Mori, T., Takeuchi, Y., Tada, S., Gemmei-Ide, M., Yokoyama, Y., and Tanaka, M. (2005). "Structure of water incorporated in sulfobetaine polymer films as studied by ATR-FTIR." *Macromolecular Bioscience*, 5(4), 314–321.

Krishnan, S., Weinmanb, C. J., and Oberb, C. K. (2008). "Advances in polymers for anti-biofouling." *Journal of Material Chemistry*, 18, 3405–3413.

Lee, H., Dellatore, S. M., Miller, W. M., and Messersmith, P. B. (2007). "Mussel-inspired surface chemistry for multifunctional coatings." *Science*, 318, 426–430.

Lee, S., and Elimelech, M. (2006). "Relating organic fouling of reverse osmosis membranes to intermolecular adhesion forces." *Environmental Science*, 40, 980–987.

Lewis, A. L. (2000). "Phosphorylchlorine-based polymers and their use in the prevention of biofouling." *Colloids Surf B: Biointerfaces*, 18, 261–275.

Mansouri, J., Harrison, S., and Chen, V. (2010). "Strategies for controlling biofouling in membrane filtration systems: challenges and opportunities." *Journal of Material Chemistry*, 20, 4567–4586.

McCutcheon, J. R., McGinnis, R. L., and Elimelech, M. (2006). "Desalination by ammonia-carbon dioxide forward osmosis: Influence of draw and feed solution concentrations on process performance." *Journal of Membrane Science*, 278, 114–123.

McGinnis, R. L., and Elimelech, M. (2008). "Global challenges in energy and water supply: the promise of engineered osmosis." *Environmental Science & Technology*, 42, 8625–8629.

McGinnis, R. L., and Elimelech, M. (2007). "Energy requirements of ammonia-carbon dioxide forward osmosis desalination." *Desalination* , 207, 370–382.

Mi, B., and Elimelech, M. (2010). "Organic fouling of forward osmosis membranes: Fouling reversibility and cleaning without chemical reagents." *Journal of Membrane Science*, 348, 337–345.

Mi, B., and Elimelech, M. (2008). "Chemical and physical aspects of organic fouling of forward osmosis membranes." *Journal of Membrane Science*, 320, 292–302.

Miller, D. J., Araújo, P. A., Correia, P. B., Ramsey, M. M., Kruithof, J. C., Loosdrecht, M. C. M., Freeman, B. D., Paul, D. R., Whiteley, M., and Vrouwenvelder, J. S. (2012). "Short-term adhesion and long-term biofouling testing of polydopamine and poly(ethylene glycol) surface modifications of membranes and feed spacers for biofouling control." *Water Research*, 46, 3737–3753.

Miller, J. E., and Evans, L. R. (2006). *Forward osmosis: a new approach to water purification and desalination.* Sandia National Laboratories, USA.

Ng, H. Y., Tang, W., and Wong, W. S. (2006). "Performance of forward (direct) osmosis: Membrane structure and transport phenomenon." *Environmental Science and Technology*, 40, 2408–2413

Nguyen, A., Azari, S., and Zou, L. (2013) *Coating zwitterionic amino acid l-DOPA to increase fouling resistance of forward osmosis membrane. Desalination*, 312, 82–87.

Pan, J. R., Su, Y. C., Huang, C., and Lee, H. C. (2009). "Effect of sludge characteristics on membrane fouling in membrane bioreactors." *Journal of Membrane Science*, 349, 287–249.

Pasmore, M., Todd, P., Smith, S., Baker, D., and Silverstein, J. (2001). "Effects of ultrafiltration membrane surface properties on Pseudomonas aeruginosa biofilm initiation for the purpose of reducing biofouling." *Journal of Membrane Science*, 194(1), 15–32.

Phillip, W. A., Yong, J. S., and Elimelech, M. (2010). "Reverse draw solute permeation in forward osmosis: Modeling and experiments." *Environmental Science & Technology*, 44, 5170–5176.

Roosjen, A., Van derMei, H. C., Busscher, H. J., and Norde, W. (2004). "Microbial adhesion to poly(ethylene oxide) brushes: influence of polymer chain length and temperature." *Langmuir*, 20(25), 10949–10955.

Sagle, A. C., Van Wanger, E. M., Ju, H., McCloskey, B. D., and Freeman, B. D. (2009). "PEG-coated reverse osmosis membranes: desalination properties and fouling." *Journal of Membrane Science*, 340, 92–108.

Saravia, F., Zwiener, C., and Frimmel, F. (2006). "Interactions between membrane surface, dissolved organic substances and ions in submerged membrane filtration." *Desalination*, 192, 280–287.

Su, Y. L., Cheng, W., Li, C., and Jiang, Z. (2009). "Preparation of antifouling ultrafiltration membranes with poly(ethylene glycol)-graft-polyacrylonitrile copolymers." *Journal of Membrane Science*, 329, 246–252.

Tang, W., and Ng, H. Y. (2008). "Concentration of brine by forward osmosis: Performance and influence of membrane structure." *Desalination*, 224, 143–153.

Wilf, M., and Alt, S. (2000). "Application of low fouling RO membrane elements for reclamation of municipal wastewater." *Desalination*, 132, 11–19.

Wang, K. Y., Ong, R. C., and Chung, T. -S. (2010). "Double-skinned forward osmosis membranes for reducing internal concentration polarization within the porous sublayer." *Industrial and Engineering Chemistry Research*, 49, 4824–4831.

Wavhal, D. S., and Fisher, E. R. (2002). "Hydrophilic modification of polyethersulfone membranes by low temperature plasma-induced graft polymerization." *Journal of Membrane Science*, 209, 255–269.

Xi, Z. Y., Xu, Y. Y., Zhu, L. P., and Wang, Y. (2009). "A facile method of surface modification for hydrophobic polymer membranes based on the adhesive behavior of poly(DOPA) and poly(dopamine)." *Journal of Membrane Science*, 327, 244–253.

Yang, Y. F., Li, Y., Li, Q. L., and Wan, L. S. (2010). "Surface hydrophilization of microporous polypropylene membrane by grafting zwitterionic polymer for anti-biofouling." *Journal of Membrane Science*, 362, 255–264.

Yu, H. Y., Liu, L. Q., Tang, Z. Q., Yan, M. G., Gu, J. S., and Wei, X. W. (2008). "Surface modification of polypropylene microporous membrane to improve its antifouling characteristics in an SMBR: Air plasma treatment." *Journal of Membrane Science*, 311, 216–224.

Zhao, S., and Zou, L. (2011). "Effects of working temperature on separation performance, membrane scaling and cleaning in forward osmosis desalination." *Desalination*, 278, 157–164.

Zhao, S., Zou, L., and Mulcahy, D. (2011). "Effects of membrane orientation on process performance in forward osmosis applications." *Journal of Membrane Science*, 382, 308–315.

Zhu, L., Yu, J., Xu, Y., Xi, Z., and Zhu, B. (2009). "Surface modification of PVDF porous membranes via poly(DOPA) coating and heparin immobolisation." *Colloids Surf. B: Biointerfaces*, 69, 152–155.

Zou, S., Wang, Y. -N., Wicaksana, F., Aung, T., Wong, P. C. Y., Fane, A. G., and Tang, C. Y. (2013). "Direct microscopic observation of forward osmosis membrane fouling by microalgae: Critical flux and the role of operational conditions." *Journal of Membrane Science*, 436, 174–185.

CHAPTER 11

Membrane Fouling in Forward Osmosis Processes

Seungkwan Hong
Ho Kyong Shon

11.1 BACKGROUND

11.1.1 What is Membrane Fouling?

Since the world has been suffered from water scarcity, membrane filtration process has been emerging as the most promising technology in the water treatment system and has been developed significantly over the past few decades. To date, a number of filtration systems have been applied to advanced water treatment (seawater desalination and wastewater reuse) and even conventional water treatment processes. Although the membrane filtration systems are expected as the most feasible alternative for water treatment, several problems still remain for efficient operation of membrane filtration. Among the problems, fouling which is the deposition of undesirable materials on the membrane surface and/or pores and which reduces the permeate flux and operation efficiency, is one of the major challenges that should be overcome in membrane technology. Membrane fouling can not only cause severe flux decline but also affect the quality of the water produced. Severe fouling often requires intense chemical cleaning or membrane replacement in order to maintain membrane performance which leads to increase of operating costs in a treatment plant.

Membrane fouling can be considered to be a two-step process. The first step is foulant adsorption on the clean membrane surface, which is defined as membrane-foulant interaction (i.e., initial rapid fouling). The fouling rate of the first step is decisive in overall efficiency. The next step of the fouling mechanism is its accumulation on the membrane surface, which is defined as foulant-foulant interaction. This step takes up the greater part of the fouling mechanism (long-term gradual fouling). Between the two steps of membrane fouling, foulant adsorption on the membrane surface, named as the membrane-foulant interaction, is mainly affected by the membrane surface property and usually determines the initial fouling rate. However the foulant–foulant interaction is not clearly distinguishable with respect to the membrane surface property as the clean membrane surface has already been covered by a number of foulants due to the initial membrane-foulant interaction.

There are various types of foulants which cause membrane fouling: colloidal (clays, flocs), biological (bacteria, fungi), organic (oils, polyelectrolytes, humics) and scaling (mineral precipitates) (Baker 2004). The fouling type is determined by each of those foulants as colloidal fouling, biofouling, organic fouling and inorganic fouling. Moreover, based on the attachment strength of foulants to the membrane surface, fouling can be divided into reversible and irreversible fouling. Reversible fouling can be removed by physical cleaning such as a strong shear force of backwashing or increased cross flow velocity. Formation of a strong matrix of fouling layer with solutes during a continuous operation of membrane filtration may transform reversible fouling into an irreversible fouling layer. Irreversible fouling, a strong attachment of particles, cannot be removed by physical cleaning; it requires chemical cleaning (Choi et al. 2005).

11.1.2 Fouling Properties of Forward Osmosis Processes

During FO, fouling mechanism steps are identical, foulant adsorption on the clean membrane surface which is defined as membrane-foulant interaction and its accumulation on the fouled membrane surface which is defined as foulant-foulant interaction. Similarly, because of its identical fouling behavior steps between RO and FO, those factors which affect membrane fouling during RO process also affect FO membrane fouling.

It was revealed that surface characteristics such as functional groups, roughness, surface charge, hydrophobicity and heterogeneity of the FO membrane surface are important factors that affect membrane fouling rate and extent (Hoek et al. 2003; Mi and Elimelech 2010; Gu et al. 2013). However, since RO membrane is a polyamide thin-film composite membrane and FO membrane is a cellulose based membrane, fouling trend could be different between RO and FO due to different membrane surface characteristics, which will be discussed in the next section specifically.

Fouling studies on FO were investigated in terms of the effect of solution chemistries on FO fouling (Zhu and Elimelech 1995; Faibish et al. 1998; Singh and Song 2005, 2007). They found that feed solution (FS) chemistry such as pH and divalent cations is important factors for membrane fouling, especially natural organic matter (NOM) fouling. Furthermore, hydrodynamic condition such as initial water flux and cross-flow velocity is a crucial factor during FO fouling formation (Lay et al. 2010; Tang et al. 2010) because it affects membrane fouling extent and rate. However, caused by distinction of fundamental process mechanisms between FO and RO, there are some differences of fouling mechanisms. The distinction of membrane process mechanisms leads to different fouling rate and extent between FO and RO. Therefore, it is significant to study the fundamental process mechanisms of both processes.

First, each process has different driving forces. The driving force of FO is osmotic pressure difference between FS and DS, while RO is hydraulic pressure driven process. It is very important because one of the main forces that build up fouling layer on membrane surface is driving force for water permeation. Since osmotic pressure is chemical pressure, it is possible that its impact on fouling formation may be different from hydraulic pressure that is physical pressure.

The second difference of the process is solute flux direction. During RO operation, salt is dissolved in the FS and passes through the membrane as identical direction of water flux. During FO process, on the contrary, salt passes through the membrane from DS to FS due to its higher ion concentration of the DS then the FS (Phillip et al. 2010). It is called reverse salt diffusion (or reverse solute flux, RSF). As a result, these two major differences (driving force and salt flux direction) could be related to the difference of flux decline rate and extent and lead to different fouling mechanisms between RO and FO. For example, it is confirmed that accelerated cake-enhanced osmotic pressure (A-CEOP) in FO is mostly due to reverse salt diffusion from the DS which is accumulated between membrane surface and fouling layer then leads to much thicker layer which results in more flux decline compared to RO. Consequently, it reduces the net driving force for water permeation by increase of fouling layer resistance in FO. Moreover, it is found that the structure of the organic fouling layer is also determined by the applied hydraulic pressure (Lee et al. 2010b). FO fouling is revealed that fouling is reversible in FO while irreversible in RO. Since FO does not require hydraulic pressure, deformable foulants make loose and sparse fouling layer that can easily be removed by physical cleaning.

11.1.3 Fouling Properties of Reverse Osmosis Processes

Reverse osmosis (RO) process has been considered as one of the most promising technologies in hydraulic pressure-driven membrane process. However, RO also has suffered from fouling, especially in wastewater reuse and desalination. It was found that major factors affecting RO membrane fouling are membrane surface characteristics, feed water solution chemistry and hydrodynamic conditions.

Surface Characteristics

Table 11.1 shows general characteristics of membrane and its effects on fouling. Surface characteristics of membrane are of paramount importance in RO fouling. Membrane materials are known to affect foulant-membrane interactions, thereby influencing the initial stage of membrane fouling and the efficiency of membrane cleaning. From these studies, it can be concluded that RO membrane should have the smoother, more negatively charged, and less hydrophobic properties for reducing fouling potential of RO membranes.

Solution Chemistry

With regard to the solution chemistry, pH and divalent cations have been reported to be key factors affecting natural organic matter (NOM) fouling (Hong and Elimelech 1997; Lee and Elimelech 2006). Organic matters usually have functional groups such as carboxyl (—COOH) and phenolic groups (—OH) which are directly affected by the feed water pH and divalent cation concentration. It has

Table 11.1. Membrane characteristics and their fouling contributions

Parameter	*Remark*
• Functional groups	RO fouling is greatly affected by the characteristics of the functional groups of the membrane surface and foulants. Membrane with higher adhesive interaction force with foulants shows greater fouling potential due to high attractive force and vice versa (Yang et al. 2010).
• Roughness	Roughness plays an important role on colloidal fouling (Vrijenhoek et al. 2001, Elimelech et al. 1997). The membranes with rougher surface are liable to colloidal fouling as the valleys created by the rough surface produce wells of low interaction energy in which colloidal particles preferentially deposit (Hoek et al. 2003).
• Surface charge	RO membranes with the negatively charged surface exhibit low fouling tendency. Most foulants are negatively charged and thus the electrostatic repulsion between negatively charged foulants and the negatively charged membrane surface prevents foulant adhesion (Childress and Deshmukh 1998; Shim et al. 2002; Brant et al. 2006a).
• Hydrophobicity	Hydrophobic membranes suffer more severely from membrane fouling than hydrophilic membranes due to the strong hydrophobic interaction, which could allow thick fouling layers on the membrane surface (Jucker and Clark 1994; Brant and Childress 2004; Boussu et al. 2007).
• Chemical heterogeneity	Heterogeneously charged RO membrane exhibits high fouling tendency because membrane surface chemical heterogeneities may affect the foulant transport and adsorption on the membrane surface (Sakuma et al. 1989; Lee et al. 2010a; Brant et al. 2006b; Lee et al. 2011).

been known that organic fouling is usually accelerated with decreasing pH and increasing divalent cation (i.e., calcium ions) concentration (Jones and O'Melia 2000; Seidel and Elimelech 2002; Lee et al. 2005; Lee et al. 2006). In these conditions (i.e., low pH and high divalent cation concentration), charge property of organic matters diminishes through the neutralization of functional groups as well as organic-calcium complexation. These results in the accelerated accumulation of organic matters on the membrane surface enhance the foulant-membrane interaction as well as foulant-oulant interactions (Hong and Elimelech

1997; Seidel and Elimelech 2002; Lee et al. 2005; Lee and Elimelech 2006). Especially, under the typical surface water ionic strength condition (e.g., 10 mM), organic fouling was significantly affected by the solution chemistry (e.g., pH and calcium concentration), showing more rapid flux decline with decreasing pH and increasing calcium concentration. However, these visible impacts of feed water pH and calcium concentration on organic fouling almost disappeared at the condition of high ionic strength such as the TDS of seawater. These results obviously showed that the influence of solution chemistry such as feed water pH and calcium concentration would have much less impacts on organic fouling of RO membranes during seawater desalination (Yu et al. 2010).

Hydrodynamic Condition

Hydrodynamic conditions such as initial permeate flux and cross flow velocity also affect RO membrane fouling. This indicates that an increased applied pressure (and water flux) and a reduced cross flow velocity can cause faster flux decline (Wu et al. 1999). In the case of initial flux, higher initial flux tends to increase severe membrane fouling as a result of increased permeate drag force in addition to the enhanced concentration polarization. After athe long-term period operation, the fluxes for all membranes became slowly changed, and the fluxes are considered pseudo stable. Unlike initial flux, there is a limiting value for the pseudo stable flux under a given condition, and increasing initial flux beyond the limiting value does not lead to increase of the pseudo stable flux. Therefore, membrane having fluxes larger than limiting flux tends to reach the limiting value. On the other hand, membrane having fluxes below limiting flux seems to be liable to membrane fouling (Tang and Leckie 2007; Wang and Tang 2011; Tang and Leckie 2007). In the case of cross flow velocity, it affects the shear induced inertial force and lateral migration. Therefore, so it may affect membrane fouling tendency. However, the shear induced inertial force and lateral migration depend strongly on the particle size and are of paramount importance in relatively large particles (size $\geq \sim\sim$ or >100 nm) (Bacchin et al. 1995; Bacchin et al. 2006). Thus shear induced inertial force and lateral migration are less important for the small particle sizes ($\ll 100$ nm) (Bacchin et al. 2006; Wang and Tang 2011; Bacchin et al. 2006).

11.2 FORWARD OSMOSIS FOULING

11.2.1 Reverse Salt Flux and Accelerated Cake-enhanced Osmotic Pressure

A-CEOP phenomenon is that the cake layer (one of the fouling types) prevents the back diffusion of salt feed bulk solution thereby resulting in increased osmotic pressure adjacent the membrane surface, thus water flux decreases. Lee et al. (2010b) reported a relationship between RSF and fouling formation, now it is well known as A-CEOP phenomenon that is caused by solute from DS and it contributes to much thicker cake layers compared to RO except bovine serum albumin (BSA) fouling. The relationship between RSF and fouling is proved by

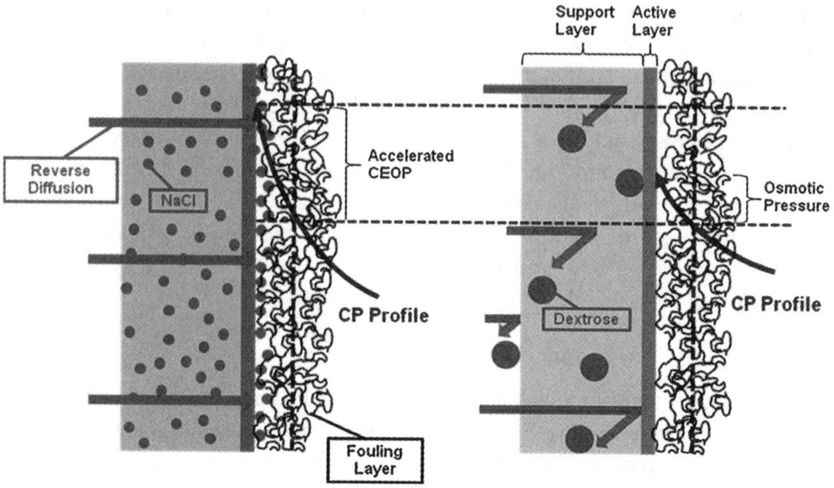

Figure 11.1. A conceptual illustration of the effect of draw solute reverse diffusion on A-CEOP
SOURCE: Adapted from Lee et al. 2010b; reproduced with permission from Elsevier

comparing water flux decline when NaCl and dextrose as draw solute. Dextrose was effectively retained by the membrane because of its much larger hydrated radius than that of NaCl, leading to negligible reverse solute flux of dextrose from the draw to the feed side therefore A-CEOP phenomena were only observed for NaCl solution (Figure 11.1).

The colloidal fouling is also affected by RSF. In addition to RSF, colloidal particle size contributes to fouling severity (Lee et al. 2010b; Boo et al. 2012). In both papers, when small silica particle, approximately 20 nm, was used as foulant, water flux decline was moderate compared to large particle, 139 nm and 300 nm respectively. This is because the small colloidal particles have large back diffusion compared to the convective permeate flow, thereby resulting in a thin cake layer which has short pathway of back diffusion, thus A-CEOP is insignificant.

11.2.2 Membrane Orientation

Since FO uses no hydraulic pressure, it is possible that operating FO process on two membrane orientations that the membrane active layer is placed against the FS (FO mode), and the membrane active layer is against the DS (PRO mode). Mi and Elimelech (2008) observed that membrane orientation has different effects on the organic foulants such as alginate, BSA, and humic acid (HA). It is interesting to note that fouling propensity depends on foulant type and membrane orientation. For alginate, fouling is not affected by membrane orientation, with a similar flux decline obtained in FO and PRO modes. In contrast, for BSA and HA, the flux decline is more severe in the PRO mode than in the FO mode (Figure 11.2). This observation is most notable with HA. The effects of different membrane orientation suggest that different organic foulants have different fouling mechanisms.

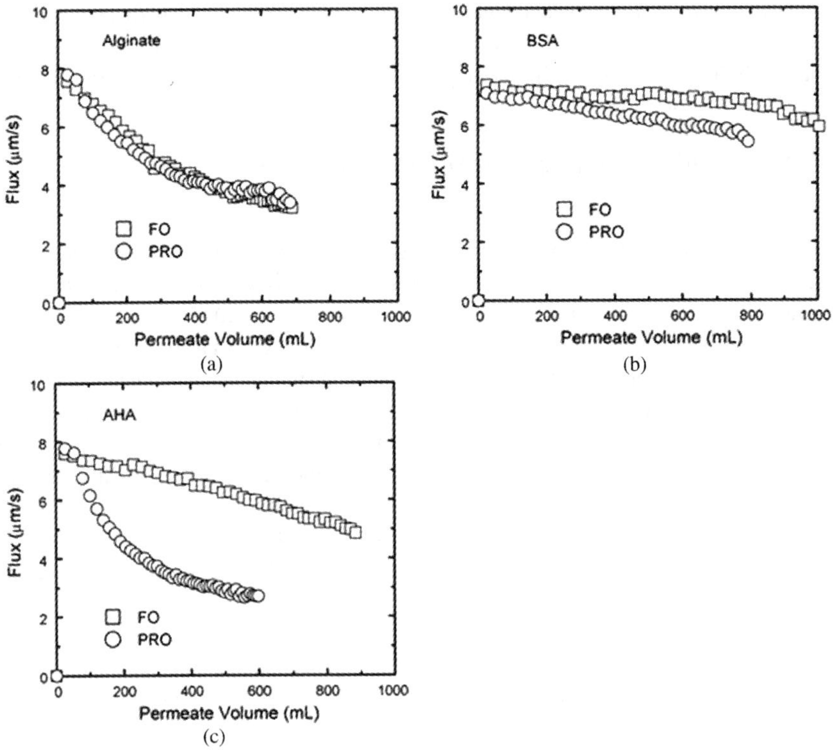

Figure 11.2. Effect of membrane orientation on FO membrane fouling by alginate, BSA, and HA
SOURCE: Adapted from Mi and Elimelech 2010; reproduced with permission from Elsevier

Since the FO membrane has an asymmetric structure, an active layer on top of a porous support layer, membrane fouling occurs on different surfaces in FO and PRO modes. For instance, in FO mode, foulant deposition (or accumulation) occurs on top of the active layer. On the other hand, foulant deposition takes place within the porous structure of the membrane in PRO mode. The biggest difference between two modes is that cross-flow velocity vanishes within the porous support layer, thus the influence of hydrodynamic shear forces is absent at the initial stage of fouling in PRO mode. As the two membrane orientations provide different hydrodynamic conditions during membrane fouling, the membrane orientation also should be considered as a key factor of membrane fouling in FO.

Figure 11.2 shows that water flux variation for alginate is identical on both modes, which means that hydrodynamic interactions do not play crucial role in alginate fouling. Chemical interactions, like calcium binding, play a dominant role because highly built gel layer is relatively unaffected by hydrodynamic condition change. However, in the case of BSA and HA, hydrodynamic condition change influences fouling formation significantly by comparing water flux variation on both FO and PRO modes. In addition, the remarkable flux decline with HA on the

PRO mode is attributed to cake layer formation due to lack of shear force as well as hindered back diffusion of HA aggregates in the porous structure (Mi and Elimelech 2008). These results are in good agreement with previous studies (Tang et al. 2010; Parida and Ng 2013).

11.2.3 Solution Chemistry (pH and Calcium Binding)

pH

Solution chemistry has a prominent effect on the electrokinetic properties of colloidal particles and on the rate of fouling in membrane separation (Zhu and Elimelech 1995; Faibish et al. 1998; Singh and Song 2005, 2007). Boo et al. (2012) investigated the relationship between solution pH on colloidal fouling propensity and both monodisperse and polydisperse colloidal suspensions. Regardless of pH, there were no water flux differences when monodisperse colloidal particles were used as foulants. On the other hand, at high pH (pH 9), water flux for the polydisperse suspension was suddenly dropped while moderate water flux decline was observed at pH 4 for identical polydisperse suspension (Figure 11.3). This is because hydrodynamic radius of the large particles, 139 nm, significantly increased for high pH (pH 9) and high ionic strength (1 M of NaCl) solutions. Increased

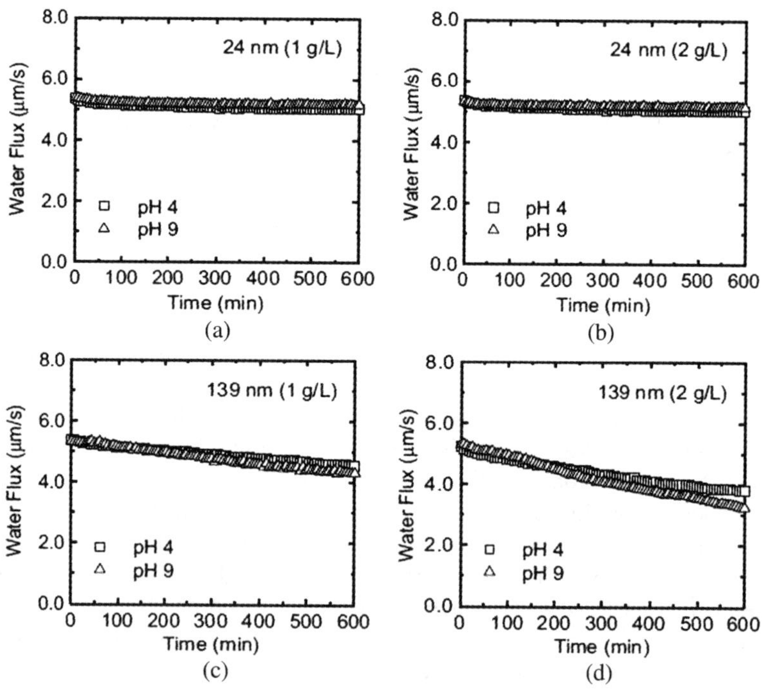

Figure 11.3. Flux decline curves obtained during the FO colloidal fouling runs with silica particles at pH 4 and 9

SOURCE: Adapted from Boo et al. 2012; reproduced with permission from Elsevier

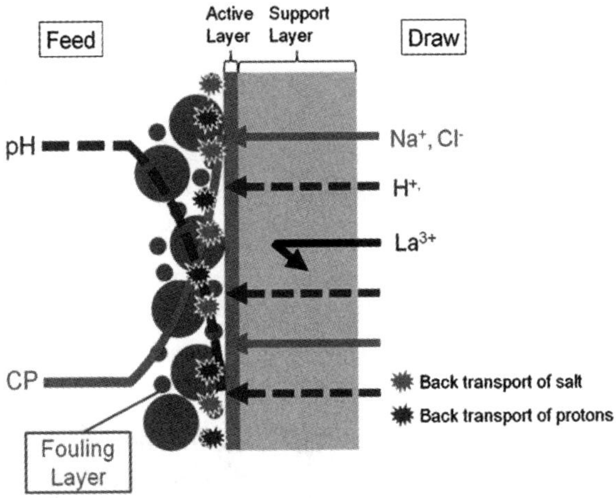

Figure 11.4. A conceptual illustration for the effect of reverse diffusion of DS on the concentration polarization and pH profiles within the fouling layer
SOURCE: Adapted from Boo et al. 2012; reproduced with permission from Elsevier

hydrodynamic radius of particles tends to aggregate each other, resulting in fouling formation.

These phenomena were proven by comparing NaCl and $LaCl_3$ as a draw solute. Water flux was dropped when NaCl was used as a draw solute while no water flux changed when $LaCl_3$ was used. This may be due to RSF. In addition to RSF, it was observed that pH of FS decreased due to reverse diffusion of protons generated by the hydrolysis of $LaCl_3$. Since low pH makes colloidal particles stable, such pH decreasing effect by $LaCl_3$ solution also helps low fouling propensity (Figure 11.4).

Calcium Binding

It is well known that calcium binds with carboxylic acid groups in NOM and causes accelerated fouling. In FO process, fouling severity is different from membrane orientation. Parida and Ng (2013) reported Ca^{2+} effect in FO process. When the FO active layer was subject to a FS containing calcium ion and organic foulants, the flux remained nearly constant throughout 20 h of the experimental run (Figure 11.5). However, in the PRO mode, the water flux sharply dropped within the first 1–2 h operation, followed by a gradual drop. According to the results, the previous tests without Ca^{2+} addition showed that the water flux in PRO mode after 20 h dropped by 36% (from 33.0 to 24.0 LMH) for 30 ppm of TOC in the FS, and by 45% (from 33.0 to 18.0 LMH) in the case with 50 ppm of TOC. With the inclusion of Ca^{2+}, as high as 85% reduction in water flux (from 33.0 to 4.8 LMH) was observed in the PRO mode. It shows that not only Ca^{2+} binding but also membrane orientation is a key parameter in FO process because loose and rough structure of porous layer exacerbates fouling on PRO mode.

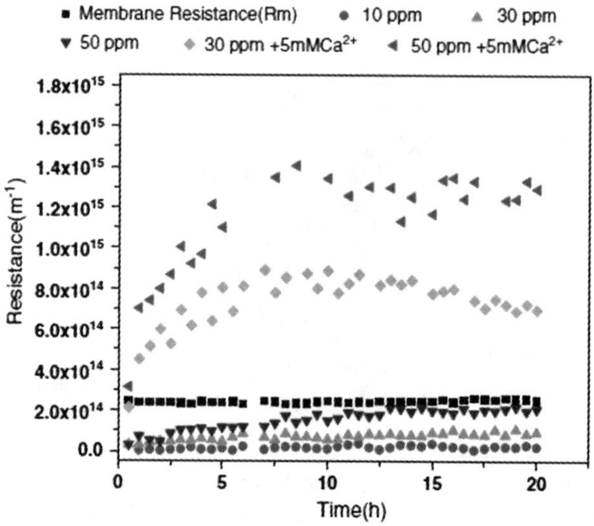

Figure 11.5. Resistance (m^{-1}) versus experimental time in the absence and presence of Ca^{2+} (5 mM) in both FO and PRO modes

SOURCE: Adapted from Parida and Ng 2013; reproduced with permission from Elsevier

On contrary, no severe fouling was observed in FO mode. Hence, although calcium-foulant interactions could cause fouling, the smooth structure of the dense layer of the FO membrane impeded fouling. Furthermore, high cross flow velocity (50 cm/s) was enough to wash off foulants from the membrane surface, thus reducing the effects of external concentration polarization on the densely fouled membrane surface.

11.2.4 Initial Flux

As FO is an osmotically driven membrane process, the DS concentration plays a critical role in FO performance (Mi and Elimelech 2008; Tang et al. 2010; Lay et al. 2010). The effects of initial flux on organic fouling in PRO mode are shown in Figure 11.6. In the case of BSA, the higher initial flux caused higher flux decline. Since the influence of initial flux on membrane fouling is principally attributed to the permeation drag resulting from convective flow toward the membrane, the results indicate that the stronger permeation drag exerted on BSA molecules results in the formation of a cake layer. The large difference in the fouling behavior with BSA for low and high permeation drags reflects a transition in the fouling layer from a loose fouling layer structure to a much more compact cake layer. For alginate and HA, the effect of initial flux is relatively small because the fouling layers under both conditions are already in the form of a cake layer. Once a cake layer forms, fouling becomes much less sensitive.

Wang et al. (2010) studied the effect of DS concentration for FO particle fouling on PRO mode. Apparently, the DS concentration critically affects FO fouling. While particle deposition was negligible at the low DS concentration at

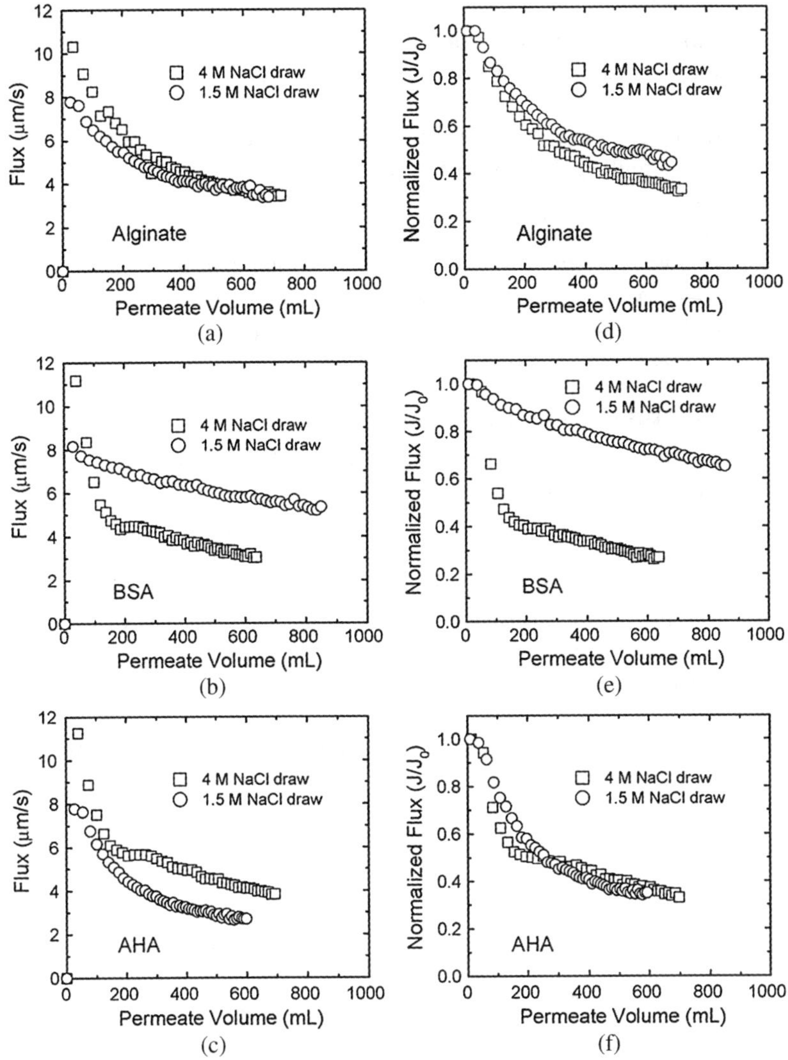

Figure 11.6. Effect of initial flux on membrane fouling in the PRO mode
SOURCE: Adapted from Mi and Elimelech 2010; reproduced with permission from Elsevier

0.5 M, small patches were observable for 2 M DS. At higher DS concentrations, the surface coverage was extensive (about 40% for 3 M DS and about 70% for 4 M DS at 70 min). Corresponding to the dramatic particle deposition at higher DS concentrations, the flux decline was also more severe for the 3 and 4 M concentration of DS. The greater fouling propensity at higher DS concentrations may be explained by the higher flux levels as a result of increased FO driving force. For pressure-driven membrane processes, it is well-known that elevated flux levels inevitably lead to severe permeability loss (Bacchin et al. 2006; Tang et al. 2007).

In contrast, fouling is minimal when flux is below some threshold values, which is known as the critical flux behavior (Bacchin et al. 2006; Tang et al. 2007). The study demonstrated that the critical flux concept is also applicable to the osmotically driven FO process.

The strong dependence of the critical flux value on DS chemistry is a unique feature for FO fouling. In this regard, a concept of DS type-dependent critical concentration may also be of practical interest. When there was no Mg^{2+} ions in the original feed water, the critical concentration of $MgCl_2$ was below 0.5 M while that for NaCl was between 2 M and 4 M for algae fouling in the present case. Zou et al. (2011) suggests a strong interplay between membrane flux and DS chemistry on FO fouling in addition to the well documented interplay between flux and solution FS chemistry (Figure 11.7).

11.2.5 Membrane Material Type

Mi and Elimelech (2010) compared the cellulose triacetate (CA) and polyamide (PA) (or TFC: thin film composite) FO membranes in the RO mode to elucidate the effects of membrane materials on alginate fouling and cleaning behavior (Figure 11.8). Compared to CA membrane, PA membrane has a lower average adhesion force which is measured by atomic force microscopy but those 18% of its adhesion forces are distributed in the range of 1.2–2.0 mN/m while the maximum adhesion force of the CA membrane is 1.2 mN/m. Consequently, these small but sticky sites on the PA membrane may cause more severe fouling at the initial stage.

Gu et al (2013) also compared CA FO membrane and TFC PA FO membrane in terms of flux performance and foulant mass deposition during FO fouling

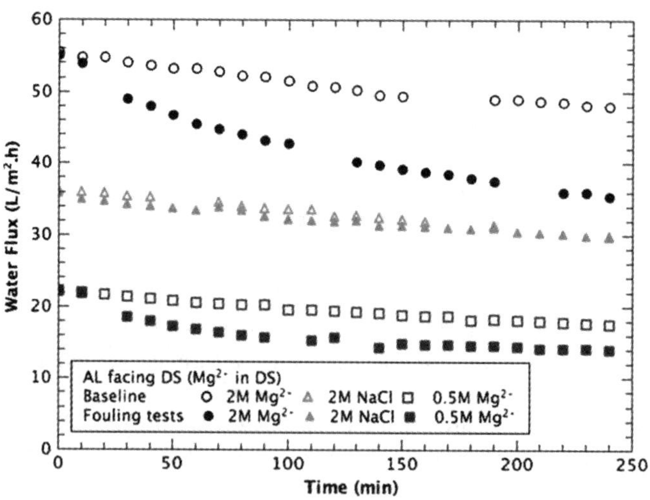

Figure 11.7. Effect of DS type (NaCl vs. $MgCl_2$) on FO fouling

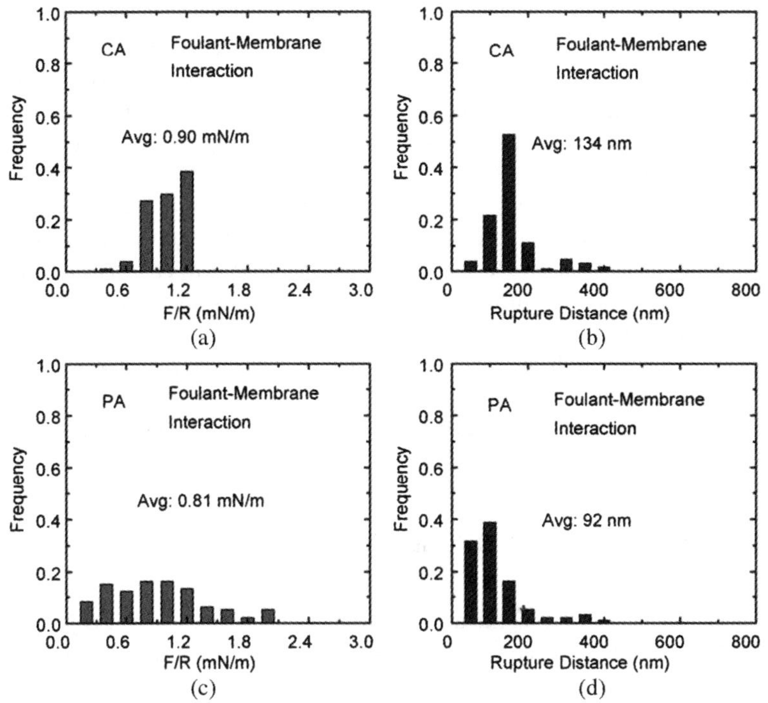

Figure 11.8. Adhesion force measurements of foulant-membrane interactions for the cellulose acetate (CA) and polyamide (PA) membranes

SOURCE: Adapted from Mi and Elimelech 2010; reproduced with permission from Elsevier

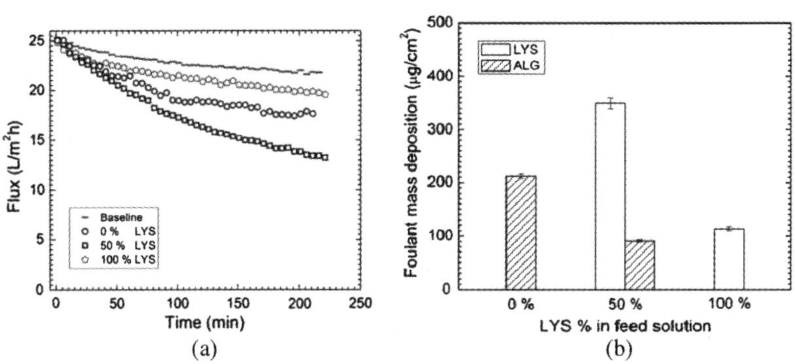

Figure 11.9. Effect of foulant composition in FS on FO fouling of TFC membrane

SOURCE: Adapted from Gu et al. 2013; reproduced with permission from Elsevier

(Figure 11.9). For FS containing either only sodium alginate (ALG) (0% Lysozyme (LYS)) or only LYS (100% LYS), it showed clearly that the TFC membrane was more likely to have severe fouling compared to the CA membrane. However, when two foulants were mixed, LYS and ALG (50% LYS in feed), both the CA and the

TFC membranes experienced severe flux decline and the final flux values towards the end of fouling tests were similar (~15 LMH for CA and 13 LMH for TFC). Moreover, the deposited foulant mass and composition for the two membranes were nearly identical. In both cases, the unfavorable interaction between the foulant species in the FS (i.e., the electrostatic attraction between LYS and ALG) caused foulants deposition on the membrane surface, thus severe flux decline. Results from the current study suggest that: (1) under mild FO fouling conditions, TFC FO membranes can have greater fouling tendency compared to CA FO membranes due to their greater surface roughness (the surface roughness plays a more dominant role over surface hydrophilicity in membrane fouling, noting that the TFC membrane was more hydrophilic than the CA membrane); (2) under severe FO fouling conditions, membrane surface properties play a less important role; and (3) although FO is believed to have superior fouling resistance in the AL-FS orientation, fouling can occur even at moderate flux levels, especially for TFC membranes or for unstable FS.

11.3 FOULING MITIGATION METHODS

In this section, various strategies that mitigate fouling formation in FO process are discussed in terms of spacer, cross flow velocity, air scouring, cross flow type and osmotic backwashing.

11.3.1 Spacer

Inserting spacer on the feed side is the easiest and effective way to reduce fouling on membrane surface because it generates shear force that promotes turbulence near the membrane surface on the feed side. In Figure 11.10, the feed channel spacer enhances mass transfer in the boundary layer near the membrane surface, thereby resulting in less accumulation of foulants on the membrane surface Boo et al. (2012). The flux decline is notably diminished during fouling runs with spacers. It clearly demonstrates the less accumulation of organic foulants on the membrane surface. Similarly, the amount of algal biomass deposition was reduced by the inclusion of feed spacer in the PRO mode (Zou et al. 2013).

Meanwhile, it was reported that feed spacer induces accumulation of particles near and under the spacer filament (Wang et al. 2010). Such particle accumulation was likely due to the low shear force region caused by spacer, physical blockage, due to the relatively large particle sized latex used in the study. However, approximately 10% flux decline was observed without feed spacer for 2 h fouling run, whereas no considerable flux loss with regard to the baseline was observed during fouling run with the feed spacer in the FO cell. The paper may hypothesize that the membrane area covered with the spacer has relatively marginal contribution to the overall permeate water so that local deposition of particles underneath the spacer has little effect on the flux stability.

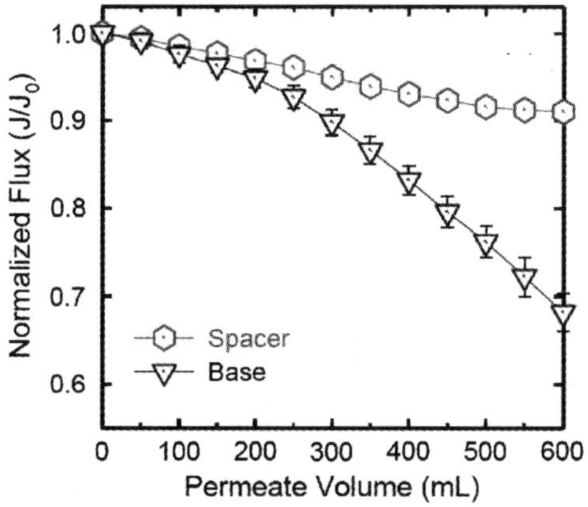

Figure 11.10. Impact of hydrodynamic control strategies on FO fouling mitigation using feed channel spacer

SOURCE: Adapted from Boo et al. 2012; reproduced with permission from Elsevier

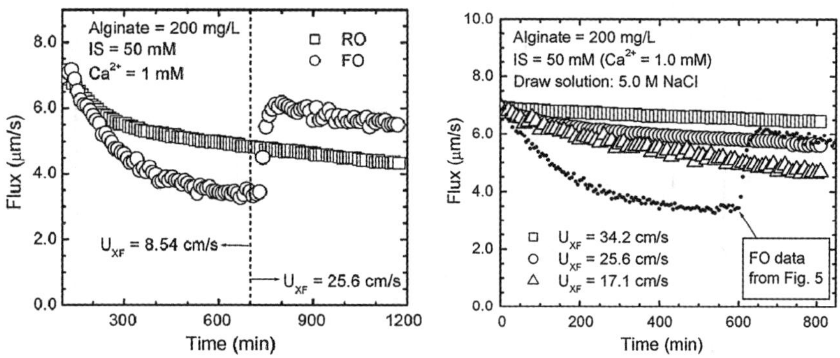

Figure 11.11. Impact of cross flow velocity on fouling behavior

SOURCE: Adapted from Lee et al. 2010b; reproduced with permission from Elsevier

11.3.2 Cross Flow Velocity

The main advantage of using FO is that it has lower irreversible fouling than pressure-driven membrane processes due to the lack of applied hydraulic pressure. Lee et al. (2010b) compared the fouling behaviors in regard to effect of cross flow velocity during FO and RO (Figure 11.11). Although flux decline is much severer in FO than RO, but the reduced flux is readily recovered when the cross flow velocity increases during the FO fouling run and that the following flux decline rate lessens after increasing the cross-flow velocity. It should be noted that this simple physical cleaning by increasing cross flow velocity is not effective in RO fouling. Another result shows that fouling propensity decreases as cross flow velocity increases. Table 11.2 shows the effect of fouling mitigation by increasing cross flow velocity.

Table 11.2. Effect of fouling mitigation by increasing cross flow velocity

Ref	*Foulants*	*pH*	*Cross flow velocity*	*Duration*	*Efficiency*
Li et al. (2012)	Seawater pre-filtered by a 10 mm capsule filter	8.2	4.2 cm/s to 16.7 cm/s	Fouling run	Flux decline reduced
Boo et al. (2012)	1 g/L of 139 nm silica	4	8.5 cm/s to 25.6 cm/s	Not mentioned	Fully reversible
		9			80%
Boo et al. (2012)	75 mg/L of alginate, SRNOM, and BSA	6.5–7.5 of feed, 7.5–8.5 of draw	10.7 cm/s to 32.1 cm/s	Fouling run	Flux decline reduced
Mi and Elimelech (2010)	200 mg/L alginate + 0.5 mM $CaCl_2$	5.8	8.5 cm/s to 21 cm/s	15 min	Fully reversible
Liu and Mi (2012)	Gypsum scaling	7.5	8.5 cm/s to 21 cm/s with air bubbles	20 min	Fully reversible
	Combined gypsum-alginate				75%

11.3.3 Air Scouring

Air scouring has been proved to be an effective way to reduce membrane fouling in membrane bioreactors. Mi and Elimelech (2010) investigated the fouling reversibility of FO, introducing air bubbles into the DI water stream for air scouring. As shown in Figure 11.12, using air bubble combined with increasing cross flow shortens cleaning time. The enhanced cleaning efficiency by air bubbles is primarily due to the shear force increased overall on the membrane and to more turbulence generated by the introduction of bubbles (Ndinisa et al. 2006). Yangali-Quintanilla et al. (2011) obtained the same result introducing air scouring that the FO membranes were hydraulically cleaned for 30 min, operating with real seawater and secondary wastewater effluent in FO. For the 28% flux decline, the reversible fouling was 98.8%. On the other hand, Liu and Mi (2012) found that flux recovery by air scouring for combined gypsum and alginate fouling was low. The synergistic fouling effect is mainly a result of the intensified gypsum scaling with alginate molecules. Alginate molecules act like nuclei in gypsum crystal growth, thus significantly increasing the size of gypsum crystal and accelerating crystallization kinetics.

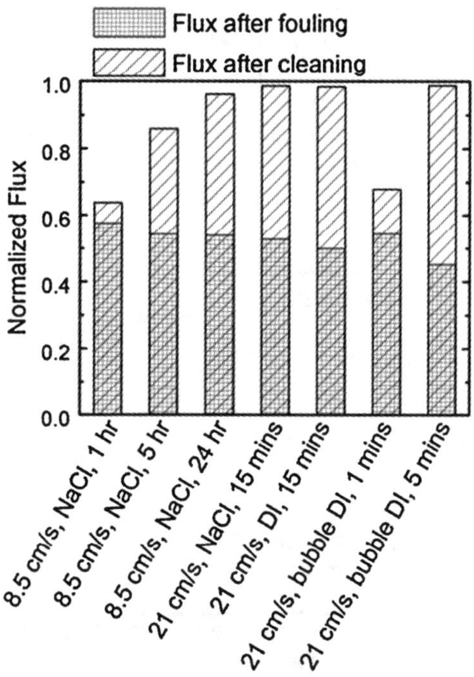

Figure 11.12. Cleaning efficiency under various crossflow, bubble and DS conditions

SOURCE: Adapted from Mi and Elimelech 2010; reproduced with permission from Elsevier

11.3.4 Cross Flow Type

Boo et al. (2012) investigated novel method for mitigating fouling. Pulsed flow has potential to remove foulants from the membrane surface. In Figure 11.13, the notable effect of pulsed flow on organic fouling mitigation is clearly demonstrated. The mechanisms by which pulsed flow controls membrane fouling include: (1) creating a highly turbulent flow regime, which leads to lower foulant concentration polarization; (2) increasing the overall shear force near to the membrane surface, which hinders foulant accumulation on the membrane surface; and (3) inducing membrane movement, which dislodges foulants from the membrane surface.

11.3.5 Osmotic Backwashing

Osmotic backwashing has been developed to remove foulants on membrane surface. When DS is replaced by DI water, the osmotic pressure gradients is in reverse and the permeate (i.e. backwash water) flows from draw (DI water) to feed sides. Therefore, the deposited foulants are possibly separated by this opposite flow and then washed away by the cross flow.

Kim et al. (2012) applied osmotic backwashing to clean FO membranes after being fouled by organic and inorganic foulants. Various backwashing conditions under two different membrane orientations were also studied for better cleaning efficiency. For equivalent permeation drag force in both membrane orientation, initial water flux was adjusted to be the same.

Although water flux decline showed difference depending on foulants type and membrane orientation, overall efficiency of osmotic backwashing is remarkably higher. The difference may be caused by various fouling parameters that are discussed above. For example, humic acid (HA) is effectively removed by shear force on FO mode. On the other hand, HA captures in porous structure of support

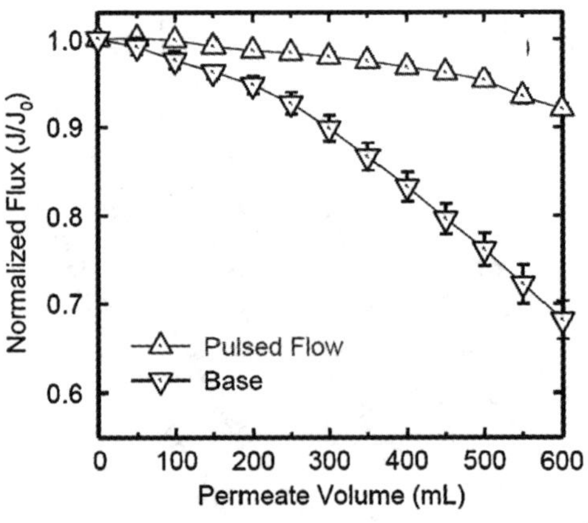

Figure 11.13. Impact of hydrodynamic control strategies on FO fouling mitigation: pulsed flow

SOURCE: Adapted from Boo et al. 2012; reproduced with permission from Elsevier layer in PRO mode and then is not removed effectively through the osmotic backwashing. It can be interpreted in a way that hydrophobicity of HA strengthens chemical interaction with FO membrane surface and possibly increased their attachments to the inside pores of PRO mode.

In the case of experiments with 20 nm particles, the flux decline on PRO mode is more severe than that on FO mode. The shear force generated by cross flow is not able to effectively flush small colloidal particles entrapped inside pores of support layer. However, a noticeable flux recovery on PRO mode was found after osmotic backwashing, although flux was not completely recovered. In contrast, fouling behaviors of 100 nm SiO_2 particles are almost the same during FO mode and PRO mode runs. These findings can be explained by A-CEOP at the presence of particle cake layer formed by large silica colloidal particles. Furthermore, large colloids are less removed from the membrane surface because of small diffusion coefficient of large particles. However, achieving a flux recovery is over 85% at the end of osmotic backwashing. The results are summarized in Table 11.3.

Table 11.3. Fouling characteristics in terms of foulant types, orientation and back washing

Foulants	*Characteristic*	*Orientation*	*Flux decline*	*Back washing efficiency*	*Cause*
Alginate	Hydrophilic	FO mode	Mild	High	Less chemically interacted
		PRO mode	Mild	High	Less chemically interacted
Humic acid	Hydrophobic	FO mode	Mild	Low	Chemically interacted
		PRO mode	Severe	Low	Chemically interacted Attachment to inside pore
Silica	20 nm	FO mode	Mild	Low	Shear force
		PRO mode	Severe	Medium	Entrapped inside pore
	100 nm	FO mode	Severe	Medium	A-CEOP Small diffusion coefficient
		PRO mode	Severe	Medium	A-CEOP Small diffusion coefficient

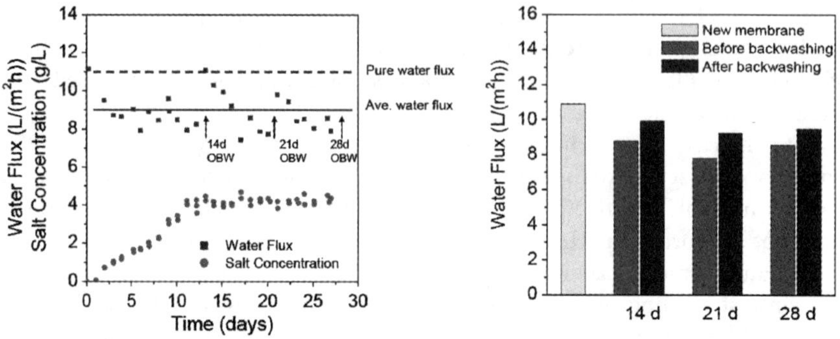

Figure 11.14. Comparison between water flux for the new membrane, and before/after osmotic backwashing at 14, 21, and 28 d
SOURCE: Adapted from Holloway et al. 2007; reproduced with permission from Elsevier

The osmotic backwashing is applied to an osmotic membrane bioreactor (OsMBR) system that utilizes a submerged FO membrane in the bioreactor. FO offers advantage of lower fouling propensity than pressure driven systems, and therefore requires less frequent backwashing compared to the MF or UF process.

Figure 11.14 shows the relationship between water flux and osmotic backwashing. The flux decreased by nearly 20% from the initial flux of the virgin membrane after 14 d. Osmotic backwashing was able to recover approximately 50% of this flux loss, leaving an overall 10% lower flux due to irreversible fouling. Similar trend was observed after 21 and 28 d. After an initial phase of irreversible fouling (occurring in the first 14 days), later phases of fouling were more reversible and the water flux remains a constant value over time. These results were in agreement with previous studies (Achilli et al. 2009; Lay et al. 2012).

On the other hand, Valladares Linares et al. (2013) produced conflicting results that osmotic backwash did not help recover flux. This research showed the cleaning efficiency of several methods to remove NOM-fouling on the active layer of FO membrane submerged in a municipal secondary wastewater effluent. This research suggests hypothesis regarding ineffective osmotic backwashing: (1) as replacing the feed and draw with 4% NaCl and DI respectively, the interaction of the salt in the cleaning solution with the foulants might have enhanced the pore blockage and thus the water flux kept decreasing; and (2) salt accumulation in the fouling layer reduces the osmotic pressure gradient and then eventually reduces the flux through the membrane.

11.4 ABBREVIATIONS/NOMENCLATURE

A-CEOP accelerated cake-enhanced osmotic pressure
ALG sodium alginate
BSA bovine serum albumin

CA	cellulose triacetate
CP	concentration polarization
DI	Deionsed water
DS	draw solution
FO	forward osmosis
FS	feed solution
HA	humic acid
LYS	Lysozyme
NOM	natural organic matter
PA	polyamide
PRO	pressure retarded osmosis
RO	reverse osmosis
RSF	reverse solute flux
TDS	total dissolved solid
TFC	thin film composite

References

Achilli, A., Cath, T. Y., Marchand, E. A., and Childress, A. E. (2009). "The forward osmosis membrane bioreactor: A low fouling alternative to MBR processes." *Desalination*, 239(1-3): 10–21.

Bacchin, P., Aimar, P., and Field, R. W. (2006). "Critical and sustainable fluxes: Theory, experiments and applications." *J. Membr. Sci.*, 281(1-2), 42–69.

Bacchin, P., Aimar, P., and Sanchez, V. (1995). "Model for colloidal fouling of membranes." *AIChE Journal*, 41(2), 368–376.

Baker, R. W. (2004). *Membrane Technology and Applications*, England: John Wiley & Sons Ltd.

Boo, C., Elimelech, M., and Hong, S. (2013). "Fouling control in a forward osmosis process integrating seawater desalination and wastewater reclamation." *Journal of Membrane Science*, 444(0): 148–156.

Boo, C., Lee, S., Elimelech, M., Meng, Z., and Hong, S. (2012). "Colloidal fouling in forward osmosis: Role of reverse salt diffusion." *J. Membr. Sci.*, 390-391, 277–284

Boussu, K., Belpaire, A., Volodin, A., Van Haesendonck, C., Van der Meeren, P., Vandecasteele, C., and Van der Bruggen, B. (2007). "Influence of membrane and colloid characteristics on fouling of nanofiltration membranes." *J. Membr. Sci.*, 289(1-2), 220–230.

Brant, J. A. and Childress, A. E. (2004). "Colloidal adhesion to hydrophilic membrane surfaces." *J. Membr. Sci.*, 241(2), 235–248.

Brant, J. A., Johnson, K. M., and Childress, A. E. (2006a). "Examining the electrochemical properties of a nanofiltration membrane with atomic force microscopy." *J. Membr. Sci.*, 276(1-2), 286–294.

Brant, J. A., Johnson, K. M., and Childress, A. E. (2006b). "Characterizing NF and RO membrane surface heterogeneity using chemical force microscopy." *Colloids and Surfaces A: Physicochemical and Engineering Aspects*, 280(1-3), 45–57.

Childress, A. E. and Deshmukh, S. S. (1998). "Effect of humic substances and anionic surfactants on the surface charge and performance of reverse osmosis membranes." *Desalination*, 118(1-3), 167–174.

Choi, H., Zhang, K., Dionysiou, D. D., Oerther, D. B., and Sorial, G. A. (2005). "Effect of permeate flux and tangential flow on membrane fouling for wastewater treatment." *Separation and Purification Technology*, 45(1), 68–78.

Elimelech, M., Xiaohua, Z., Childress, A. E., and Seungkwan, H. (1997). "Role of membrane surface morphology in colloidal fouling of cellulose acetate and composite aromatic polyamide reverse osmosis membranes." *J. Membr. Sci.*, 127(1), 101–109.

Faibish, R. S., Elimelech, M., and Cohen, Y. (1998). "Effect of interparticle electrostatic double layer interactions on permeate flux decline in crossflow membrane filtration of colloidal suspensions: An experimental investigation." *Journal of Colloid and Interface Science*, 204(1), 77–86.

Gu, Y., Wang, Y.-N., Wei, J., and Tang, C. Y. (2013). "Organic fouling of thin-film composite polyamide and cellulose triacetate forward osmosis membranes by oppositely charged macromolecules." *Water Research*, 47(5), 1867–1874.

Hoek, E. M. V., Bhattacharjee, S., and Elimelech, M. (2003). "Effect of Membrane Surface Roughness on Colloid-Membrane DLVO Interactions." *Langmuir*, 19(11), 4836–4847.

Holloway, R. W., Childress, A. E., Dennett, K. E., and Cath, T. Y. (2007). "Forward osmosis for concentration of anaerobic digester centrate." *Water Research*, 41, 4005–4014.

Hong, S. and Elimelech, M. (1997). "Chemical and Physical Aspects of Natural Organic Matter (NOM) Fouling of Nanofiltration Membranes." *J. Membr. Sci.*, 132, 159–181.

Jones, K. L. and O'Melia, C. R. (2000). "Protein and humic acid adsorption onto hydrophilic membrane surfaces: effects of pH and ionic strength." *J. Membr. Sci.*, 165(1), 31–46.

Jucker, C. and Clark, M. M. (1994). "Adsorption of aquatic humic substances on hydrophobic ultrafiltration membranes." *J. Membr. Sci.*, 97(0), 37–52.

Kim, C, Lee, S., and Hong, S. (2012). "Application of osmotic backwashing in forward osmosis: mechanisms and factors involved." *Desalination and Water Treatment*, 43, 314–322

Lay, W. C., Chong, T. H., Tang, C. Y., Fane, A. G., Zhang, J., and Liu, Y. (2010). "Fouling propensity of forward osmosis: investigation of the slower flux decline phenomenon." *Water Sci. Technol.*, 61(4), 927–936.

Lay, W. C. L., Zhang, J., Tang, C., Wang, R., Liu, Y., and Fane, A. G. (2012). "Factors affecting flux performance of forward osmosis systems." *Journal of Membrane Science*, 394–395(0): 151–168.

Lee, E., Lee, S., and Hong, S. (2010a). "A new approach to the characterization of reverse osmosis membrane by dynamic hysteresis." *Desalination and Water Treatment*, 18, 257–263.

Lee, S. and Elimelech, M. (2006). "Relating Organic Fouling of Reverse Osmosis Membranes to Intermolecular Adhesion Forces." *Environmental Science & Technology*, 40(3), 980–987.

Lee, S., Ang, W. S., and Elimelech, M. (2006). "Fouling of reverse osmosis membranes by hydrophilic organic matter: implications for water reuse." *Desalination*, 187(1-3), 313–321.

Lee, S., Boo, C., Elimelech, M., and Hong, S. (2010b). "Comparison of fouling behavior in forward osmosis (FO) and reverse osmosis (RO)." *J. Membr. Sci.*, 365(1-2), 34–39.

Lee, S., Cho, J., and Elimelech, M. (2005). "Combined influence of natural organic matter (NOM) and colloidal particles on nanofiltration membrane fouling." *J. Membr. Sci.*, 262(1-2), 27–41.

Lee, S., Lee, E., Elimelech, M., and Hong, S. (2011). "Membrane characterization by dynamic hysteresis: Measurements, mechanisms, and implications for membrane fouling." *J. Membr. Sci.*, 366(1-2), 17–24.

Liu, Y. and Mi, B. (2012). "Combined fouling of forward osmosis membranes: Synergistic foulant interaction and direct observation of fouling layer formation." *Journal of Membrane Science*, 407–408(0), 136–144.

Mi, B. and Elimelech, M. (2008). "Chemical and physical aspects of organic fouling of forward osmosis membranes." *Journal of Membrane Science*, 320(2008), 292–302.

Mi, B. and Elimelech, M. (2010). "Organic fouling of forward osmosis membranes: Fouling reversibility and cleaning without chemical reagents." *J. Membr. Sci.*, 348(1-2), 337–345.

Ndinisa, N., Fane, A., Wiley, D., Fletcher, D. (2006). "Fouling control in a submerged flat sheet membrane system. Part II. Two-phase flow characterization and CFD simulations." *Sep. Sci. Technol.*, 41, 1411–1445.

Parida, V. and Ng, H. Y. (2013). "Forward osmosis organic fouling: Effects of organic loading, calcium and membrane orientation." *Desalination*, 312(0), 88–98.

Phillip, W. A., Yong, J. S., and Elimelech, M. (2010). "Reverse draw solute permeation in forward osmosis: modeling and experiments." *Environmental Science & Technology*, 44(13), 5170–5176.

Sakuma, S., Ohshima, H., and Kondo, T. (1989). "Charge distribution in poly(N, N-l-lysinediylterephthaloyl) microcapsule membranes." *Journal of Colloid and Interface Science*, 133(1), 253–256.

Seidel, A. and Elimelech, M. (2002). "Coupling between chemical and physical interactions in natural organic matter (NOM) fouling of nanofiltration membranes: implications for fouling control." *J. Membr. Sci.*, 203(1-2), 245–255.

Shim, Y., Lee, H.-J., Lee, S., Moon, S.-H., and Cho, J. (2002). "Effects of natural organic matter and ionic species on membrane surface charge." *Environmental Science & Technology*, 36(17), 3864–3871.

Singh, G. and Song, L. (2005). "Quantifying the effect of ionic strength on colloidal fouling potential in membrane filtration." *Journal of Colloid and Interface Science*, 284(2), 630–638.

Singh, G. and Song, L. (2007). "Experimental correlations of pH and ionic strength effects on the colloidal fouling potential of silica nanoparticles in crossflow ultrafiltration." *J. Membr. Sci.*, 303(1-2), 112–118.

Tang, C. Y. and Leckie, J. O. (2007). "Membrane independent limiting flux for ro and nf membranes fouled by humic acid." *Environmental Science & Technology*, 41(13), 4767–4773.

Tang, C. Y., She, Q., Lay, W. C. L., Wang, R., and Fane, A. G. (2010). "Coupled effects of internal concentration polarization and fouling on flux behavior of forward osmosis membranes during humic acid filtration." *J. Membr. Sci.*, 354(1-2), 123–133.

Valladares Linares R., Li, Z., Abu-Ghdaib, M., Wei, C. H., Amy, G., and Vrouwenvelder, J. S. (2013). "Water harvesting from municipal wastewater via osmotic gradient: An evaluation of process performance." *J. Membr. Sci.*, 447, 50–56.

Vrijenhoek, E. M., Hong, S., and Elimelech, M. (2001). "Influence of membrane surface properties on initial rate of colloidal fouling of reverse osmosis and nanofiltration membranes." *J. Membr. Sci.*, 188(1), 115–128.

Wang, Y. -N. and Tang, C. Y. (2011). "Protein fouling of nanofiltration, reverse osmosis, and ultrafiltration membranes-The role of hydrodynamic conditions, solution chemistry, and membrane properties." *J. Membr. Sci.*, 376(1-2), 275–282.

Wang, Y., Wicaksana, F., Tang, C. Y., and Fane, A. G. (2010). "Direct Microscopic Observation of Forward Osmosis Membrane Fouling." *Environmental Science & Technology*, 44(18), 7102–7109

Wu, D., Howell, J. A., and Field, R. W. (1999). "Critical flux measurement for model colloids." *J. Membr. Sci.*, 152(1), 89–98.

Yang, J., Lee, S., Yu, Y., Kuk, J., Hong, S., Lee, S., and Min, K. (2010). "Role of foulant-membrane interactions in organic fouling of RO membranes with respect to membrane properties." *Separation Science and Technology*, 45(7), 948–955.

Yangali-Quintanilla, V., Li, Z., Valladares, R., Li, Q., and Amy, G. (2011). "Indirect desalination of Red Sea water with forward osmosis and low pressure reverse osmosis for water reuse." *Desalination* 280(1-3), 160–166.

Yu, Y., Lee, S., and Hong, S. (2010). "Effect of solution chemistry on organic fouling of reverse osmosis membranes in seawater desalination." *J. Membr. Sci.*, 351(1-2), 205–213.

Zhu, X. and Elimelech, M. (1995). "Fouling of reverse osmosis membranes by aluminum oxide colloids." *J. Environmental Engineering*, 121(12), 884–892.

Zou, S., Gu, Y., Xiao, D., and Tang, C. Y. (2011). "The role of physical and chemical parameters on forward osmosis membrane fouling during algae separation." *Journal of Membrane Science*, 366(1-2), 356–362.

Zou, S., Wang, Y.-N., Wicaksana, F., Aung, T., Wong, P. C. Y., Fane, A. G., and Tang, C. Y. (2013). "Direct microscopic observation of forward osmosis membrane fouling by microalgae: Critical flux and the role of operational conditions." *Journal of Membrane Science*, 436(0), 174–185.

CHAPTER 12

Biofouling in Osmotic Membrane Bioreactor

In S. Kim
Lan Hee Kim
Sung-Jo Kim
Kyoung-Yeol Kim

In this chapter, the most important aspects of biofouling in the FO membrane process are reviewed and discussed. Especially, the biofouling phenomenon shows quite different characteristics particularly in reverse osmosis (RO) and forward osmosis (FO) membrane. Several biofouling studies accomplished with the osmotic membrane bioreactor (OMBR), forward osmosis-microbial fuel cell (FO-MFC) processes and lab-scale FO filtration units are introduced and reviewed. In addition, biofouling by transparent exopolymer particles (TEP), which is an emerging algae organic matter, is also discussed. Results of these studies would throw light on advanced and innovative biofouling research in the future.

12.1 INTRODUCTION

12.1.1 Membrane Biofouling

Membrane-based water treatment systems including desalination and wastewater treatment have been considered a promising technological field for water supply in the face of predicted future shortages. Membrane fouling continues to present a serious technical challenge though, as it significantly reduces membrane performance efficiency (Kim et al. 2011). The membrane can be considered the heart of a desalination plant where the cost of each individual membrane unit is about 20–25% of the total capital cost. Consequently, familiarity with all factors involved in reduction of membrane performance and longevity is crucial, and in particular membrane fouling (Al-Amoudi and Lovitt 2007). Fouling types can be categorized into inorganic fouling, organic fouling, and biofouling (Sohrabi et al. 2011). Inorganic fouling is the deposition of inorganic scales such as $BaSO_4$, $CaSO_4$, and $CaCO_3$ on a membrane. Organic fouling is caused by natural organic material

(NOM) in the process stream. Biofouling is caused by the undesired development of microbial layers on surfaces (Flemming 2002; Al-Amoudi and Lovitt 2007).

Biofouling is recognized as a major contributing factor to more than 45% of all membrane fouling (Nguyen et al. 2012). Because fouling can be controlled by eliminating foulants from the liquid phase, in the case of biofouling, even if 99.9–99.99% of all bacteria are eliminated by pretreatment, a few will enter the system to become protected, adhere to surfaces and multiply at the expense of biodegradable substances (Flemming 2002). Biofouling can have one or more of several adverse effects on membrane systems, as listed below (Nguyen et al. 2012):

- Membrane flux decline due to the formation of a low permeability biofilm on the membrane surface;
- Increased differential pressure and feed pressure being needed to maintain the same production rate due to biofilm resistance;
- Membrane biodegradation caused by acidic by-products which are concentrated at the membrane surface. For example, cellulose acetate membrane has been found to be more susceptible to being biodegraded;
- Increased salt passage through membrane and reduced quality of the product water due to the accumulation of dissolved ions in the biofilm at the membrane surface thus increasing the degree of concentration polarization; and
- Increased energy consumption due to higher pressure being required to overcome the biofilm resistance and the flux decline.

12.1.2 Biofouling Formation

The process by which bacterial fouling is formed on the surface can be divided into three main phases: (1) transport of the organisms to the surface; (2) attachment to the substratum; and (3) growth of bacteria. At first, reversible adsorption is an instantaneous attraction that holds bacteria near the surface. Physical forces including Brownian motion, electrostatic interaction, gravity, and van-der-Waals forces essentially govern the process of bacterial adsorption. The second irreversible attachment phase is made by bacteria that produce extracellular bridging polymer substances (Abarzua and Jakubowski 1995). Factors affecting the adhesion of microorganisms to membrane surfaces are summarized in Table 12.1 (Nguyen et al. 2012).

Biofouling is mainly the result of bacteria and biopolymers. Groups of biopolymers are mostly termed extracellular polymeric substances (EPS) or soluble microbial products (SMP). EPS are the first cell enveloped structures encountering a solid surface in primary contact. Thus, they are of great interest for the understanding of the adhesion process. EPS may play an important role in adhesion. EPS can determine the chemistry of the cell surface and act as a "polymer bridge" between a cell and the surface. It can contribute to the adhesion and aggregation of bacterial cells and enzyme activity. In addition, it can function as the nutrient source.

Table 12.1. Factors affecting microorganism adhesion to membrane surfaces

Microorganism	*Surface*	*Feed water*
Species	Chemical composition	Temperature & pH
Composition of mixed population	Surface charge	Dissolved organic matter
Population density	Surface tension	Dissolved inorganics
Growth phase	Hydrophobicity	Suspended matter
Nutrient status	Conditioning film	Viscosity
Hydrophobicity	Roughness	Shear forces
Charges	Porosity	Boundary layer
Physiological responses		Flux

SOURCE: Adapted from Nguyen et al. 2012

SMPs are defined as soluble cellular components released during substrate metabolism and cell lysis. In biological treatment processes, effluent contains a portion of organic matter known as effluent organic matter (EfOM) which mainly comes from the SMP. SMP are closely related with the EPS that is bound to bacterial cells. EPS are of biological origin, participate in the formation of microbial aggregates, and consist of insoluble materials (sheaths, capsular polymers, condensed gel, loosely bound polymers, and attached organic material). In contrast SMP are considered soluble EPS (soluble macro-molecules, colloids, and slimes) (Jang et al. 2007). Recently, the terms biopolymers or biopolymeric clusters (BPC) have also come into use.

12.2 BIOFOULING ANALYSIS

12.2.1 Biomass Quantification

12.2.1.1 Measurement of Bacteria

Cell numbers can be determined by total direct cell counts (TDC values, cells/cm^2) and heterotrophic plate counts (HPC values, CFU/cm^2). Active bacteria were determined by measuring the adenosine triphosphate (ATP) concentration (Vrouwenvelder et al. 2008). TDC value is determined via fluorescence microscopy using dyes such as SYTO9, acridine orange, and 4'-6-diamidino-2-phenylindole (DAPI). However, a problem is encountered in that these fluorochromes stain all bacterial cells in the samples. To differentiate live and dead cells, cell viability can be measured using a BacLight LIVE/DEAD staining kit (Molecular Probes Inc.). A bacterial viability kit employs two nucleic acid stains: green fluorescent SYTO9 stain; and red fluorescent propidium iodide (PI) stain. SYTO9 stain labels both live and dead bacteria. Alternatively, PI penetrates only

Table 12.2. Quantification methods of bacterial activity and cell numbers

Parameters	*Features*
ATP	• Provides a measure for the amount of active biomass within minutes • Utilizes an enzymatic reaction using the luciferine-luciferase assay • Can be applied on site • Testing volume range 0.1 ~ 10 mL based on the predicted number of bacteria • Bacterial cells should be lysed to release ATP • Improper separation of bacterial ATP from the nonbacterial ATP
TDC	• Epifluorescence microscopic cell count (cells/cm^2) using acridine orange, DAPI, SYTO 9 or other nucleic acid-bound fluorophores • The total amount of fluorescent cells is measurable • Not currently possible: discrimination between active and inactive cells • Microscopic observations give information about the variety and appearance of microbes present. • Large standard deviation in the presence of large clusters of thousands of cells
Cell viability	• Differentiate between live cells and dead cells • Microscopic observation after staining with SYTO9 and PI
HPC	• Cultivation on R2A medium (enumerated as colony-forming units) • Only a small proportion of the total microbial community • A minor part of TDC present (<1%)

bacteria with damaged membranes, reducing SYTO9 fluorescence when both dyes are present. Thus, live bacteria with intact membranes fluoresce green, while dead bacteria with damaged membranes fluoresce red. HPC can be obtained by incubating the sample on R2A plates at 20 or 28°C for 5–7 days to enumerate more of the heterotrophic bacteria found in the biomass suspension samples (Table 12.2) (Lee and Kim 2011).

12.2.1.2 Biopolymers

An analysis of biopolymers is needed to understand the biofouling phenomenon. Many EPS extraction methods have been developed and summarized in physical and chemical methods; a combined technique which has a higher output capacity of EPS and a lower level of damage is more effective than a single method (Table 12.3)).

Table 12.3. Comparison of extraction methods of EPS

Methods	*Operation methods and features*
Physical method:	
Ultrasonication	Samples are ultrasonicated then centrifuged. The amount of cell disruption is comparatively small and the extracellular polymer yield is relatively high; no chemical reagent contamination.
Heating	Samples are in phosphate buffer conical flask. Heating at 80 °C and then centrifuge to remove cells. Cells are easy to damage; the income of proteins and polysaccharides is not high.
Steaming	Samples are steamed in an autoclave and then centrifuged while still hot. During centrifugation the temperature is reduced in order to reduce the disruptive effects on the cells of boiling or autoclaving.
Centrifugation	Samples are ultrasonicated: accelerated particle settling velocity; under mild conditions; extraction efficiency is low; damage is small.
Mixing	Samples are ultrasonicated; deionized water is added to the original sample volume; stirred at high-speed supernatant; centrifugation. Extraction efficiency is low; Cell disruption level is low.
Chemical methods:	
NaOH	Centrifuge and then discard the supernatant; add NaOH to the original sample volume; centrifuge and discard the pellet. Extraction efficiency and the level of cell disruption are high. Operation period is long.
H_2SO_4	H_2SO_4 is added to samples; after centrifugation twice, the supernatant is filtered; the supernatant is taken after dialysis. EPS is highly soluble; cell disruption level is high.
Formaldehyde	Sodium chloride solution of formaldehyde is added to samples; supernatant extracted after centrifugation. It is propitious to fix cells, but it destroys the protein structure.
Formaldehyde-NaOH	Add formaldehyde and mix, and then add NaOH and mix. Centrifuge twice and filter supernatant; obtain extract after dialyzing. High extraction efficiency.
EDTA	Add EDTA into sample, and keep 3 h. After centrifugation, filter the supernatant. High extraction efficiency and high rate of cell damage.
PBS	Add quantitative PBS into the sample and mix 2–4 h in a water bath. Low extraction efficiency.

(Continued)

Table 12-3. Comparison of extraction methods of EPS (Continued)

Methods	*Operation methods and features*
CER	Add extraction buffer and ion-exchange resin into sample and mix: high extraction efficiency; mild operating conditions; complex operation.
Ethanol	Add ethanol into sample; keep still 1–16 m in an enclosed place: centrifuge 10 min.; low extraction efficiency; long cycle.
Tris/HCl	Add Tris/HCl buffer; keep still 1h in thermostat water bath of 80°C. Low extraction efficiency.
Glutaricdialdehyde	Add glutaric dialdehyde into a sample; mix over night at 4°C. Low extraction efficiency; long cycle.
Electric field chemical agent	The extraction efficiency of electric field-chemical agent existed was better than one factor existed. High extraction efficiency and low cell damage rate.

Source: Adapted from Yu et al. 2012; reproduced with permission from Elsevier.

EPS and SMP consist of polysaccharides (PS), proteins, lipids, and nucleic acids, which can originate from cell lysis, microbial metabolites or un-metabolized wastewater components. Usually, PS and proteins are assumed to be the major fractions that contribute to fouling. Therefore, the determination of EPS or SMP concentration relies almost exclusively on PS and protein measurements (Drews 2010). Li et al. (2011) analyzed biopolymers in fouled FO membranes. EPS was observed using an optical microscope with bright illumination and UV epifluorescence after membrane staining with Alcian Blue 8GX and DAPI solution (Li et al. 2011).

12.2.2 Interaction of Biopolymer and Membrane Surface

Various methods have been adopted for studying protein-surface interactions, including X-ray based techniques, vibrational and electronic spectroscopy, atomic force microscopy, surface plasmon resonance, dual polarization waveguide interferometry, and quartz crystal microbalance measurements (Hall et al. 2013).

Quartz crystal microbalance with dissipation (QCM-D) has emerged as one of the most convenient and highly sensitive instruments that can measure the structural properties of adsorbed protein layers on the membrane and detect the mass of proteins adsorbed as small as 5 ng/cm^2 (Liu and Kim 2009). When the quartz crystal is immersed in gas or an aqueous solution, it is excited to oscillate at a fundamental resonant frequency (5 MHz). Once the substrate is adsorbed on the crystal surface, the resonant frequency would be decreased with the increase in the amount of mass adsorbed on the crystal. The QCM-D is used in dairy protein adsorption onto the polyethersulfone (PES) polymeric membrane surface, and the results are interpreted with the Kevin-Voigt viscoelastic model to obtain the thickness, viscosity and elasticity of the protein layer.

12.2.3 Microbial Composition

There is a high demand for information on the most efficient and effective methods leading to the prevention of biofouling. For a microbial analysis of biofoulants, processes such as real-time polymerase chain reaction (PCR), denaturing gradient gel electrophoresis (DGGE), restriction fragment length polymorphism (RFLP), and fluorescence in situ hybridization (FISH) can be utilized (Table 12.4). Real-time PCR can be used to quantify specific bacteria genes of interest, and DGGE and RFLP are methods used to identify and quantify mixed bacteria. However, these protocols include time-consuming and labor intensive steps (e.g., DNA extraction, PCR amplification, restriction enzyme digestion and DNA fragment separation by gel electrophoresis); restrictions limiting their application are yet to be overcome (Lee and Kim 2011).

12.2.4 Characterization of Membrane Surface

Many studies have been reported to identify biofilm structure. They include atomic force microscopy (Tansel et al. 2006), optical coherence tomography (OCT) (Haisch and Niessner 2007), scanning electron microscopy (SEM) (Schneider et al. 2005), magnetic resonance imaging (MRI) (Manz et al. 2005), attenuated total internal reflection spectroscopy (ATR-Spectroscopy), fourier transform infrared (FTIR) spectroscopy (Lee et al. 2009), confocal laser scanning microscopy (CLSM) (Chae et al. 2008) and transmission electron microscopy (TEM) (Lawrence et al. 2003) (Table 12.5).

To observe the morphology of the foulants on the membrane, CLSM and SEM were applied. In addition, a FTIR spectrophotometer was applied used for the characterization of the composition of biopolymers in the cake layer (Zhang et al. 2012b). The spectra of the foulants can be easily distinguished from the spectra of the membrane material. ATR-FTIR spectroscopy also can indicate the presence of inorganic foulants as well as FTIR spectra describing various functional groups such as amides, carbo acids and nitro compounds that may explain the accumulation of protein and carbohydrate onto the membrane surface. A peak at 2931 cm^{-1} (carbon acid, 3100 ~ 2900 cm^{-1} or sulfur, 3000 ~ 2800 cm^{-1}), a peak at 2855 cm^{-1} (alkanes, 2882 ~ 2862 cm^{-1}), a peak at 1644 cm^{-1} (aldehyde, 1670 ~ 1645 cm^{-1}), a peak at 1541 cm^{-1} (amide, 1570 ~ 1515 cm^{-1}), a peak at 1251 cm^{-1} (nitro, 1300 ~ 1250 cm^{-1}) and a peak at 1045 cm^{-1} (alcohols, 1075 ~ 1000 cm^{-1}) were observed. The FTIR spectra of two fouled membranes were very similar; it was therefore speculated that the two membranes had similar organic characteristics (Lee and Kim 2011). Identification of specific species deposited onto membrane surfaces can also be carried out using matrix assisted laser desorption ionization mass spectroscopy (MALDI-MS) (Goosen et al. 2005).

From the CLSM image in Figure 12.1(a), biofilm thickness is shown to be ~18 μm. The FTIR spectrum (Figure 12.1(b, c, d)) shows two sharp adsorption peaks at 1635 cm^{-1} and 1543 cm^{-1} that are characteristics of protein secondary

Table 12.4. Microbiological methods for biofoulants analysis

*Methods/Ref**	*Strengths*	*Limitations*
DGGE/[1]	• The total community composition among samples can be directly compared at a glance, and the individual components can be identified by excision and sequencing of bands. • High reliability, good repetition and easy operation	• Preparation of denaturing gels with chemical denaturing agents is not always reproducible.
T-RFLP (Terminal restriction fragment length polymorphism)/[2]	• Direct reference to the sequence database can be made. • Nucleic acid sequencing technology has considerable greater resolution than the electrophoretic systems than DGGE. • The output is digital.	• Time consuming, labor intensive, often requires use of radioactive materials. (Alternatively, detection sensitivity of EtBr is low.) • Complicated by incomplete digestion with the restriction enzymes, which leads to underestimation of the percentage of heteroplasmy. • Nonspecific PCR products many interfere with the analysis.
FISH	• It allows for the in situ localization and the study of spatial organization of cells as they occur in their natural habitat or biofilm. • No needs for cell cultivations before analysis. • Cells need not be alive.	• The process of preparing probes is complex due to the fact that it is necessary to tailor the probes to identify specific sequences of DNA. • It is difficult to count total numbers in probe-stained clusters of cells.

Real-time PCR	• Quantification of interest gene is possible. • 96 well or 384 well plate formats available in a shorter time	• Multiple-targeting. Currently 5-flex (5 colors) is possible. However, developments continue.
Microarray/[3]	• Parallel, high-throughput detection and quantification of many nucleic acids.	• Moderate or high price for developments
Enzyme-Linked Immunosorbent Assay (ELISA)	• Simple and flexible • Well published methodology/techniques • Inexpensive • Amenable to automation • Widely accepted format	• Only single-plex • Low sensitivity
Luminex (bead-based suspension array)	• Easy to learn • Multiplexing flexibility • Inexpensive • High sensitivity	• Automation
Pyrosequencing/[4]	• Real-time DNA sequencing technique generating short reads rapidly and inexpensively. • 1,000,000 high-qualities read per run and read lengths of 400 bases. Extended read lengths of 1000 bases will be approached soon. (genome sequencer FLX system, www.454.com)	• Limitation in the read length • Limitation in complete incorporation of nucleotides in homopolymeric regions • Non-linear light response to the sequential incorporation of more than 5-6 identical nucleotides.

*References: [1] = Balogh et al. (2004); [2] = Marsh (1999); [3] = Bodrossy and Sessitsch (2004); [4] = Ronaghi and Elahi (2002).
SOURCE: Modified from Lee and Kim 2011; reproduced with permission from Elsevier

Table 12.5. Methods for evaluation of membrane structure

Instruments	*Advantages/disadvantages*
CLSM/[1]*	• Not only the presence and the viability of the biofilm consortium but also biofilm/substrate interactions can be investigated. • Visualize the 3D structure of biofilms • Optical sectioning of thick biological structures with no out-of-focus disturbance • Fixation step is needed to enhance the permeability of the cell for the gene probes. • Thereby the 3D information is partly lost due to shrinking of the biofilm. • Without staining, imaging is not possible so is less suitable for online monitoring.
AFM/[2]	• The tip of the probe gives information on the surface contours of the sample. • Direct visualization of intact hydrated specimens at high magnification. • It cannot elucidate internal biofilm structures.
OCT/[3]	• Optical signal acquisition and processing method allowing extremely high-quality, micrometer-resolution, 3D image from within optical scattering media (or reflecting structure) • Non-invasive • It does not have the perspective to observe and distinguish single cells as it is routinely done by CLSM.
SEM/[4]	• The surface topology of the biofilms can be examined at a high magnification. • Shrinking problem due to dehydration and fixation of samples
TEM/[5]	• Cross-section gives useful information about the spatial relationships of microorganism within the biofilm matrix. • Sample preparations result in the inclusion of artifacts.
MRI/[6]	• Sequences of NMR measurements in the presence of magnetic fields with linear gradient are carried out and a certain space resolution is derived. Finally the generation of 2D images of a sample is possible.
ATR-Spectroscopy	• Limited to a thin layer

Table 12-5. Methods for evaluation of membrane structure (Continued)

Instruments	*Advantages/disadvantages*
FTIR Spectroscopy/[7]	• Scanning of the IR absorption in a sample at different wavelengths yields a characteristic "fingerprint" of certain wavelengths in which the radiation has been absorbed stronger than in other ones. • Sufficient to monitor the formation of biofilms thicker than 0.05 mm. • Reported work was limited to pure cultures of *Pseudomonas fluorescens*

*References: [1] = Chae et al. (2008); [2] = Tansel et al. (2006); [3] = Haisch and Niessner (2007); [4] = Schneider et al. (2005); [5] = Lawrence et al. (2003); [6] = Manz et al. (2005); [7] = Lee et al. (2009).

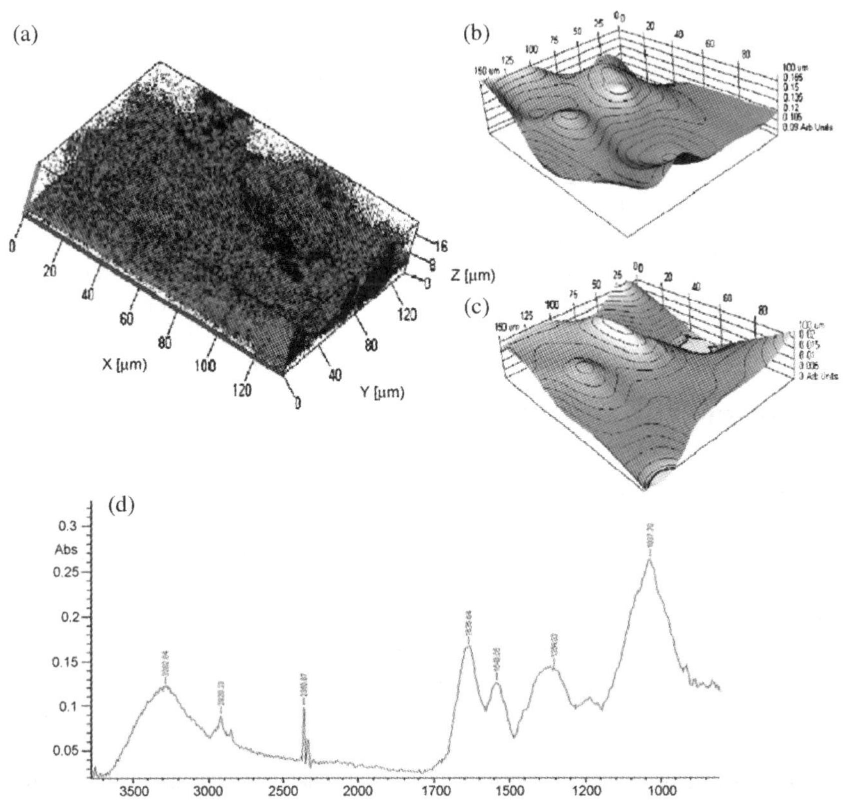

Figure 12.1. (a) CLSM image showing the surface of a fouled membrane. (b, c) 3D FTIR spectra distribution map of protein-like EPS and polysaccharide-like EPS. (d) FTIR spectra of biofilm surface

SOURCE: Adapted from Zhang et al. 2012b; reproduced with permission from Elsevier

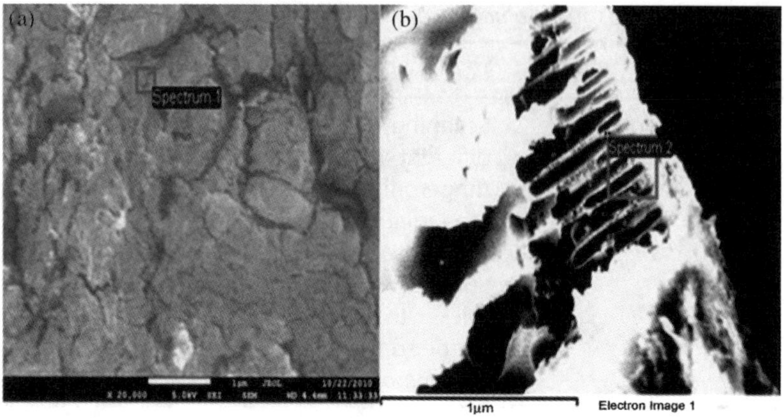

Figure 12.2. SEM image of fouling surface and cross section of FO membrane: (a) cake layer of the fouled FO membrane; and (b) cross section image and EDX scan region of fouled FO membrane
SOURCE: Adapted from Zhang et al. 2012b; reproduced with permission from Elsevier

structures. The broad peak at $\sim 1037\,cm^{-1}$ is assigned to polysaccharides or polysaccharide-like substances.

SEM results (Figure 12.2(a)) confirmed the presence of a surface fouling layer on the FO support. The cross-sectional SEM image (Figure 12.2(b)) showed mild internal pore clogging in this FO membrane, which implies that, the fouling occurred mainly on the membrane surface. In this FO process, the presence of combined structures of bacteria and EPS covered the membrane substrate surface and plugged the porous membrane support layer (Zhang et al. 2012b).

12.3 MEMBRANE BIOFOULING IN FO

Membrane biofouling is regarded as a severe problem in the field of FO when compared to other types of fouling. It is reported that fouling caused by the accumulation of colloidal, inorganic and organic foulants in FO could be easily removed by the application of hydraulic shear force, e.g. an increase of cross-flow velocity (CFV) (Lee et al. 2010; Mi and Elimelech 2010). Mi et al. (2010) investigate the fouled FO membrane surface and cross-section images using SEM. Thick and loosely combined fouling layers were formed on the FO membrane surface compared to that of RO due to the absence of applied pressure on FO.

The propensity for FO biofouling was systematically evaluated recently using lab-scale membrane filtration systems. In this study, flux decline occurred as a result of biofouling was never improved upon from an 8-fold increase of CFV (Figure 12.3(a)). In addition, chemical cleaning such as the use of chlorine, which is one of commonly used anti-fouling agents, is required in order to restore the

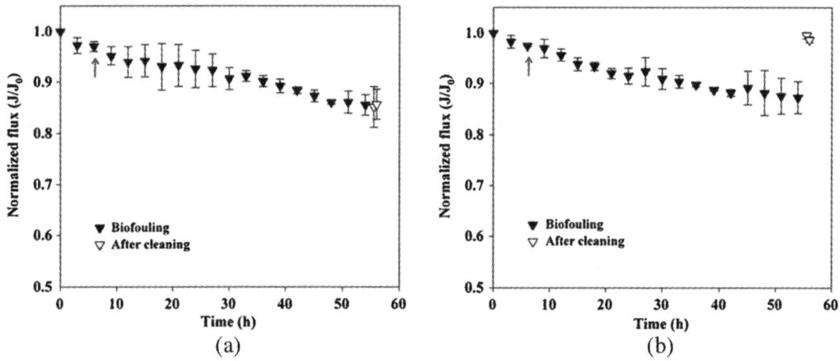

Figure 12.3. Normalized flux behavior of FO under biofouling conditions: (a) Physical cleaning for 1 h with hydraulic stress and (b) chemical cleaning for 1 h with chlorine (biofouling test–feed solution: 10 mM NaCl, 1 mM $CaCl_2$ and TSB 0.1%, draw solution: 4 M NaCl, initial flux: 22.5 ± 1 L/m^{2h}, initial Pseudomonas aeruginosa *PAO1 GFP concentration in feed: about 10^7CFU/mL, 25°C, CFV: 4 cm/s; cleaning test–CFV: 33 cm/s, (a) feed reservoir: DI water, draw reservoir: 4 M NaCl, (b) feed reservoir: chlorine 100* mg/L *at pH 7, draw reservoir: 4 M NaCl)*
SOURCE: Adapted from Yoon et al. 2013; reproduced with permission from Elsevier

permeate flux (Figure 12.3(b)) (Yoon et al. 2013). In this way, biofouling in FO could affect the efficiency by causing a decrease in permeate flux, and necessitating an increase in frequency of chemical cleaning of the membrane, thereby eventually shortening the membrane lifetime. Systematic and mechanical research of membrane biofouling is therefore an indispensable area for the development and application of FO.

Despite the significance of biofouling to FO, at this date there appear to be but few researches in this area, likely due to the difficulty of an accepted systematic approach to membrane biofouling. Major researches of biofouling in FO are performed in terms of OMBR, FO-MFC, and biofouling by microbial and algae excreta (e.g. TEP). Interactions between biofouling and FO membranes and the effects of increased salinity within a bioreactor on biofouling (due to high retention of FO membranes and reverse solute flux) are mainly investigated. Detailed explanations of each research are listed as follows.

12.3.1 Osmotic Membrane Bioreactor (OMBR)

A membrane bioreactor (MBR) involves the combination process of an activated sludge process and physical membrane separation process. Before the introduction of MBR, a conventionally activated sludge process was widely used, involving wastewater treatment technology in which aerobic bacteria ingest organic substances such as sugars and protein and decompose them to produce water and carbon dioxide. However, there are a range of disadvantages within this activated sludge process including instability, settling problems, large space requirements,

Table 12.6. Advantages and disadvantages of MBR process

Advantages	*Remarks*
1. High effluent quality	Membranes can retain all bacteria cells of a high molecular weight, soluble compounds such as proteins, starch, casein and precipitated phosphorus. These are good conditions for bacteria that have relatively much lower growth rate than aerobic microorganisms such as nitrifying bacteria; these conditions prolong their residence time, therefore improving the opportunity for biodegradation in the biological reactors. (Chiemchaisri and Yamamoto 1993).
2. High resistance to shock loads	MBR absorbs variations and fluctuations of hydraulic and organic loads to the system and the problem of bulking and foaming does not need to be considered. It is reported that an MBR could cope with a short-term fluctuation of inflow up to three-fold (Ueda and Hata 1999).
3. Low sludge production	MBR systems produce less sludge than conventionally activated sludge systems due to the relatively low F/M ratio and high sludge age. Autolysis and hydrolysis, which decrease the efficiency of the conventional systems, are favourable to MBR systems (Yamamoto et al. 1989).
4. Simple and compact system	There is no need to develop flocs that are needed in conventional aerobic activated sludge systems to settle. It is possible to operate with biomass concentrations between 10,000 and 20,000 mg/L, and up to 50,000 mg/L in zero-sludge wastage, compared with 1,500–3,500 mg/L for a conventional activated sludge bioreactor (Muller et al. 1995).
Disadvantages	*Remarks*
1. Membrane fouling	Concentration polarization and other membrane fouling problems can reduce MBR process efficiency, and increase the need for chemical cleaning for membrane regeneration. Reducing membrane biofouling is one of the main research areas in the MBR process (Le-Clech et al. 2006).

Table 12-6. Advantages and disadvantages of MBR process (Continued)

Disadvantages	*Remarks*
2. High operation and capital cost	Capital cost of membranes is relatively high compared to activated sludge process (Gander et al. 2000). Membrane lifetimes estimated by manufacturers range from 3–5 years for polymeric membranes, but membrane filters can blind very quickly due to membrane fouling, irreversibly in certain circumstances. Consequently, it also leads to high operation costs for MBR process.
3. Decreased biomass viability	A long biomass retention time by membrane results in decreased biomass viability. Retention and accumulation of non-reactive compounds (e.g. inorganic salt) in the bioreactor by the membrane could lead to microbial inhibition or toxicity, fouling of the membrane surface (Lay et al. 2010).
4. Low oxygen transfer	Because of higher concentrations of biomass, there is a high consumption of oxygen due to the oxidation processes and high minimum maintenance energy is needed, while the oxygen transfer in bioreactor membrane system is low.

and difficulty in scaling (Judd 2010). In order to minimize the problems of the activated sludge process, membrane modules (microfiltration and ultrafiltration grades) were applied and came to fulfill the role of clarifier for suspended biomass or conventional separation devices in the MBR process. As membrane technology has progressed, the MBR process has gained considerable attention in wastewater treatment, reclamation and reuse.

Table 12.6 shows a summary of advantages and disadvantages of the MBR process. The MBR process provides various advantages such as high effluent quality, high resistance to shock loads, low quantities of sludge production, and is a compact system. In contrast, MBR also produces a variety of problems including membrane fouling and high operational costs. In conclusion, it is desirable that future research in MBR enhances the strengths and supplements the weakness of the conventional MBR process.

According to these approaches, OMBR, the innovative and novel concept of MBR, was developed. Figure 12.4 indicates the submerged type of OMBR developed by Achilli et al. (2009). OMBR employ naturally occurring osmotic pressure difference between feed and draw solutions as the driving force for

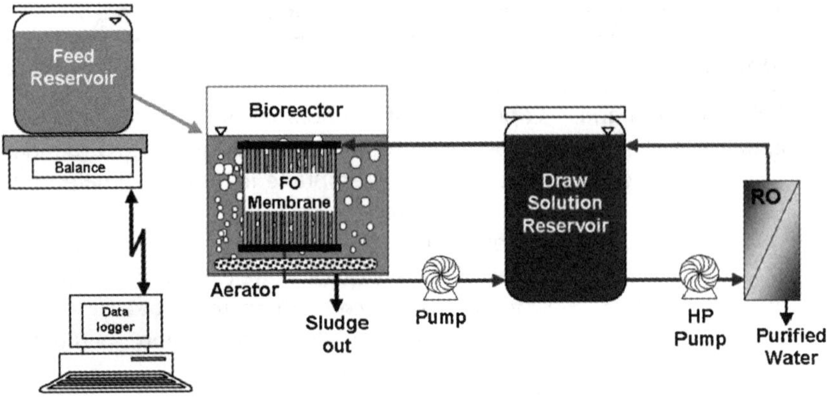

Figure 12.4. Schematic diagram of submerged OMBR
SOURCE: Adapted from Achilli et al. 2009; reproduced with permission from Elsevier

membrane filtration. Diluted draw solution was regenerated by post-treatment unit process such as RO and membrane distillation (MD). The primary differences between OMBR and conventional MBR are the utilization of other membranes and the introduction of draw solution. In this study (Figure 12.4), cellulose triacetate (CTA) FO membrane and sodium chloride were used as the model membrane and the draw solute, respectively. Recently, various FO membranes (type and materials) with high flux and low reverse solute flux (RSF) are available to improve OMBR performance. In addition, various high concentrate draw solutions (inorganic salt or organic materials) are also used to extract pure water from the wastewater in the bioreactor. Except for the FO membrane and the draw solution, other components are the same as those used in other MBR.

With usage of a bioreactor and an FO membrane, OMBR has great benefits compared to the conventional MBR. First of all, OMBR possesses a lower membrane fouling propensity which helps resolve the paramount problem within the MBR process. It is believed that no hydraulic pressure in the FO processes would reduce the compacting of a fouling layer and allow simple cleaning without chemical treatment. Figure 12.5 demonstrates that OMBR performed well at the low or no flux decline and salinity pattern during the long operation time (maximum 73 days) (Lay et al. 2011). Secondly, OMBR could be operated with minimum energy to produce high quality effluent in comparison with conventional MBR. This can be attributed to two factors: energy and fouling reduction. The operation of OMBR requires less energy than other MBR processes; further, membrane fouling reduction offers significant contributions in energy savings. Lastly, OMBR can lessen the load of the post treatment process of bioreactors by producing higher quality effluent. Since the currently developed FO membrane could reject all particles and colloidal materials, organic (partial micropollutants) and partial ions, permeate from the FO membrane shows a low fouling potential for post treatment units. That's why many researchers are paying attentions to

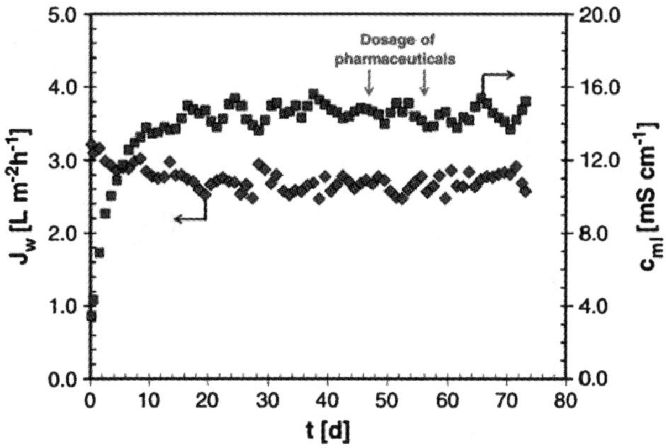

Figure 12.5. Permeate water flux and salinity of OMBR (experimental conditions–feed solution: synthetic wastewater, draw solution: 0.5 M NaCl, 20–22°C, cross-flow velocity: 0.5 L/min)

SOURCE: Adapted from Lay et al. 2011; reproduced with permission from Elsevier

OMBR in the application of wastewater reclamation and reuse (Alturki et al. 2012).

There are two major research streams in OMBR. One is to investigate the relationship between permeate flux and several operational parameters (draw solution type, concentration, CFV, membrane orientation and HRT) of OMBR (Lay et al. 2011; Alturki et al. 2012; Zhang et al. 2012a). These researches are mainly performed using lab-scale membrane bioreactor systems. The other research stream elucidates the influences of accumulated salts in a bioreactor on membrane fouling and performance of OMBR (Nawaz et al. 2013; Xiao et al. 2011). Computational modeling and experimental approaches are simultaneously carried out for the research.

12.3.1.1 Configuration of OMBR

Conventional MBR systems can be summarized as falling into two types, including cross-flow and submerged type MBR. In cross-flow MBR, the shear stress produced by cross-flow velocity breaks bacterial cells and induces a discharge of granulated matter such as glycogen and cell wall fragments (Shimizu et al. 1996). However, cross-flow MBR requires higher energy consumption than that of submerged MBR. The average power consumption of cross-flow MBR is about 3–4 kWh/m^3 compared to the 2.0 kWh/m^3 of treated wastewater for a submerged MBR (Ueda et al. 1996). On the other hand, submerged MBR immersed membrane pores may be prone to clogging by various foulants. Therefore, several parameters affecting the performance of MBR should be considered when determining MBR configuration.

OMBRs are also divided into two configurations. Figure 12.6 demonstrates two representative OMBR configurations, cross-flow and submerged OMBR in

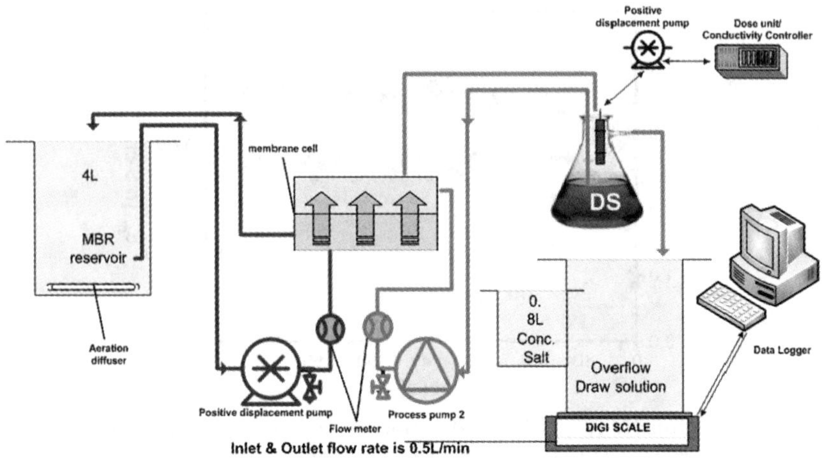

(a)

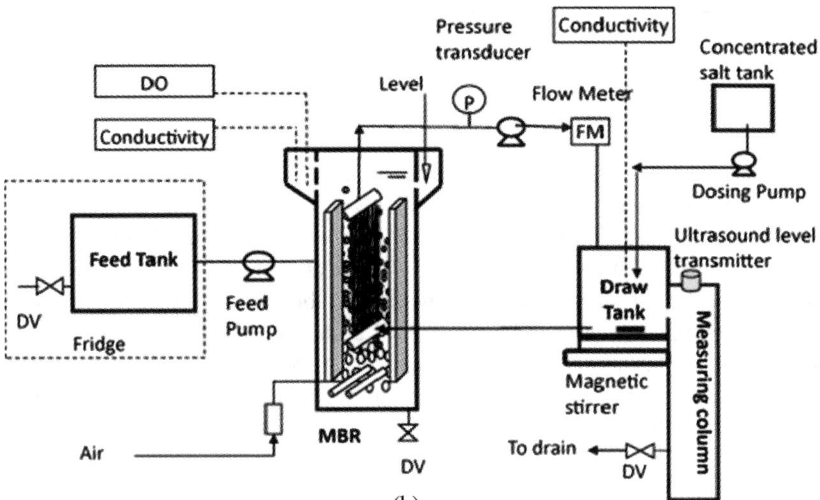

(b)

Figure 12.6. Schematic diagram of OMBR: (a) Cross-flow type and (b) submerged type

SOURCE: Adapted from (a) Alturki et al. (2012) and (b) Zhang et al. (2012b); reproduced with permission from Elsevier

accordance with the position of membrane cells. Cross-flow type features the effect of fouling reduction, while submerged type possesses a higher effective membrane area than other configurations. Several sensors (pH, conductivity, weight, water level, etc.,) are installed for the monitoring of water and other molecules transported in OMBR systems. Recently, the device for supplemental draw solute supply is also required to maintain constant draw solution concentration during long term experiments, which can eliminate the over estimation of flux decline.

12.3.1.2 Factors Affecting the Membrane Biofouling in OMBR

Membrane biofouling in MBR is attributed to the physicochemical interactions between the biofluid (soluble materials suspended solids) and membrane. Membrane fouling is the most serious problem affecting the membrane separation performance in MBR. Membrane fouling leads to permeate flux decline, and more frequent membrane cleaning and replacement, which increases operating costs. Chang et al. (2002) summarized major three factors for MBR fouling: membrane characteristics (configuration, hydrophobicity, porosity and pore size), biomass characteristics (mixed liquor suspended solids (MLSS), EPS, floc size and structure and dissolved matter) and operating conditions (configuration, CFV, aeration, hydraulic retention time (HRT), solids retention time (SRT) and trans-membrane pressure (TMP). Because of the heterogeneous nature in the bioreactor system and several factors which are correlated with others, it is extremely difficult to determine the relationship between them. Therefore, simplification of these factors within integrated viewpoints is urgently needed to better understand biofouling in MBR (Jang 2006).

Table 12.7 summarizes operational factors affecting membrane fouling in OMBR. In this chapter, various parameters including the configuration, feed/draw solution, membrane (type and material), CFV, SRT, HRT and average flux are summarized with fouling status of each study. Through the table, it is concluded that the OMBR process equipped with intermittent cleaning units can be operated under the high potential condition of biofouling without severe flux decline. However, average water fluxes in the table are extremely low (in the range of 2.7 to 9 LMH) compared to other MBR systems. Numerous efforts to enhance the performance of OMBR including the application of high flux membranes or draw solute, which has high osmotic pressure, are needed for the commercialization of OMBR. In addition, the effects of increased salinity in the bioreactor on microbial activity and overall performance should be systematically elucidated with various draw solutes such as inorganic and organic salts.

12.3.1.3 Salt Accumulation of OMBR

As mentioned above, salt accumulation in a bioreactor could exert adverse effects on the overall performance of an OMBR. In general, elevated salt concentration in the mixed liquor could hinder the growth and metabolism of microorganisms and cause the membrane scaling which can result in OMBR performance decline (Lay et al. 2010). Moreover, increased salinity leads to reduction of overall osmotic pressure difference, the driving force of FO filtration systems. To resolve this problem researchers are focused on a computer simulation for salt transport within the bioreactor, and finding alternative draw solutes which are suitable for OMBR application (Xiao et al. 2011; Bowden et al. 2012; Nawaz et al. 2013). Here, three previous studies in OMBR are introduced with experimental results in the following sections.

Novel computational modeling research about salinity in OMBR bioreactors was published by Xiao et al. (2011). Figure 12.7 demonstrates the conceptual

Table 12.7. Summary of operational parameters affecting membrane biofouling in OMBR

Ref.	*Stream type*	*Feed/Draw solution*	*Membrane*	*CFV*	*SRT*	*HRT*	*Average Flux*[a] *(LMH)*	*Fouling conditions*
Cornelissen et al. 2008	Submerged	Activated sludge (AS)/ four salts (NaCl, $MgSO_4$, $NaNO_3$, $ZnSO_4$)	Flat sheet, three membrane (TFC-NF, TFC-RO, CA-FO)	5.5 L/min	–	7-8 h	5.1-6.2 (0.5 M NaCl)	No irreversible membrane fouling occurred with AS for 7–8 h operation
Achilli et al. 2009	Submerged	Synthetic wastewater/ NaCl	Plate-and-frame, CA-FO	1.5 L/min	15 d	3.5 d	9(50 g NaCl/L)	RSF plays a substantial role in flux decline over the first 14 d. After 14 d, there is no irreversible fouling with osmotically backwashing.
Qin et al. 2010	Cross-flow	Domestic sewage/ two salts (NaCl, $MgSO_4$)	Flat sheet,CA-FO	9.0 L/min	12 d	6.3 h	4.3 (0.5 M NaCl)	Low or no flux decline (W/ air scouring operation), 30% flux decline ($5.9 \rightarrow 4.3$) (W/O air scouring)
Lay et al. 2011	Submerged	Synthetic wastewater/ NaCl	Flat sheet, CA-FO	0.5 L/min	20 d	–	2.7 (0.5 M NaCl)	No deterioration in membrane performance was detected and fouling with gel-like layer was mild during 73 days operation.

Alturki et al. 2012	Cross-flow	Synthetic wastewater/ NaCl	Flat sheet,CA-FO	2.0 m/s	–	–	3(1 M NaCl)	Continues deterioration of flux (12 → 3, after 4 days operation) was due to coupled effects of biofouling and increased salinity. (No intermittent cleaning lead to severe flux decline)
Zhang et al. 2012b	Submerged	Synthetic wastewater/ NaCl	Hollow fiber, TFC-FO	0.75 m/s	10 d	7 h	8(0.5 M NaCl)	FO fouling occurred by the combination of biofilm formation and inorganic scaling. No fouling was observed at AL-FS and tap water intermittent flushing

[a]Measured values under the fouling conditions

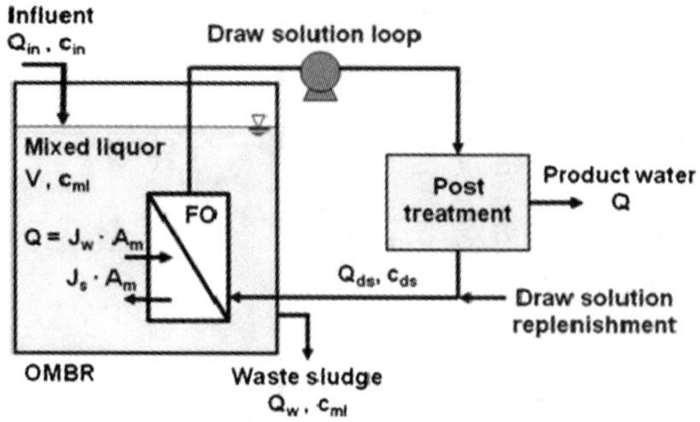

Figure 12.7. A conceptual diagram of an OMBR system
SOURCE: Adapted from Lay et al. 2012; reproduced with permission from Elsevier

diagram of the submerged OMBR system used in this study. The properties of an FO membrane and operational conditions of OMBR are suggested as the major factors determining the concentration of salt in OMBR. Monitored membrane properties are water permeability (A), salt permeability (B), mass transfer coefficient (K_m), and membrane orientation, while operational parameters include initial salt concentration of feed solution, draw solution concentration, HRT and SRT, etc., which are commonly referred to within the previous MBR and OMBR studies (Le-Clech et al. 2006; Lay et al. 2012).

According to the results of this study, there are two major factors influencing the salt accumulation within OMBR: the initial influent salt concentration; and the draw solution RSF. It turned out that the accumulated salt in OMBR could be accurately predicted by comparing the two factors. In particular, the relative importance of the two factors is decided in terms of the selectivity of an FO membrane; i.e., in the case of a high selectivity membrane ($B/A \ll$ osmotic pressure of feed solution), an RSF draw solution can be negligible for this estimation (Lay et al. 2012). In addition, the change of salt concentration in OMBR is linearly proportional to the volumetric concentration factor (the SRT/HRT ratio). In conclusion, an accurate prediction of the fate and transport of salt in OMBR systems is achievable by monitoring membrane properties (B/A) and the volumetric concentration factor.

In general, high salinity in a bioreactor system greatly affects the physical and biological properties of activated sludge and alters the production rate of EPS and SMP (Reid et al. 2006). Consequently, the permeate water flux and overall performance of bioreactors significantly decreased. Figure 12.8 indicates the change of cell growth under salt accumulation conditions with commonly used inorganic salt. Eight inorganic draw solutes (NaCl, $CaCl_2$, KCl, $MgCl_2$, K_2SO_4, NH_4HCO_3, Na_2SO_4 and $(NH_4)_2SO_4$) were employed in this study. The purpose of this study was to evaluate the effects of reverse diffused inorganic salts on bacterial cell growth (Nawaz et al. 2013). As shown in Figure 12.8, *Escherichia coli* growth patterns were significantly different depending on the draw solute. Increases in cell growth patterns were observed within results with regard to

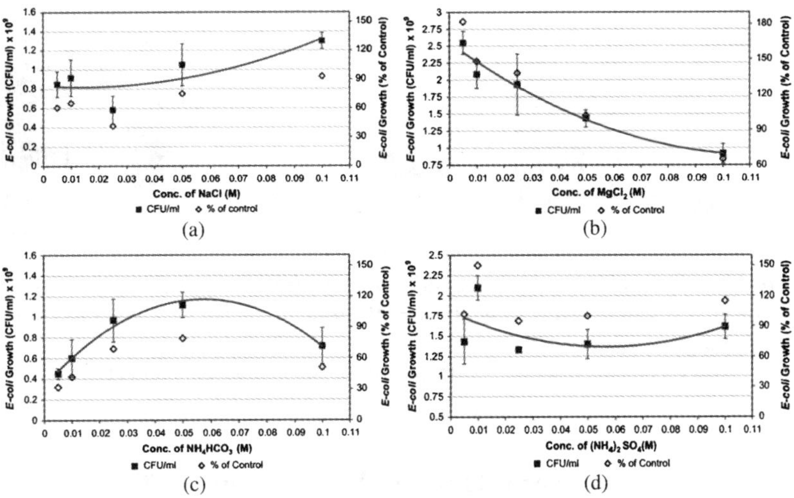

Figure 12.8. Effects of draw solutes on E.coli *growth (a) NaCl, (b) $MgCl_2$, (c) NH_4HCO_3, (d) $(NH_4)_2SO_4$*
SOURCE: Adapted from Nawaz et al. 2013; reproduced with permission from Elsevier

NaCl, $CaCl_2$, KCl, Na_2SO_4 and $(NH_4)_2SO_4$. On the other hand, decreasing patterns were detected within experiments with $MgCl_2$ and K_2SO_4 (experimental concentration range = 0–0.11M). This is the first published research paper related to the ecological evaluation of reversely diffused salts in the OMBR area. Supplementary studies about the application of multi-cultural systems and the correlation between accumulated inorganic salts and microbial products (EPS, SMP) are crucial to further develop an understanding of the OMBR process.

Finding alternative draw solutes could be the solution to salt problems in OMBR. Some researches related to the evaluation of several draw solutes in FO were performed mainly to determine the ideal draw solutes which have high osmotic pressure and easy regeneration methods (Achilli et al. 2010; Ge et al. 2012a). In the case of OMBR, the amount of RSF from draw solutions should be added as additional criterion. Recently, specific salt fluxes of four organic ionic salts and representative inorganic salts (NaCl, $MgCl_2$) were compared. As shown in Figure 12.9, two organic ionic salts (magnesium acetate ($Mg(C_2H_3O_2)_2$) and sodium propionate ($Na(C_3H_5COO)$)) have five or six times lower specific salt flux than NaCl or $MgCl_2$ (<0.2 g/L) (Bowden et al. 2012). Considering other properties such as the reconcentration performance, biodegradation potential and replacement cost, magnesium acetate and sodium propionate also perform impressively. Furthermore, organic anions degraded in a bioreactor could provide carbon sources and support nutrient removal. It is expected that the high-performance and sustainable OMBR process will continue to develop with various draw solutes.

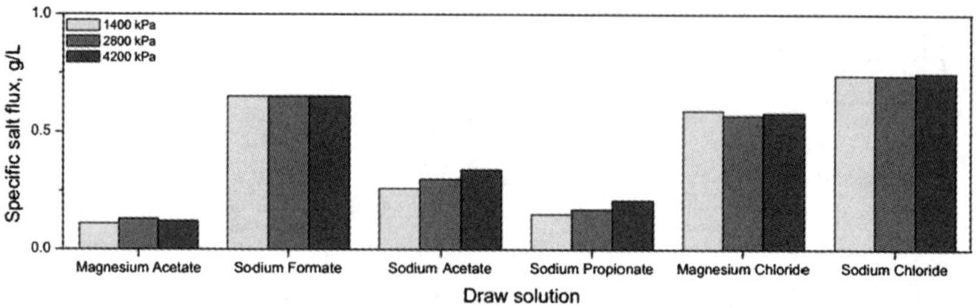

Figure 12.9. Comparison of reverse salt diffusion (reverse salt flux (J_S)/water flux (J_W)) for all the organic salts at each osmotic pressure tested with sodium chloride and magnesium chloride

SOURCE: Adapted from Bowden et al. 2012; reproduced with permission from Elsevier

12.3.2 FO-MFC (Forward Osmosis-Microbial Fuel Cell)

Microbial fuel cells (MFCs) have been studied over recent decades in an effort to recover energy from the degradation of organic compounds in wastewater. Recent developments have seen the FO membrane integrated MFC (labeled osmotic MFC or FO-MFC) developed to simultaneously achieve wastewater treatment (organic compound degradation), water extraction (producing effluent from wastewater using osmotic pressure), and seawater desalination (if the draw solution is seawater) (Zhang et al. 2011; Ge et al. 2012b). In osmotic MFCs (Figure 12.10), an FO membrane is installed between the anode and cathode chamber instead of an ion exchange membrane that is generally utilized as a separator in many MFC studies. Due to the difference of osmotic pressure between the anode and cathode chamber, water can be extracted from the anode to cathode chamber and this water permeation through the FO membrane can influence the electricity generation in osmotic MFC by enhancing the proton or ion transport through the FO membrane (Zhang et al. 2011). Biofouling is a crucial factor in the osmotic MFC system, because the biofouling in osmotic MFC can affect not only water flux decreases as in other membrane processes but also electricity generation efficiencies by influencing the ion or proton transport through the FO membrane, which is the one of decisive factors for electricity generation (Rabaey and Verstraete 2005). Osmotic MFC research is still in its infancy now as there appear to be as yet only a few studies conducted for osmotic MFC; in this section, we would like to establish the osmotic MFC and understand the effects of biofouling on MFC's performance.

Osmotic microbial fuel cell (OsMFC) is a new MFC technology that can extract product water from wastewater using osmotic pressure between the anode and cathode chambers. The history of OsMFC study is relatively short; the first paper introducing the concept of OsMFC was published in 2011 (Zhang et al. 2011). In general given MFCs, the ion exchange membrane (cation or anion) is used to separate anode and cathode, but recently, various separators have been employed in MFCs in order to lower the cost of the separator and internal

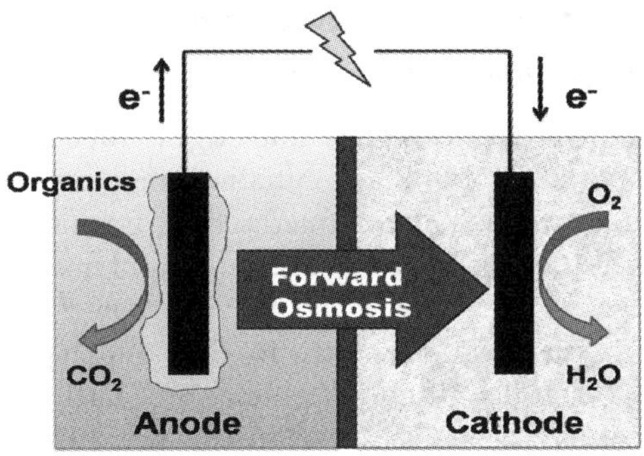

Figure 12.10. A conceptual diagram of osmotic MFCs
SOURCE: Adapted from Zhang et al. 2011; reproduced with permission from Elsevier

resistances that are caused by low proton or ion transport through the separator or add more functions such as a high-quality effluent production (Biffinger et al. 2007; Kim et al. 2007; Kim et al. 2013). With the FO membrane integration in MFCs, a higher maximum power density was achieved than cation exchange membrane (CEM) or anion exchange membrane (AEM) integrated MFCs (Werner et al. 2013). Higher power density of OsMFC was related to a lower internal resistance (54 Ω for FO membrane, 63 Ω for AEM, 73 Ω for CEM), and these low internal resistances were consistent with easy transport of protons through the FO membrane with water extraction (permeation) which has not occurred in ion exchange membrane integrated MFCs. The pH of the draw solution increased during the operation due to the reduction reaction at the cathode ($O_2 + 4\,H^+ + 4e^- \rightarrow 2H_2O$), thereby the pH adjustment by acid or utilization of the draw solution that has a buffering capacity is needed for long term operation (Ge and He 2012).

12.3.2.1 Effect of Biofouling on Osmotic MFCs

In OsMFC, protons or other dissolved ion transports through the FO membrane with the water permeation are crucial factors which can determine system performance. A study for biofouling phenomena which can influence the ion transport (and water flux) in OsMFC is important. In CEM integrated MFCs, biofouling on the CEM surface led to lower electricity generation due to the slight increase of membrane electrical resistance (MER, virgin CEM:15.65 Ω cm^2; fouled CEM:19.1 Ω cm^2) (Choi et al. 2011). According to literature on OsMFC, water flux decreased due to biofouling during the OsMFC operation. But unfortunately, biofouling behavior and its effect on the performance of OsMFC is as yet still not fully understood due to the short history of study. Considering the importance of biofouling processes within OsMFCs, further biofouling studies should be completed in order to understand the water flux decrease during operation and ion transport behavior through the FO membrane.

12.3.3 Biofouling by Microbial Excreta in FO

12.3.3.1 TEP and Formation of Biofilm

TEP, microscopic organic particles present in large numbers in marine and fresh water, play an important role in the initiation and development of aquatic biofilm (Berman et al. 2011). TEP are microscopic organic particles ranging in size from about 0.4 to 100–200 microns. They appear in many forms; amorphous blobs, filaments, clumps and sheets, and they are sometimes recognizable as debris from broken plankton. Similarly to EPS, these particles are mainly composed of polysaccharides although proteins and nucleic acids may also be present. Biofilm cohesiveness was found to have a strong correlation with polysaccharide rather than protein concentrations (Villacorte et al. 2009). TEPs have large, negatively charged, surface areas that make them very sticky because that area is composed of highly surface-active polymers containing fucose and rhamnose. They are about 2 to 4 orders of magnitude more sticky than phytoplankton or mineral particles, having a high probability of attachment upon collision and therefore are likely to play an important role in coating surfaces.

Figure 12.11 shows the progress of biofilm formation induced by TEP. Bar-Zeev et al. (2012) proposed the new paradigm of aquatic biofilm formation induced by TEP. They introduced the term "protobiofilm" to refer to TEPs with extensive microbial outgrowth and colonization. The proposed progress of development of early stage biofilm was as follows (Bar-Zeev et al. 2012):

1. Conditioning of surface begins upon contact with seawater and organic polymers, colloids and large particles such as TEP and protobiofilm were adhered (Figure 12.11(a, b, c)). They provide a favorable substrate for the attachment of bacteria and additional microgel particles. Although single bacteria can contact with surface, they do not become firmly attached (Figure 12.11(d)).
2. Within 30 min of seawater/surface interaction, further TEP and protobiofilm particles adhere firmly to the surface. Single bacteria attach irreversibly to the preconditioned surface (Figure 12.11(e)).
3. The matured biofilm was derived mainly from TEP and protobiofilm, and becomes established within a few hours (Figure 12.11(f)). Bacterial populations associated with the attached microgels and also single bacteria adhering to the surface begin to grow out and proliferate EPSs.

In the case of membrane systems, TEP creates the perfect agent to bond with the electropositive surface of membranes and element components to become the initial cause of biofouling (Komlenic 2010). Once TEP adheres to the membrane surface, they can provide a nutritious substrate for microbial growth and the establishment of biofilm. Further fouling is then stimulated by the continuous stream of TEP being trapped at the biofilm surface (Berman and Holenberg 2005).

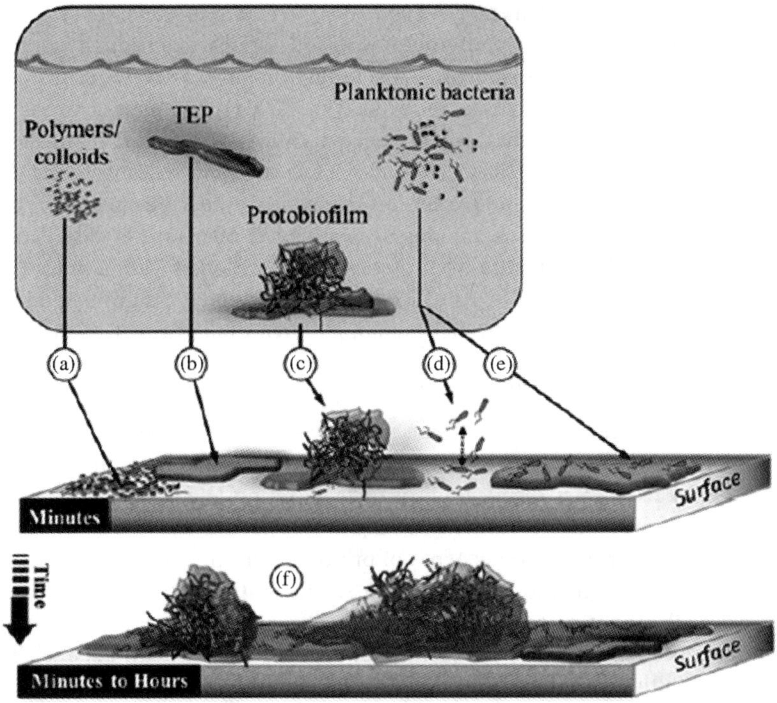

Figure 12.11. Schematic illustrations showing the involvement of organic polymers and colloids, TEPs, and protobiofilm in the initial stages of aquatic biofilm formation. Immediately upon exposure to seawater, organic polymers and colloids (A), and microgels such as uncolonized TEPs (B) and protobiofilm (C) begin to attach to pristine surfaces. Single cells of planktonic bacteria also attach reversibly (D) or irreversibly (E) to conditioned surfaces. With time (minutes to hours), a continuous coverage of mature biofilm (F) develops
SOURCE: Adapted from Bar-Zeev et al. 2012; reproduced with permission from PNAS

12.3.3.2 Visualization and Quantification of TEP

The number abundance and sizes of TEP can be analyzed using the microscopic enumeration method. Samples are prepared for microscopic enumeration by filtering the water samples through 0.4 or 0.2 μm polycarbonate filters. The TEP are stained directly on the filters with Alcian Blue (a copper phthalocyanin with four (methylene-tetramethyl-cisthiouronium-chloride) side chains). Alcian Blue is a hydrophilic cationic dye that complexes with anionic carboxyl or halfester-sulfate groups of acidic polymers (Tokarczyk et al. 1995).

To quantify TEP, the dye bound to the particles is re-dissolved in 80% sulfuric acid and can be measured using the colorimetric method. This method is semi-quantitative because of the amount of this specificity; the amount of dye bound to TEP is standardized using Gum Xanthan. TEP concentration is expressed in μg Gum Xanthan equivalent liter^{-1} (μg Xeq. l^{-1}) (Berman et al. 2011; Passow 2002). However, the spectrophotometric method for the determination of TEP proposed is labor-intensive and time-consuming, involving several critical steps which lowered precision. Arruda Fatibello et al. (2004) developed a simple and rapid spectrophotometric method for the determination of TEP in freshwater samples. TEP reacts with excess of Alcian Blue solution yielding a non-ionic pigment of low solubility, and after centrifugation, the concentration of the remaining dye in the supernatant was determined at 602 nm (Arruda Fatibello et al. 2004). Recently, to quantitatively assess the deposition of TEP in RO systems, the specific deposition rate (SDR) of TEP can be calculated assuming a uniform TEP deposition throughout the RO membrane area (Villacorte et al. 2009).

12.3.3.3 Strategies to Reduce TEP

To remove TEP from seawater, some pretreatment methods using membranes were applied. However, Linares et al. (2012) showed that TEP cannot be easily controlled by feed water disinfection or microfiltration (Valladares Linares et al. 2012). Ultrafiltration pre-treatment was the most effective way to remove particulate TEP (>0.4 μm) than other types of pre-treatment investigated (microfiltration, conventional method (coagulation-sedimentation-filtration)). However, colloidal TEP cannot be removed effectively with either MF/UF or conventional pre-treatment methods (<0.4 μm). At the Adom desalination plant in Israel, it was found that although pre-treatment by sand filtration followed by cartridge filtration significantly lowered the Silt Density Index (SDI) and turbidity of water reaching the RO membranes, TEP levels were usually much less affected (Berman 2010).

Also, TEP has been considered a biofouling formation factor in FO systems. When secondary wastewater effluent (SWWE) was applied as feed water and seawater was applied as draw solution, the support layer contact with seawater presents TEP fouling. Figure 12.12 shows the cleaning of TEP on an active layers and the support layer of an FO membrane with two kinds of cleaning agents. Figure 12.12 a shows a picture of the support layer of a used FO membrane after the dying process. Because the support layer and active layer were faced with seawater and SWWE, respectively, TEP was observed only on the support layer (Figure 12.12(a, b)). As mentioned above, the filtration methods would not remove TEP effectively. In this study, the cleaning potential of the two solutions was tested to analyze the best option for TEP removal on the support side of the FO membrane. The first solution contained 0.8% sodium ethylenediaminetetraacetic acid (EDTA) and 1% Alconox, a detergent composed of sodium dodecylbenzenesulfonate (10–30%), sodium carbonate (7–13%), tetrasodium pyrophosphate (10-30%) and sodium phosphate (10–30%) (solution 1). The second one contained 1% aqueous solution of NaOCl (pH 7.6) (solution 2). After *in situ* cleanings, the TEP concentration was significantly reduced, but, small particles of TEP remained when cleaned with solution 1 (Figure 12.12©). Unlike solution 1, almost no TEP could be found on the membrane when treated with solution 2 (Figure 12.12(d)) (Valladares Linares et al. 2012).

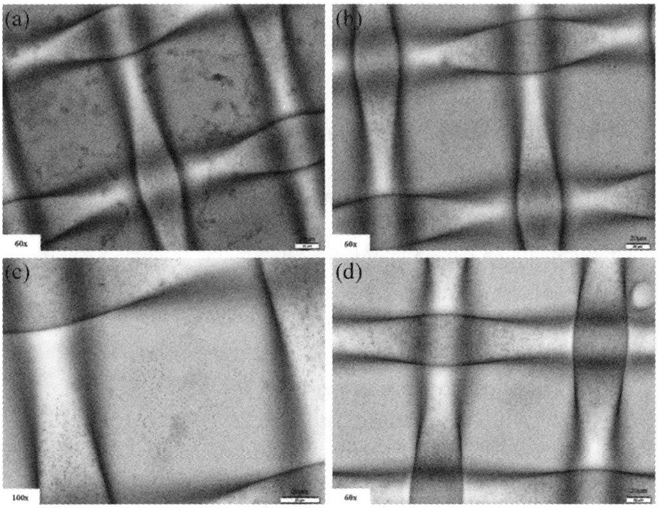

Figure 12.12. Bright light microscopy images of a fouled FO membrane stained with Alcian Blue dye. (a) TEP fouling on the support layer (60×); (b) active layer (60×); (c) support layer of a cleaned FO membrane (100×) with Solution 1; (d) support layer of a cleaned FO membrane (60×) with Solution 2
(Source: Adapted from Linares et al. 2012; reproduced with permission from Elsevier)

12.4 CONCLUDING REMARKS

The FO process currently is regarded as an innovative and alternative water treatment technology which could hold the potential to help solve the water scarcity problem throughout the world. As continued progress on membrane technology makes feasible the FO process, the development of other unit processes such as devices for the draw solute regeneration would lead to the commercialization of FO. However, the advancement of operations and maintenance (O&M) technology is also important as well. Membrane fouling is the one of the most crucial aspects of O&M research including FO process. In particular, biofouling originated by microbial products is more difficult to control than other fouling and this leads to sever performance decline. Therefore, the researches associated with the prevention and control of biofouling is greatly required in FO membrane process.

12.5 ABBREVIATIONS/NOMENCLATURE

A	water permeability
AEM	anion exchange membrane
AFM	atomic force microscopy

ATP	adenosine triphosphate
ATR	attenuated total internal reflection
B	salt permeability
BPC	biopolymeric clusters
CEM	cation exchange membrane
CFV	cross-flow velocity
CLSM	confocal laser scanning microscopy
CTA	cellulose triacetate
DAPI	4'-6-diamidino-2-phenylindole
DGGE	denaturing gradient gel electrophoresis
EDTA	ethylenediaminetetraacetic acid
EfOM	effluent organic matter
ELISA	enzyme-linked immunosorbent assay
EPS	extracellular polymeric substances
FISH	fluorescence in situ hybridization
FO	forward osmosis
FO-MFC	forward osmosis-microbial fuel cell
FTIR	Fourier transform infrared
HPC	heterotrophic plate counts
HRT	hydraulic retention time
J_s	reverse salt flux
J_w	water flux
K_m	mass transfer coefficient
LMH	L/m^{2h}
MALDI-MS	matrix assisted laser desorption ionization mass spectroscopy
MBR	membrane bioreactor
MD	membrane distillation
MER	membrane electrical resistance
MFCs	microbial fuel cells
MLSS	mixed liquor suspended solids
MRI	magnetic resonance imaging
NOM	natural organic material
OCT	optical coherence tomography
OMBR	membrane bioreactor
O&M	operations and maintenance
OsMFC	osmotic microbial fuel cell
PCR	polymerase chain reaction
PES	polyethersulfone
PI	propidium iodide
PS	polysaccharides
QCM-D	quartz crystal microbalance with dissipation
RFLP	restriction fragment length polymorphism
RO	reverse osmosis
RSF	reverse solute flux
SDI	silt density index

SDR	specific deposition rate
SEM	scanning electron microscopy
SMP	soluble microbial products
SRT	solids retention time
SWWE	secondary wastewater effluent
TDC	total direct cell counts
TEM	transmission electron microscopy
TEP	transparent exopolymer particles
TMP	trans-membrane pressure
T-RFLP	terminal restriction fragment length polymorphism

References

Abarzua, S., and Jakubowski, S. (1995). "Biotechnological investigation for the prevention of biofouling. I. Biological and biochemical principles for the prevention of biofouling." *Mar Ecol Prog Ser.*, 123, 301–312.

Achilli, A., Cath, T. Y., and Childress, A. E. (2010). "Selection of inorganic-based draw solutions for forward osmosis applications." *Journal of Membrane Science*, 364, 233–241.

Achilli, A., Cath, T. Y., Marchand, E. A., and Childress, A. E. (2009). "The forward osmosis membrane bioreactor: A low fouling alternative to MBR processes." *Desalination*, 239, 10–21.

Al-Amoudi, A., and Lovitt, R. W. (2007). "Fouling strategies and the cleaning system of NF membranes and factors affecting cleaning efficiency." *Journal of Membrane Science*, 303, 4–28.

Alturki, A., McDonald, J., Khan, S. J., Hai, F. I., Price, W. E., and Nghiem, L. D. (2012). "Performance of a novel osmotic membrane bioreactor (OMBR) system: Flux stability and removal of trace organics." *Bioresource Technology*, 113, 201–206.

Arruda Fatibello, S. H., Henriques Vieira, A. A., and Fatibello-Filho, O. (2004). "A rapid spectrophotometric method for the determination of transparent exopolymer particles (TEP) in freshwater." *Talanta*, 62, 81–85.

Balogh, K., Patócs, A., Majnik, J., Rácz, K., and Hunyady, L. (2004). "Genetic screening methods for the detection of mutations responsible for multiple endocrine neoplasia type 1." *Molecular genetics and metabolism*, 83, 74–81.

Bar-Zeev, E., Berman-Frank, I., Girshevitz, O., and Berman, T. (2012). "Revised paradigm of aquatic biofilm formation facilitated by microgel transparent exopolymer particles." *Proceedings of the National Academy of Sciences*, 109, 9119–9124.

Berman, T. (2010). "Biofouling: TEP-a major challenge for water filtration." *Filtration & Separation*, 47, 20–22.

Berman, T., and Holenberg, M. (2005). "Don't fall foul of biofilm through high TEP levels." *Filtration & Separation*, 42, 30–32.

Berman, T., Mizrahi, R., and Dosoretz, C. G. (2011). "Transparent exopolymer particles (TEP): A critical factor in aquatic biofilm initiation and fouling on filtration membranes." *Desalination*, 276, 184–190.

Biffinger, J. C., Ray, R., Little, B., and Ringeisen, B. R. (2007). "Diversifying biological fuel cell designs by use of nanoporous filters." *Environmental Science and Technology*, 41, 1444–1449.

Bodrossy, L., and Sessitsch, A. (2004). "Oligonucleotide microarrays in microbial diagnostics." *Current opinion in microbiology*, 7, 245–254.

Bowden, K. S., Achilli, A., and Childress, A. E. (2012). "Organic ionic salt draw solutions for osmotic membrane bioreactors." *Bioresource Technology*, 122, 207–216.

Chae, K.-J., Rameshwar, T., Jang, A., Kim, S. H., and Kim, I. S. (2008). "Analysis of the nitrifying bacterial community in BioCube sponge media using fluorescent in situ hybridization (FISH) and microelectrodes." *Journal of Environmental Management*, 88, 1426–1435.

Chang, I.-S., Le Clech, P., Jefferson, B., and Judd, S. (2002). "Membrane fouling in membrane bioreactors for wastewater treatment." *Journal of environmental engineering*, 128(11), 1018–1029.

Chiemchaisri, C., and Yamamoto, K. (1993). "Biological nitrogen removal under low temperature in a membrane separation bioreactor." *Water Science & Technology*, 28, 325–333.

Choi, M. J., Chae, K. J., Ajayi, F. F., Kim, K. Y., Yu, H. W., Kim, C. W., and Kim, I. S. (2011). "Effects of biofouling on ion transport through cation exchange membranes and microbial fuel cell performance." *Bioresource Technology*, 102, 298–303.

Cornelissen, E. R., Harmsen, D., de Korte, K. F., Ruiken, C. J., Qin, J.-J., Oo, H., and Wessels, L. P. (2008). "Membrane fouling and process performance of forward osmosis membranes on activated sludge." *J. Memb. Sci.*, 319, 158–168.

Drews, A. (2010). "Membrane fouling in membrane bioreactors—Characterisation, contradictions, cause and cures." *Journal of Membrane Science*, 363, 1–28.

Flemming, H. C. (2002). "Biofouling in water systems- cases, causes and countermeasures." *Applied Microbiology and Biotechnology*, 59, 629–640.

Gander, M., Jefferson, B., and Judd, S. (2000). "Aerobic MBRs for domestic wastewater treatment: a review with cost considerations." *Separation and purification Technology*, 18, 119–130.

Ge, Q., Su, J., Amy, G. L., and Chung, T.-S. (2012a). "Exploration of polyelectrolytes as draw solutes in forward osmosis processes." *Water Research*, 46, 1318–1326.

Ge, Z., and He, Z. (2012). "Effects of draw solutions and membrane conditions on electricity generation and water flux in osmotic microbial fuel cells." *Bioresource Technology*, 109, 70–76.

Ge, Z., Ping, Q., Xiao, L., and He, Z. (2012b). "Reducing effluent discharge and recovering bioenergy in an osmotic microbial fuel cell treating domestic wastewater." *Desalination*, 312, 52–59.

Goosen, M., Sablani, S., Al-Hinai, H., Al-Obeidani, S., Al-Belushi, R., and Jackson, D. (2005). "Fouling of reverse osmosis and ultrafiltration membranes: a critical review." *Separation science and technology*, 39, 2261–2297.

Haisch, C., and Niessner, R. (2007). "Visualisation of transient processes in biofilms by optical coherence tomography." *Water Research*, 41, 2467–2472.

Hall, S. A., Covert, P. A., Blinn, B. R., Shakeri, S., and Hore, D. K. (2013). "Rapid and Sensitive Polarization Measurement for Characterizing Protein Adsorption at the Solid-Liquid Interface." *The Journal of Physical Chemistry C*, 117, 1796–1803.

Jang, N. (2006). Membrane bio-fouling in terms of extracellular polymeric substances (EPS) and soluble microbial products (SMP) in the submerged membrane bioreactor (SMBR), Gwangju Institute of Science and Technology.

Jang, N., Ren, X., Kim, G., Ahn, C., Cho, J., and Kim, I. S. (2007). "Characteristics of soluble microbial products and extracellular polymeric substances in the membrane bioreactor for water reuse." *Desalination*, 202, 90–98.

Judd, S. (2010). *The MBR Book: Principles and Applications of Membrane Bioreactors for Water and Wastewater Treatment*, Elsevier.

Kim, J. R., Cheng, S., Oh, S. E., and Logan, B. E. (2007). "Power Generation Using Different Cation, Anion, and Ultrafiltration Membranes in Microbial Fuel Cells." *Environmental Science and Technology*, 41, 1004–1009.

Kim, K. Y., Chae, K. J., Choi, M. J., Yang, E. T., Hwang, M. H., and Kim, I. S. (2013). "High-quality effluent and electricity production from non-CEM based flow-through type microbial fuel cell." *Chemical Engineering Journal*, 218, 19–23.

Kim, L. H., Jang, A., Yu, H.-W., Kim, S.-J., and Kim, I. S. (2011). "Effect of chemical cleaning on membrane biofouling in seawater reverse osmosis processes." *Desalination and Water Treatment*, 33, 289–294.

Komlenic, R. (2010). "Rethinking the causes of membrane biofouling." *Filtration & Separation*, 47, 26–28.

Lawrence, J. R., Swerhone, G. D. W., Leppard, G. G., Araki, T., Zhang, X., West, M. M., and Hitchcock, A. P. (2003). "Scanning Transmission X-Ray, Laser Scanning, and Transmission Electron Microscopy Mapping of the Exopolymeric Matrix of Microbial Biofilms." *Appl. Environ. Microbiol.*, 69, 5543–5554.

Lay, W. C., Liu, Y., and Fane, A. G. (2010). "Impacts of salinity on the performance of high retention membrane bioreactors for water reclamation: a review." *Water Research*, 44, 21–40.

Lay, W. C., Zhang, J., Tang, C., Wang, R., Liu, Y., and Fane, A. G. (2012). "Analysis of salt accumulation in a forward osmosis system." *Separation science and technology*, 47, 1837–1848.

Lay, W. C., Zhang, Q., Zhang, J., McDougald, D., Tang, C., Wang, R., Liu, Y., and Fane, A. G. (2011). "Study of integration of forward osmosis and biological process: Membrane performance under elevated salt environment." *Desalination*, 283, 123–130.

Le-Clech, P., Chen, V., and Fane, T. A. (2006). "Fouling in membrane bioreactors used in wastewater treatment." *Journal of Membrane Science*, 284, 17–53.

Lee, J., Jung, J.-Y., Kim, S., Chang, I. S., Mitra, S. S., and Kim, I. S. (2009). "Selection of the most problematic biofoulant in fouled RO membrane and the seawater intake to develop biosensors for membrane biofouling." *Desalination*, 247, 125–136.

Lee, J., and Kim, I. S. (2011). "Microbial community in seawater reverse osmosis and rapid diagnosis of membrane biofouling." *Desalination*, 273, 118–126.

Lee, S., Boo, C., Elimelech, M., and Hong, S. (2010). "Comparison of fouling behavior in forward osmosis (FO) and reverse osmosis (RO)." *Journal of Membrane Science*, 365, 34–39.

Li, Z.-Y., Yangali-Quintanilla, V., Valladares, R., Li, Q., Zhan, T., and Amy, G. (2011). "Flux patterns and membrane fouling propensity during desalination of seawater by forward osmosis." *Water Research*.

Liu, S. X., and Kim, J.-T. (2009). "Application of Kevin—Voigt Model in Quantifying Whey Protein Adsorption on Polyethersulfone Using QCM-D." *Journal of the Association for Laboratory Automation*, 14, 213–220.

Manz, B., Müller, K., Kucera, B., Volke, F., and Leubner-Metzger, G. (2005). "Water uptake and distribution in germinating tobacco seeds investigated in vivo by nuclear magnetic resonance imaging." *Plant Physiology*, 138, 1538–1551.

Marsh, T. L. (1999). "Terminal restriction fragment length polymorphism (T-RFLP): an emerging method for characterizing diversity among homologous populations of amplification products." *Current opinion in microbiology*, 2, 323–327.

Mi, B., and Elimelech, M. (2010). "Organic fouling of forward osmosis membranes: Fouling reversibility and cleaning without chemical reagents." *Journal of Membrane Science*, 348, 337–345.

Muller, E., Stouthamer, A., van Verseveld, H. W. v., and Eikelboom, D. (1995). "Aerobic domestic waste water treatment in a pilot plant with complete sludge retention by cross-flow filtration." *Water Research*, 29, 1179–1189.

Nawaz, M. S., Gadelha, G., Khan, S. J., and Hankins, N. (2013). "Microbial toxicity effects of reverse transported draw solute in the forward osmosis membrane bioreactor (FO-MBR)." *Journal of Membrane Science*, 429, 323–329.

Nguyen, Thang, Roddick, Felicity, A., and Fan, L. (2012). "Biofouling of Water Treatment Membranes: A Review of the Underlying Causes, Monitoring Techniques, and Control Measures." *Membranes*, 2, 804–840.

Passow, U. (2002). "Transparent exopolymer particles (TEP) in aquatic environments." *Progress in Oceanography*, 55, 287–333.

Qin, J.J., Kekre, K.A., Oo, M.H., Tao, G., Lay, C.L., Lew, C.H., Cornelissen, E.R., Ruiken, C.J. (2010). "Preliminary study of osmotic membrane bioreactor: effects of draw solution on water flux and air scouring on fouling." *Water Science & Technology*, 62, 1353–1360.

Rabaey, K., and Verstraete, W. (2005). "Microbial fuel cells: novel biotechnology for energy generation." *Trends in Biotechnology*, 23, 291–298.

Reid, E., Liu, X., and Judd, S. (2006). "Effect of high salinity on activated sludge characteristics and membrane permeability in an immersed membrane bioreactor." *Journal of Membrane Science*, 283, 164–171.

Ronaghi, M., and Elahi, E. (2002). "Pyrosequencing for microbial typing." *Journal of Chromatography B*, 782, 67–72.

Schneider, R. P., Ferreira, L. M., Binder, P., Bejarano, E. M., Góes, K. P., Slongo, E., Machado, C. R., and Rosa, G. M. Z. (2005). "Dynamics of organic carbon and of bacterial populations in a conventional pretreatment train of a reverse osmosis unit experiencing severe biofouling." *Journal of Membrane Science*, 266, 18–29.

Shimizu, Y., Okuno, Y.-I., Uryu, K., Ohtsubo, S., and Watanabe, A. (1996). "Filtration characteristics of hollow fiber microfiltration membranes used in membrane bioreactor for domestic wastewater treatment." *Water Research*, 30, 2385–2392.

Sohrabi, M. R., Madaeni, S. S., Khosravi, M., and Ghaedi, A. M. (2011). "Chemical cleaning of reverse osmosis and nanofiltration membranes fouled by licorice aqueous solutions." *Desalination*, 267, 93–100.

Tansel, B., Sager, J., Garland, J., Xu, S., Levine, L., and Bisbee, P. (2006). "Deposition of extracellular polymeric substances (EPS) and microtopographical changes on membrane surfaces during intermittent filtration conditions." *Journal of Membrane Science*, 285, 225–231.

Thang Nguyen, Felicity, A. Roddick, Fan, L. (2012). "Biofouling of Water Treatment Membranes: A Review of the Underlying Causes, Monitoring Techniques and Control Measures." *Membranes*, 2, 804–840.

Tokarczyk, R., Tait, V., Poulin, M., Green, C., Newman, K., Gschwend, P., and Macfarlane, J. (1995). "A dye-binding assay for the spectrophotometric measurement of transparent exopolymer particles (TEP)." *Fresenius Z Anal Chem*, 327, 198–204.

Ueda, T., and Hata, K. (1999). "Domestic wastewater treatment by a submerged membrane bioreactor with gravitational filtration." *Wat. Res.*, 33, 2888–2892.

Ueda, T., Hata, K., and Kikuoka, Y. (1996). "Treatment of domestic sewage from rural settlements by a membrane bioreactor." *Water Sci. Technol.*, 34, 189–196.

Valladares Linares, R., Yangali-Quintanilla, V., Li, Z., and Amy, G. (2012). "NOM and TEP fouling of a forward osmosis (FO) membrane: Foulant identification and cleaning." *Journal of Membrane Science.*

Villacorte, L. O., Kennedy, M. D., Amy, G. L., and Schippers, J. C. (2009). "The fate of transparent exopolymer particles (TEP) in integrated membrane systems: removal through pre-treatment processes and deposition on reverse osmosis membranes." *Water Research*, 43, 5039–5052.

Vrouwenvelder, J. S., Manolarakis, S. A., van der Hoek, J. P., van Paassen, J. A. M., van der Meer, W. G. J., van Agtmaal, J. M. C., Prummel, H. D. M., Kruithof, J. C., and van Loosdrecht, M. C. M. (2008). "Quantitative biofouling diagnosis in full scale nanofiltration and reverse osmosis installations." *Water Research*, 42, 4856–4868.

Werner, C. M., Logan, B. E., Saikaly, P. E., and Amy, G. L. (2013). "Wastewater treatment, energy recovery and desalination using a forward osmosis membrane in an air-cathode microbial osmotic fuel cell." *Journal of Membrane Science*, 428, 116–122.

Xiao, D., Tang, C. Y., Zhang, J., Lay, W. C. L., Wang, R., and Fane, A. G. (2011). "Modeling salt accumulation in osmotic membrane bioreactors: Implications for FO membrane selection and system operation." *Journal of Membrane Science*, 366, 314–324.

Yamamoto, K., Hiasa, M., Mahmood, T., and Matsuo, T. (1989). "Direct solid-liquid separation using hollow fiber membrane in an activated sludge aeration tank." *Water Science & Technology*, 21, 43–54.

Yoon, H., Baek, Y., Yu, J., and Yoon, J. (2013). "Biofouling occurrence process and its control in the forward osmosis." *Desalination*, 325, 30–36.

Yu, D.-y., Yin, X., Zhang, J., Wang, G., and Cao, S.-x. (2012). "Research Progress of Biofouling Extracellular Polymeric Substances in Cooling Water System." *Energy Procedia*, 16, 1671–1677.

Zhang, F., Brastad, K. S., and He, Z. (2011). "Integrating forward osmosis into microbial fuel cells for wastewater treatment, water extraction and bioelectricity generation." *Environmental Science and Technology*, 45, 6690–6696.

Zhang, H., Ma, Y., Jiang, T., Zhang, G., and Yang, F. (2012a). "Influence of activated sludge properties on flux behavior in osmosis membrane bioreactor (OMBR)." *Journal of Membrane Science*, 390, 270–276.

Zhang, J., Loong, W. L. C., Chou, S., Tang, C., Wang, R., and Fane, A. G. (2012b). "Membrane biofouling and scaling in forward osmosis membrane bioreactor." *Journal of Membrane Science*, 403, 8–14.

CHAPTER 13

Forward Osmosis Hybrid Processes for Water Reclamation and Reuse

Melvin Tang
Liu Xiao
How Yong Ng

With diminishing supplies and an escalating demand for drinking water, motivation for societies to recycle wastewater and desalinate brackish or seawater becomes stronger over the years. However, current methodologies for such technological implementations are energy intensive and are not congruent with current global concerns of energy and climate change. With the advancement of membrane material sciences in the recent decades, interests in applying forward osmosis (FO) for water reclamation and reuse are mounting. FO utilizes a semi-permeable membrane that separates the feed stream, which is usually rather low in osmotic pressures, from a body of highly concentrated salt solution called the draw solution. The water concentration differences between the two bodies of water allow water molecules to be drawn into the salt solution by passing through the membrane. Osmosis is a natural, spontaneous process and hydraulic pressures are no longer required for permeate production. Hence, FO provides for an energy efficient way from which water can be extracted from impaired water sources and seawater streams by leveraging on osmotic pressure differences that are established across the semi-permeable FO membrane. However, the FO process produces diluted salt solutions that are not directly usable downstream, and this is its main difference in comparison with traditional membrane processes running on vacuum suction or high pressures. Thus, they cannot be standalone systems and will have to exist in hybrid configurations. The second components after FO in these hybrid systems are usually technologies that provide a mean to recover a drinking water product from the diluted FO draw solution. These secondary appendages can involve well-established, conventional pressurized filtrations such as RO and NF systems. With deepening interests in and rapid development of novel draw solutes, alternatives to such traditional systems can be utilized to provide the required conditions or stimuli for water extraction downstream. Therefore, the purpose of this chapter is to provide an overview of the fundamental terminologies, concepts, performance discussions and concerns of FO hybrid systems for water reclamation and reuse.

13.1 INTRODUCTION

Availability of fresh water across the globe has become increasingly inadequate to meet demands. Many communities are presently facing water scarcity woes stemming from the issues of population growth, periodic droughts, depletion and contamination of water sources, environmentally unsustainable manners of economic development and lastly, increased per capita consumption of goods and services (Vorosmarty et al. 2000; Miller 2006; Shaffer et al. 2012). The current water demand estimates presented a total global water demand dominated by agriculture (70% of all global water use), followed by industrial use (19%) and domestic use (11%) (UNESCO 2012).

While there is still a substantial uncertainty of how water demands will scale in the future, burgeoning global population and the associated demand for food from irrigated agriculture are expected to increase considerably and put acute pressure on the already strained water supply (Foley et al. 2011). In addition, climate change will be likely to further aggravate the intensity of water scarcity in the future with more extreme weather conditions and imbalanced water distribution (Vorosmarty et al. 2000). In view of these challenges, water agencies and communities across the globe have been propelled to search for new sources of water and become progressively more motivated to look into alternatives, such as recycling and reusing treated wastewater and/or desalinate seawater or brackish water (Shannon et al. 2008). Seawater desalination with membrane treatment processes such as reverse osmosis (RO) and nanofiltration (NF) has become commonplace to support the growing demand for water in areas with access to the ocean; however, these processes are generally energy-intensive and the water recovery is often limited. In addition to energy concerns, high susceptibility to membrane fouling and scaling, increased treatment cost, shortened membrane life time and finally, brine disposal with its accompanying adverse environmental impacts are also some of the critical drawbacks of membrane desalination technologies (Farooque et al. 2008; Cath et al. 2010).

Water reclamation is defined as the treatment or processing of wastewater to make it reusable with definable treatment reliability and meeting water quality criteria. On the other hand, water reuse is the application of treated water for beneficial reuses such as agricultural irrigation and industrial cooling (Metcalf and Eddy 2004). Some other terms related to water reclamation and reuse which are frequently used are summarized in Table 13.1, adapted from Metcalf & Eddy (2004).

Water reclamation and reuse attempt to alleviate water stresses by posing as viable and reliable water sources to create high-quality water supplies. In fact, the use of highly treated wastewater effluents are already an important element in water resources planning and the future potential for reclaimed treated effluent is

Table 13.1. Definition of common terms for water reclamation and reuse

Terms	*Definitions*
Beneficial uses	The many ways water can be used, either directly by people, or for their overall benefit. Examples include municipal water supply, agricultural and industrial applications, navigation, fish and wildlife
Direct potable reuse	A form of reuse by the incorporation of reclaimed water directly into a potable water supply system, often implying the blending of reclaimed water with potable water
Direct reuse	The use of reclaimed water which has been transported from the wastewater reclamation plant to the water reuse site without intervening discharge to a natural body of water, including such uses as agricultural and landscape irrigation
Indirect potable reuse	Potable reuse by incorporation of reclaimed water into a raw water supply, allowing mixing and assimilation by discharge into an impoundment or natural body of water, such as in a domestic water supply of reservoir or ground water
Indirect reuse	Use of reclaimed water indirectly by passing through a natural body of water or use of groundwater that has been recharged with reclaimed water
Non-potable water reuse	All reuse applications that do not involve either indirect or direct potable use
Potable water reuse	An augmentation of drinking water supplies directly or indirectly by reclaimed water that is highly treated to protect public health
Reclaimed water	Water that, as a result of wastewater treatment, is suitable for a direct beneficial use or a controlled use that would not otherwise occur
Recycled water	Reclaimed water that has been used beneficially. The term recycled water is used synonymously with reclaimed water
Water reclamation	Treatment or processing of wastewater to make it reusable. Also this term is often used to include delivery of reclaimed water to place of use and its actual use
Water reuse	The use of treated wastewater for a beneficial use, such as agricultural irrigation and industrial cooling

enormous (Metcalf and Eddy 2004; Miller 2006). Some of the key benefits associated with water reclamation and reuse are listed below:

1. A sustainable, alternative water supply;
2. Less energy demanding than water import;
3. Avoidance of construction impact associated with new supply development; and
4. Reduction or elimination of the quantity of treated wastewater discharged to sensitive or impaired surface waters.

The most noteworthy value of water reclamation and reuse is the providence of a high quality water supply to serve a variety of beneficial usages (UNESCO 2012). There are seven principal categories of water reclamation and reuse application discussed and highlighted by Metcalf and Eddy (2004), including:

1. Agricultural irrigation;
2. Landscape irrigation;
3. Industrial recycling and reuse;
4. Groundwater recharge;
5. Recreational uses;
6. Non-potable urban uses; and
7. Potable reuses.

Nevertheless, even though water reclamation and reuse is practiced in many countries, the existing levels of reuse constitute only a small fraction of the total volume of municipal and industrial wastewater effluent that is generated. Therefore, much of the potential largely remains untapped on as it is the consequence of a multitude of impediments, which include the lack of governmental support and the public resistance to planned indirect potable usages based on psychological fears.

Technologies that are commonly applied for water reclamation and reuse include the following:

1. Membrane separation technologies;
2. Carbon adsorption;
3. Advanced oxidation; and
4. Ion exchange, etc.

Membrane processes have been pinpointed to be one of the most significant developments because the final products have the flexibility to be tailored according to needs based on considerations such as nature of treatment applications and water quality criteria, to name a few. Membranes had almost exclusively been limited to desalination usages previously but they are increasingly being investigated for their potential in wastewater applications, which will allow for the production of high quality effluents that are suitable for reclamation and reuse. The added benefit of enhanced contaminant removal not only enables the product water to be fit for quality reuse applications but also lessen health risks. As the need for indirect potable reuse intensifies to augment existing water supplies, membranes are expected to be one of the predominant treatment technologies of the future (Metcalf and Eddy 2004; Farooque et al. 2008).

13.2 FO HYBRID SYSTEMS

Due to the fact that FO produces diluted salt solutions that are not directly usable or potable downstream of the membrane unit, FO is typically combined with other processes to create hybrid systems for reasons of practicality and applicability. In the following sections, several FO hybrid systems such as the FO dilution-reverse osmosis (FODN-RO) and FO-NF hybrid will be examined.

13.2.1 FODN-RO Hybrid Systems in Water Reclamation and Reuse

RO and NF membrane processes are commonly employed for desalination and have been tested and utilized since the late 1970s for advanced treatment of reclaimed water leading to indirect potable reuse. Provided that adequate pretreatment is implemented, these processes are capable of producing high grade potable water from reclaimed waters, surface or ground waters compromised by discharge from wastewater treatment plants (Xu et al. 2005; Bellona and Drewes 2007). Nevertheless, inherent drawbacks remain as these processes are often energy intensive, prone to membrane fouling and the rejection of salts, nutrients, and organic micropollutants by many semi-permeable membranes is incomplete.

One way to moderate energy consumption during RO desalination is via the dilution of the highly saline seawater feed stream since dilution will reduce the innate osmotic pressures that must be overcome to produce RO or NF permeates (Glueckstern and Priel 1998). The relatively low salinity of most impaired and reclaimed waters makes them good candidates for such dilution purposes (Lew et al. 2005); still, direct dilution/combination of the streams may contaminate and alter the chemistry of the feed stream to the desalination process, possibly aggravating membrane fouling and subsequently lowering product water quality. Consequently, pretreatment of impaired water before desalination of diluted saline water becomes a necessity. Yet again, many of such advanced pretreatment processes may still be chemically and/or energy intensive and thus does not offer any new advantages over existing desalination systems (Lew et al. 2005).

Recently, researchers started to explore and assess the treatment performance and economics of the FODN-RO hybrid process to co-treat impaired or reclaimed water and seawater for reuse (Cath and Childress 2005; Cath et al. 2009, 2010; Yangali-Quintanilla et al. 2011; Hancock et al. 2012). FO, as the core concept running through the entire book, operates based on the osmotic pressure differences across a semi-permeable membrane rather than hydraulic pressure differentials, such as in the case for RO and NF, to extract clean water from the feed. The FO-NF hybrid process presented here utilizes the chemical potential of a saline stream (pretreated seawater) as the driving force for osmotic dilution in which water from an impaired stream (reclaimed water) spontaneously diffuses through a semipermeable membrane (FO membrane) and into this saline stream. The resultant diluted saline stream, or more appropriately, brackish water stream, will go through a RO stage downstream to produce high quality product water for water reuse. By using this hybrid process, several merits related to energy consumption and product water quality can be realized:

1. Seawater would be diluted before desalination, which reduces the energy cost of desalting the seawater;
2. Pretreatment of impaired water reduces the fouling propensity of water in the RO stage;
3. Multi-barrier protection of product water since contaminants present in the impaired water are prohibited from entering the product water through two established barriers - FO membrane and then RO membrane units; and
4. An opportunity for safe reuse of impaired water can be realized.

Moreover, unlike the conventional FO or osmotic membrane bioreactor (OMBR), the hybrid system, operating in the osmotic dilution mode, does not require closed loop draw solution reconcentration; instead, the process can operate with both the feed and draw solution in a once through flow configuration (Cath et al. 2006). This unique operating configuration not only offers the same benefits of FO such as low energy operation and high resistance to irreversible fouling, but also eradicates the energy costs associated with re-concentrating the draw solution and reducing operational complexities (Hancock and Cath 2009).

An example of the hybrid process is shown in Figure 13.1 below. An impaired water stream in ODN 1 osmotically dilutes pretreated seawater as shown in the

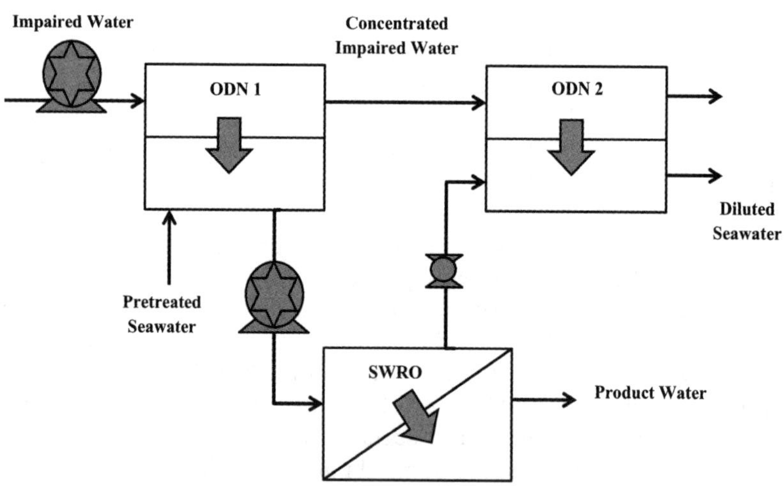

Figure 13.1. A schematic of FODN-RO Hybrid System for co-treating seawater and reclaimed water

SOURCE: Adapted from Cath et al. 2010; reproduced with permission from Elsevier figure. The diluted seawater, now a brackish water stream, will go through seawater RO stage to produce high quality product water for reuse at reduced energy consumption. Before discharge, the concentrated SWRO brine can be osmotically diluted again in the ODN2 by the concentrated impaired water, reducing the environmental impacts associated with discharging concentrated brines back to the ocean. The concentrated impaired water from ODN 2 can be used for other beneficial purposes such as undergoing further dewatering to produce high grade fertilizer or being returned to a wastewater treatment plant for retreatment (Holloway et al. 2007).

The FODN-RO hybrid system is still at its early stage of development to date. Bench-top and pilot-scale studies have been conducted to examine the viability and feasibility of the system based on considerations of parameters such as water flux, membrane fouling, transport of solutes and economic viability (Cath et al. 2009, 2010; Yangali-Quintanilla et al. 2011; Hancock et al. 2012). A comprehensive life cycle assessment of the hybrid process compared to conventional seawater reverse osmosis treatment had also been provided by Hancock et al. (2012).

13.2.1.1 Water Flux

Water flux through the FO membrane was generally low, for example Cath et al. (2009) reported approximately a water flux of 5.6 LMH through the FO membrane at 35 g/L sea salt draw solution concentration in the FODN-RO hybrid system. The relatively slow water permeation was attributed to the low osmotic driving forces generated by the synthetic seawater (Cath et al. 2010). Operation at higher osmotic pressures can produce greater water flux, but water fluxes in FO are still lower than those observed in most RO applications with equivalent hydraulic pressure driving forces. Table 13.2 is adapted from Cath et al. (2010), and summarizes FO water permeability results with varying draw solution concentration. The FO membrane used was provided by HTI, orientated with the active layer of membrane facing the feed solution.

Table 13.2. FO water permeability results with varying draw solution concentration

Draw Solution concentration (g/L)	Water flux (L/m^2-h)	Pi (MPa)	Specific water flux (L/m^2-h-MPa)
5	1.1	0.35	3.14
10	2.1	0.70	3.00
15	3.0	1.05	2.86
20	3.6	1.39	2.59
25	4.2	1.74	2.41
30	4.9	2.09	2.34
35	5.7	2.45	2.33
70	8.3	4.90	1.69

The results indicated that FO water flux is highly dependent on the osmotic driving forces and draw solution concentration. In addition, there is a diminishing return on providence of additional driving force. At higher draw solution concentration, the water flux increment is also observed to deviate from linearity. The specific water flux declines due to dilutive internal concentration polarization (ICP) phenomenon, where the draw solution concentration in the porous support layer of the membrane is lowered by the water molecules diffusing from the feed into the draw solution, which reduces the effective osmotic driving force (McCutcheon and Elimelech 2006). The phenomenon of dilutive ICP was previously reported in other literatures as well (Gray et al. 2006; Tan and Ng 2008; Tan and Ng 2010b) where the draw solutions flow against the support layer (active layer facing feed water). Although the AL-FS configuration will incur ICP during osmosis, it is still the preferred orientation as it allows for reduced membrane fouling propensities and thus allows a less significant loss of flux under operation over protracted periods of time, as compared to the configuration where active layer faces the draw solution (Gray et al. 2006). Lastly, it was also reported that feed water quality had a minimal effect on water flux through the FO membrane (Cath et al. 2010).

Water flux magnitudes are also dependent on the rate of draw solution recirculation speeds where the higher the recirculation rates, the higher the obtained water fluxes. A study reported that the fluxes were mainly provided by osmotic pressure differences (as the pressure drops across the inlet and outlet of the flow cell is negligible), but no further explanation was put forth for the flux increments that corresponded to the increased draw solute recirculation rates (Tan and Ng 2008). As a suggestion, the increased turbulence by higher recirculation rates could have helped mitigated the severity of the ECP phenomenon and helped improve water fluxes at a given bulk osmotic pressure difference. Additionally, faster recirculation rates could have helped maximize the water concentration differences across the membrane and help support higher fluxes over extended operational periods.

13.2.1.2 Membrane Fouling

Membrane fouling is one critical aspect in the operation of any membrane-based system, and it is of special importance and great concern in the present FODN-RO hybrid system in virtue of employing impaired water as the feed stream. Fouling can change membrane characteristics and thus affecting membrane water flux and contaminant rejection. Severe fouling may even permanently damage the membrane and reduce the overall productivity of a membrane system, compromising the economic feasibility of the hybrid system (Cath et al. 2009).

Both bench- and pilot-scale fouling tests had been conducted and reported for the hybrid system in the literatures (Cath et al. 2009, 2010; Yangali-Quintanilla et al. 2011). Generally, for short-term bench-scale experiments, minimal fouling of the FO membrane was observed. Water flux remained relatively constant and the authors attributed this observation to several factors, including low osmotic pressures of the feed solution and limited transport of foulants to membrane due to lower water flux (Cath et al. 2010). However, the combination of a hydrophilic FO membrane and low operating pressure allegedly plays a more prominent role in reducing the rate of membrane fouling during bench-scale experiments (Mi and Elimelech 2010). It was also reported that pressure-driven membrane processes under comparable experimental set up conditions (i.e., water flux, temperature, and feed characteristics) experienced more substantial membrane fouling and flux decline (Agenson and Urase 2007).

However, fouling was reported to be more conspicuous for long-term pilot-scale experiments. The FO water fluxes declined more rapidly when the hybrid process was tested with a continuous supply of impaired waters, such as secondary effluents that contained and deposited more suspended solids on the FO membrane. It was claimed that the components of natural organic matter (NOM) present in impaired waters such as secondary (wastewater) effluents are the most important foulants in water reuse facilities based on membrane operations (Zheng et al. 2010). During the FO osmotic dilution stage, interactions between the membrane and the NOM present in the impaired water cause membrane fouling and therefore, a decrease of the membrane flux (due to increased water mass transfer resistance) besides a flux decrement due to dilutive ICP. The accelerated reduction of water flux observed in pilot-scale tests after the period of sustained higher flux was ascribed to be a result of reverse diffusion of solutes from the draw solution that are becoming immobilized by the cake layer on the feed side of the membrane. This phenomenon may cause a foulant-enhanced external concentration polarization (ECP) layer where the draw solution solutes accumulate on the feed side of the membrane, reducing the effective osmotic pressure difference and subsequently decreasing the water flux (Cath et al. 2009).

In a bid to better understand the fouling phenomenon so as to further reduce FO fouling, the effect of membrane cell orientation on fouling was also investigated (Cath, et al. 2010). The membrane cell was reported to orientate in such a way that the feed stream flowed either above or below the membrane, despite the fact that membrane orientation remained the same with the active layer always facing the impaired feed water feed.

Figure 13.2 shows the water flux as a function of time for two experiments with different membrane cell orientations. The results demonstrated that with the new configuration where feed flowed below the membrane, water flux declined minimally as compared to the previously observed decline of approximately 10% per day when the feed channel was above the membrane. The reported data well corroborated the hypothesis put forth by the authors, whereby the shear flow and gravity settling was postulated to prevent accumulation of solids on the membrane surface and thus avoiding flux decline due to membrane fouling. The solids accumulated in the feed channel were observed to settle onto the side-plate of the membrane cell rather than the membrane itself, resulting in lower membrane fouling. Lastly, the authors (Cath et al. 2009, 2010) also mentioned that minimal membrane fouling was observed for the downstream SWRO, nor did the RO membranes received or required cleaning during the three-month deployment

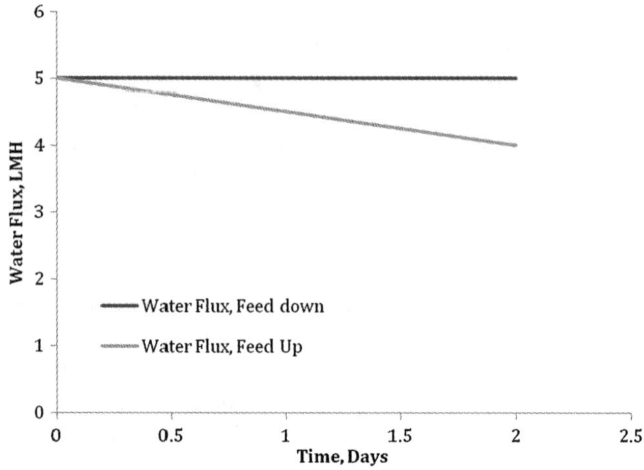

Figure 13.2. Water flux as a function of time with different FO membrane cell orientations

and operation of the hybrid pilot system. The reported observation nicely demonstrated the precise advantage of osmotic dilution as an effective pretreatment stage for the co-treatment of a highly impaired water stream prior to SWRO.

13.2.1.3 Transport of Solutes

For purposes of water reclamation and reuse, the removal and elimination of traditional and emerging contaminants are major concerns to regulators, utilities, and the general public. The concept of offering a multi-barrier treatment scheme for water reuse has been gaining nods from regulators and used as a measure for public assurance. Although FO membranes are relatively new and the rejections of such membranes towards contaminants present in impaired water have not been extensively documented, the membranes are dense, semi-permeable and are fabricated from similar polymers as used in RO membrane productions. As a result, it is expected that the hybrid FODN-RO process would offer an additional level of rejection for solutes.

In FO processes utilizing a closed draw solution loop in conjunction with RO reconcentration, loss of the osmotic agent, such as sodium chloride, into the feed solution and into the RO permeate can make the process uneconomical, or require treatment of the feed solution before discharge. For the contemporary FODN-RO hybrid system, in virtue of the high concentration of solutes present in the draw solution, in this case seawater, transportation of inorganic compounds occurs mostly from the draw solution into the impaired feed stream (i.e., reverse solute diffusion). This may give rise to a reduction in the driving force within the osmotic dilution process. Nevertheless, since the draw solution will not be operated in a close loop configuration, reverse diffusion of TDS is unlikely to impinge on the process efficiency and replenishment of draw solutes is also not required. In the

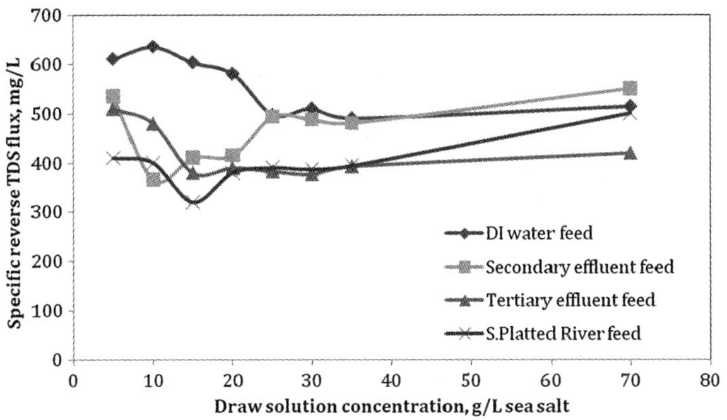

Figure 13.3. Specific reverse TDS flux versus draw solution concentration with varying feed stream

Table 13.3. Rejection values for ammonia, nitrate and TOC by the system

Parameter	*FO membrane % Rejection*	*RO membrane*	*Total hybrid system*
Ammonia	74	77	94
Nitrate	78	82	97
TOC	85	99	>99.9

pilot studies conducted by (Cath et al. 2009, 2010), the reverse diffusion of TDS through the FO membrane, though greater than that through the RO membrane, was within the levels reported by other studies. Sodium and chloride ions accounted for the major portion of the diffusing TDS (97% of the solute mass flux), arguably because they had exhibited the highest diffusion gradient. Figure 13.3 (Cath et al. 2010) presents the specific reverse TDS flux as a function of draw solution concentration to various feed streams. The results revealed that the feed solution chemistry has a more profound influence on solute flux at lower draw solution concentrations, and that TDS flux for the different feed solutions converges at higher draw solution concentrations. This is purportedly caused by the faster diffusion of water molecules at higher draw solution concentrations, which consequentially hinder the reverse diffusion of salts through the membrane.

Transport of ammonia, nitrate and TOCs are of special interests as well since they are generally present in the impaired water streams and membrane treatment generally shows moderate rejection towards ammonia and nitrate. Table 13.3 was developed from Cath et al. (2009), showcasing the results of the individual processes and entire hybrid system rejection towards ammonia, nitrate and TOC.

Table 13.4. Concentrations of selected micropollutants at various treatment stages in the hybrid system

	Feed (ng/L)	*Draw solution (ng/L)*	*RO permeate (ng/L)*	*Rate of transport through the FO membrane (ng/m^2-h)*
Diclofenac	155	65	<1	1.3
Gemfibrozil	960	1650	<2	32.3
Ibuprofen	385	255	<5	5
Naproxen	435	360	<1	7
Salicylic acid	260	85	<5	1.7
TCEP	800	625	<20	12.2

The presented data indicate that both FO and RO membranes provided moderate rejection of ammonia and nitrate, but fortunately, the overall system rejection was relatively high. The overall multiple-barrier approach for both ammonia and nitrate removal was demonstrated to be effective. Similar to ammonia and nitrate, UV absorbing compounds, which were used in the study as a surrogate for TOC compounds, were moderately rejected by the FO membrane, more highly rejected by the RO membrane and their presence in the FO/RO system was almost at the detection limits of the analytical equipment. This suggested that organic matter, especially those containing aromatic moieties, could be efficiently removed by the present FODN-RO hybrid system.

Transportation of selected organic micropollutants, including commonly occurring trace organic compounds present in secondary effluent, was also studied and presented at both bench- and pilot-scale levels in the literature reports (Cath et al. 2009, 2010). Table 13.4 summarizes the concentrations of trace organic chemicals before and after an 8-day operation of the pilot system using 35 g/L sea-salt draws solution and secondary effluent feed. The results provided affirmation that the FO membrane provided relatively high rejection of most of the targeted organic micropollutants, and the remainder was well removed by the RO membrane since four out of six compounds exhibited concentrations below the detection limit in the RO permeate stream.

13.2.1.4 Energy Saving and Economic Viability

The feasibility of the hybrid system in co-treating impaired water and seawater strongly depends on economics of the system. In pressure-driven membrane processes such as RO or NF, economic feasibility is very much dependent on the optimal recovery of pure water from the source water. At higher than optimal recoveries, reject concentration rises, and this increases energy requirements. On the other hand, lower recovery RO, while using less energy, inefficiently and unnecessarily wastes a larger portion of energy on pretreatment of the feed water stream. In an osmotically-driven membrane process such as FO, the maximum economically feasible recovery depends on the osmotic pressure and flow rates of the (draw and feed) streams. The draw solution is diluted when recovery increases. This dilution, however, will decrease the osmotic driving force and lower the water flux. Consequently, this increases the required membrane surface area to achieve the additional water production (Cath et al. 2009).

In order to determine the maximum economically feasible recovery for the FODN-RO hybrid system, Cath et al. (2010) developed a model where the impaired water recovery can be increased incrementally, and system feasibility is determined at each unit increment in impaired water recovery. Water flux and performance data from reports (Cath et al. 2009, 2010) were used to configure the model. In the constant flux model, the RO influent stream was maintained at constant flow rate and the flow rate of water across the FO membrane was increased in steps of one percent of the RO influent flow rate. At each step, the incremental capital costs of the FO system were calculated, along with the incremental energy savings due to dilution of the RO influent. Cath et al. (2010) reported the benefit cost ratio (BCR) and cumulative cost of implementing the osmotic dilution as a function of the fraction of impaired water recovered by the process. According to the model, the dual barrier hybrid process is estimated to achieve favorable economic returns (i.e., BCR greater than unity) during operation, with up to 63% recovery from the impaired water stream. Beyond 63% recovery, the capital costs associated with increasing required membrane area for osmosis will counter-balance the saving from reduced energy consumption of the SWRO process. The cumulative cost savings of the coupled process was also found to decrease initially because the reduction in energy cost for operation of the SWRO units offsets the cost of installing additional FO osmotic membrane units. However, as the water recovery increases in the FODN unit, the lower driving force for water flux demands increasingly more membrane surface area to achieve the same water production. At the optimal recovery, the additional cost associated with the incremental amount of membrane area required to achieve higher recovery of impaired water balanced the energy cost savings realized by same deducted amount of water recovery in the SWRO system nicely.

In a separate study (Yangali-Quintanilla et al. 2011) where a FODN-LPRO system was tested, it was estimated from literatures that the energy consumption for conventional methods of desalting water with high pressure RO membranes lies between 2.5 and 4 kWh/m^3 (Fritzmann et al. 2007), while existing water reuse installations using membrane filtration (microfiltration or ultrafiltration) and RO have an overall energy demand of 1.5–1.7 kWh/m^3 (de Koning et al. 2008). The energy consumption associated with the proposed FO-LPRO concept was revealed to lie within a conservative estimated range of 1.3–1.5 kWh/m^3 for desalinating the diluted seawater. Thus it can be demonstrated that hybrid FO-LPRO system is an attractive consideration from an energy-saving stand point of view. The comparison of FO-LPRO to RO is presented in Figure 13.4.

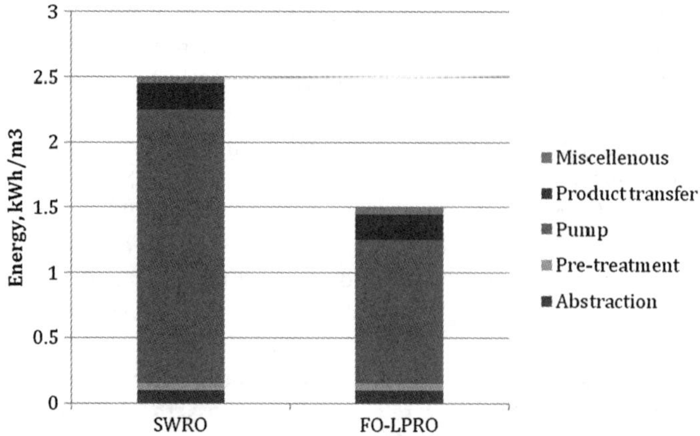

Figure 13.4. Comparative estimation of energy cost for SWRO and FO-LPRO

13.2.1.5 Alternative Direct Water Reuse after FO Process

It is possible to have direct water reuse application with low salinity pretreated seawater that was diluted osmotically using impaired water with a preceding FO process. It was stated by Yangali-Quintanilla et al. (2011) that low-salinity water, in the range of 15 g/l of TDS, can be achieved by the FO stage in the FODN-RO hybrid system, using seawater as draw solution and impaired water as feed solution. They suggested that this salinity could be further reduced to 6–10 g/L via the following: 1) using a reduced volume of seawater at the beginning of each batch FO cycle; and 2) using more FO membrane area.

Although the USEPA guideline (Metcalf and Eddy 2004; USEPA 2012) suggested that the TDS levels of water for irrigation reuse or other purposes should be less than 450 mg/L. In detail, high salinity levels may lead to a decrease in productivity for certain crops and the destabilization of soil structure (Hussain and Al-Saati 1999). One possibility for direct reuse is the use of the low salinity water as feed water for aquaculture. Low salinity (4–10 g/L) water for shrimp farming has also been practiced in Thailand widely (Szuster and Flaherty 2001). On another note, in the country of Saudi Arabia, there are also incentives to move from seawater aquaculture to brackish water aquaculture (shrimps), by employing partial desalination to produce the diluted seawater using FO process (Hussain and Al-Saati 1999).

13.2.2 FODN-RO Hybrid System in FOMBR for Water Reclamation and Reuse

As the water and wastewater treatment industry moves into an era of energy and climate concerns, the energy consumption and consequently, carbon footprint of treatment processes will be under close scrutiny. Having lower consumptions and/or better energy efficiencies will be crucial in determining the makeup of the individual units in wastewater treatment trains of the future. Wastewaters in general, such as those from domestic and industrial sources possess lesser osmotic pressure but elevated fouling tendencies in comparison with seawater. Since reduced fouling propensity is one of the prominent advantages of FO, it will be good to look into the application and integration of FO process into existing wastewater treatment units to create novel systems that have the potential of demonstrating lowered energy consumptions to tackle global energy and climate anxieties. In short, FO holds great promise in improving the future of the wastewater treatment industry.

The novel integration of a FO membrane process into the conventional MBR aeration tank resulted in the FO membrane bioreactor (FOMBR), as pioneered back in 2009 by the research group of Achilli et al. from the University of Nevada. While limited, several papers on the FOMBR have been published since then and have helped advance the academic knowledge on this exciting new system (Lay et al. 2011; Lay et al. 2012).

As shown in Figure 13.5, an FOMBR is a hybrid of the FO membrane process (for water extraction), aeration process (for the cultivation of activated sludge to treatment wastewater contaminants) and lastly, a pressurized filtration stage such as RO or NF system (for drinking water recovery and draw solute regeneration). Water was extracted from the impaired water source, in this case, domestic wastewater, and causing dilution of the draw solution. Using a pressurized filtration stage such as RO or NF system in the downstream, product water is obtained and draw solution is regenerated as well. FOMBRs differ from conventional MBRs in the manner which water is being extracted from the mixed liquor (inside the aeration tank). While conventional systems utilize hydraulic pressures and vacuum suction for permeate production, FOMBRs leverage on osmotic pressure differences that exist across the FO membrane, between the mixed liquor and the draw solution (Achilli et al. 2009; Lay et al. 2011; Lay et al. 2012). The impaired water source here is domestic wastewater that will be treated biologically by the activated sludge.

The non-porous nature of the semi permeable FO membranes has allowed Achilli et al. (2009) to reject 98% of total organic carbon (TOC) in the FOMBR

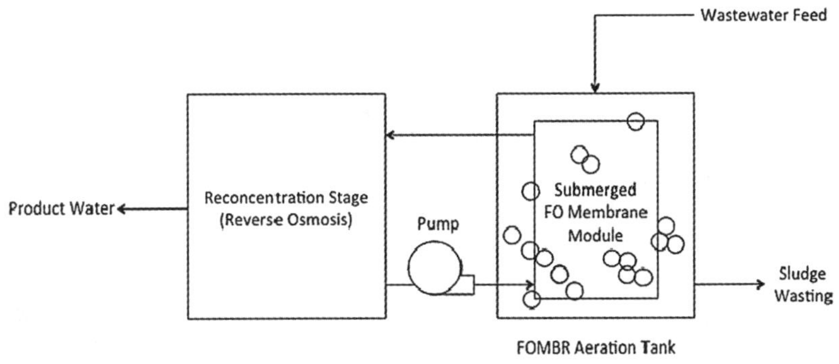

Figure 13.5. Schematic diagram of an FOMBR system
SOURCE: Adapted from Achilli et al. 2009; reproduced with permission from Elsevier

Table 13.5. TOC and NH_4^+-N removal efficiencies at various stages of the FOMBR system

	Rejection ratio of FO membrane	*Removal ratio of FOMBR*	*Total rejection ratio (FOMBR+RO)*
Total Organic Carbon (mg/L)	0.979	0.998	0.998
Ammonium-N (mg/L)	0.9	0.977	0.994

SOURCE: Data from Achilli et al. 2009.

system as presented in Table 13.5. The high removal of TOC would have reduced the amount of soluble organics within the diluted draw solution, which were reported to be the contaminants majorly responsible for the fouling of the RO process downstream (Barger and Carnahan 1991; Vrouwenvelder and van der Kooij 2001). In addition, the presence of RO downstream allowed for a polishing effect on the effluent quality as it improved the overall TOC and NH_4^+-N removal due to the multi-barrier effect offered by the two established membrane barriers in hybrid systems. Hence, it has been definitively proven that the FO-RO hybrid system employed in the FOMBR system is a better candidate than conventional MBRs that are employing micro-porous membranes for water recovery from impaired water sources with high fouling propensities due to the multi-barrier advantage offered by hybrid configurations, and importantly, the high rejection performance of the FO membranes.

The presence of FO membranes presents challenges that are unheard of for conventional MBRs. The high rejection performance of FO membranes gives rise to the problem of acclimatizing activated sludge to high total dissolved solids (TDS) accumulation within the mixed liquor (Lay et al. 2011). Yet this is a double-edged sword whereby high retention performance is also a highly sought after character especially for treatment trains in areas that recycle used water over protracted periods of time. High membrane rejection lengthens the retention time of small molecular weight (M_w) contaminants that will accumulate within recycled water bodies, and thus allowing sufficient time for biodegradation by the activated sludge systems. MBRs employing high retention membranes such as FO membranes, NF membranes or membrane distillation (MD) membranes are termed high retention membrane bioreactors or HRMBRs (Rautenbach and Mellis 1994; Choi et al. 2002; Phattaranawik et al. 2008; Lay et al. 2011). HRMBRs are novelties that introduce the issue of salinity stresses on the activated sludge population and this is an area that is yet to be studied. Nonetheless, clues can be obtained from high salinity studies performed on conventional systems. Moderate to high mixed liquor TDS values have been acknowledged to confer adverse impacts onto bacterial populations that were not acclimatized to highly saline environments. Salt concentrations exceeding 1% have demonstrated the ability to invoke bacterial plasmolysis and/or loss of metabolic activities (Kargi and Dincer 1996; Kargi and Dinçer 1998; Dinçer and Kargi 2001). With genetic variation amongst any given bacterial population, high salinities enforce a strong selection pressure for bacteria capable of withstanding osmotic stresses while the unadapted ones die off, giving rise to a lower genetic diversity in the steady state stage of reactor operation.

Another study by Reid et al. (2006) discovered that salinity shocks (in the range that was studied by them) were likely to pose more challenges to activated sludge viability than those thriving in MBRs with steady salinity levels. This finding sheds important insights on the impact of TDS accumulation within the FOMBR mixed liquor, as salinity increases are not introduced as spikes but a very gradual elevation of salinity within the aeration tank contents (directly as a result of high membrane rejection). Therefore, while there are no papers currently discussing this aspect of FOMBR, it can be extrapolated from existing wisdom that the FOMBR sludge innoculum will have an arduous task of adapting to the unstable (and ever-rising) salinities and will perhaps, produce unprecedented interesting trends in data.

13.2.3 FO-NF Hybrid System in Water Reclamation and Reuse

FO processes are fundamentally different from conventional membrane-based water extraction methods because clean water is not obtainable downstream of the membrane unit. FO processes leverage on the osmotic pressures exerted by a highly concentrated salt solution that serves as the draw solution for water extraction from the feed stream across the FO membrane. The product that is directly downstream of the FO unit is a stream of diluted salt solution that is usually not usable directly (unless it is designed as direct potable reuse by using glucose as draw solution as seen in hydration bags). Thus, additional recovery treatment will be required to be performed on the diluted salt stream. The FO-RO hybrid is the most commonly used system applied for draw solute reconcentration and product water recovery. However, the associated intense energy consumption for RO filtration is in stark incongruence with the very reason for the rising popularity of FO processes- lower energy usage.

NF or low pressure RO offers a lower energy-consuming alternative to RO filtration for draw solute recovery and the various benefits attainable for the FODN-RO configuration (as discussed in the preceding section) can also be achieved by a hybrid FO-NF process (Tan and Ng 2010a; Zhao et al. 2012). Thus, it is expected that FO-NF hybrids will have lower energy consumption than FO-RO systems for the same feed streams. The diluted draw solution produced by the water molecules passing through the FO membrane and into the salt stream will be reconcentrated by an NF unit downstream, allowing the recovery of a drinking water product and regeneration of the draw solution, as depicted in the schematic within Figure 13.6.

The catch for the application of lower energy consuming NF units is the restriction on the types of draw solutes that can be used. While the larger pores of

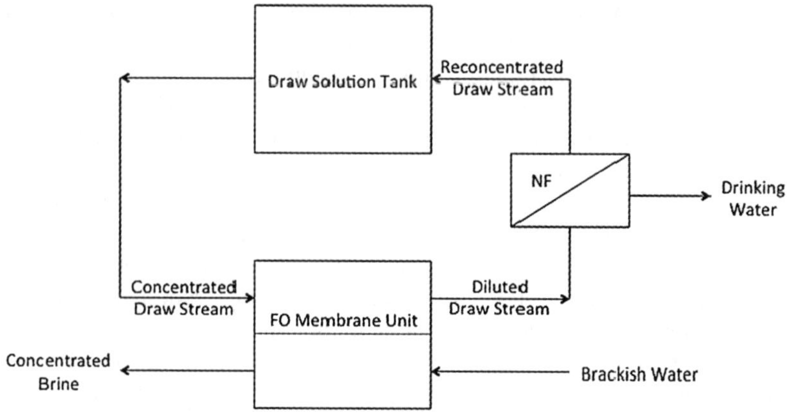

Figure 13.6. Schematic diagram of the hybrid FO–NF system configuration
SOURCE: Adapted from Zhao et al. 2012; reproduced with permission from Elsevier

the NF membranes allow for reduced energy consumption, it also entails a less than desired rejection of smaller solute ions such as sodium and chloride ions. Commercially available sodium sulphate salt has been successfully applied in a recent research for the purpose of brackish water desalination using the FO-NF configuration (Zhao et al. 2012). In another similar research running simulated seawater (0.6 M sodium chloride solution) with laboratory scale FO and NF systems, the hybrid FO-NF process with single-stage two-pass NF regeneration has been concluded to be of good performance (Tan and Ng 2010a). Both papers have suggested sodium sulphate and magnesium sulphate as appropriate draw solutes for FO-NF hybrid systems as the larger ionic radius of the sulphate ions will allow for better solute rejection performance by either loose RO or dense NF membranes (via principles of size and charge exclusion). Although sodium ions are much smaller (than sulphate ions), they will be all rejected together with the sulphate ions for the maintenance of charge neutrality in the source and product water stream.

The advantages of a hybrid FO-NF configuration are even more evident against a standalone RO configuration (instead of FO-RO hybrid) for the purpose of brackish water desalination. It was discovered that the hybrid arrangement offers many benefits over the stand-alone RO process (Zhao et al. 2012). The benefits consist of the following:

1. Lowered energy consumption from application of lower hydraulic pressures;
2. Lower fouling propensities and consequently, lower flux declines due to fouling;
3. Better flux recovery upon membrane cleansing;
4. Elimination of pretreatment for source water;
5. Better effluent qualities; and
6. Elimination of the need for chemical cleaning.

In conclusion, with the correct selection of salt as the draw solute, FO-NF hybrid systems can be an improvement over FO-RO hybrids and also, excel and outperform traditional standalone RO systems and therefore, should be the direction of future studies on FO hybrid configurations.

13.2.4 FO-MD Hybrid System in Water Reclamation and Reuse

Membrane distillation (MD) is a membrane process driven by thermal energies to power the transportation of (water) vapor through a porous hydrophobic membrane. The feed stream, which is usually seawater, that is undergoing treatment by the MD process must be in direct contact with one side of the membrane and should not cause pore wetting. The separation process fails once the pore is wet as the feed stream can pass through the membrane directly. Henceforth it is of utmost importance that the MD membrane is highly hydrophobic with a high contact angle and small maximum pore size (Qu et al. 2009). The high hydrophobicity will ensure that the water surface tensions are of appropriate magnitudes to prevent liquid solutions from the feed side of the membrane from entering the pores. More specifically, MD is a thermal process that requires the establishment of a temperature difference between the feed and permeate stream. Therefore, the driving force empowering the MD processes is the vapor pressure difference (between the feed and permeate stream) that is created due to temperature disparities. Under the premise that the feed stream does not contain any volatile solutes, MD systems theoretically enable only water vapor to be transferred across the membrane pores, thus allowing for the production of a pure water permeate stream. Thus MD process has a high salt rejection ratio and additionally, the process is also less dependent on the initial feed salinity. Consequentially, MD is often proposed for the purpose of desalination (Hsu et al. 2002; Peng et al. 2005; Qu et al. 2009), especially for seawaters of salinity higher than the usual 35 g/L NaCl concentration, such as those found in the Middle Eastern seas.

On the other hand, FO is a spontaneous process driven by osmotic differences that exists across the FO membrane between the feed and draw stream. Thus in contrast with conventional membrane processes, FO systems do not require hydraulic pressures for operation and is therefore, more energy saving by nature. As the driving force for permeate production is induced by the osmotic pressures offered by a highly concentrated salt solution, only an undesirable diluted salt stream is obtainable downstream of the membrane unit. Hence, FO processes have to exist as hybrids by being coupled to other techniques downstream to allow it to become a sustainable methodology of water recovery from impaired water sources and seawater. Since MD processes operate independently of the feed salinity, their traditional application as a seawater desalination technique can be ported over for the reconcentration and extraction of water from the diluted draw stream of FO processes. The result of this integration is the FO-MD hybrid system where water extraction from the impaired water source is performed by the FO unit, and the drinking water product is produced from the diluted draw solution via the coupled MD process.

The FO-MD hybrid system was proposed earlier on as a means of water reclamation in space for astronaut consumption (Cath et al. 2007). More recently,

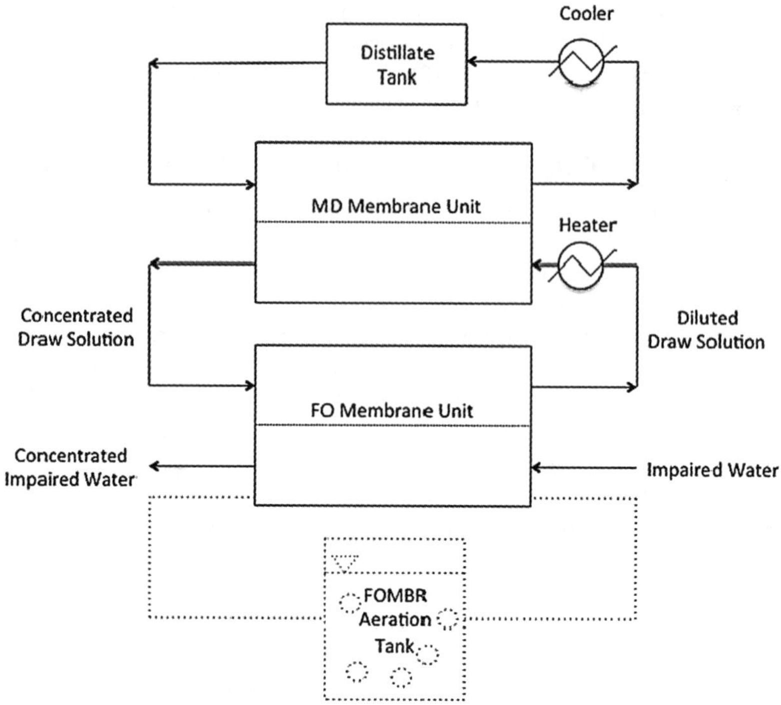

Figure 13.7. Schematic demonstrating the FO-MD concept and process flow. Drawings done in solid lines represent the design flows as adapted and inspired from existing literatures. Dotted lines represent a possible manner that the FO-MD concept can be integrated into the FOMBR system for water reclamation from wastewater sources, as envisioned by the authors of this book chapter

a FO-MD system was successfully applied for the concentration of proteins (Wang et al. 2011). The FO-MD hybrid system has been evaluated as a promising system for concentration of proteins and pharmaceuticals because of the low temperature and pressure needs. Additionally, process repeatability, controllability and predictability have also been exhibited by the studied FO-MD hybrid setup, making it a potential candidate as the mainstream FO hybrid configuration of the future.

A schematic of a FO-MD hybrid process design, as inspired by the aforementioned readings, is presented in Figure 13.7. The original process flow in the literatures have been adapted and further developed to put forth suggestions as to how the FO-MD hybrid can be incorporated into the FOMBR (FO membrane bioreactor) system, as envisioned by the authors of this book chapter.

As presented in Figure 13.7, the mixed liquor within the aeration tank of the FOMBR can be recirculated out into an external FO filtration unit. This will be a better candidate than the FO-NF hybrid to perform reconcentration of draw solutions on the basis of smaller ions such as sodium chloride. Where sodium and chloride ions could leak out and lead to poor rejection performance for FO-NF systems, FO-MD systems will provide a much better solute rejection (100% rejection theoretically) to produce product water of much higher qualities. Despite the numerous good qualities and advantages of FO-MD systems that have been discussed till now, the involvement of heat in this hybrid system poses some limitations and challenges during operation. As in the case for concentration of proteins, the possibility of protein denaturation due to heat is an important consideration. Similarly, the elevated temperatures of draw solutions due to the reconcentration process via MD can be transferred across the FO membrane and into the mixed liquor (in the case of integrating FOMBR into the FO-MD hybrid). Slightly higher temperatures in the aeration tank might be beneficial for bacterial growth but detrimental when it exceeds the optimal range. This necessarily places an upper limit on the operational temperatures at the MD unit, which consequentially, limits the obtainable operational flux from the FO-MD hybrid.

13.2.5 FO-Thermal Hybrid System in Water Reclamation and Reuse

The actual makeup of FO hybrid systems is dependent on numerous factors, and one in particular, the type of draw solute that is being utilized. For instance, the use of solutes with smaller ionic radii like NaCl dictates that FO-NF hybrid configurations will be inappropriate for good water recovery and solute separation, which is very much due to the larger pore sizes (as compared to RO) in loose RO and/or dense NF membranes. Furthermore, the development of a good draw solute that allow for an energy efficient manner through which drinking water and draw solutes can be recovered is one of the most important factors deciding the viability of FO processes becoming mainstream in the future. An ideal draw solute should satisfy the criteria listed below (Cath et al. 2006):

1. Low energy consumption for regeneration;
2. Complete separation from drinking water product;
3. Low or no toxicity; and
4. Not chemically reactive with FO membranes.

A recent potential candidate to improve FO feasibility is the hydrogel polymer, which, under the influence of external stimuli, such as thermal and pressure exertion, can undergo reversible volume changes. Hydrogels possess a three dimensional, insoluble cross-linked network of polymers that can allow for a capacity to absorb a large water volume due to the hydrophilicity and flexibility of the polymeric chains (Gutowska et al. 1994; Muniz and Geuskens 2001). Their swelling upon water uptake (due to the dissociation of ionic species within the polymer) allows them to induce a higher osmotic pressure internally, which can be then tapped on to drive water extraction across an FO membrane. Hence, coupled with their ability to undergo reversible volume changes when external stimuli such as pH, temperature, electric field and mechanical stress are applied (Dai et al. 2009), it positioned them favorably as a novel draw solute that can meet the criterion of an ideal draw solute as laid out previously.

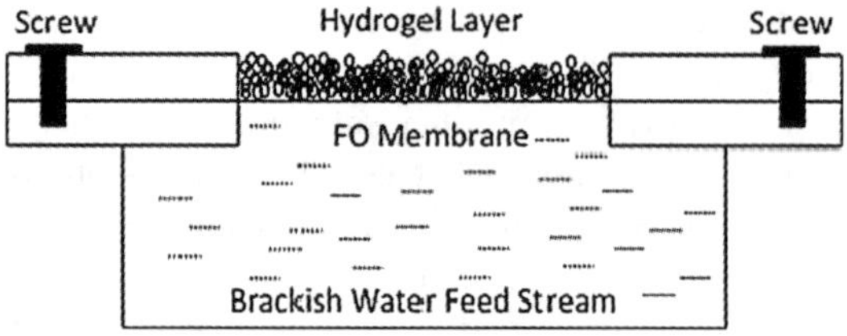

Figure 13.8. Schematic the apparatus for the FO flux measurements based on thermal responsive hydrogels

Since hydrogels have a gel-like or powder consistency, conventional existing hybrid systems, such as FO-NF and FO-RO, will be unsuitable for use with FO systems employing hydrogels. The downstream stage of the hybrid system should consist of a process that provide the required stimulus for swollen hydrogels to release the absorbed water within their polymer networks. It has been reported that hydrogel dewatering can be achieved using temperatures as low as 40 to 50°C (Li et al. 2011; Cai et al. 2013), which makes thermal stimuli responsive hydrogel polymers such as semi-IPN hydrogels an excellent candidate to study and improve on.

A schematic of a thermally responsive hydrogel setup is presented in Figure 13.8. Thermal stimulus was not incorporated directly into the experimental setup and required manual dewatering in a batch routine. For instance, Cai et al. (2013) reported that the hydrogel disc was immersed into 40°C water to release the absorbed water after each cycle of FO experiment. Thus for existing literatures on thermal water recovery from hydrogels, dewatering methods are ambiguous, heavily dependent on manual effort and/or non-continuous in nature. Therefore, in addition to efforts placed on the development of better performing hydrogels, the understanding will never be complete without investment into the development of appropriate water recovery design for such novel draw solutes. Future works should thus put in more effort in developing simple and effective systems that can incorporate controlled providence of external stimuli (e.g., pH dosing, turning on of magnetic fields, thermal irradiation, etc) to allow for better process control in scaled up system.

References

Achilli, A., Cath, T. Y., Marchand, E. A., and Childress, A. E. (2009). "The forward osmosis membrane bioreactor: A low fouling alternative to MBR processes." *Desalination*, 239 (13), 10–21.

Agenson, K. O., and Urase, T. (2007). "Change in membrane performance due to organic fouling in nanofiltration (NF)/reverse osmosis (RO) applications." *Separation and Purification Technology*, 55(2), 147–156.

Barger, M., and Carnahan, R. P. (1991). "Fouling prediction in reverse osmosis processes." *Desalination*, 83(1–3), 3–33.

Bellona, C., and Drewes, J. E. (2007). "Viability of a low-pressure nanofilter in treating recycled water for water reuse applications: A pilot-scale study." *Water Research*, 41(17), 3948–3958.

Cai, Y., Shen, W., Loo, S. L., Krantz, W. B., Wang, R., Fane, A. G., and Hu, X. (2013). "Towards temperature driven forward osmosis desalination using Semi-IPN hydrogels as reversible draw agents." *Water Research*, 47(11), 3773–3781.

Cath, T. Y., and Childress, A. E. (2005). "System and Methods for Forward Osmosis Assisted Desalination of Liquids." U.S. Patent 7914680 B2, Dec 6, 2004.

Cath, T. Y., Childress, A. E., and Elimelech, M. (2006). "Forward osmosis: Principles, applications, and recent developments." *Journal of Membrane Science*, 281(12), 70–87.

Cath, T. Y., Childress, A. E., and Elimelech, M. (2006). "Forward osmosis: Principles applications, and recent developments." *Journal of Membrane Science*, 281(1–2), 70–87.

Cath, T. Y., Childress, A. E., and Martinetti, C. R. (2007). *Combined membrane-distillation forward-osmosis systems and methods of use*, W. I. P. Organization, Board of Regents of the Nevada System of Higher Education, on Behalf of The University of Nevada, Reno.

Cath, T. Y., Drewes, J. E., and Lundin, C. D. (2009). *A Novel Hybrid Forward Osmosis Process for Drinking Water Augmentation using Impaired Water and Saline Water Sources*, Water Research Foundation.

Cath, T. Y., Hancock, N. T., Lundin, C. D., Hoppe-Jones, C., and Drewes, J. E. (2010). "A multi barrier osmotic dilution process for simultaneous desalination and purification of impaired water." *Journal of Membrane Science*, 362(1–2), 417–426.

Choi, J. -H., Dockko, S., Fukushi, K., and Yamamoto, K. (2002). "A novel application of a submerged nanofiltration membrane bioreactor (NF MBR) for wastewater treatment." *Desalination*, 146(1–3), 413–420.

Dai, S., Ravi, P., and Tam, K. C. (2009). "Thermo- and photo-responsive polymeric systems." *Soft Matter*, 5(13), 2513–2533.

de Koning, J., Bixio, D., Karabelas, A., Salgot, M., and Schäfer, A. (2008). "Characterisation and assessment of water treatment technologies for reuse." *Desalination*, 218(1–3), 92–104.

Dinçer, A. R., and Kargi, F. (2001). "Performance of rotating biological disc system treating saline wastewater." *Process Biochemistry*, 36(8–9), 901–906.

Metcalf, and Eddy (2004). *Wastewater Engineering, Treatment and Reuse*, 4th Ed., McGraw-Hill.

Farooque, A. M., Jamaluddin, A. T. M., Al-Reweli, A. R., Jalaluddin, P. A. M., Al-Marwani, S. M., Al-Mobayed, A. A., and Qasim, A. H. (2008). "Parametric analyses of energy consumption and losses in SWCC SWRO plants utilizing energy recovery devices." *Desalination*, 219(1–3), 137–159.

Foley, J. A., Ramankutty, N., Brauman, K. A., Cassidy, E. S., Gerber, J. S., Johnston, M., Mueller, N. D., O'Connell, C., Ray, D. K., West, P. C., Balzer, C., Bennett, E. M., Carpenter, S. R., Hill, J., Monfreda, C., Polasky, S., Rockström, J., Sheehan, J., Siebert, S., Tilman, D., and Zaks, D. P. M. (2011). "Solutions for a cultivated planet." *Nature*, 478 (7369), 337–342.

Fritzmann, C., Löwenberg, J., Wintgens, T., and Melin, T. (2007). "State-of-the-art of reverse osmosis desalination." *Desalination*, 216(1–3), 1–76.

Glueckstern, P., and Priel, M. (1998). "A hybrid SWRO/BWRO desalination plant." *Desalination*, 115(2), 203–209.

Gray, G. T., McCutcheon, J. R., and Elimelech, M. (2006). "Internal concentration polarization in forward osmosis: role of membrane orientation." *Desalination*, 197(1–3), 1–8.

Gutowska, A., Bae, Y. H., Jacobs, H., Feijen, J., and Kim, S. W. (1994). "Thermosensitive interpenetrating polymer networks: Synthesis, characterization, and macromolecular release." *Macromolecules*, 27(15), 4167–4175.

Hancock, N. T., Black, N. D., and Cath, T. Y. (2012). "A comparative life cycle assessment of hybrid osmotic dilution desalination and established seawater desalination and wastewater reclamation processes." *Water Research*, 46(4), 1145–1154.

Hancock, N. T., and Cath, T. Y. (2009). "Solute coupled diffusion in osmotically driven membrane processes." *Environmental Science and Technology*, 43(17), 6769–6775.

Holloway, R. W., Childress, A. E., Dennett, K. E., and Cath, T. Y. (2007). "Forward osmosis for concentration of anaerobic digester centrate." *Water Research*, 41(17), 4005–4014.

Hsu, S. T., Cheng, K. T., and Chiou, J. S. (2002). "Seawater desalination by direct contact membrane distillation." *Desalination*, 143(3), 279–287.

Hussain, G., and Al-Saati, A. J. (1999). "Wastewater quality and its reuse in agriculture in Saudi Arabia." *Desalination*, 123(2–3), 241–251.

Kargi, F., and Dincer, A. R. (1996). "Effect of salt concentration on biological treatment of saline wastewater by fed-batch operation." *Enzyme and Microbial Technology*, 19(7), 529–537.

Kargi, F., and Dinçer, A. R. (1998). "Saline wastewater treatment by halophile-supplemented activated sludge culture in an aerated rotating biodisc contactor." *Enzyme and Microbial Technology*, 22(6), 427–433.

Lay, W. C. L., Zhang, J., Tang, C., Wang, R., Liu, T., and Fane, A.G. (2012). "Factors affecting flux performance of forward osmosis systems." *Journal of Membrane Science*, 394–395, 151–168.

Lay, W. C. L., Zhang, Q., Zhang, J., McDougald, D., Tang, C., Wang, R., Liu, Y., and Fane, A. G. (2011). "Study of integration of forward osmosis and biological process: Membrane performance under elevated salt environment." *Desalination*, 283, 123–130.

Lew, C. H., Hu, J. Y., Song, L. F., Lee, L. Y., Ong, S. L., Ng, W. J., and Seah, H. (2005). "Development of an integrated membrane process for water reclamation." *Water Science and Technology*, 51(6–7), 455–463.

Li, D., Zhang, X., Yao, J., Simon, G. P., and Wang, H. (2011). "Stimuli-responsive polymer hydrogels as a new class of draw agent for forward osmosis desalination." *Chemical Communications*, 47(6), 1710–1712.

McCutcheon, J. R., and Elimelech, M. (2006). "Influence of concentrative and dilutive internal concentration polarization on flux behavior in forward osmosis." *Journal of Membrane Science*, 284(1–2), 237–247.

Mi, B., and Elimelech, M. (2010). "Organic fouling of forward osmosis membranes: Fouling reversibility and cleaning without chemical reagents." *Journal of Membrane Science*, 348(1–2), 337–345.

Miller, G. W. (2006). "Integrated concepts in water reuse: managing global water needs." *Desalination*, 187(1–3), 65–75.

Muniz, E. C., and Geuskens, G. (2001). "Compressive elastic modulus of polyacrylamide hydrogels and semi-ipns with poly(n-isopropylacrylamide)." *Macromolecules*, 34(13), 4480–4484.

Peng, P., Fane, A. G., and Li, X. (2005). "Desalination by membrane distillation adopting a hydrophilic membrane." *Desalination*, 173(1), 45–54.

Phattaranawik, J., Fane, A. G., Pasquier, A. C. S., and Bing, W. (2008). "A novel membrane bioreactor based on membrane distillation." *Desalination*, 223(1–3), 386–395.

Qu, D., Wang, J., Fan, B., Luan, Z., and Hou, D. (2009). "Study on concentrating primary reverse osmosis retentate by direct contact membrane distillation." *Desalination*, 247(1–3), 540–550.

Rautenbach, R., and Mellis, R. (1994). "Waste water treatment by a combination of bioreactor and nanofiltration." *Desalination*, 95(2), 171–188.

Reid, E., Liu, X., and Judd, S. J. (2006). "Effect of high salinity on activated sludge characteristics and membrane permeability in an immersed membrane bioreactor." *Journal of Membrane Science*, 283(1–2), 164–171.

Shaffer, D. L., Yip, N. Y., Gilron, J., and Elimelech, M. (2012). "Seawater desalination for agriculture by integrated forward and reverse osmosis: Improved product water quality for potentially less energy." *Journal of Membrane Science*, 415, 1–8.

Shannon, M. A., Bohn, P. W., Elimelech, M., Georgiadis, J. G., Marĩas, B. J., and Mayes, A. M. (2008). "Science and technology for water purification in the coming decades." *Nature*, 452(7185), 301–310.

Szuster, B. W., and Flaherty, M. (2001). "Inland Low-salinity Shrimp Farming in the Central Plains Region of Thailand." *In Proceedings of the international conference: the Chao Phraya Delta: historical development, dynamics and challenges of Thailand's rice bowl, December 12-15, 2000*, Kasetsart University, Bangkok.

Tan, C. H., and Ng, H. Y. (2008). "Modified models to predict flux behavior in forward osmosis in consideration of external and internal concentration polarizations." *Journal of Membrane Science*, 324(1–2), 209–219.

Tan, C. H., and Ng, H. Y. (2010a). "A novel hybrid forward osmosis—nanofiltration (FO-NF) process for seawater desalination: Draw solution selection and system configuration." *Desalination and Water Treatment*, 13(1–3), 356–361.

Tan, C. H., and Ng, H. Y. (2010b). "Modeling of external and internal concentration polarization effect on flux behaviour of forward osmosis." *Water Science and Technology: Water Supply*, 8(5), 533–539.

Cath, T. Y., Drewes, J. E., and Lundin, C. D. (2009). *A Novel Hybrid Forward Osmosis Process for Drinking Water Augmentation using Impaired Water and Saline Water Sources*, Water Research Foundation.

UNESCO. (2012). *The United Nations World Water Development Report 4: Managing Water Under Uncertainty*, Paris, France.

USEPA. (2012). *Guidelines for Water Reuse*, Chapter 8 Water Reuse Outside the U.S.

Vorosmarty, C. J., Green, P., Salisbury, J., and Lammers, R. B. (2000). "Global water resources: Vulnerability from climate change and population growth." *Science*, 289 (5477), 284–288.

Vrouwenvelder, J. S., and van der Kooij, D. (2001). "Diagnosis, prediction and prevention of biofouling of NF and RO membranes." *Desalination*, 139(1–3), 65–71.

Wang, K. Y., Teoh, M. M., Nugroho, A., and Chung, T.-S. (2011). "Integrated forward osmosis membrane distillation (FO-MD) hybrid system for the concentration of protein solutions." *Chemical Engineering Science*, 66(11), 2421–2430.

Xu, P., Drewes, J. E., Bellona, C., Amy, G., Kim, T. U., Adam, M., and Heberer, T. (2005). "Rejection of emerging organic micropollutants in nanofiltration-reverse osmosis membrane applications." *Water Environment Research*, 77(1), 40–48.

Yangali-Quintanilla, V., Li, Z., Valladares, R., Li, Q., and Amy, G. (2011). "Indirect desalination of Red Sea water with forward osmosis and low pressure reverse osmosis for water reuse." *Desalination*, 280(1–3), 160–166.

Zhao, S., Zou, L., and Mulcahy, D. (2012). "Brackish water desalination by a hybrid forward osmosis-nanofiltration system using divalent draw solute." *Desalination*, 284, 175–181.

Zheng, X., Ernst, M., Huck, P. M., and Jeke, M. (2010). "Biopolymer fouling in dead-end ultrafiltration of treated domestic wastewater." *Water Research*, 44(18), 5212–5221.

CHAPTER 14

Forward Osmosis Processes for Desalination: Evolution, Case Studies, and Future Directions

Tian C. Zhang
Ye Yuan
Rao Y. Surampalli

Over the past century, human consumption of freshwater has increased by tenfold, and is still expected to double every 20 years in the future. This increase is mostly driven by population growth and rising consumption per capita in the developed world (Zhang et al. 2012). It has been noted that in developing countries more than one billion people are deprived of improved sanitation services (Tan 2012). Fresh water is necessary for human life and for industrial activities. Although about 70 percent of the Earth's surface is water-covered, 97% of the water is sea/ocean water, which has a salt content averaging 35,000 mg/L. The 3% freshwater is not easily available for potable uses because 2% is like ice at the polar caps, 0.6% is trapped in groundwater aquifer, 0.3% is in the atmosphere, and only 0.1% is in rivers and lakes (Frenkel 2011). Therefore, eventually seawater could be one of the most desirable long-term reliable sources of drinking water for future generations.

Engineers and scientists were challenged by the American President, John F. Kennedy in April 1961 in his well-known speech, in which he said, "If we could ever competitively, at a cheap rate, get freshwater from salt water, that would be in the long-range interest of humanity and would dwarf any other scientific accomplishments." Desalination refers to any of several processes that remove some amounts of salt and other minerals from saline water. Generally, desalination may also refer to the removal of salts and minerals in a solution. For decades, Reverse Osmosis (RO) technologies are widely utilized for desalination. However, RO processes need high hydraulic pressure to force water flow through a membrane, resulting in primary drawbacks, such as a) intensive consumption of energy, b) severe membrane fouling, and c) brine discharge. Nowadays, research is being focused on tackling desalination with new technologies. One of such new technologies is forward (or direct) osmosis (FO) processes. Although the FO concept was proposed for several decades, most of the studies conducted in the last century are at bench scales. However, as the quality of FO membranes are improved, we see more and more successful field-scale demonstration projects of using FO processes for desalination (FOD). As a result, there is a need to summarize these projects for promoting more applications of FO processes in the future. This chapter introduces the principle of FOD and historical evolution of desalination technologies from RO to FO, followed by the in-depth presentation of different case studies on FOD processes. Through the presentation, materials are summarized by FO operation conditions, performance, and key issues related to design/management, future trend and research directions of FO for desalination.

14.1 BACKGROUND

14.1.1 FO Processes

Principles of RO, FO and Pressure-Retarded Osmosis (PRO)

FO processes may be used to desalinate saline water sources at a notably reduced cost. In RO, FO and PRO processes, water transports across a semi-permeable membrane that is impermeable to salt. In RO processes, hydraulic pressure (ΔP) is used to overcome osmotic pressure ($\Delta\pi$) to force freshwater transport through the membrane (Figure 14.1). In the FO process, the osmotic pressure gradient is used to let freshwater flow from the feed solution chamber to the draw solution chamber (Figure 14.1). Osmotic driving forces in FO can be significantly greater than hydraulic driving forces in RO, potentially leading to higher water flux rates and recoveries. In PRO, ΔP drives the water to the turbine for power generation (Zhang and Surampalli 2012).

FO Systems

As an emerging and novel technology for desalination, FO technology has advantages over the RO process in terms of energy consumption and fouling resistance. Unlike the RO process, where high hydraulic pressure is required, a FO

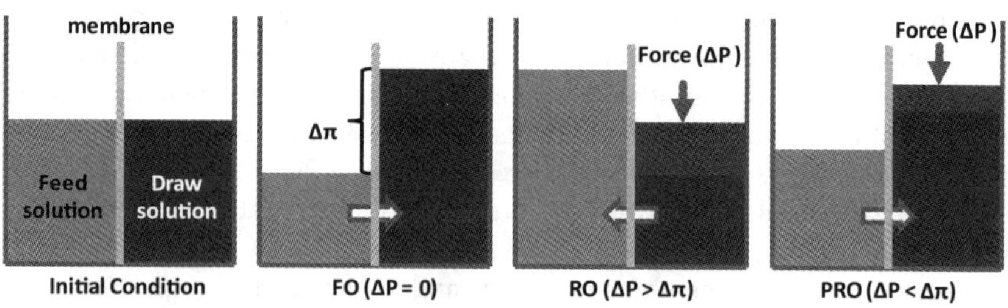

Figure 14.1. Water flows in FO, RO, and PRO

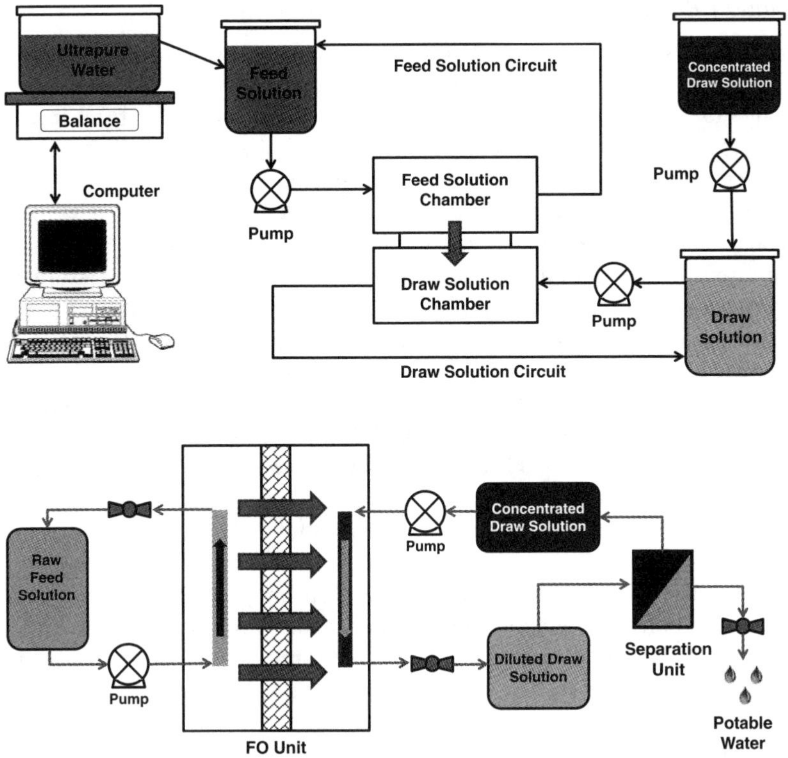

Figure 14.2. Schematic diagram of a widely used flat-sheet bench-scale FO system (top), in which the balance is used to quantify product water passing the FO membrane, and a FO desalination unit + a draw solution/product water separation unit (bottom), which can be RO, nanofiltration (NF), distillation, magnetic, or precipitation units, depending on draw solute used
SOURCE: Adapted from Liu et al. (2009) and Achilli et al. (2010)

process simply uses the intrinsic osmotic pressure difference to produce freshwater. In principle, the FO unit can incorporate spiral wound or hollow fiber membrane modules, and the feed water and draw solution often flow tangent to the membrane in a cross-flow mode (top panel of Figure 14.2). To yield potable water, the diluted draw solution is sent to a separation unit, which will use some cost-effective technologies to separate the freshwater and reconcentrate the draw solution to be recycled in the FO unit (bottom panel of Figure 14.2).

14.1.2 Evolution of FOD Technology

Historically, the evolution of using FOD processes centers around two-fold: i) development of new FO membranes and processes to overcome FO self-limiting flux behavior; and ii) development of new efficient draw solutions. The major progresses in the two aspects are briefly described below.

Evolution Related to FO Processes and Membranes

Considerable studies have been conducted on FO membranes since the late 1960s. Early work concerning FOD included Kravath and Davis (1975), and Kessler ad Moody (1976); both groups suggested that FO could be used for batch seawater FOD, especially as an emergency source of water in emergency lifeboats. Yaeli (1992) applied the FO process as a pretreatment step to a low pressure RO (LPRO) process, which is more energy efficient than a single RO process for desalination. In 2005, Hydration Technology Innovations (HTI) developed the FOPRO (Forward Osmosis Pressurized Reverse Osmosis) which uses osmotic pressure from the FO element to push the piston to drive a RO process for producing portable water without external energy input and with a better quality than that obtained by similar RO devices (Tan et al. 2012). Recently, many research groups reported the development of hybrid FO-RO or FO-nanofiltration (NF) systems for FOD (see below). For example, Tan and Ng (2010) reported that the hybrid FO-NF/NF system was capable of desalinating seawater with >25% of the energy saved as compared with the 2 two-stage RO process (Table 14.1); the product water quality met the drinking water standard with two NF for the post treatment (Figure 14.3).

Theoretically, any dense, non-porous, selectively permeable material can be used as a FO membrane. However, RO membranes are very different from FO membranes (Table 14.2). In the past, several FO studies were conducted by using Loeb-Sourirajan (L-S) type RO membranes or using L-S type RO membranes with the support fabric bonded to the porous support layer being removed, and the results were affected by switching the location of feed solution and draw solution (Sidney 1997). McCutcheon and Elimelech (2008) reported that membrane support layer hydrophobicity significantly hinders water flux in osmosis-driven

Table 14.1. Comparison between FO-NF/NF and 2-stage RO systems

Process	*2-stage RO*	*FO-NF-NF*
Production capacity, m^3/d	1000	1000
Overall recovery (%)	45	45
Pass-one and pass two feed pressure, bar	60 and 20	40 and 10
High power pump efficiency (%)	88.5	88.5
Energy recovery, Pelton wheel, (%)	92	92
FO feed pump power consumption, kW	–	4.36
Stage-one and stage-two feed pump power consumption, kW	183.55 and 27.53	174.37 and 14.53
Total pump power consumption, kW	211.08	193.26
Total recovered power, kW	81.88	97.90
Net energy consumption, kW	129.19	95.30
Specific work, kWh/m^3	3.10	2.29

SOURCE: Tan et al. (2012); copyright ASCE

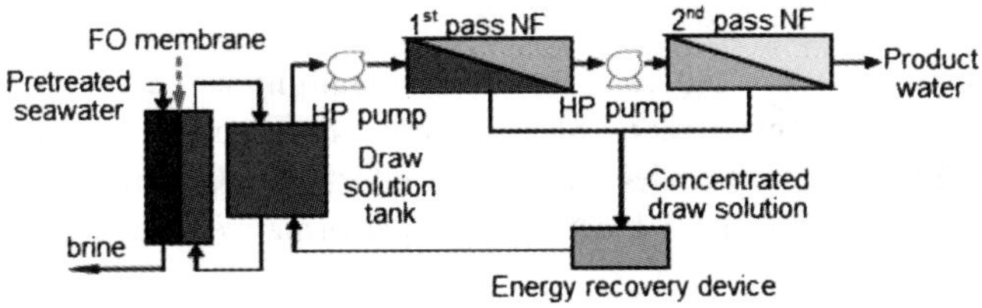

Figure 14.3. A hybrid FO-NF/NF process for desalination
SOURCE: Tan et al. 2012; copyright ASCE

Table 14.2. Comparison of requirements between RO and FO membranes

Requirements/ properties	*RO*	*FO*
• Compression resistance	• Be able to withstand large pressure	• Don't need to resist compaction
• The porous supporting layer	• Need a thick supporting layer, but no special requirement for the layer	• Can be a thick layer, but must be made of loose fibers to allow water to diffuse easily
• Importance of skin and supporting layer	• Skin layer is more important	• Both are important

membrane processes. Membranes with very thin dense layers for rejecting the salt and very thick porous support layers for supporting the thin layer under the hydraulic pressures required for operation is not suitable for FO or PRO (Liu et al. 2009). In a FO process, when permeate flows through the FO membrane, it will dilute the draw solution closely adjacent to the membrane (called dilutive ICP or ECP), resulting in a decrease in the osmotic pressure near the membrane. Therefore, water in the feed solution near the membrane will never contact the bulk draw solution, but rather the diluted draw solution as soon as FO starts. This is so-called self-limiting flux behavior (Zhou et al. 2012; McCutcheon and Elimelech 2006). Considerable studies are focusing on making more effective FO membranes that can ease the ICP effect as well as membranes that are semi-permeable and as thin as possible without a porous support layer. Currently, two strategies are used in FO membrane development: 1) introduce a backbone or support into the thin film and 2) add to the thin film a highly porous and very thin support layer. Using strategy 1, HTI coats cellulose triacetate (CTA) onto polyester screen mesh (thickness <50 μm). HTI membranes are widely used in research and practices (Ling et al. 2010). However, new FO membranes have been developed, such as thin-film composite (Yip et al. 2010), hollow-fiber and double-skinned FO membranes (Wang et al. 2010).

Evolution Related to Draw Solutes

Several attempts to use FO as a means of desalting saline waters have been reported in the past four decades. Many different draw solutes have been studied (Cath et al. 2006). Here, the relevant previous efforts for FOD-mostly presented as patents with limited technical details and performance data-are summarized below.

Batchelder (1965) described a process of adding volatile solutes, such as sulfur dioxide, to seawater or freshwater to create a solution which may be used in a FO process to extract water from seawater. The suggested membrane to be used in this process was cellulosic in nature. Other examples in the patent described the use of carrot root as a membrane material. The process is carded out until the draw solution is sufficiently dilute, at which point the volatile solute is removed by heating and/or air stripping. The patent, however, only determined that positive water flux occurred in the experiments and did not quantify the flux or salt rejection.

Glew (1965) expanded on this idea by describing a FO method using a mixture of water and another gas or liquid. This mixture is intended to lower the "activity" of the water solution to the point that a net flow of potable water will be induced from the seawater, after which the miscible solid or liquid will be removed from the fresh water by "standard means." Sulfur dioxide and aliphatic alcohols are suggested as the components of the draw solution. This is also the first FO technique that suggests the removal and recycle of the draw solute in the overall FO process.

Frank (1972) described a FO method using a precipitable salt, in this case aluminum sulfate, as the draw solution solute. Following osmosis of water across the membrane, the diluted draw solution is dosed with calcium hydroxide, leading to the precipitation of aluminum hydroxide and calcium sulfate. The precipitate is removed by standard methods, leaving the fresh product water. Excess calcium hydroxide from the precipitation step can be removed by dosing with sulfuric acid or carbon dioxide, which produces calcium sulfate and calcium carbonate precipitates, respectively. This step requires additional solid removal and leads to neutral pH in the product water. The membrane used in the patent is of similar characteristics to the Loeb-Sourirarjan type cellulose acetate membrane. Water flux and salt rejection data were not presented.

Kravath and Davis (1975) described a process of seawater desalination achieved by FO of water across a cellulose acetate membrane. Initial tests were run with a dialysis cell with glucose as the draw solute and seawater as the feed. Additional tests were run with glucose dissolved in seawater as a draw solution. Emergency lifeboats were considered as a possible use of the process in which seawater would be brought aboard a lifeboat and glucose would be added. Additional seawater would be passed through a dialysis unit leading to osmosis and a dilution of the seawater/glucose draw solution. Upon dilution, the salinity would be reduced to a level where ingestion is possible for short term consumption. The flat-sheet cellulose acetate membranes did not perform well in terms of salt rejection. Hollow fiber membranes were also tried and results improved. Draw solute removal was not considered as it was intended for ingestion.

Stache (1989) expanded the idea of Kravath and Davis by using a membrane bag to desalinate small volumes of seawater while simultaneously creating a nutritious drink. This batch process is accomplished by placing a concentrated fructose solution in a semi-permeable membrane bag. Fructose is used because of its relatively high "osmotic efficiency." The bag is submersed in seawater and water diffuses into the concentrated fructose draw solution until it is suitably dilute. Separation of the draw solute is unnecessary since the solute is intended for ingestion. It is a batch process, and for repeated use, additional concentrated fructose solution is needed.

Yaeli (1992) expanded yet again on the idea of using sugar as a draw solution by combining FO and low pressure RO into a continuous process. Using a concentrated sucrose solution as a draw solution, water is extracted from seawater by osmosis. The diluted glucose solution is fed to a RO unit, where a low pressure RO membrane separates potable water from the sucrose draw solution. A loose RO membrane can be used because of the relatively large size of the sucrose molecules. The concentrated sucrose solution is fed back to the FO membrane module for reuse. Recovery in this process is limited, however, due to a relatively low osmotic efficiency of sucrose.

McGinnis (2002) described a FO method using a combination of draw solutions across several semi-permeable membranes. This patent combined the ideas of draw solution recycle with an osmotically efficient draw solution to increase recovery. The two-stage FO process takes the advantage of the temperature dependent solubility of the solutes, in this case potassium nitrate (KNO_3) and sulfur dioxide (SO_2). Seawater is heated and fed to the FO membrane unit where a heated solution of saturated KNO_3 serves as the draw solution. The diluted draw solution is sent to a new chamber where it is cooled by incoming seawater, which is simultaneously heated to the appropriate feed temperature. Upon cooling, a significant portion of the KNO_3 precipitates out of the solution, reducing the osmotic pressure. Next, the diluted KNO_3 solution is drawn to another FO unit, where dissolved SO_2 acts as the draw solution. The dilute KNO_3 solution has a low osmotic pressure in comparison with the saturated SO_2 solution, and water diffuses across the membrane while the KNO_3 is rejected. The sulfur dioxide is then removed through standard means, leaving potable water. All solutes are recycled in the process. A system using SO_2 as the only draw solution is not considered since a saturated SO_2 solution does not have enough osmotic pressure to realize as high a recovery as a saturated KNO_3.

In 2004, McGinnis proposed a process that uses an ammonium bicarbonate draw solution to extract water from saline feed water across a semi-permeable polymeric membrane. Water is extracted from seawater and dilutes the ammonia–carbon dioxide draw solution. Upon moderate heating (~58°C), the draw solution decomposes to ammonia and carbon dioxide. The fresh product water can be separated from the diluted draw solution by several separation methods, e.g. column distillation or membrane distillation (MD). The degasified solution left behind is pure product water, and the distillate is a reconcentrated draw solution available for reuse. Experiments with a laboratory-scale FO unit utilizing a flat-sheet cellulose tri-acetate membrane have demonstrated high product water flux and relatively high salt rejection.

In recent bench-scale studies, it was demonstrated that when using a suitable FO membrane (e.g., the FO CTA membrane) and a strong draw solution (highly soluble ammonia and carbon dioxide gases), seawater can be efficiently desalinated with FO. The draw solution was formed by mixing together ammonium carbonate and ammonium hydroxide in specific proportions (McCutcheon et al. 2005, 2006). The salt species formed include ammonium bicarbonate, ammonium carbonate, and ammonium carbonate. Analysis of the process has shown that an osmotic pressure driving force ($\Delta\pi$) as high as 238 bar for a feed water with a salt concentration of 0.05 M NaCl, and as high as 127 bar for a feed water with a salt concentration of 2 M NaCl, can be achieved with the ammonia/carbon dioxide draw solution. This is a rather high driving force considering that 2 M NaCl is equivalent to brine from seawater desalination at ~70% water recovery.

Bench-scale FO data demonstrates that the ammonia-carbon dioxide FO process is a viable desalination process. Salt rejections greater than 95% and fluxes as high as 25 $L/m^2 h$ were achieved with the FO CTA membrane with a calculated driving force of more than 200 bar (McCutcheon et al. 2005). Although this is a relatively high flux, much greater flux is actually expected for such a high driving force. Further analysis of the results has indicated that the performance ratio (defined as experimental water flux divided by theoretical water flux) of the FO CTA membrane used was at most 20%, and on average between 5% and 10%. This lower-than-expected flux is attributed to ICP. The system saves 72 to 85% energy, but still requires considerable amounts of energy to boil away NH_3 and CO_2 in the draw solution. Moreover, the maximum contaminant level for NH_3 in drinking water is <2 mg/L, which makes the NH_4HCO_3 draw solution face potential safety drawbacks. Furthermore, NH_4HCO_3, NaCl and $MgCl_2$ can reversely diffuse into the feed solution chamber, which may jeopardize the FO system.

Ling et al. (2010) used highly water-soluble magnetic nanoparticles (MNPs) as novel draw solutes in FO; however, they didn't report how long the coated MNPs can be used. In addition, the complexity of reusing MNPs still exists, and should be addressed. Liu et al. (2011) used $Al_2(SO_4)_3$ draw solution to draw fresh water from wastewater across a semipermeable FO membrane, and then, in another tank, used negatively-charged and magnetic nanoparticles to mimic a natural "destabilization" phenomenon to economically separate $Al_2(SO_4)_3$ draw solute. Liu et al (2011) claimed that this draw solution separation process does not require any intensive energy input, but this conclusion needs more real experimental data to support. Yen et al. (2010) reported using 2-methylimidazole-based organic compounds as draw solutes; some compounds can produce higher fluxes with lower reverse diffusion. The larger molecular size enhances the potential in membrane-based solute recovery (e.g., use RO); however, it may induce more ICP. Li et al. (2011a & b) used environmentally-sensitive polyelectrolytes (ESP) or composite polymers as a new class of draw solutes for FOD (Figure 14.4).

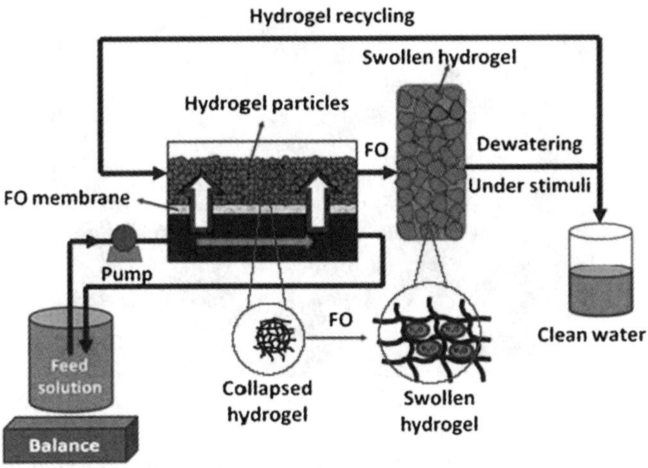

Figure 14.4. Schematic diagram of the ESP hydrogel-based FO process with pure or composite hydrogels as draw solutes

Hydrogels are able to draw pure water through FO membranes, and the product water can be removed by pressure and/or heating (e.g., solar energy). The incorporation of light-absorbing particles leads to natural, enhanced heating and dewatering of the composites. Water fluxes of these hydrogels range from 1.5 L/m^2-h (LMH) at the beginning to ~0.1–0.2 LMH after 24 h, and may be improved by adjusting the hydrogels' compositions. These studies opened a new window for using hydrogels as draw solute to overcome the two major drawbacks of the FO processes.

14.2 CASE STUDIES

In this section, several case studies are introduced to discuss important achievements, the engineering challenges and the commercialization of the FOD process.

14.2.1 Case Study A–FO Processes for Wastewater Reclamation

This case study is about wastewater reclamation with FO and Red Sea seawater. The study was conducted by the research group at the King Abdullah University of Science and Technology (KAUST), Saudi Arabia. In this case, the first objective was to achieve a new design of an efficient FO configuration; the second objective was to study feed waters and draw solutions to achieve an optimal coupling of FO and RO. To approach the objectives, a custom-made plate and frame configuration for the membranes (Figure 14.5) was designed, implemented and studied (Yangali et al. 2011).

Two types of HTI membranes were used, i.e., HTI-1 (from a pouch) and HTI-2 (from a cartridge). The FO unit was used to run dilution experiments for

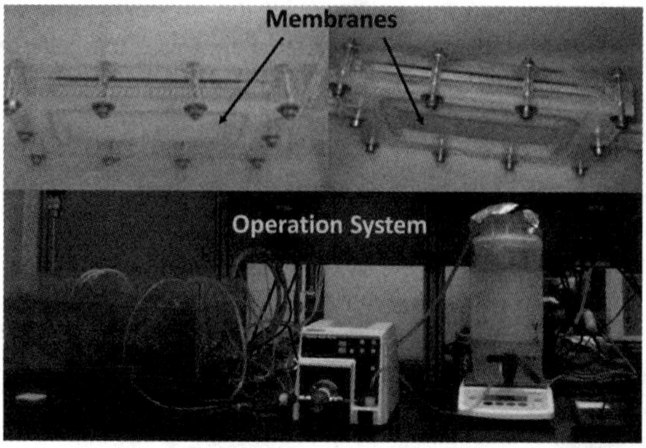

Figure 14.5. Experimental setup of the FO system. (Top) FO units with two flat-sheet HTI membranes (each with a total area of 404 cm²). (Bottom) the FO system. The membranes are immersed in a water tank, and are connected to a draw solution reservoir. A pump is used to continuously recirculate the DS. The FO setup is connected to a cross-flow filtration unit containing an RO membrane (139 cm²) (not shown), thus forming a single hybrid unit
SOURCE: Adapted from Yangali et al. 2011

24 hours, starting with 1 L of draw solution and increasing its volume due to osmosis produced by recirculation of draw solution in the cells. These experiments were used to calculate average fluxes. During experiments with continuous recirculation of draw solution, the RO system was operated in a recirculation mode, with the concentrate returning to the DS reservoir. A Dow-Filmtec low pressure RO membrane (BW30) was used for desalination of the DS and final recovery of wastewater. The performance of the entire hybrid system was evaluated using deionized water and treated domestic wastewater (shown in Table 14.3) as feed water, with either sodium chloride (0.5 M, 29.22 g/L) or Red Sea seawater (filtrated with 0.45-μm filters, Table 14.3) as the draw solution.

The results of these experiments show an average flux of 3.5 LMH in 9 h, and 2.6 LMH in 24 h. After a dilution experiment of 1 L of draw solution during 24 h, the average flux of HTI-1 membranes was 2.4 LMH with DS (0.5 M NaCl) with a recirculation flow of 150 mL/ min, when using DI water as feed to the FO membrane. The average flux of HTI-2 membranes was slightly greater than HTI-1, 2.6 LMH with DS (0.5 M NaCl) and a recirculation flow of 150 mL/ min, using DI water as feed to the FO membrane. When seawater was used as draw solution and DI water as feed, the average flux increased by 20% for HTI-1 (2.88 LMH) membranes and 27% for HTI-2 membranes (3.36 LMH).

HTI-2 membranes were used with the secondary wastewater effluent as the feed and Red Sea water (filtrated) as a draw solution. The FO unit resulted in a flux variation shown in Figures 14.6 and 14.7. The conductivity of the FO feed, draw solution and RO permeate was also monitored during the 4 days. RO was performed at a flux of around 6.3 LMH and at a pressure of 30 bar (Figure 14.6), with a recovery of 1.5%. The RO membrane area, the permeate flow and the

Table 14.3. Wastewater effluent and seawater characteristics

Wastewater:					
Conductivity (μS/cm)	3300		DOC (mg/L)		5.3
pH	7.3		DO (mg/L)		6.3
Seawater Parameter	*Raw*	*After filtration*	*Parameter*	*Raw*	*After filtration*
Conductivity (μS/cm)	57700	57500	Sodium (mg/L)	12470	n.a.
Temperature (°C)	20.5	20.5	Strontium (mg/L)	7	n.a.
pH	7.8	7.8	Bicarbonate (mg/L)	141	n.a.
TDS (mg/L)	40600	40500	Boron (mg/L)	2.2	n.a.
SDI	2	Not available (n.a.)	Carbonate (mg/L)	8	n.a.
Barium (mg/L)	0.01	n.a.	Chloride (mg/L)	23073	n.a.
Calcium (mg/L)	571	n.a.	Fluoride (mg/L)	1.5	n.a.
Magnesium (mg/L)	1458	n.a.	Sulfates (mg/L)	2400	n.a.
Potassium (mg/L)	488	n.a.			

crossflow velocity (40 cm/s) in the RO unit determined the operating RO flux (6.3 LMH). Conductivity readings resulted in salt removals of more than 99%. The flux of the FO unit during a continuous experiment (~4.3 LMH, Figure 14.6) is greater than the average flux of a dilution experiment (~2.6 LMH at 24 h). As a major advantage, the RO unit of the hybrid FO-RO system can be operated at pressures <15 bar. The RO unit continuously renovates the DS strength, making it possible for the FO unit to maintain a relatively high flux (4.3 LMH).

It should be pointed out that the new FO configuration (Figure 14.5) is different from the FO membrane contactors described elsewhere. HTI provides plate and frame modules for treatment of landfill leachate. However, KAUST custom-made FO plate and frame (KOPF) configuration is very different from HTI. The KOPF works immersed in a tank, and resembles the Kubota MBR membrane configuration. The HTI plate and frame module configuration does not work immersed in a water tank, and is more similar to a membrane contactor working as a direct osmotic concentration unit. The two different configurations may behave differently in membrane fouling. For example, after 4 days of

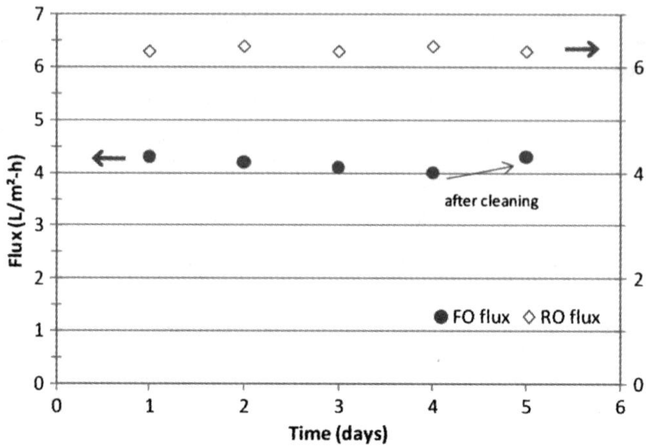

Figure 14.6. Flux vs. time for FO and RO. The results correspond to a configuration where the DS is being recirculated, and the concentrated DS is being continuously pumped from the RO unit into the draw solution tank
SOURCE: Adapted from Yangali et al. 2011

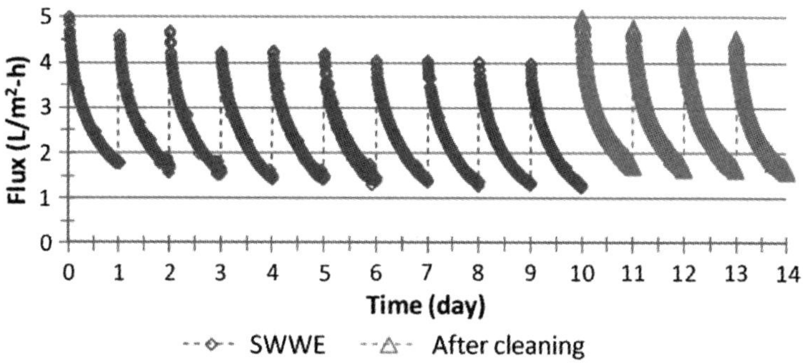

Figure 14.7. FO flux vs. time for FO. Results are obtained with the DS being replaced every day and the concentrated DS is discharged to the sea. SWWE = Secondary wastewater effluent (shown in Table 14.3)
SOURCE: Adapted from Yangali et al. 2011

continuous operation, the flux declined 7% due to fouling of the membrane (Figure 14.6). Cleaning the membranes by soaking and air scouring them with tap water could restore the initial membrane flux. A similar result was obtained after 10 days of batch operation, cleaning with air scouring restored 98.8% of the initial flux (Figure 14.7). This is quite interesting in that the natural osmotic movement of water from the feed tank to the draw solution fouled the membrane in such a way that it could be easily cleaned by hydraulically flushing the membrane with clean water and air. In contrast, a FO membrane contactor configuration may potentially increase fouling of the membranes when compared to these results.

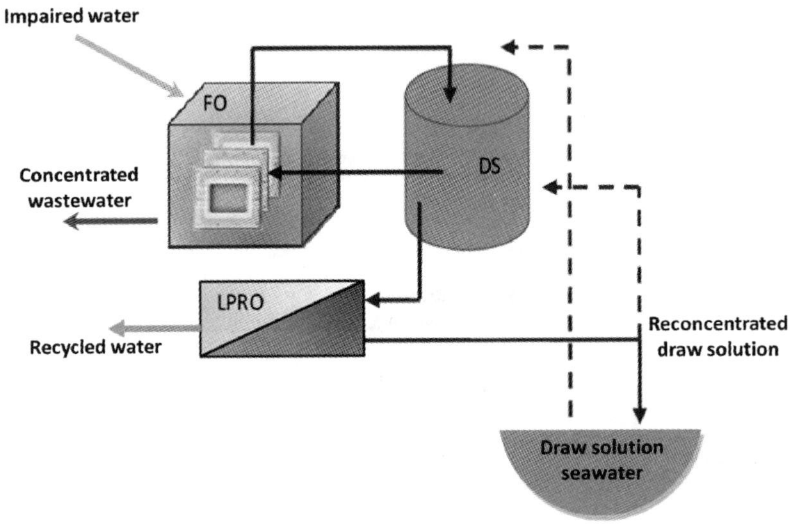

Figure 14.8. Wastewater reclamation with FO and LPRO
SOURCE: Adapted from Yangali et al. 2011

Physical cleaning with a water jet and chemical cleaning are necessary in order to restore the flux of a FO membrane fouled in a membrane contactor configuration.

This case study demonstrates that type HTI-2 FO membranes are suitable for water reclamation with the hybrid FO-RO system as shown in Figure 14.8. One alternative (shown by dash lines) is that a single injection of pre-treated seawater to the DS/RO feed tank is necessary. Later, the DS will be continuously recirculated through the FO cells to extract water from the impaired feed. At the same time, the reconcentrated DS (reject of RO membrane) originating from the LPRO system will be returned to the DS/RO feed tank. An alternative configuration considers batch supplies of fresh DS. In this way only the diluted DS is used as feed of the LPRO, and the concentrate from LPRO is discharged to the sea (shown by solid lines). The novelty of this approach, when compared to previous research, is the new osmosis membrane configuration. The plate and frame osmosis cell that functions immersed in a tank presents the following advantages: i) reduced fouling of osmosis membranes and ii) low energy consumption of draw solution recirculation. In a FO membrane contactor configuration, counter cross-flows of 1.5 L/s have been reported in both directions of the cells (counter flows), which will demand a high energy consumption when up-scaling the systems.

14.2.2 Case Study B–FO for Making up Cooling Water

System Description

Nicoll et al. (2011) outlined the progress made to date on the development and commercial deployment of a FO process applied to evaporative cooling tower make-up water. The process (shown in Figure 14.9) is very simple in concept. To

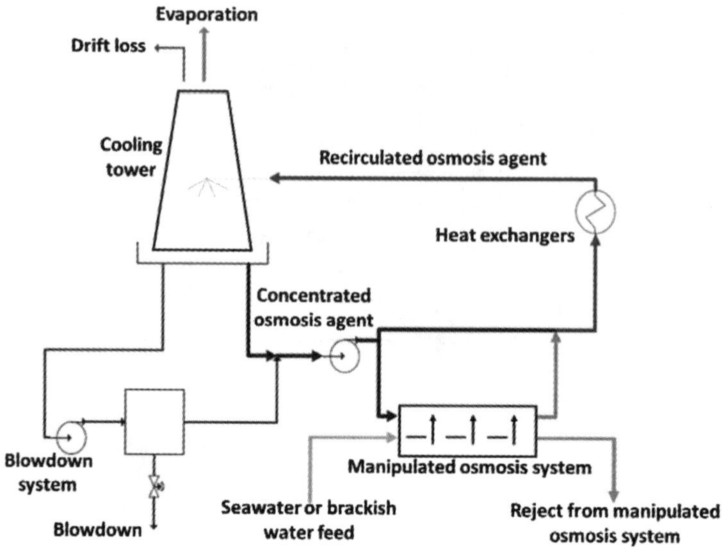

Figure 14.9. Using FO process to make up cooling water. Osmotic Agent = draw solute
SOURCE: Adapted from Nicoll et al. 2011

replace water that was lost by evaporation drift and blowdown, one can change the cooling water chemistry by adding chemical reagents so as to increase its osmotic pressure, and use the high osmotic pressure solution as draw solution in a FO unit with seawater, brackish water or treated sewage effluent being used as feed solution. The FO process makes pure water that continuously compensates the lost water in the process.

Modern Water made a demonstration unit (Nicoll et al. 2011). A pre-requisite is that the design of the demonstration unit should not interrupt the operation of an existing cooling water system, so that there is little or no risk to the client. Therefore, the demonstration unit has an external packaged evaporative cooling tower with a nominal cooling capacity of 50 kW. Feed water can be treated sewage effluent or seawater. A separate heat exchanger is installed between the heat load supplied by the host (Table 14.4). Hence, all aspects of the process can be demonstrated on a client site. The demonstration unit has been tested in a petrochemical plant in Sohar, Oman (Figure 14.10).

Energy and Cost

Nicoll et al. (2011) analyzed the energy and cost of the FO and RO systems used for making up cooling water with several assumptions (Table 14.5). Figure 14.11 shows a significantly better power consumption and cost advantage of FO when compared to RO. In the case of FO without using the osmotic agent (draw solute) recovery system on the blowdown stream, the power consumption is the lowest and is constant vs. feed TDS. However, the cost of draw solute may be higher than

Table 14.4. Properties of major system components of the demonstration unit

Component	*Description*
Size	Housed in a 20-foot container, can be trialed at different client site
FO membranes	Specifically made for Modern Water; operated at 2–3 bar with improved permeability; Recovery ≈ the RO system; with resistance to oxidizing biocides (e.g., chlorine)
Draw solute	A safe, economical, readily available commodity chemical; not corrosive to stainless steel; corrosion rates for carbon steel are reduced by adding a corrosion inhibitor (a blend phosphonates and carboxylic acids); may inhibit microbes to grow in cooling systems (e.g., *Legionella peneumophila* can't grow there)
Blowdown	Used to recover the draw solute. The blowdown system is a loose membrane system
Cooling tower	An external packaged evaporative cooling tower with a nominal cooling capacity of 50 kW
Pre-treatment	A full pre-treatment system for the raw water, based on multi-media filtration

Figure 14.10. The demonstration plant located in Sohar, Oman
SOURCE: Adapted from Nicoll et al. 2011

the energy saving. These figures are conservative for FO because the conversion (make-up water to feedwater ratio) of the process is fixed to 30% across the range of feed water TDS, whereas the process actually would have a similar conversion to RO. These comparisons do not consider the other advantages of the FO system,

Table 14.5. Parameters assumed/used in energy and cost estimation

Parameter	*Value*	*Parameter*	*Value*
Feedwater temperature	25 °C	Pre-treatment required	Same for FO and RO
Pump efficiency	70%	Energy recovery efficiency	70%
Pressure loss across pre-treatment and the FO units conservative)	3 bar	Max. conversion of FO (make-up water to feed water ratio)	Limited to 30% (very
Conversion for RO	80–40%, feed TDS sensitive	Cooling tower concentration ratio	5
Power cost	US0.075/kWh (low)	Osmotic agents (= draw solutes)	US75/tonne

SOURCE: Data from Nicoll et al. (2011)

including an increased availability, fewer membrane replacement and lower chemical cleaning costs.

Summary of Case Study B

Therefore, it can be concluded that the FO process can be used as a robust and cost-effective process for making up cooling water. However, more research is needed to commercialize the process.

14.2.3 Case Study C–Modern Water's FO Practices

Since 2007, Modern Water has been developing FOD processes for drinking water production on a commercial. In September 2008, Modern Water successfully commissioned the world's first FOD plant located in Gibraltar on the Mediterranean Sea, and since May 1, 2009, water has been exported contentiously and put into the public water supply. In September 2009, a larger FOD plant (100 m^3/d) was installed in the Sultanate of Oman at the Public Authority of Electricity and Water's (PAEW) site at Al Khaluf. PAEW selected this site because of the extremely challenging seawater, taken from a very shallow open seawater intake which was sometimes exposed at low tide. The FOD plant shares a common pre-treatment with an existing similarly-sized seawater RO (SWRO) plant, thus providing a unique opportunity to trial the two technologies on a like-for-like basis (Nicoll et al. 2011). In June 2011, Modern Water was awarded the contract worth US$ 650,000 to build and operate a 200 m^3/d FOD plant by Oman's PAEW. Modern Water says it is the world's first commercial FO desalination plant and the group's second plant in Oman (D&WR 2012). In 2012, Modern Water signed a

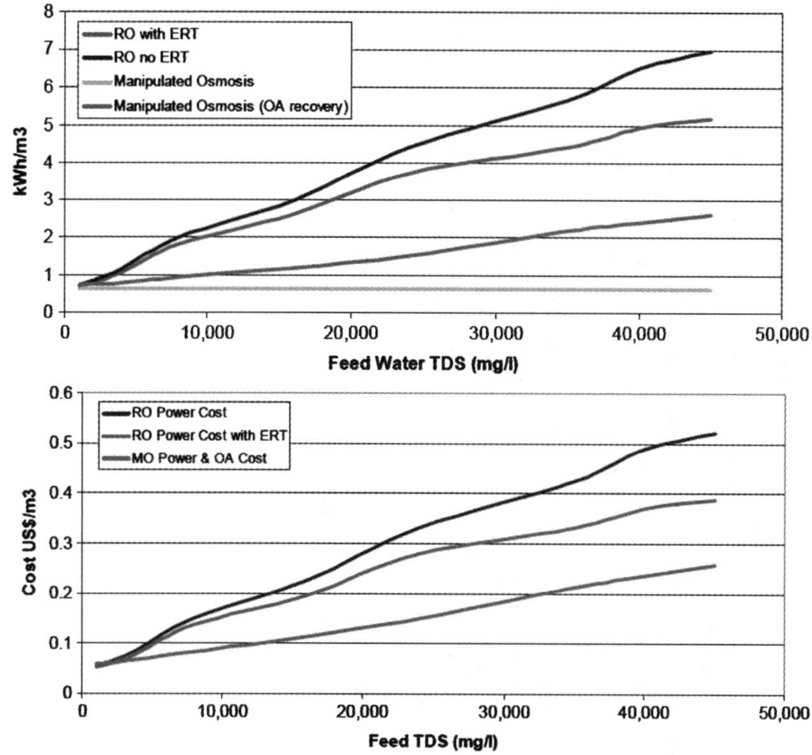

Figure 14.11. Comparison of power (top) and operational cost (bottom) of FO with RO. Note: ERT = energy recovery turbine; Manipulated Osmosis = MO = FO; and OA = Osmotic agent = draw solute
SOURCE: Adapted from Nicoll et al. 2011

Framework Agreement with Hangzhou Development Center of Water Treatment Technology in the People's Republic of China. Sources estimating the industry in China will see new investment of up to RMB20 billion (US$3.2 billion) over the next four years for desalination (WaterWorld 2012). This case study describes the key information on these processes and operational results.

FOD Facilities and Process Components

The FOD development process from the laboratory to a commercial facility is best illustrated by Table 14.6 and Figures 14.12–14.15 (Thompson and Nicoll 2011). To make high quality water, Modern Water's FOD process integrates the FO system in a single cycle with a regeneration system. The major FOD components include i) FO membranes, ii) regeneration system, iii) draw solute, and iv) pre-treatment of feed water. The characteristics of these components are shown in Table 14.7. Details of Modern Water's trail and production facilities are listed in Tables 14.8 and 14.9 below. Comparison between the FOD and the SWRO facility are shown in Figure 14.16 and Table 14.10.

Table 14.6. Three Modern Water's key FOD facilities

Facility	*Description*
Lab test rig (Figure 14.12)	First test facility owned the Centre for Osmotic Research and Applications (CORA) at the University of Surrey; used by CORA to obtain design parameters and develop the concept of FOD; paving the way for implementing the system outside of the laboratory environment
Trial facility (Figure 14.13)	The first field-scale FOD system, located at Gibraltar, Mediterranean Sea; delivering water to the local drinking water system since May 1, 2009. Key features are as follows: • housed in a 20 ft ISO shipping container, the system has i) extensive automated measurements of flowrates, pressures, temperatures, pH and conductivity; ii) a Programmable Logic Controller (PLC) + PC running SCADA software for data recording; iii) pumps, tanks and pipework with much large capacity to permit future testing; iv) interconnections to external services, including a pressurized seawater supply, a pressurized brine discharge system, permeate from the adjacent plant (for initial fill etc.), and product export line and power supply • with several FO membrane elements (membranes are 3rd generation) • uses either RO or nano-filtration-based systems (each several elements) for draw solute regeneration

Production facility (Figure 14.14)	A field-scale FOD system (design capacity = 100 m^3/d), located in Al Khaluf, Sultanate of Oman; providing product water into the public supply in Nov. 2009. Key features and test site conditions below: • system features: a) housed in a 40 ft high-cube ISO shipping container, the system has key features of the facility similar as those in trial facility with a 415 VAC 3-phase power supply; b) with several FO membrane elements (membranes are 3rd generation); and c) uses either RO or nano-filtration-based systems (each several elements) for draw solute regeneration • test site conditions: a) the site owned by the PAEW; b) there is a RO water plant (100 m^3/d) comprised with i) 2 borewells (one not operational), ii) temporary seawater intake, iii) feedwater chlorination, iv) media filter, v) feedwater storage, vi) main process hall containing two RO vessels (8″ spiral wound type), vii) dosing pumps, etc., viii) permeate storage, ix) boron removal equipment, x) chlorine dosing (product water), xi) product water storage, and xii) tanker filling station

SOURCE: Data from Thompson and Nicoll 2011

Figure 14.12. Laboratory Test Rig
SOURCE: Adapted from Thompson and Nicoll 2011

Figure 14.13. Gibraltar Trial Facility
SOURCE: Adapted from Thompson and Nicoll 2011

Figure 14.16 shows that there is a 30% decline in the output from new membranes in just five months in the existing SERO plant, despite repeated cleaning of the membranes. In contrast, the Modern Water's FOD system over the same period has remained relatively unchanged, despite the fact that no chemical

Figure 14.14. The FOD plant at Al Khaluf
SOURCE: Adapted from Thompson and Nicoll 2011

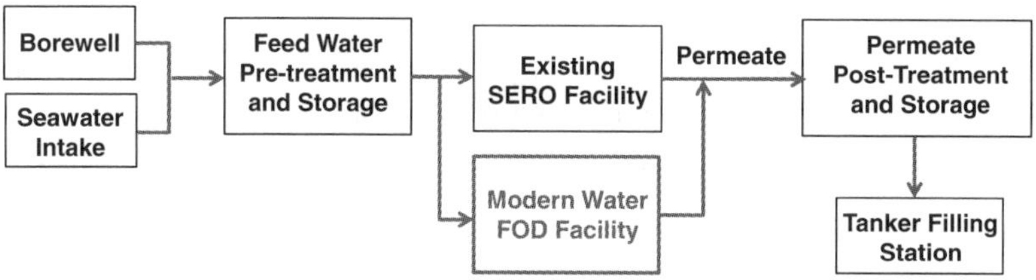

Figure 14.15. The SERO and FOD plant
SOURCE: Adapted from Thompson and Nicoll 2011

cleaning has been performed, and operation has been at a higher seawater recovery (Table 14.10). Despite the atrocious feed water conditions at Al Khaluf, the FO membranes have not required cleaning in over a year of operation. This demonstrates that the performance of the FOD process is reliable in particularly challenging feed water. Furthermore, the FOD system can be operated at about 60% of the energy consumption of the competing SWRO facility (Table 14.10) even though the FOD facility tested at the PAEW water site was not designed with optimum energy consumption in mind.

Summary of Case Study C

Over the course of a 4-year period, Modern Water has successfully taken the FOD process from the laboratory to a full-scale commercial facility. The results from

Table 14.7. Properties of major components of Modern Water's FOD systems

Component	*Description*
FO membranes	Specifically made for Modern Water; with resistance to oxidizing biocides (e.g., chlorine); FO membranes are arranged according to the feedwater conditions and the required plant performance; the draw solute is typically fed in parallel to every membrane, banks of membranes may be arranged in parallel or in a tapered array on the feedwater side
Draw solute	A safe, economical, readily available commodity chemical; non-corrosive; suitable for use in drinking water facilities
Regeneration system	Based on semi-permeable or loose membranes; the energy consumption is minimized by selecting the draw solute and operational conditions
Pre-treatment	A full pre-treatment system for the raw water, based on multi-media filtration; if no pre-treatment, an increase in FO membrane areas or draw solute concentrations is needed

Modern Water's FOD system are significantly better than expectations, with the following key benefits of the FOD being demonstrated:

- Lower fouling propensity/lower energy consumption/lower O/M cost;
- Lower seawater contaminants (e.g., boron)/reduced caper/opex; and
- Lower capital cost/easier fabrication due to using low pressure pipework/fittings.

14.3 RESEARCH NEEDS AND FUTURE TREND

Current Status of FO Research

Although many applications of FOD have been demonstrated; the existing FOD processes have two major drawbacks that need immediate fundamental research to overcome (Cath et al. 2006; Zhang et al. 2012):

1. the presence of dilutive and concentrative external concentration polarization (ECP) and internal concentration polarization (ICP) in the currently used FO membranes significantly reduces the available osmotic driving force; and
2. the product water has to be separated subsequently from the draw solution with another RO or energy intensive heating/cooling process (e.g., column or membrane distillation), and the draw solution needs to be concentrated before its reuse.

Table 14.8. Major conditions of and performance observations of Gibraltar Trial Facility

Units/parameter	*Observations/benefits*
Feedwater	pre-treated water with $SDI_{15} = 3–4$
Production water	typically TDS < 200 mg/l; boron = 0.6–0.8 mg/L with optimization or 1–1.5 mg/L without optimization
FO membrane	3rd generation, with 30% performance improvement as compared with the last generation
RO for draw solution regeneration	the RO membranes performed much better than a conventional seawater RO (SWRO) plant: the recovery of the RO unit in a FO-RO system = 27× that of SWRO[a] higher flux rates achieved due to a much better quality of the RO feed (draw solution)
Properties of draw solute and product water	can be modified by making relatively small operational or chemical changes (e.g., pH adjustment), which then will allow the modification of product quality
Boron control	can achieve 40% reduction of boron by a simple modification to the operating flow conditions

[a] Calculated based on the statement "6 or 7 membrane elements in a pressure vessel are needed to achieve a 40% recovery in a conventional SWRO plant, whereas in the RO of the Gibraltar system a 50% conversion needs only 3 elements" that is modified from Thompson and Nicoll (2011).

Table 14.9. Major conditions and performance observations of production facility

Units/parameter	*Observations/benefits*
Feedwater	from a shallow open seawater intake; very low water quality
Pre-treatment	feed water is pre-treated with one dual media filter (filtration velocity = 16–25 m/h), and then a simple cartridge filtration to produce the water with $SDI_5 = 5$
FOD system and production water	the FOD system has been operating successfully with a seawater recovery of 35%; the product water meets the Omani Standard, typically TDS < 200 mg/l; boron = 0.6–0.8 mg/L

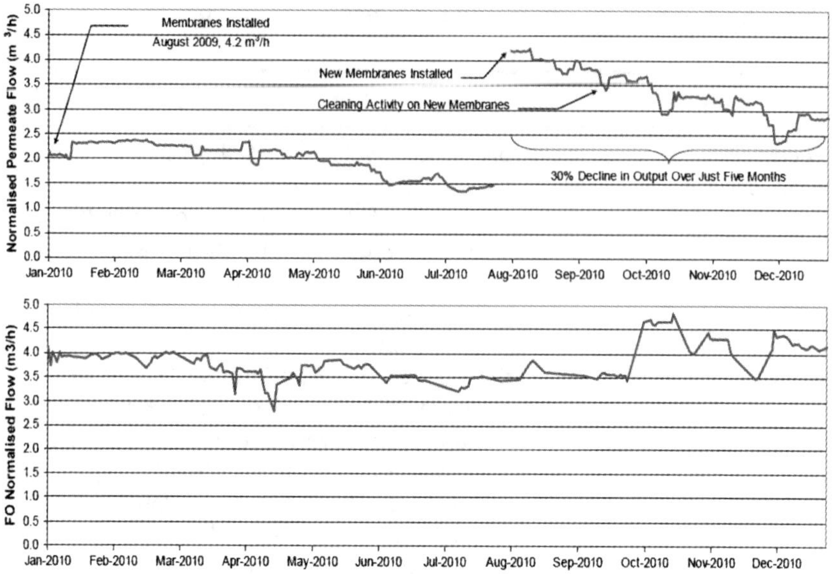

Figure 14.16. Normalized output from SWRO Facility (top) and FOD system (bottom)

Most of the current practices in the field-scale FOD processes and cutting-edge research are focusing on overcoming these two drawbacks.

For shortcoming 1). Models to predict flux behavior in FO under the influence of ECP and ICP have been reported by McCutcheon and Elimelech (2006) and modified by Tan and Ng (2008). However, the contribution of ICP and ECP to the flux reduction is not clear. Currently, it is commonly accepted that asymmetric FO membranes would induce significant ICP due to the thick and dense porous support layer. Developing more efficient FO membranes to eliminate ICP is highly desirable.

Su et al. (2011) conducted fundamental research on molecular design of FO membranes. They fabricated and then studied four groups of FO membranes: i) FO membranes made by various cellulose esters on different substrates; ii) novel double-dense structured FO membrane; iii) the thin-film composite (TFC) FO membranes; and iv) dual-layer hollow fiber membranes (HFMs) with a very thin selective outer skin layer and a highly porous support layer. They reported that the formation mechanism of cellulose ester membranes at the interface between polymer and casting substrate strongly depends on their physicochemical properties. Besides, the novel TFC FO membranes casted upon polyethersulfone (PES)/sufonated polysulfone (SPSf) blend substrates have a water flux of 47.5 LMH using a 2.0 NaCl draw solution in the pressure retarded osmosis (PRO) mode. It is feasible to create two dense layers through thermal annealing on the FO membranes prepared from CA. With the two dense layers, i.e. one being dense

Table 14.10. SWRO and FOD performance at Al Khaluf, Sultanate of Oman

Technology	*SWRO*	*FOD*	*Technology*	*SWRO*	*FOD*
Permeate Extraction from Feed water			*Draw Solution (DS) Regeneration*		
Feed water recovery (%)	5	35	DS recovery (%)	NA	47
Feed water supply (m^3/h)/pressure(bar)	11.9/65	11.9/4	Dilute DS feed(m^3/h)/ pressure (bar)	NA	8.9/65
Feed water pump eff. (%)/kW absorbed	85/25.3	85/1.6	DS regen. pump eff. (%)/kW abs.	NA	85/18.8
Product water flow (m^3/d)	71.4	100			
	Overall Plant: Specific energy consumption (kWh/m^3)			8.5	4.9

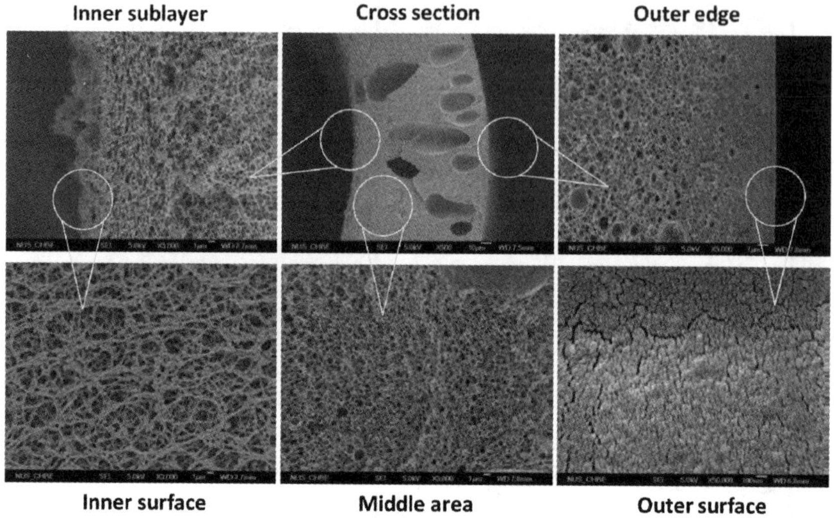

Figure 14.17. Morphology of the cross section, inner and outer surface of a dual-dense structured CA HFM
SOURCE: Adapted from Su et al. 2011

and acting as the selective layer and another one being less dense and showing increased rejection to solutes (Figure 14.17), the effect of ICP may be effectively reduced as evidenced by the little difference between water fluxes from PRO and FO modes. The dual-layer CA HFM shows a water flux of 36.7 LMH with a very low reverse salt flux of 0.21 g/m^2-h using a 2.0 M $MgCl_2$ draw solution (Su et al. 2011). In addition, several other research groups tried to make HFMs for FO processes. A research group made a polybenzimidazole (PBI) nanofiltration hollow fibre for FO applications, but PBI material was found to be very fragile. An improvement has been made by a new type of FO membrane (#C-FO) (Gadêlha and Hankins 2011).

In addition, some studies have been focused on modeling of water flux and the ICP factor under different conditions (e.g., draw solutions, configurations, and operational conditions) (McCutcheon and Elimelech 2006; Tan and Ng 2008; Gadêlha and Hankins 2011). Membrane properties and configuration play an important role in determining waster flux. The FO module and flow conditions have been evaluated as well. As indicated in Case Study A, the plate and frame osmosis cell that functions immersed in a tank presents is better than the flat-sheet counter cross-flow FO cell. Moreover, HFMs are better than flat-sheet ones as HFMs are self-supported and have a flow pattern necessary for FO processes.

Furthermore, membrane vibration has been examined as a new way to control the phenomena of CP in FO processes. Yuan and Zhang (2013) developed a batch reactor for dynamic studies on FO's flux behavior. The performance of "fresh water" production and reverse diffusion in the FO process were evaluated

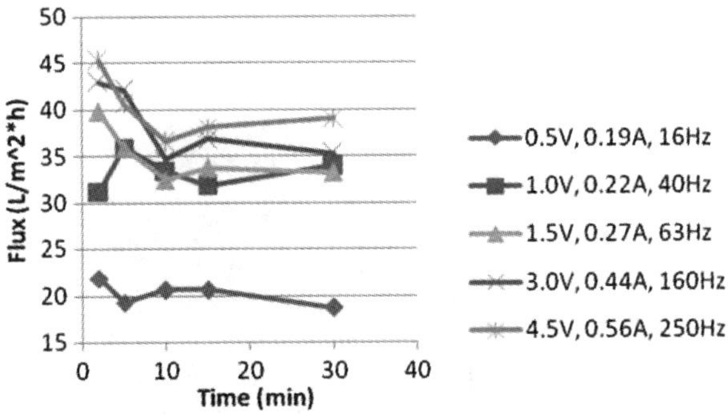

Figure 14.18. Effect of vibration frequency on FO flux with NaCl (1 M) as draw solution

SOURCE: Adapted from Yuan and Zhang (2013); copyright ASCE

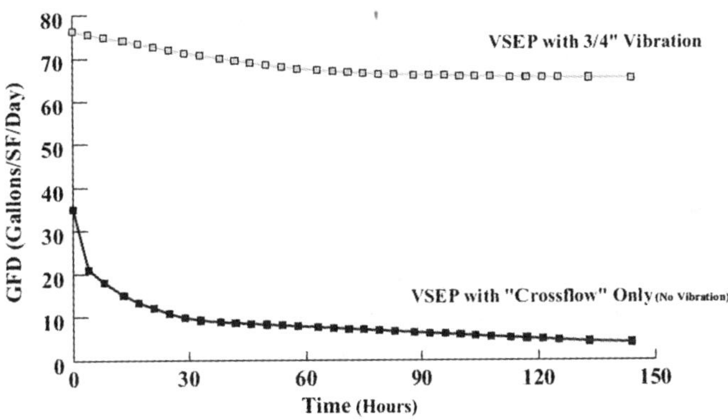

Figure 14.19. Effect of vibration on RO performance. Test conditions: 400 psi, 25°C, saturated calcium sulfate slurry

SOURCE: Adapted from NLR 2013

by vibrating the FO membrane under different experimental conditions. Results indicate that, with a NaCl solution (concentration = 1.0 M) as the draw solution, the flux was 20–50 LMH for the membrane vibration (Figure 14.18), which are much higher than that generated by the magnetic stirring method (1–6 LMH) as reported by Zhou et al. (2012). The energy consumption of the method was relatively high, but was lower than that of a RO process. Therefore, membrane vibration could increase the flux of fresh water production, and thus, may be used to control the CP better than the methods being used currently.

It is interesting to note that New Logics Research (NLR) developed a Vibratory Shear Enhanced Process (VSEP) RO system to reduce the fouling of RO membranes and to increase fluxes (Figure 14.19). The system employs torsional oscillation (50 times/second and the membrane displacement = $\frac{3}{4}$" peak to peak at the perimeter) of the membrane stack and the resulted shear is 150,000

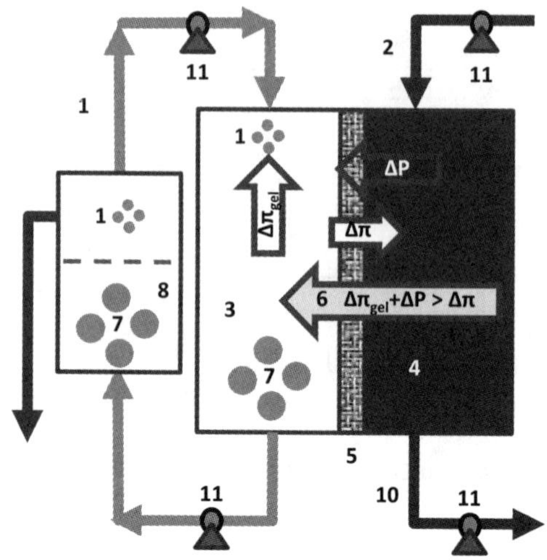

Figure 14.20. Schematic of an ESP hydrogel-based FO system. Draw solution (= shrunken hydrogels (1) + pure water) and feed solution (e.g., seawater) (2) are fed to the draw solution chamber (3) and the feed solution chamber (4), respectively. Through the FO membrane (5), permeate flows into chamber (3) by the overall draw force (6) and then is sorbed by hydrogels due to strong osmotic pressure gradient ($\Delta\pi_{gel}$) created by shrunken gels. Swollen gels (7) flow to the gel contraction unit (8), where, due to environmental condition changes, phase transition occurs to shrink gels back to (1) for their reuse. The squeezed-out permeate (9) and finished feed solution (10) flows out of the system. Pumps (11) are used to maintain a pressure difference (ΔP) between (4) and (3)

inverse seconds (ten times greater than the shear in crossflow systems). In VSEP, the diffusion layer is limited in size due to the gravitational pull (G forces) of the vibrating membrane. In addition, the vibration and oscillation of the membrane surface itself inhibits crystal formation (NLR 2013).

For shortcoming 2). Considerable studies have been conducted on finding new suitable FO draw materials (e.g., inorganic salts, large molecular compounds, magnetic nanoparticles, dendrimers and micelles close to the Kraft point, fertilizers) (Zhang 2013). In an application, a high osmolarity fertilizer solution is used to extract pure water from seawater in a FO process. The end result is a diluted fertilizer solution for direct use without the need for separation of draw solutes from the produced water (Phuntsho et al. 2011). More attention is being given to easy and efficient separation/recovery of the draw solutes. For example, dendrimers may be a good draw solute as they can provide high osmotic pressure and be regenerated by ultrafiltration (UF). Colloidal systems such as micellar solutions and free hydrophilic polyelectrolytes are also recommended due to their several advantages (Gadêlha and Hankins 2011). As another example, the environmentally-sensitive gel-based FO system (Figure 14.20) and the FO systems with polymer coated magnetic nanoparticles (MNPs) as draw solute were proposed (Zhang et al. 2012; Zhang 2013). The EPS hydrogels can release water with a minimal energy requirement during phase transition, and MNPs can be separated from product water once a magnetic field is applied.

In the FO system proposed in Figure 14.20, the slurry of shrunken hydrogels is fed into the draw solution chamber, and hydrogels are in good contact with both the FO membrane and other hydrogels. Permeate flows through the FO membrane because of $\Delta\pi_{gel}$ and from hydrogels near the FO membrane to hydrogels far away from it, again, because of $\Delta\pi_{gel}$. In the draw solution chamber, water has a tendency to flow into hydrogels and back to the feed solution chamber (thus, a $\Delta\pi$ exists). However, as long as $\Delta P + \Delta\pi_{gel} > \Delta\pi$, water will flow from feed solution to draw solution. The overall driving forces of three different membrane systems are:

$$\text{For the RO system: } J_w = k_w(\Delta P - \Delta\pi) \tag{14.1}$$

$$\text{For the conventional FO system: } J_w = k_w(\Delta P + \Delta\pi) \tag{14.2}$$

$$\text{For the proposed FO system: } J_w = k_w(\Delta P + \Delta\pi_{gel} - \Delta\pi) \tag{14.3}$$

where, k_w = water mass transfer coefficient; ΔP = average applied hydraulic pressure difference between the two chambers; $\Delta\pi$ = osmotic pressure difference defined in Figure 14.1; and $\Delta\pi_{gel}$ = the osmotic pressure gradient between the inside and outside of the ESP hydrogels. Because of self-limiting flux behavior, $\Delta\pi$ in both the conventional FO system and that shown in Figure 14.20 is much lower than the theoretical values. Therefore, the introduction of $\Delta\pi_{gel}$ can increase the overall driving force of the proposed FO system as proved by Li et al. (2011a & b), and thus, save energy. The potential of the proposed FO system stems from $\Delta\pi_{gel}$ which can be increased by several orders of magnitude depending on the hydrogels' structure/characteristics (Xie et al. 2008). Tests are being conducted to evaluate the FO system shown in Figure 14.20.

Moreover, modeling of draw solution behavior in the FO process and the protocol for the selection of optimal draw solutions for FO applications have been developed to determine the most appropriate draw solutions under different conditions (Achilli et al. 2010; Sagiv and Semiat 2011). The protocol usually includes a desktop screening process and laboratory and modeling analyses (Figure 14.21), which can be used for draw solution selection as new FO membranes and draw solutions are developed and applied to new FO applications.

PRO

It seems that desalination technologies evolved from RO to FO and, nowadays, extend to PRO (Sourirajan 1970; Li et al. 2008; Frenkel 2011; Zhang et al. 2012). The concept of PRO was proposed in the 1970s (Loeb 1974), and the principle of using a conventional PRO process to generate blue power is shown in Figure 14.22.

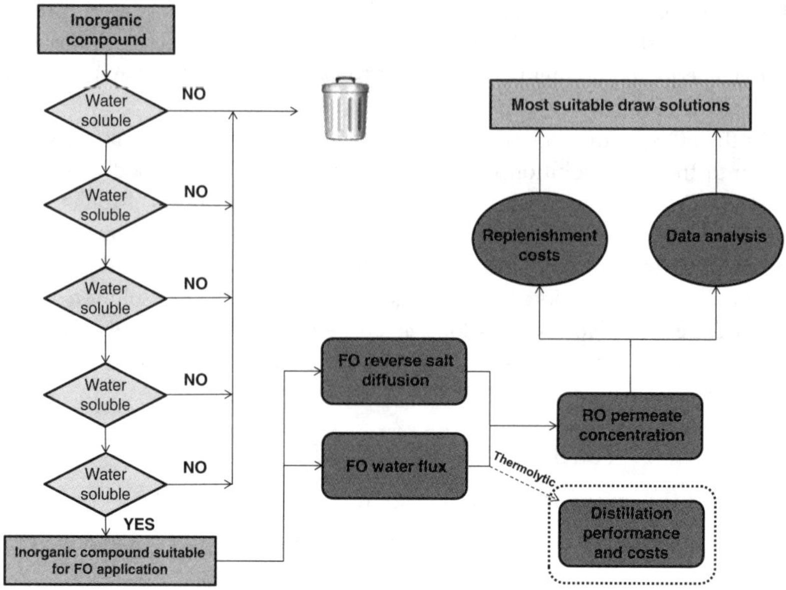

Figure 14.21. Flow chart for draw solution selection
SOURCE: Adapted from Achilli et al. 2010

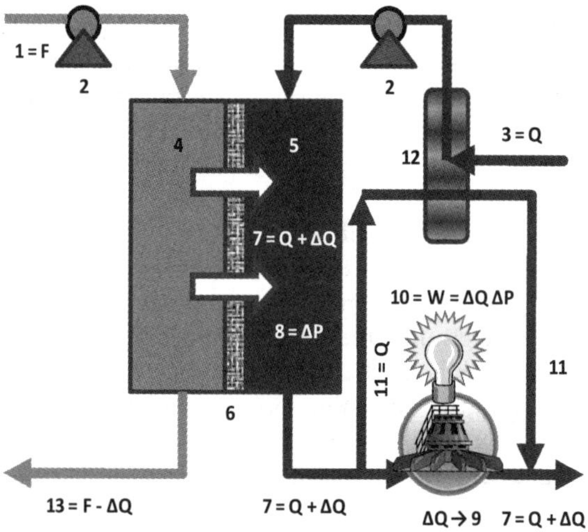

Figure 14.22. A PRO system. Fresh water (1) at a flow rate F and seawater (3) at a Q are pumped (2) into a feed (4) and draw (5) chamber separated by a FO membrane (6), which only allows water to pass through but stops salts. Osmosis transports water (ΔQ, m^3/s) from (4) to (5); the pressure difference between (4) and (5) is ΔP (8). ΔQ passes through a turbine (9) to generate power (10) (= W). The stream (11) passes through a pressure exchanger (12) to pressurize seawater (3)

In a PRO process, the power density, W (W/m^2) and water permeation flux, J_w (L/m^2-h) can be expressed as follows:

$$W = J_w \Delta P = A(\Delta\pi - \Delta P)\Delta P \tag{14.4}$$

$$J_w = A(\sigma\Delta\pi - \Delta P) \tag{14.5}$$

where J_w = volumetric flux of water; ΔP = hydrostatic pressure of the salt solution (= hydrostatic pressure of the feed, P_f – hydrostatic pressure of the permeate, P_p); A = water permeation coefficient; $\Delta\pi$ = osmotic pressure of the salt solution (= osmotic pressure in the feed, π_f – osmotic pressure in the permeate). σ = the reflection coefficient (= 1 for an impermeable solute). Notice that W_{max} can be achieved at $\Delta P = \sigma\Delta\pi/2$.

There is no detailed report about using a full-scale PRO process for blue power generation. Statkraft, an European company accelerated the PRO development in 2007 has been operating the world's first prototype PRO power plant in Norway since 2009 (Skråmestø et al. 2009); however, no detailed information about the test is available. The PRO system is limited to the estuary where seawater is used as a natural draw solute. According to Statkraft, the river-into-sea PRO system is ~66–132/MWh (at year 2015), which is comparable and competitive with the other new renewable energy sources (e.g., 82–106/MWh for wind offshore; 80–90/MWh for biomass, and 116–150/MWh for wave and tidal) (Skråmestø et al. 2009). The global potential of conventional osmotic power using seawater is about 1600–1700 TWh/year–equivalent to China's electricity consumption in 2002 (Skråmestø et al. 2009). If the additional osmotic power generated by the proposed PRO system from brackish water or fresh water is considered as suggested by Zhang and Surampalli (2012), the contribution of PRO systems to the generation of clean, renewable energy would be very significant.

Currently, the PRO system is not considered to be cost-effectively compatible with fossil fuel systems because of some major problems: a) dilution of draw solution which requires a constant replacement of draw solution; b) salt leakage from the draw chamber to the feed chamber, which reduces $\Delta\pi$; and c) CP across the salt-rejecting skin that reduces $\Delta\pi$ to some effective value $\Delta\pi_{eff}$ (Post et al. 2007; Zhang and Surampalli 2012). For most of the membranes, $\Delta\pi_{eff}/\Delta\pi$ is ~5–90% at a water flux of 2 gal/ft^2-d and 0.5–10% at a water flux of 200 gal/ft^2-d (an economically viable flux) (Lee et al. 1981). Since the maximum PRO power $W_{max} = A(\Delta\pi_{eff})^2/4$, a small reduction of $\Delta\pi_{eff}$ can reduce PRO power tremendously.

Current research about the PRO systems focuses on 1) membrane development, 2) improvement of draw solutes, and 3) enhancement of the performance of pressure recovery devices (Lee et al. 1981; Cath et al. 2006; Post et al. 2007; Peinemann et al. 2008). The break even value for the membrane performance is 5 W/m^2. Currently, CA membranes can achieve a performance of 1.3 W/m^2. TFC membranes and inner coated TFC HFMs have been tested. Yang et al. (2009) developed a dual-layer hollow fiber FO membrane with a water flux of

33.8 L/m^2-h and a salt flux <1.0 g/m^2-h using 5 M $mgCl_2$ as the draw solution. Yip et al. (2010) reported a TFC membrane with a water flux >18 LMH in a 1.5 M NaCl draw solution and a pure water feed. Modeling and tests have been conducted about the reverse diffusion of draw solute across an asymmetric membrane in FO; results indicate that the reverse flux selectivity, the ratio of the permeate flux to the reverse solute flux, is a key parameter in the design of osmotically driven membrane processes (Phillip et al. 2010).

Future Trend

Future research will focus on FO and/or PRO processes and their applications, such as good FO and PRO membranes, good FO and PRO draw solutions, combination of UF, NF, RO, FO, PRO and MBR (membrane bioreactor) for drinking water production, wastewater treatment/reuse, and blue energy generation. Some of these applications will be beyond the scope of desalination. For example, if desalination is not needed, but turbidity and bacteria in feed solution need to be removed, one can use UF or MF to replace the FO membrane in a hydrogel-based FO system (Figure 14.20). In such a system, the water flux $J_w = k_w (\Delta P + \Delta\pi_{gel})$ as UF or MF can't generate $\Delta\pi$. Zhang (2013) reported that when the HTI FO membrane was replaced by a Kubota membrane (type H-203, KUBOTA Membrane USA Co., Redmond, WA), the same hydrogels can generate a flux 80 times higher than that of the FO system shown in Figure 14.20 at a water column height of 1 m. Thus, adding hydrogels may significantly lower the ΔP that otherwise is needed in a conventional MF system. In such a new system, the function of the UF or MF is to remove unwanted materials (e.g., particles, bacteria) larger than the pore size of the membrane, while hydrogels are for lowering the energy consumption. RO or FO membranes will be used only as needed to remove salt or micro-pollutants (e.g., emerging contaminants). Thus, this new membrane process could have much wider applications (e.g., water treatment, wastewater reclamation, etc.).

14.4 CONCLUSIONS

When compared to a RO system, there are many advantages of a FO system. While RO systems involve high pressures, and therefore, high energy inputs and exotic materials, the FO process takes place at low pressures and therefore does not require the same energy input or high strength materials. However, FO systems do have additional complications when compared to RO systems. In particular, the FO process does not provide high quality water for use in a single step as the permeate is mixed with the draw solution. These are the major obstacles of the FOD process. Although the novel concept of FO was developed as early as 1968, it has not been able to advance mainly due to the lack of suitable draw solution and lack of suitable FO membranes. Therefore, future research is needed.

14.5 ABBREVIATIONS/NOMENCLATURE

CP	concentration polarization
CTA	cellulose triacetate
DO	dissolved oxygen
DOC	dissolved organic carbon
DS	draw solution
ECP	external concentration polarization
FO	forward osmosis
FOD	forward osmosis (processes) for desalination
HFM	hollow fiber membrane
HTI	Hydration Technology Innovations
ICP	internal concentration polarization
KAUST	the King Abdullah University of Science and Technology
KOPF	KAUST custom-made FO plate and frame
LMH	L/m^2-h
LPRO	low pressure reverse osmosis
MF	microfiltration
NF	nanofiltration
PRO	pressure retarded osmosis
RO	reverse osmosis
TFC	thin film composite
ΔP	hydraulic pressure
$\Delta\pi$	osmotic pressure difference
$\Delta\pi_{gel}$	osmotic pressure gradient between inside and outside of hydrogels

References

Achilli, A., Cath, T. Y., and Childress, A. E. (2010). "Selection of inorganic-based draw solutions for forward osmosis applications." *J. Memb. Sci.*, 364, 233–241.

Batchelder, G. W. (1965). *Process for the demineralization of water*, US Patent 3171799.

Cath, T. Y., Childress, A. E., and Elimelech, M. (2006). "Forward osmosis: principles, applications, and recent developments." *J. Membr. Sci.*, 281, 70–87.

D&WR. (2012). "Small FO desalination plant commissioned in Oman." Posted on September 6, 2012, the International Desalination & Water Reuse Quarterly industry website ⟨www.desalinationbiz/news⟩ (accessed March 2013).

Frank, B. S. (1972). "Desalination of sea water." US Patent 3670897.

Frenkel, V. S. (2011). "Reverse osmosis, forward osmosis, pressure retarded osmosis. What is next?" *IDA World Congress/Perth Convention and Exhibition Centre (PCEC), Perth, Western Australia, Sept. 4–9, 2011*, REF: IDAWC/PER11-009.

Gadêlha, G. A. F., and Hankins, N. P. (2011). "Assessment of draw-solution performance in combination with asymmetric membranes for forward osmosis." *IDA World Congress/ Perth Convention and Exhibition Centre (PCEC)*, Perth, Western Australia, September 4–9, 2011, REF: IDAWC/PER11-024.

Glew, D. N. (1965). "Process for liquid recovery and solution concentration." US Patent 3216930.

Kessler, J. O., and Moody, C. D. (1976). "Drinking water from seawater by forward osmosis." *Desalination*, 18, 297–306.

Kravath, R. E., and Davis, J. A. (1975). "Desalination of seawater by direct osmosis." *Desalination*, 16, 151–155.

Lee, K. L., Baker, R. W., and Lonsdale, H. K. (1981). "Membranes for power generation by pressure-retarded osmosis." *J. Membrane Sci.*, 8, 141–171.

Li, D., Zhang, X., Yao, J., Zeng, Y., Simon, G. P., and Wang, H. (2011a). "Stimuli-responsive polymer hydrogels as a new class of draw agent for forward osmosis desalination." *ChemComm.*, 47, 1710–1712.

Li, D., Zhang, X., Yao, J., Zeng, Y., Simon, G. P., and Wang, H. (2011b). "Composite polymer hydrogels as draw agents in forward osmosis and solar dewatering." *Soft Matter*, 7, 10048–10056.

Li, N. N., Fane, A. G., Ho, W. S. W., and Matsuura, T. (2008). *Advanced Membrane Technology and Applications*, John Wiley & Sons, Hoboken, New Jersey, 2008.

Ling, M. M., Wang, K. Y., and Chung, T.-S. (2010). "Highly water-soluble magnetic nanoparticles as novel draw solutes in forward osmosis for water reuse." *Ind. Eng. Chem. Res.*, 49, 5869–5876.

Liu, L., Wang, M., Wang, D., and Gao, C. (2009). "Current patents of forward osmosis membrane process." *Recent Patents on Chemical Eng.*, 2, 76–82.

Liu, Z., Bai, H., Lee, J., and Sun, D. D. (2011). "A low-energy forward osmosis process to produce drinking water." *Energy & Environ. Sci.*, 4, 2582–2585.

Loeb, S. (1974). "Osmotic power plants." *Science*, 189, 350.

McCutcheon, J. R., McGinnis, R. L., and Elimelech, M. (2005). "A novel ammonia-carbon dioxide forward (direct) osmosis desalination process." *Desalination*, 174, 1–11.

McCutcheon, J., and Elimelech, M. (2006). "Influence of concentrative and dilutive internal concentration polarization on flux behavior in forward osmosis." *J. Membrane Science*, 284, 237–247.

McCutcheon, J. R., McGinnis, R. L., Elimelech, M. (2006). "Desalination by a novel ammonia–carbon dioxide forward osmosis process: influence of draw and feed solution concentrations on process performance." *J. Membr. Sci.*, 278, 114–123.

McCutcheon, J. R., and Elimelech, M. (2008). "Influence of membrane support layer hydrophobicity on water flux in osmotically driven membrane processes." *J. Memb. Sci.*, 318, 458–466.

McGinnis, R. L. (2002). "Osmotic desalination process." US Patent 6391205 B1.

Nicoll, P. G., Thompson, N. A., and Bedford, M. R. (2011). "Manipulated osmosis applied to evaporative cooling make-up water–revolutionary technology." *IDA World Congress/ Perth Convention and Exhibition Centre (PCEC), Perth, Western Australia, September 4–9, 2011*, REF: IDAWC/PER11-119.

NLR (New Logic Research). (2013). "Membrane filtration of landfill leachate: Advances in module designs enhance landfill effluent treatment." Technical article, New Logic Research, Emeryville, CA, USA.

Peinemann, K.-V., Gerstandt, K., Skilhagen, S. E., Thorsen, T., and Holt, T. (2008). "Membranes for power generation by pressure retarded osmosis." *Ch. 9 in Membranes for Energy Conversion*, Vol. 2, K.-V. Peinemann and S. P. Pereira-Nunes (eds.), Wiley-VCH Verlag GmbH & Co. KGaA, p. 263–273.

Phillip, W. A., Yong, J. S., and Elimelech, M. (2010). "Reverse draw solute permeation in forward osmosis: modeling and experiments." *Environ. Sci. Technol.*, 44(13), 5170–5176.

Phuntsho, S., Shon, H. K., Hong, S., Lee, S., and Vigneswaran, S. (2011). "A novel low energy fertilizer driven forward osmosis desalination for direct fertigation: Evaluating the performance of fertilizer draw solutions." *Journal of Membrane Science*, 375, 172–181.

Post, J. W., Veerman, J., Hamelers, H. V. M., Euverink, G. J. W., Metz, S. J., Nymeijer, K., and Buisman, C. J. N. (2007). "Salinity-gradient power: evaluation of pressure-retarded osmosis and reverse electrodialysis." *J. Membrane Sci.*, 288, 218–230.

Sagiv, A., and Semiat, R. (2011). "Finite element analysis of forward osmosis process using NaCl solutions." *Journal of Membrane Science*, 379(2011), 86–96.

Sidney, L., Leonid, T., Emmanuel, K., and Joseph, F. (1997). "Effect of porous support fabric on osmosis through a Loeb-Sourirajan type asymmetric membrane." *J. Memb. Sci.*, 129, 243–249.

Skråmestø, Ø., Skilhagen, S. E., and Nielsen, W. K. (2009). "Power production based on osmotic pressure." *Waterpower*, 16, 1–10. Available at ⟨www.statkraft.com⟩ (accessed Aug. 2010).

Sourirajan, S. (1970). *Reverse osmosis*, Academic Press, New York.

Stache, K. (1989). "Apparatus for transforming sea water, brackish water, polluted water or the like into a nutrious drink by means of osmosis." US Patent 4879030.

Su, J., Wang, K. Y., Zhang, S., Ong, R. C., and Chung, T.-S. (2011). "Molecular design of advanced forward osmosis membranes for water reuse." *IDA World Congress/Perth Convention and Exhibition Centre (PCEC), Perth, Western Australia, September 4–9, 2011*, REF: IDAWC/PER11-031.

Tan, C. H., and Ng, H. Y. (2008). "Modified models to predict flux behavior in forward osmosis in consideration of external and internal concentration polarizations." *J. Membr. Sci.*, 324, 209–219.

Tan, C. H., and Ng, H. Y. (2010). "A novel hybrid forward osmosis — nanofiltration (FO-NF) process for seawater desalination: Draw solution selection and system configuration." *Desalination and Water Treatment*, 13(2010), 356–361.

Tan, C. H., Lefebvre, O., Zhang, J., Ng, H. Y., and Ong, S. L. (2012). "Membrane process for desalination: Overview." *Ch. 10 in Membrane Technology and Environmental Applications*, T. C. Zhang, R. Y. Surampalli, S. Vigneswaran, R. D. Tyagi, S. L. Ong and C. M. Kao (eds.), ASCE, Reston, Virginia, 2012, 298–330.

Thompson, N. A., and Nicoll, P. G. (2011). "Forward osmosis desalination: A commercial reality." *IDA World Congress/Perth Convention and Exhibition Centre (PCEC), Perth, Western Australia, September 4–9, 2011*, REF: IDAWC/PER11-198.

Wang, K. Y., Ong, R. C., and Chung, T. S. (2010). "Double-skinned forward osmosis membranes for reducing internal concentration polarization within the porous sublayer." *Ind. Eng. Chem. Res.*, 49, 4824–4831.

WaterWorld. (2012). "Forward osmosis: Is China next to shake up the desalination sector?" Posted on Dec. 4, 2012, WaterWorld ⟨www.waterworld.com⟩ (accessed March 2013).

Xie, J.-J., Liu, S., Li, S., and He, X.-J. (2008). "Preparation and characterization of superbasorbents from AMPS by solution polymerization." *J. Central South Univ. of Forestry & Tech.*, 28(2), 32–36.

Yaeli, J. (1992). "Method and apparatus for processing liquid solutions of suspensions particularyly useful in the desalination of saline water." US Patent 5098575.

Yang, Q., Wang, K. Y., and Chung, T.-S. (2009). "Dual-layer hollow fibers with enhanced flux as novel forward osmosis membranes for water production." *Environ. Sci. Technol.*, 43, 2800–2805.

Yangali, V., Li, Z., Valladares, R., and Amy, G. (2011). "Wastewater reclamation with forward osmosis and red sea seawater." *IDA World Congress/Perth Convention and Exhibition Centre (PCEC), Perth, Western Australia, September 4–9, 2011*, REF: IDAWC/PER11-253.

Yen, S. K., Haja, N. F. M., Su, M., Wang, K. Y., and Chung, T.-S. (2010). "Study of draw solutes using 2-methylimidazole-based compounds in forward osmosis." *J. Membr. Sci.*, 364, 242–252.

Yip, N. Y., Tiraffrri, A., Phillip, W. A., Schiffman, J. D., and Elimelech, M. (2010). "High performance thin-film composite forward osmosis membrane." *Environ. Sci. Technol.*, 44 (10), 3812–3818.

Yuan, Y., and Zhang, T. C. (2013). "Control of concentration polarization in forward osmosis processes by vibration and ultrasonic cavitation." *Proc. of World Environ. & Water Resources Congress 2013, May 20–24, 2013*, Cincinnati, OH.

Zhang, T. C. (2013). *Fundamentals of Environmentally-Sensitive Hydrogel-Based Forward Osmosis*, A proposal submitted to National Sci. Foundation, Feb., 2013.

Zhang, T. C., and Surampalli, R. Y. (2012). "Novel membrane-separation techniques and their environmental applications." *Ch. 23 in Membrane technology and environmental applications*, T. C. Zhang, R. Y. Surampalli, S. Vigneswaran, R. D. Tyagi, S. L. Ong and C. M. Kao (eds.), ASCE, Reston, Virginia, 2012, 696–726.

Zhang, T. C., Surampalli, R. Y., Vigneswaran, S., Tyagi, R. D., Ong, S. L., and Kao, C. M. (eds.) (2012). *Membrane Technology and Environmental Applications*, ASCE, Reston, Virginia, 2012.

Zhou, A. J., Zhang, T. C., and Yuan, Y. (2012). "Performance of forward osmosis process under different operating conditions and draw solutes." *Proceedings of World Environmental & Water Resources Congress 2012 (CD-ROM), May 20–24, 2012*, Albuquerque, NM.

CHAPTER 15

Treatment of High Salinity Waste Water from Shale Gas Exploitation by Forward Osmosis Processes

Xue-Mei Li
Gang Chen
Ho Kyong Shon
Tao He

15.1 INTRODUCTION

Hydraulic fracturing has been a key technology in producing shale gases an affordable addition to the United States' energy supply. Hydraulic fracturing is a rather water intensive process which requires 2 million to 5 million gallons of water for a horizontal shale gas well depending on the basin and formation characteristics (Ground Water Protection Council 2009). After fracturing, the hydraulic fluid begins to flow back through the well casing to the well head. This produced water contains various dissolved constituents and organic matters. Its treatment and recycling has drawn wide attention because of its health, environmental and ecological impacts. Because of the complexity in composition, high TDS, limited footprint and cost issues, new water treatment technologies are needed that can recycle the water as fracturing make-up water, or irrigation water, and in some cases pure process water.

Forward osmosis (FO) is an osmotically driven membrane process, where a chemical potential difference acts as the driving force for transferring of water across the membrane from a dilute feed solution to a concentrated draw solution (Cath et al. 2006). The semipermeable FO membrane can block the transfer of a broad range of contaminants including organic matter, dissolved solids, and suspended solids with potential applications in treatment of domestic and industrial wastewater, concentration of beverages and pharmaceutics, and controlled drug release. The most significant characteristics of FO are low energy input, low fouling propensity, high water recovery rate, and highly tolerance to high salinity water streams. FO could potentially provide a new perspective to the

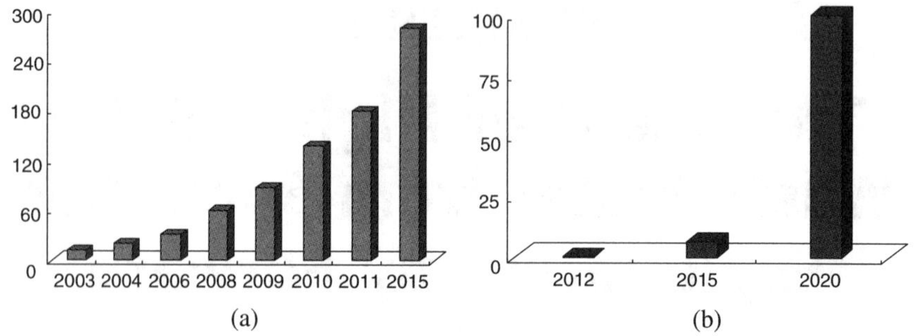

Figure 15.1. Shale gas productivity in the USA (a) and China (b) (in Billion M^3*)*

disposal of the special wastewater containing high total dissolved solids (TDS) (Shaffer et al. 2013).

This chapter reviews the state-of-the-art of the treatment of shale gas produced water with the focus on the treatment of shale gas flow-back water (SGW) by FO. A brief introduction to the origin and chemical/physical characteristics of the SGW are given, and the advantages and limitations of potential treatments methods are analyzed. The process parameters, selection of membrane and draw solutions are summarized. Finally, the potential of utilization of the FO process for the treatment of SGW in a large scale are discussed.

15.2 WATER MANAGEMENT IN SHALE GAS EXPLOITATION

15.2.1 Generation, Health and Environmental Impacts

Shale gas is an important unconventional natural resource for the energy thirsty, and its exploitation activities has been increasing. Based on the US EIA data in 2011, the reservation of the shale gas in US was about $2.44 \times 10^{4\ BM^3}$ and that in China is $3.6 \times 10^{4\ BM^3}$ (He et al. 2012). As shown in Figure 15.1, the projection of the shale gas productivity in the US will be 280 Billion cubic meter by 2015 in America and to 100 Billion cubic by 2020 in China. Between 2003 and 2010, there has been a quick and steady growth of the shale gas output in the US. It is expected that the shale gas production in China follows an even more drastic increase in the coming 10 years.

The shale gas resources in many areas had been overlooked because the production economical feasibility was not attractive enough until the development of combination of sequenced hydraulic fracture treatments and horizontal well completions for shale gas drilling. During the hydraulic fracturing process, a fracturing fluid under high pressure is pumped into a shale formation to generate fractures or cracks in the shale layer. The natural gas flows out of the shale to the well. Water and sand make up over 98% of the fracture fluid, with the rest consisting of various chemical additives that improve the effectiveness of the fracturing process. The main compositions of the fracturing fluid consist of 90.60% water, ~9% sand and other additives. The additives include biocides (sodium hypochlorite or sodium hydroxide), corrosion inhibitors, scavengers, friction reducers, surfactants, etc. The exact chemical components are the secret of the oil/gas service companies, thus not known in public. The amount of water needed to drill and fracture a horizontal shale gas well generally ranges from about 2 million to 5 million gallons of fresh water, depending on the basin and formation characteristics (Colorado School of Mines 2009).

After a hydraulic fracture treatment and relief of the pumping pressure from the well, the water-based fracturing fluid, mixed with any natural underground water, begins to flow back through the well casing to the wellhead. The time for recovering the majority of fracturing fluid ranges from several hours to a couple of weeks. In various basins and shale gas plays, the volume of produced water may account for 15–40% of the original fracture fluid volume. In some cases, flowback of fracturing fluid in produced water can continue for several months after gas production has begun. If not directly treated, the flow back water is stored in a man-made pond before further treatment or tankering. Figure 15.2 shows a typical site for shale gas mining in a remote area in the northwest China. Next to the crane, shale gas flowback water and domestic wastewater were temporarily stored in separate ponds. Both streams are of different characteristics and remains yet untreated.

The SGW contains various dissolved constituents. Initial produced water can vary from fresh (TDS < 5,000 mg/L) to varying degrees of salinity (TDS from 5,000 mg/L to 100,000 mg/L or higher). The dissolved constituents are natural compounds and vary from one shale site to the other.

The composition varies significantly as compared to the composition of produced water from Marcellus Shale drilling. The TDS in the wastewater changes

Figure 15.2. Photos of one typical shale gas exploitation site in west China. (1) Shale gas exploitation well pad; (2) domestic wastewater; (3) wastewater storage

with time as well. Initial TDS of 38500–238000 mg/L was found in the produced water from wells drilled in Marcellus Shale at 5 days post hydraulic fracturing and was in the range of 3010–261000 mg/L at 14 days of post drilling. This high TDS makes the treatment of such produced water a great challenge. In the case of southwest China, the saline content appears to be much lower than that in the US. The difference is a strong reflection of the geological variation from region to region.

It is a common concern to the public that the flowback water from shale gas drilling is a major environmental issue. Compatibility of the land use is the first concern. Contamination of the surface water and ground water as well as the release of toxic pollutants is critical to the environment. Most of the shale gas sites, particularly in southwest China, are densely inhabited farming fields. Although no specific regulations yet for the shale gas, the discharge of industrial wastewater guidelines are restricted as well. Therefore, a post-treatment of the flowback and produced water from shale gas exploitation is crucial for the sustainable development of the oil and gas industry.

15.2.2 Water Management in Shale Gas Exploitation

The potential targets for the treatment of the SGW are listed as follows:

1. brine volume reduction with the possibility for reuse to future fracturing purpose;
2. removal of the organic polymeric additives;
3. oil and grease control;
4. removal of suspended solids;
5. microbial control; and
6. soluble organics

Management of the shale gas wastewater depends on multidimensional criteria, e.g. the local regulation, site conditions, produced water quality, and most importantly, economic feasibility. Approaches used for the treatment of high salinity waste water include: (1) deep well injection; (2) transport and centralized treatment; (3) treatment and disposal; and (4) reuse. The selection of different approaches is not purely a technical choice. However, it is a common paradigm that deep well injection and treatment and disposal are one of the first choices. Centralized treatment is costly, and thus, the last choice if an economical reuse alternative exists. The main targets of the reuse include reduction of total suspended solids (TSS), oil and grease, hydrocarbons, hardness, iron, boron, etc.

For the removal of the TDS, evaporation, distillation and reverse osmosis are the main candidates. Although the water quality is of the highest level, cost per cubic of wastewater becomes very important. However, injection eliminates water permanently from the water cycle, and in some areas this is a critical environmental issue. Some processes are highly energy intensive and may require intensive pretreatment, leading to rather high operating expenses due to the membrane replacement. Nevertheless, membrane-based processes such as reverse osmosis, nanofiltration (NF), membrane distillation (MD), and FO are promising due to the advantages in simple/automatic operation, small foot-print and high efficiency, etc. Table 15.1 lists several typical membrane technologies which show potential applications in the shale gas wastewater treatment, along with their advantages and disadvantages.

For NF, reverse osmosis (RO) or MD, a pre-treatment facility to remove completely or partially the inorganic or organic contaminants, is necessary as listed in Table 15.2. Usually, membrane fouling is primarily caused by organics such as humic acid and alginate (Sioutopoulos et al. 2010; Zazouli et al. 2010; Peter-Varbanets et al. 2011; Ghouas et al. 2012; Kim and Dempsey 2013; Resosudarmo et al. 2013), which can reduce the efficiency of salt removal. Consequently, pre-treatment is often compulsory to control water chemistry and reduce fouling. For TDS concentrations of up to 20,000 mg/L, RO is the preferred method. MD is used for waters with TDS concentrations of 40,000–100,000 mg/L. FO technology was widely reported in treatment of brackish water and brine (Tang and Ng 2008; Zhao and Zou 2011; Li et al. 2013).

FO systems do not need external hydraulic pressure as the driving force. It has a high rejection rate of contaminates, and more importantly a lower propensity for membrane fouling, comparing to pressure driven osmotic processes (Cath and Bamaga 2011). In case the draw solution is directly reusable without any further post-treatment, the FO is essentially working at low energy consumption. Success in the utilization of FO technology has been reported in the literature (Hickenbottom et al. 2013). A summary of the technical assessment of the several membrane technologies are listed in Table 15.2. FO is a low energy and simple process for the reuse of the shale gas flowback water. Therefore, the following section will mainly focus on the application of FO for the reuse of the shale gas wastewater.

15.3 FO FOR TREATMENT OF SEG-PRODUCED WASTEWATER

15.3.1 Overview

FO is an osmotically driven membrane process, where a chemical potential difference acts as the driving force for water transfer across the membrane of a dilute feed solution to a concentrated draw solution. The semipermeable FO membrane can block the transfer of a broad range of contaminants including organic matter, dissolved solids, and suspended solids with application potentials in treatment of domestic and industrial wastewater, concentration of beverages and pharmaceutics, and controlled drug release. The most significant characteristics of FO include: no or little energy input; low fouling propensity; high water recovery rate; and highly tolerant to high salinity and TDS. FO could potentially provide a new perspective to the disposal of the special wastewater containing high concentrated TDS (Hickenbottom et al. 2013; McGinnis et al. 2013).

Table 15.1. Summary of advantages and challenges of the present treatment technologies for the shale gas flowback water

Technology	***Advantages***	***Challenges***
Deep well injection	Low cost: $1.5-2/Bbl; well established and accepted for public	limited deep injection well/capacity; transportation may be costly
Thermal distillation	Pure water obtained, reusable for industry and irrigation or for discharge	Bulky and complicated system; high cost alternative
Evaporation/ crystallization	No extra energy input, water reusable; cheap choice	Large footprint; possible for limited area
RO	Relatively pure water permeate; mature technology,	system cost high; limited salinity < 3.5%; high energy cost
Electrodialysis	Mature technology, high water recovery rate; suitable for water with high SDI, TOC; removal of heavy metals, cyanide, chloride, less fouling, scaling and chemical addition	Limit is TDS < 15000 ppm; pre-treatment required; no removal of bacteria, colloidal matters, boron, silica etc
Capacitive Deionization (CDI)	Minimum pre-treatment; low fouling scaling; low operating voltages/ pressures	Low adsorption capacity; energy loss in regeneration; suitable for TDS < 1500 mg/L
Ion exchange	High water recovery	Limit for low salinity water; generation of waste; chemical usage for regeneration; cost
MD	Very high permeate water quality; less chemical interaction between membranes and solutions; no hydraulic pressure needed	Heat sources needed; high cost in energy; membrane wetting; technology not yet mature; limited membrane supplier
FO	Tolerable to High salinity; Low energy consumption; very good quality permeated water; reusable for fracturing job	Suitable draw solute; limited membrane supplier; immature technology

These inherent advantages make FO a promising candidate for wastewater treatment. Consequently, studies were performed to examine the effectiveness of FO in the treatments of high salinity wastewater produced from shale gas exploitation.

15.3.2 Membranes

Hydration Technologies Inc. (HTI) has commercialized their CTA membranes for application in an FO process. The HTI FO membranes made of cellulose triacetate (CTA) supported by embedded polyester webs have been widely reported in desalination of brackish water (Zhao and Zou 2011; Zhao et al. 2012; Phuntsho et al. 2013) and seawater (Yangali-Quintanilla et al. 2011; Li et al. 2012; Boo et al. 2013). Few investigations were for disposal of high saline wastewater generated from Shale gas exploitation (Hickenbottom et al. 2013). In addition, thin-film composite (TFC) FO membranes were also used in desalination of high salinity wastewater derived from shale exploitation (McGinnis et al. 2013). Characteristics of different FO membranes used in shale gas exploitation wastewater treatment are summarized in Table 15.3. TFC membranes appear to show higher water flux than CTA membranes.

15.3.3 Draw Solution

Draw solutions, composing a draw solute and water, provide the driving force for the water to transfer across the membrane. Draw solution is a key component for a successful FO application. In general, an ideal draw solute should have a series of characters: (1) high solubility in water; (2) high osmotic pressure upon dissolution; (3) no reaction to the membrane; (4) highly rejected by the membrane; (5) non-toxic; and (6) low cost. It is even more desirable that the draw solutes can be recovered with an energy efficient process (Achilli et al. 2010). Various chemicals have been tested and compared as draw solutes, such as NaCl, $MgCl_2$, and Na_{2SO_4}, even fertilizers (Phuntsho et al. 2011). For SGW treatment, the general selecting rules hold, but special focus should be paid on high rejection, high osmotic pressure, scaling and cost.

Hickenbottom et al. (2013) employed sodium chloride solution (NaCl, 4.5 mol/L) as draw solution in the FO process for concentration of drilling mud and fracturing wastewater from oil and gas operations. Moreover, McGinnis et al. (2013) used ammonium bicarbonate (NH_{4HCO_3}) as draw solution in desalination of high salinity (> 70,000 ppm TDS) produced waters from shale gas exploration in order to produce clear water. The product water generated in the process was found to meet surface water discharge quality criteria (< 500 mg/L TDS, < 250 mg/L chlorides, < 10 mg/L barium, < 10 mg/L strontium) for the site of Pennsylvania. However, ammonium bicarbonate is alkaline in nature and may deteriorate membrane.

Sodium chloride, magnesium chloride ($MgCl_2$) and disodium ethylene diamine tetraacetate (EDTA) were compared as draw solutes in concentration of shale gas wastewater (SGW) using different FO membranes (HTI, CTA and

Table 15.2. Summary of technical assessment of membrane-based technologies

	Membrane-based technologies			
Criteria	***NF***	***RO***	***MD***	***FO***
TDS of feed water	500–25,000 mg/L	1000–35000 mg/L	40,000–100,000 mg/L	> 35,000 mg/L
Product water quality	High rejection (> 99%) of larger divalent ions and metals with moderate rejection (< 90%) of monovalent is expected.	94% rejection of TDS	Equal to distilled water	The product of FO is a diluted draw solution. To obtain pure water from the process a secondary system is required to exact pure water from the draw solution, and to re-concentrate the draw solution.
Recovery	75–90%	> 90%	60–90%	> 96%
Energy use	Approximately 2 kWh/kgal of energy is required to power the system's high-pressure pumps	No data is currently available	Require some energy input	Just need power to circulate solution across FO membrane.

Chemical use	Use of NaOH, Na_4EDTA, HCl, $Na_2S_2O_4$, or H_2O_2 for cleaning.	Use of NaOH, Na_4EDTA, HCl, $Na_2S_2O_4$, or H_3PO_4 for cleaning.	Use of NaOH, Na_4EDTA or HCl for cleaning.	Less cleaning frequency, possible to use the same chemicals as RO for the cleaning purpose
Pretreatment of feed water	Require pretreatment to mitigate harmful water quality constituents.	Require pretreatment to mitigate harmful water quality constituents.	Removal of constituents that may wet the hydrophobic, microporous pores of the MD membranes is required	Prefiltration is required to remove large debris; antiscalant may be required for high recovery operation.
Post-treatment of product water	Product water may require remineralization	Product water may require pH stabilization or remineralization	Product water may require remineralization and pH stabilization	Diluted draw solution requires further separation to produces pure water and reconcentrate the draw solution for reuse.
Capital and O&M costs	Capital cost: \$0.8 to \$4/gpd; O&M cost: \$0.7/kgal	Capital cost: \$4/gpd; O&M cost: \$0.7/kgal	\$3.34/gpd with operating costs \$1.4/kgal for a 1 MGD DCMD plant	unknown

Table 15.3. Summary of cellulose triacetate and polyamide TFC FO membrane mass transfer for the membrane used in treatment of wastewater from shale gas exploitation

	HTI	*CTA*[a]	*TFC*[a]	*TFC*[b]
Water Permeability A (L/m^2-hr-bar)	0.79 ± 0.07	0.97 ± 0.1	3.5	N.A.
Rejection (%)	89.06 ± 0.03	0.91	95	N.A.
B value (10^{-7} m/s)	2.69	3.25	3.16	1.76 ± 0.22
S value (10^{-4} m)	4.12	8.3	7.96	2.66 ± 0.46
Jw ($L/m^2 \cdot h$)	8.7	6.7	12	N.A.
Js/Jw (g/L)	1.17	1.04	0.3	N.A.

[a]Cellulose triacetate and thin-film composite fabricated by He's lab (He et al. 2012).
[b]Product developed by Oasys (McGinnis et al. 2013).

TFC) (He et al. 2012). CTA and TFC FO membranes were tailor-made (Li et al. 2013). In AL-DS and AL-FS test mode (Figure 15.3), TFC FO membrane showed better performance in the process of SGW concentration than that of HTI and CTA FO membranes using sodium chloride draw solution (3.26 mol/L).

However, when $MgCl_2$ draw solution (1.74 mol/L) was used, the FO fluxes of three different FO membranes differed quite much in the order of $J_{w\text{-}TFC} > J_{w\text{-}HTI} > J_{w\text{-}CTA}$ in the AL-DS mode and $J_{w\text{-}HTI} > J_{w\text{-}CTA} > J_{w\text{-}TFC}$ in AL-FS mode (Figure 15.4). In the AL-DS mode, using EDTA (1 mol/L) as draw solution, the flux of TFC FO membrane was similar to that of HTI, which was obviously higher than that of CTA FO membrane. In contrast, the TFC FO membrane showed higher flux than that of HTI and CTA FO membrane in AL-DS mode (Figure 15.5).

Difference in the FO flux between the AL-DS and AL-FS mode is mainly caused by the internal concentration polarization (ICP). In the case of AL-FS mode, the draw solute diffusivity mainly determines the extent of ICP. Overall, in terms of economy and efficiency, NaCl still appears to be the best choice as the draw solute. Moreover, the recycling of NaCl in an energy efficient manner remained a great challenge. Recently, temperature and pH sensitive draw solutes have received much attention. Temperature sensitive hydrogels have been published (Li et al. 2011), but their application in shale gas produced water treatment still needs further confirmation.

15.3.4 Operating Conditions

15.3.4.1 Feed and Draw Solution Temperatures

Several studies have investigated the effect of temperature on water flux and salts permeation (Zhao and Zou 2011; Phuntsho et al. 2012; You et al. 2012) in the FO

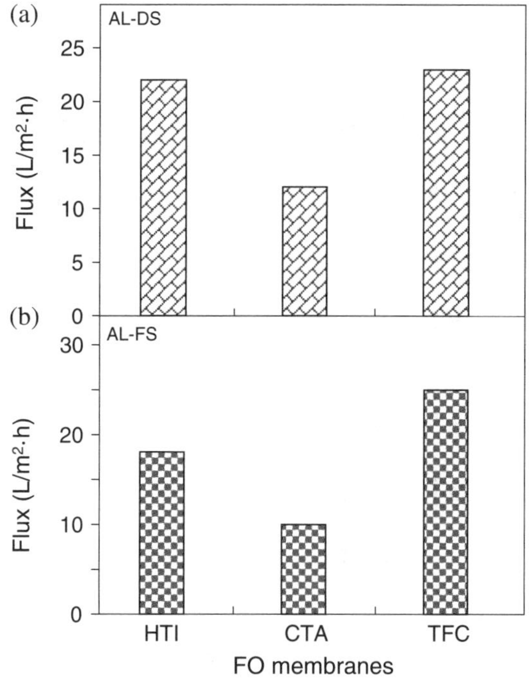

Figure 15.3. Water fluxes of three different FO membranes used in SG wastewater treatment by FO process. (a) AL-DS mode; (b) AL-FS mode. 3.26 M NaCl as the draw solution and SG waste water as the feed

process. Generally, it was concluded that water flux and salt permeation increased with increasing temperature in the FO process (McCutcheon and Elimelech 2006; Nayak and Rastogi 2010; Zhao and Zou 2011; Phuntsho et al. 2012). Recent studies have also focused on the impact of the temperature difference between the feed and draw solutions on water and draw solute permeation across FO membranes. Phuntsho et al. (2012) examined the water flux change with feed and draw solutions of different temperatures and found that water flux increased significantly by increasing draw solution temperature. You et al. (2012) proposed that the heat flux generated by the temperature difference between the feed and draw solutions could enhance the water flux due to the decrease in feed solution viscosity and the increase in water diffusivity.

Nevertheless, so far, no attention is paid to the effect of temperature on the water flux in concentration of waste stream from shale gas exploitation, which is a significant aspect to the application of the FO process in high saline wastewater reclamation. Recently, in the authors' laboratory, it was demonstrated that concentration flux of shale gas wastewater tested in AL-FS mode (Figure 15.6) increased dramatically as the system temperature increased from 15 to 45°C. In AL-FS mode, the exponential increase in the diffusivity against the temperature may effectively decrease in the ICP, resulting in high FO flux. This result indicates

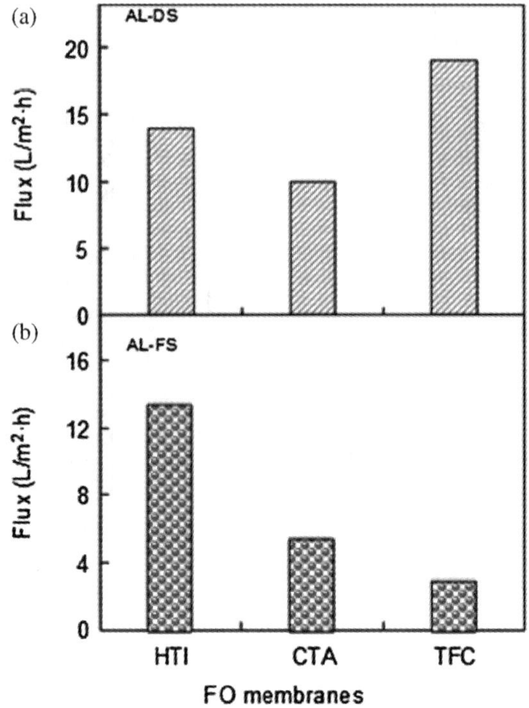

Figure 15.4. The water fluxes of three different membranes used in SG wastewater treatment by FO process. (1.74M $MgCl_2$ as the draw solution and SG waste water as the feed, (a) Water flux under AL-DS mode; (b) Water flux under AL-FS mode)

that concentration water flux of high saline wastewaters in AL-FS model can be improved by increasing system temperature of the FO process.

15.3.4.2 Flow Velocity

The flow velocity affects the hydraulic status of the FO system, which is very important for the determination of the scale of the FO plant as well as the cost. Recently, He (2013) reported the effect of the flow rate on the FO flux in concentrating shale gas wastewater in the AL-FS mode (Figure 15.7). It was found that the FO flux did not change with the flow velocity, but in AL-DS mode, an obvious increase of FO flux was observed. This phenomenon is most probably ascribed to the internal concentration polarization in the AL-FS mode where the porous support layer is facing the DS. Moreover, the flux in AL-FS mode was higher than that in AL-DS mode. Literature has shown that the AL-FS mode has shown much steady performance than in the AL-DS mode for feed solutions of complicated compositions, especially those tend to foul the membranes. Thus, for shale gas produced water treatment, the AL-FS mode should be the primary choice.

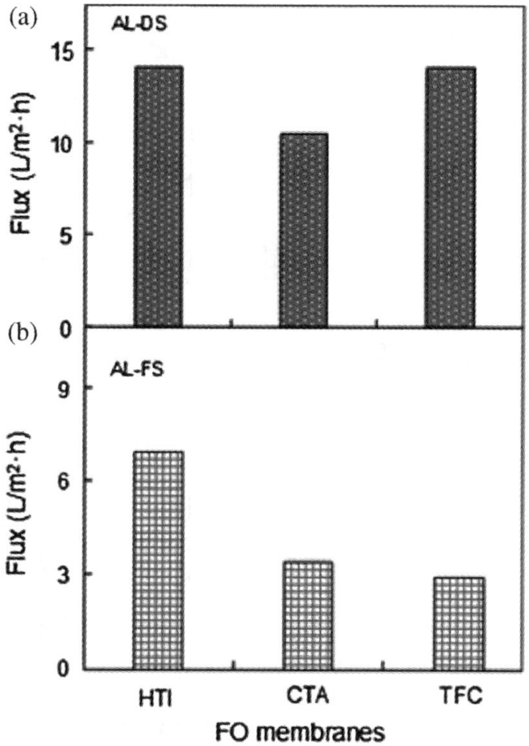

Figure 15.5. The water fluxes of three different membranes used in SG wastewater treatment by FO process. (1M EDTA as the draw solution and SG waste water as the feed, (a) Water flux under AL-DS mode; (b) Water flux under AL-FS mode)

15.3.5 Membrane Fouling

The shale gas flowback water may contain various contaminants which is particularly detrimental to the FO membrane surfaces. Organic compounds, such as oil, surfactants and other particle based colloids are the main cause for the membrane fouling. Two main factors may influence the fouling in FO are described as follows:

1. **Surface roughness may affect the fouling behavior.** A rough top surface tends to trap micro or submicrosize pollutants, especially the particulate aggregates or foulants of potential to form particulates, but not a smooth surface (Hashina et al. 2011). Moreover, once trapped, they are difficult to be washed away by the tangential flow. For a smooth surface, the possibility for the trapping of the particles is lower than the rough surface, thus less aggregation of the foulant to the membrane surface. It is thus probable that the HTI CTA membranes of a smooth surface is less accessible to the

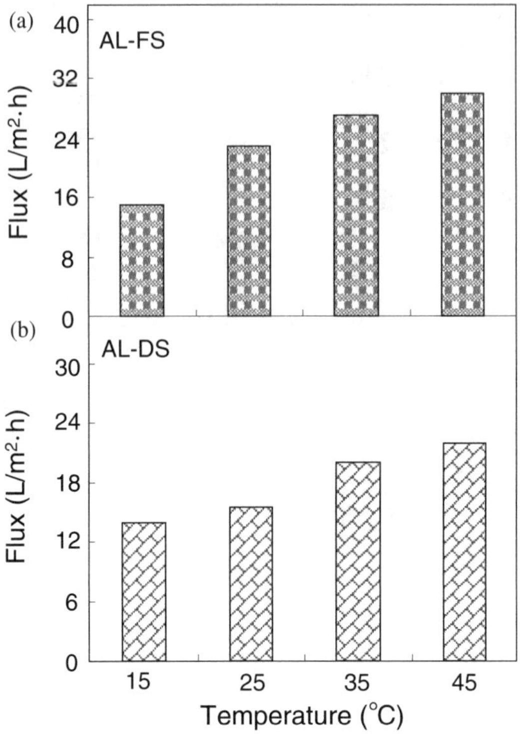

Figure 15.6. Effect of temperature on concentration of shale gas wastewater using the HTI CTA membrane. Draw solution: 20wt% KCl; Feed solution: shale gas wastewater; (a) AL-FS, (b) AL-DS

foulants in the SGW; However, for a rough membrane surface as TFC membranes, it is most likely to form a fouling layer during the FO process for the concentration of the SGW.

2. **The concentration of the particles in the feed solution.** A large amount of particles and the aggregation of the particles can result in the instantaneous decline in the flux, for feed solutions with low concentration of particulates, there may be a certain time duration before significant reduction in flux is visualized.

Figure 15.8 shows the schematic of the fouling mechanism of fouling in an FO process for smooth and rough surface membranes. The particles are illustrated as the foulants and are negatively charged. The negatively charged membrane surface may repulse the foulants away from approaching the surface, denoted as F_R. Other forces that may prevent the fouling include the tangential sweeping force given by the flow rate across the membrane surface, denoted as F_T. These two forces are the main factors to prevent the adherence of the foulants to the membrane surface. However, due to water diffusion in the FO process from the bulk to the membrane

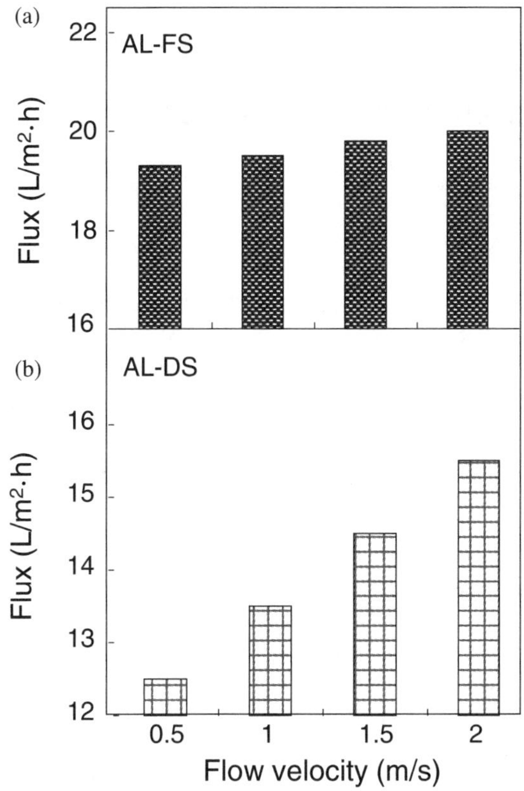

Figure 15.7. Relationship between the draw solution concentration and the FO flux with shale gas wastewater as the feed using HTI membranes at different flow velocity. Note: DS, 20wt% KCl; (a) AL-FS, (b) AL-DS

surface, there is a force which pushes the foulants towards the membrane surface, denoted as F_{FO}. This is the only force which may lead to the formation of a fouling layer during the FO process as shown in Figure 15.8(A). For a smooth membrane surface as HTI CTA membranes, both F_R and F_T act as the factors to reduce the fouling formation, and the adherence of the foulants to the membrane surface driving by F_{FO} may not be strong enough, thus, swept away by the surface flow.

In the case of a rough membrane surface, as seen in Figure 15.8(B), when the foulants follows the water diffusion direction approaching the FO membrane surface, they tend to be trapped by the valley-like surface structures. Beneath the layer adjacent to the membrane surface, the F_T is counter-balanced by the blocking microstructures in the membrane surfaces. Thus, foulants preferentially aggregate to the membrane surface, resulting in a cake layer. Once the membrane surface morphology is filled up with the foulants, the membrane may behave as a smooth surface again and no significant fouling is possible anymore. It is therefore

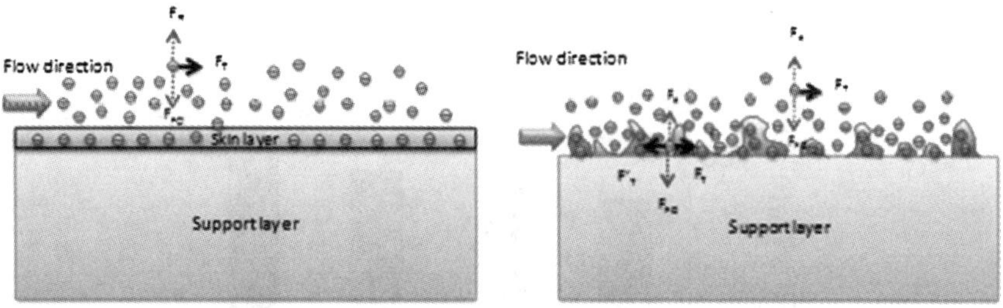

(a) (b)

Figure 15.8. Mechanisms of the deposition of the foulants on to the membrane surface; (A) smooth surface, representing an HTI membrane; (B) rough surface, representing a thin-film composite membrane. Note that the surface charges were omitted due to the high saline concentration in the feed solution. The black arrow points to the direction of the main flow, and the red dashed arrow points to the direction of water flowing to the membrane surface due to osmosis

theoretically preferred to have a membrane of smooth surface instead of a rough membrane surface in order to decrease the fouling tendency during the FO process for SGW treatment.

Similar to other FO processes, the FO performance in shale gas produced water treatment especially water flux is highly dependent on membrane orientation, operation temperature and fluid velocity. Although the FO process has the capability to treat highly concentrated waste streams such as oil and gas wastewater, membrane fouling has been observed. Some foulants absorbed on FO membrane surface can be washed off through increasing the flow rate. In most cases, the cleaning is performed without the use of chemicals. Several studies demonstrated successful restoration of water flux through FO membranes by increasing the flow rate of membrane surface (Holloway et al. 2007; Cath et al. 2009; Sagiv et al. 2010).

15.4 APPLICATION CASES

15.4.1 Osmotic Dilution

Recently, Bear Creek Services and HTI (Albany, OR) developed a new water reclamation system, Green Machine, for the deposal of oil-field related wastewaters (HTI 2010). The Green machine is a portable, modular and scalable system. It was claimed that the technology could dramatically reduce the environmentally damage and cost for tankering. The basis of this machine is osmotic dilution. In an osmotic dilution process, the draw solution is diluted by the water from the waste streams and then reused as the process liquid. It was reported that the Green machine had been operated in the Haynesville Shale exploitation play in North Louisiana and East Texas. Over 20% of the water required for hydraulic fracturing of new wells is supplied by the wastewater utilizing the osmotic dilution process (Hickenbottom et al. 2013). This operation reduces significantly the fresh water demand in the hydraulic fracturing site, and it recycles the shale gas wastewater as well.

15.4.2 FO-Distillation Processes

Although osmotic dilution can reduce the water demand in the hydraulic fracturing process, it does not generate fresh water that meets the discharge standards. Moreover, because of the operation principle, solely an individual FO membrane process can produce water that meets discharge standards (Cath et al. 2006). To purify the flowback wastewater from shale gas wastewater, a hybrid FO-MD process was investigated by McGinnis et al. (2013). They demonstrated a pilot-scale integral FO-MD concentrator (Figure 15.9) to desalinate fracturing flowback and produced waters from natural gas extraction operations in the Marcellus shale region. Compared to initial concentration of wastewater with TDS of 73,000 ± 4200 mg/L, the salt concentration of the product was 300 ± 115 mg/L TDS that meets surface water discharge quality criteria for the State of Pennsylvania.

The conventional evaporation process and the FO process combined with membrane brine concentrator are shown in Figure 15.10 for energy consumption comparison. Compared with evaporative desalination methods, producing 1 kg of

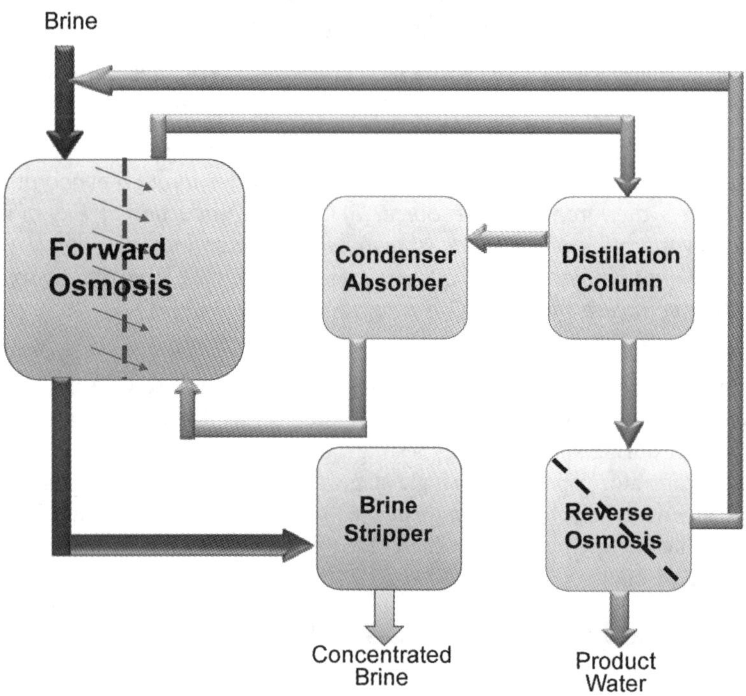

Figure 15.9. Block flow diagram of the FO membrane brine concentrator setup
SOURCE: Adapted from McGinnis et al. 2013

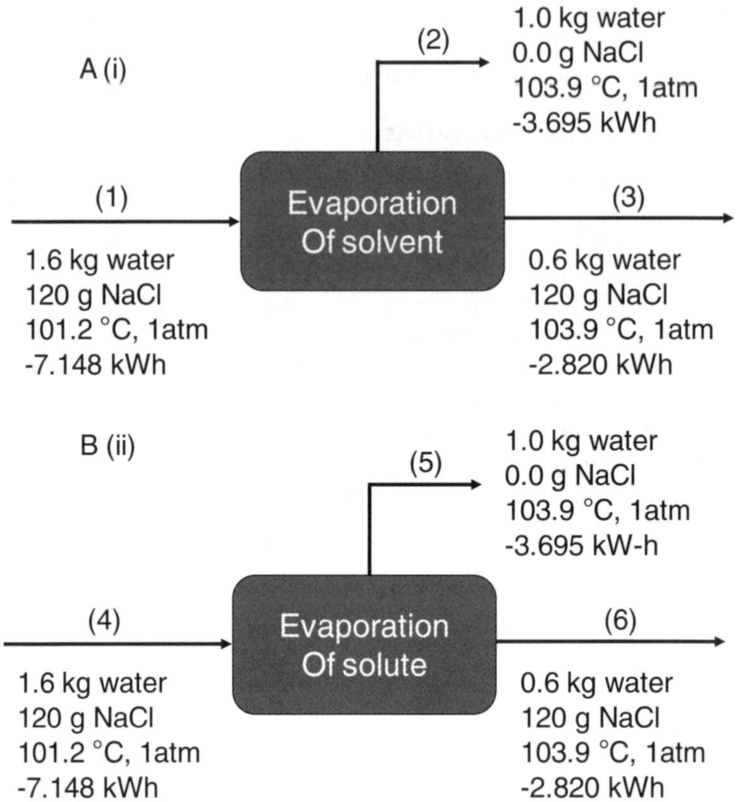

Figure 15.10. Mass and enthalpy balance diagrams based on thermophysical modeling. (A) Case (i), production of 1 kg of pure water through evaporation of the solvent (water) from a saline source. (B) Case (ii), production of 1 kg of water through evaporation of the solute (shown here as molar flows of C and N, which are equivalent to molar flows of CO_2 *and* NH_3 *in stream (5)) from a draw solution diluted by permeate from an FO membrane process*

water product by evaporating water from a 73,000 mg/L NaCl solution to a recovery of 50% (identical to the recovery of the FO membrane brine concentrator (MBC) pilot during specific energy testing), in a similarly configured evaporative brine concentrator (open cycle, single stage, no energy recovery) is estimated to require an energy input of approximately 633 kWh/m^3 of thermal energy. This is 2.3 times the energy measured in the FO MBC pilot (275 kWh/m^3), measured in the FO MBC pilot.

15.4.3 FO-vacuum Membrane Distillation

Application of FO-vacuum membrane distillation (FO-VMD) for the reuse of the shale gas wastewater has been investigated in the authors' laboratory. A schematic of the hybrid process is shown in Figure 15.11. HTI CTA membrane was used as

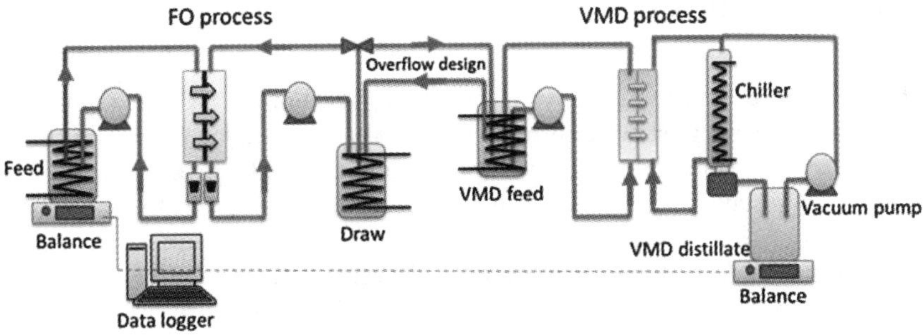

Figure 15.11. Schematic of the hybrid FO-MD hydride setup

the FO membrane, and a KCl 20 wt% solution was used as the draw solution. Before test, the feed water was pre-treated using coagulation and ultrafiltration. The quality of the permeate water was listed in Table 15.4. In comparison to the local potable water, permeate in the FO-VMD process contains much less ions. The conductivity of MD permeate was 5 μS/cm, which was similar to that of a deionized water.

In this hybrid process, the organic contaminants are removed by FO membranes, and the MD membrane faces only the single salt solution. It is therefore probable that the fouling of the MD is significantly low. The draw solute is inorganic salt, which does not decompose upon heating, and thus cannot go through the MD membrane, which guarantees a high product water quality.

The disadvantage of the hybrid process is the high energy consumption due to the evaporation stage in the MD. This process is most probably suitable for arid regions where water is more precious than energy. For regions where the solar power is abundant, it is possible to reduce the cost to an acceptable level.

Table 15.4. Comparison of water quality between the potable water and the permeation from FO-MD process treating the shale gas flowback water

	Potable water in Shanghai	***FO-MD permeate***
Conductivity (μS/cm)	43.6	5.0
pH	7.88	7.38
COD_{-Cr} (mg/L)	1.1	0.9
Turbidity (NTU)	0.09	0.07
K (mg/L)	3.74	0.48
Ca (mg/L)	0	0
Mg (mg/L)	0	0
Na (mg/L)	0.3	0.12
B (mg/L)	0.02	0
As (mg/L)	0	0
Sr (mg/L)	0	0
Mn	0	0

15.5 CONCLUSIONS

This chapter provides an overview on the application of FO as a potential technology to treat wastewater produced from shale gas exploitation. The choice of draw solution and operating conditions including membrane orientation, flow velocity are discussed. The relationship between the fouling and the membrane surface morphology is hypothetically proposed, i.e., smooth top surface tends to be more preferentially antifouling in SGW application. The advantages and limitations of osmotic dilution, FO-distillation and FO-VMD hybrid processes are analyzed. Osmotic dilution remains as the least energy intensive process for the treatment of the shale gas wastewater. Hybrid processes show better permeate water quality at higher energy cost. Overall, FO processes have shown potential in the treatment of waste waters of very complicated compositions, especially the treatment of oil and gas drilling wastewater. Further development of membranes with better performance and the search for a draw solute that can be regenerated at low cost are the future research directions.

15.6 ABBREVIATIONS/NOMENCLATURE

CDI	capacitive deionization
EDTA	disodium ethylene diamine tetraacetate
HTI	Hydration Technologies Inc.
ICP	internal concentration polarization
FO	forward osmosis
MBC	membrane brine concentrator
MD	membrane distillation
NF	nanofiltration
RO	reverse osmosis
SGW	shale gas flow-back water
TDS	total dissolved solids
TFC	thin-film composite
TSS	total suspended solids
VMD	vacuum membrane distillation

References

Boo, C., Elimelech, M., and Hong, S. (2013). "Fouling control in a forward osmosis process integrating seawater desalination and wastewater reclamation." *Journal of Membrane Science*, 444(1), 148–156.

Cath, T. Y., Childress, A. E., and Elimelech, M. (2006). "Forward osmosis: principles, applications, and recent developments." *Journal of Membrane Science*, 281, 70–87.

Colorado School of Mines (2009). *An integrated framework for treatment and management of produced water: Technical assessment of produced water treatment technologies*, First edition, November 2009.

Ghouas, H., Haddou, B., Kameche, M., Derriche, Z., and Gourdon, C. (2012). "Extraction of humic acid by coacervate: Investigation of direct and back processes." *Journal of Hazardous Materials*, 205-206(29), 171–178.

Ground Water Protection Council, ALL Consulting (2009). *Modern Shale Gas Development in the United States: A Primer Report*, April 2009.

He, T., and Li, X. M. (2012). "Forward osmosis process for recovery of highly saline wastewater." in *ADVANCED MEMBRANE TECHNOLOGY V: Membranes for Sustainable Water, Energy and the Environment*, An ECI Conference Series, Singapore, October 14-19, 2012.

He, T. (2013). "Carbon neutral process based on forward osmosis (FO) and membrane distillation (MD) for the reuse of high salinity wastewater (HSW)." *IEEC meeting 2013, Seoul. Session: Fouling Behaviour of Forward Osmosis Membrane: organic, inorganic, biofouling and its modelling.*

HTI (2010). "HBC Systems Launches Revolutionary 'GREEN MACHINE'." http://www.htiwater.com/news/press-room/content/2010/pr-hbc-green-machine-011510.pdf (accessed January 15, 2010).

Hashino, M., Katagiri, T., Kubota, N., Ohmukai, Y., Maruyama, T., and Matsuyama, H. (2011). "Effect of surface roughness of hollow fiber membranes with gear-shaped structure on membrane fouling by sodium alginate." *Journal of Membrane Science*, 366 (1–2), 389–397.

Holloway, R. W., Childress, A. E., Dennett, K. E., and Cath, T. Y. (2007). "Forward osmosis for concentration of anaerobic digester centrate." *Water Research*, 41, 4005–4014.

Hickenbottom, K. L., Hancock, N. T., Hutchings, N. R., Appleton, E. W., Beaudry, E. G., Xu, P., and Cath, T. Y. (2013). "Forward osmosis treatment of drilling mud and fracturing wastewater from oil and gas operations." *Desalination*, 312, 60–66.

Lee, C., Boo, C., Elimelech, M., and Hong, S. (2010). "Comparison of fouling behavior in Forward osmosis (FO) and reverse osmosis (RO)." *Journal of Membrane Science*, 365, 34–39.

Li, D., Zhang, X., Yao, J., Simon, G. P., and Wang, H. (2011). "Stimuli-responsive polymer hydrogels as a new class of draw agent for forward osmosis desalination." *Chemical Communication*, 47(6), 1710–1712.

Li, G., Li, X. M., He, T., Jiang, B., and Gao, C. J. (2013). "Cellulose triacetate forward osmosis membranes: preparation and characterization." *Desalination and Water Treatment*, 51 (13–15), 2656–2665

Li, W., Krantz, W. B., Cornelissen, E. R., Post, J. W., Verliefde, A. R. D., and Tang, C. Y. (2013). "A novel hybrid process of reverse electrodialysis and reverse osmosis for low energy seawater desalination and brine management." *Applied Energy*, 104, 592–602.

Li, Z. Y., Yangali-Quintanilla, V., Valladares-Linares, R., Li, Q., Zhan, T., and Amy, G. (2012). "Flux patterns and membrane fouling propensity during desalination of seawater by forward osmosis." *Water Research*, 46(1), 195–204.

Kargbo, D. M., Wilhelm, R., and Campbell, D. J. (2010). "Natural gas plays in the Marcelus Shale: challenges and potential opportunities." *Environmental Science and Technology*, 44, 5679–5684.

Kim, C., Lee, S., and Hong, S. (2012). "Application of osmotic backwashing in forward osmosis: mechanisms and factors involved." *Desalination and Water Treatment*, 43, 314–322.

Kim, H. C., and Dempsey, B. A. (2013). "Membrane fouling due to alginate, SMP, EfOM, humic acid, and NOM." *Journal of Membrane Science*, 428, 190–197.

McGinnis, R. L., Hancock, N. T., Nowosielski-Slepowron, M. S., and McGurgan, G. D. (2013). "Pilot demonstration of the NH3/CO2 forward osmosis desalination process on high salinity brines." *Desalination*, 312, 67–74.

Martinetti, C. R., Childress, A. E., and Cath, T. Y. (2009). "High recovery of concentrated RO brines using forward osmosis and membrane distillation." *Journal of Membrane Science*, 331(1–2), 31–39.

Mark, C. (2010). "Forward-osmosis technology may revolutionize water reclamation in the oil field." *Well Servicing*, May/June 68–70.

McCutcheon, J. R., and Elimelech, M. (2006). "Influence of concentrative and dilutive internal concentration polarization on flux behavior in forward osmosis." *Journal of Membrane Science*, 284(1–2), 237–247.

Nayak, C. A., and Rastogi, N. K. (2010). "Forward osmosis for the concentration of anthocyanin from GarciniaindicaChoisy." *Separation and Purification Technology*, 71(2), 144–151.

Peter-Varbanets, M., Margot, J., Traber, J., and Pronk, W. (2011). "Mechanisms of membrane fouling during ultra-low pressure ultrafiltration." *Journal of Membrane Science*, 377(1–2), 42–53.

Phuntsho, S., Hong, S., Elimelech, M., and Shon, H. K. (2013). "Forward osmosis desalination of brackish groundwater: Meeting water quality requirements for fertigation by integrating nanofiltration." *Journal of Membrane Science*, 436, 1–15.

Phuntsho, S., Vigneswaran, S., Kandasamy, J., Hong, S., Lee, S., and Shon, H. K. (2012). "Influence of temperature and temperature difference in the performance of forward osmosis desalination process." *Journal of Membrane Science*, 415–416(0), 734–744.

Phuntsho, S., Shon, H. Y., Hong, H. Y., Hong, S., Lee, S., and Vigneswaran, S. (2011). "A novel low energy fertilizer driven forward osmosis desalination for direct fertigation: evaluating the performance of fertilizer draw solutions." *Journal of Membrane Science*, 375(1–2), 172–181.

Resosudarmo, A., Ye, Y., Le-Clech, P., and Chen, V. (2013) "Analysis of UF membrane fouling mechanisms caused by organic interactions in seawater." *Water Research*, 47(2), 911–921.

Shaffer, D. L., Arias, L. H., Moshe Ben-Sasson, C., Romero-Vargas Castrillón, S., Yip, Ng. Y., and Elimelech, M. (2013). "Desalination and reuse of high-salinity shale gas produced water: drivers, technologies, and future directions." *Environ. Sci. Technol.*, DOI: 10.1021/es401966e.

Sagiv, A., Avraham, N., Dosoretz, C. G., and Semiat, R. (2010). "Parameters affecting backwash variables of RO membranes." *Desalination*, 261, 347–353.

Shapiro, S., Dinske, C., and Langenbruch, C. (2010). "Seismogenic index and magnitude probability of earthquakes induced during reservoir fluid stimulation." *Society of Exploration Geophysicists*, 29(3), 304–309.

Sioutopoulos, D. C., Karabelas, A. J., and Yiantsios, S. G. (2010). "Organic fouling of RO membranes: Investigating the correlation of RO and UF fouling resistances for predictive purposes." *Desalination*, 261(3), 272–283.

Tang, W., and Ng, H. Y. (2008). "Concentration of brine by forward osmosis: performance and influence of membrane structure." *Desalination*, 224(1–3), 143–153.

Yangali-Quintanilla, V., Li, Z., Valladares, R., Li, Q., and Amy, G. (2011). "Indirect desalination of Red Sea water with forward osmosis and low pressure reverse osmosis for water reuse." *Desalination*, 280(1–3), 160–166.

You, S.-J., Wang, X.-H., Zhong, M., Zhong, Y.-J., Yu, C., and Ren, N.-Q. (2012). "Temperature as a factor affecting transmembrane water flux in forward osmosis: Steady-state modeling and experimental validation." *Chemical Engineering Journal*, 198–199(0), 52–60.

Zazouli, M. A., Nasseri, S., and Uibricht, M. (2010). "Fouling effects of humic and alginic acids in nanofiltration and influence of solution composition." *Desalination*, 250(2), 688–692.

Zhao, S., and Zou, L. (2011). "Effects of working temperature on separation performance, membrane scaling and cleaning in forward osmosis desalination." *Desalination*, 278(1–3), 157–164.

Zhao, S., Zou, L., and Mulcahy, D. (2012). "Brackish water desalination by a hybrid forward osmosis-nanofiltration system using divalent draw solute." *Desalination*, 284, 175–181.

CHAPTER 16

Removal of Emerging Trace Organic Chemicals by Forward Osmosis

Ming Xie
Long D. Nghiem
Tao He
William E. Price

16.1 INTRODUCTION

Occurrence of emerging trace organic chemicals in sewage, secondary treated effluent and sewage impacted water bodies is a major human and environmental health concern. The full extent of the impact of these emerging trace organic chemicals on human health is still a subject of intense scientific debate. However, some of these chemicals have already been shown to cause serious adverse effects on a range of organisms at environmentally relevant concentrations (Snyder et al. 2003; Carballa et al. 2004; Schwarzenbach et al. 2006; Basile et al. 2011). Forward osmosis (FO) can potentially provide a new perspective to the removal of these emerging trace organic chemicals. In FO, a water-permeable and salt-rejecting membrane is placed between a feed solution and a concentrated draw solution with high osmotic pressure. Extraction of water is driven by the osmotic pressure difference and at the same time, it couples the rejection of the salt and contaminants in the feed solution by the FO membrane. To produce freshwater, FO is usually combined with pressure driven membrane processes, such as NF and RO (Hoover et al. 2011; Yangali-Quintanilla et al. 2011; Shaffer et al. 2012), or thermal processes, such as conventional column distillation (McCutcheon et al. 2005; McGinnis and Elimelech 2007) and membrane distillation (MD) (Cath et al. 2005; Martinetti et al. 2009). As a result, elucidating transport mechanisms of these emerging organic chemicals through an FO membrane is fundamental and critical for the further development and deployment of FO.

In this chapter, we provide a comprehensive review on the current state of knowledge on the rejection of emerging trace organic chemicals by FO processes. The review begins with a brief introduction of the occurrence of emerging trace organic chemicals in municipal wastewater effluent. Three major aspects are outlined to delineate their impacts on the rejections of trace organic chemicals: process parameters including properties of membrane and draw solution and operating conditions; membrane fouling; and FO-based hybrid processes. This review ends with several concluding remarks that will further strengthen the potential of FO in effectively dealing with emerging trace organic chemicals.

16.2 EMERGING TRACE ORGANIC CHEMICALS

16.2.1 Occurrence, Health and Environmental Impacts

The widespread occurrence of emerging trace organic chemicals in the environment has received significant scientific attention in recent years (Schwarzenbach et al. 2006). Emerging trace organic chemicals (Table 16.1) including pharmaceuticals and personal care products (PPCPs) and endocrine disrupting chemicals (EDCs) are biologically active and can threaten the aquatic environment with

Table 16.1. Examples of molecular structures of typical emerging organic chemicals (PPCPs and EDCs)

Endocrine Disrupting Chemicals (EDCs)		
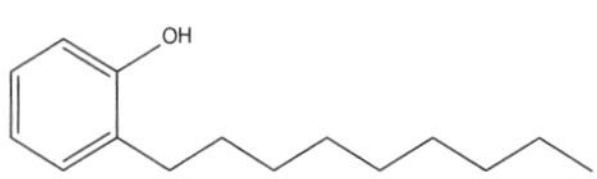 Nonylphenol	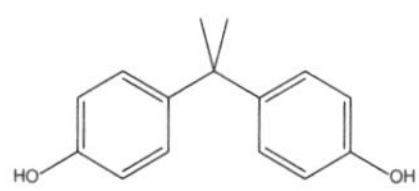 Bisphenol A	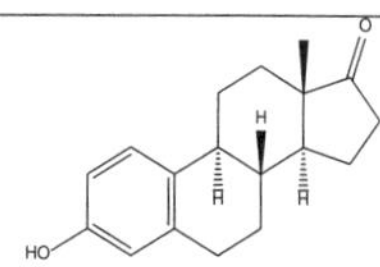 Estrone
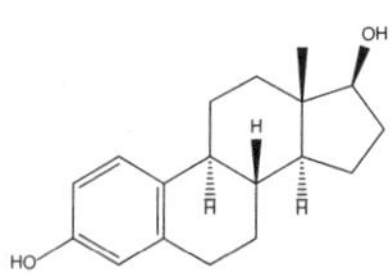 17β-Estradiol	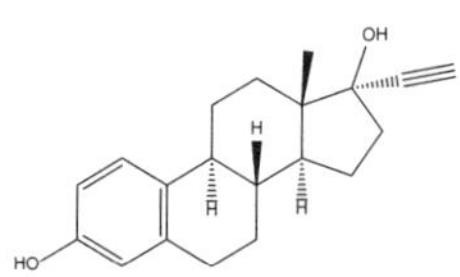 17α-Ethinylestradiol	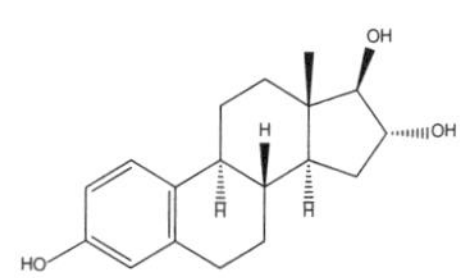 Estriol
Pharmaceuticals and Personal Care Products (PPCPs)		
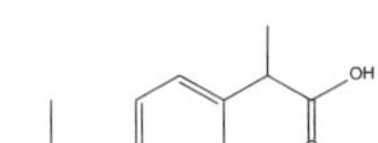 Ibuprofen	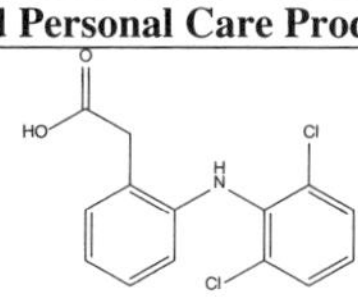 Diclofenac	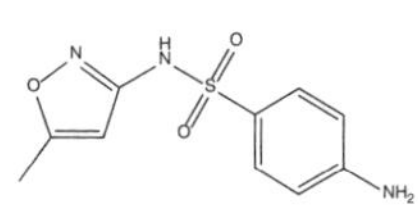 Sulfamethoxazole
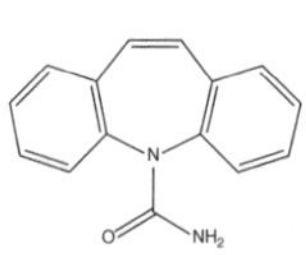 Carbamazepine	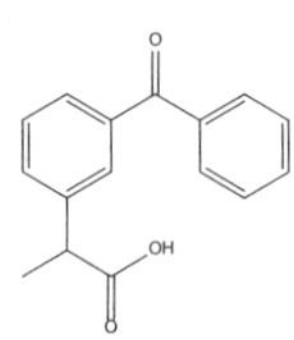 Ketoprofen	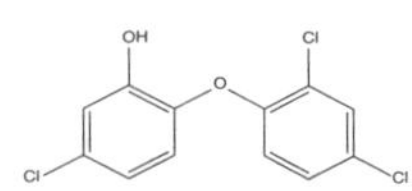 Triclosan

effects such as acute and chronic toxicity to aquatic organisms, accumulation in the ecosystem and loss of habitats and biodiversity, as well as a range of possible adverse effects on human health.

Clinching scientific evidence has revealed the potential risks associated with the presence of emerging trace organic chemicals in the aquatic environment. These contaminants can cause severe consequences to human health (Cunningham et al. 2009), such as abnormal physiological processes and reproductive impairment (Routledge et al. 1998; Rodgers-Gray et al. 2000; Sharpe et al. 2003), increased incidences of cancer (Victorin and Ahlborg 1987; Kannan et al. 2003) and development of antibiotic-resistant bacteria (Rijal et al. 2009). A further concern is the synergistic effect of mixtures of these emerging organic chemicals (Jones et al. 2005). Although effects of individual substances may be insignificant, long-term exposure to a mixture of emerging organic chemicals may have health effects on human and other biota. Furthermore, for many substances, their chronic effects on humans and aquatic ecosystems are still not well understood.

A major pathway for these emerging organic chemicals entering aquatic environment is secondary treated effluent discharged from wastewater treatment plants. This is due to the incomplete elimination of these (often polar) compounds in conventional wastewater (Carballa et al. 2004), drinking water (Huerta-Fontela et al. 2011) and water reclamation (Al-Rifai et al. 2007) treatment plants. These substances in the effluents can deteriorate water quality of receiving water bodies and will challenge water reclamation schemes that are becoming popular around the world seeking to augment the freshwater supply. Investigations carried out in Australia, China, Canada, Germany, Japan, Korea, New Zealand, Spain, Switzerland and the USA reported that these emerging organic chemicals occurred widely in secondary treated effluents at concentration level up to several μg/L (Figure 16.1). As a result, the public and water utilities have significant concerns over the quality and safety of water reclamation. Intensive treatments are required to eliminate the risks to human health and aquatic ecological wellbeing when secondary treated effluent is used as water source in water reclamation.

16.2.2 Challenges for Water Reclamation

Due to the frequent detection of emerging organic chemicals in secondary treated effluent, a multi-barrier concept is widely utilised in the design of wastewater reclamation process to ensure the quality of product water. An array of technologies, such as advanced oxidation, activated carbon adsorption, and membrane filtration, are integrated in wastewater reclamation (Council 2012). Among these, NF and RO membrane filtration is widely recognised as the leading and critical technique for wastewater reuse (American Water Works and James 2011). With well-proven separation mechanisms including size exclusion, charge repulsion and adsorption diffusion, NF and RO membrane filtration provides medium to high rejection of these emerging trace organic chemicals. Notably, in several pilot and full scale facilities, RO membranes could effectively reject up to 36 emerging

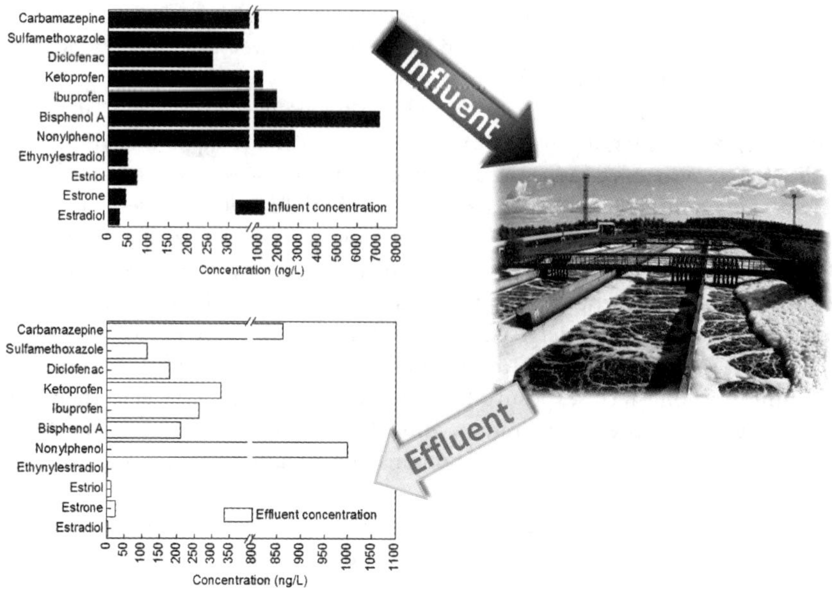

Figure 16.1. Examples of widespread occurrence of emerging organic chemicals in the influent and effluent of a conventional wastewater treatment plant. Bars represent data from the literature

SOURCE: Farré et al. 2002; Kolpin et al. 2002; Onda et al. 2003; Auriol et al. 2006; Watkinson et al. 2007; Kasprzyk-Hordern et al. 2009; Miège et al. 2009; Yu and Chu 2009

organic chemicals with varying physicochemical properties to concentrations below the detection limit (Drewes et al. 2005; Snyder et al. 2007).

Despite the prominence of NF and RO membrane in wastewater reclamation, these technologies are not without limitations. NF and RO membranes are not an absolute physical barrier against the permeation of emerging trace organic chemicals. Previous findings suggested that NF and, to a lesser extent, RO membranes can partially reject emerging trace organic chemicals (Kimura et al. 2004). Nghiem et al. (2004) reported that adsorption and subsequent diffusion of steroid hormones through a membrane polymeric matrix resulted in a much lower rejection than that would be expected based on steric hindrance. Similar observations were also reported by Steinle-Darling et al. (2010) when they examined the rejection of perfluorooctane sulphonamide by NF270 membrane and concluded that hydrophobic compounds resulted in a low rejection due to their adsorption by the membrane skin layer. As a result, the multi-barrier concept is widely accepted in the design of wastewater reclamation processes, and additional purification technologies are recommended to guarantee the wastewater reclamation if the permeation and accumulation of these emerging trace organic chemicals risk the quality of product water. Some advanced oxidation technologies, such as ultra violet (UV) radiation and radical oxidation (e.g., H_2O_2 or ClO_2), are commonly combined with NF and RO membrane filtration to further enhance the

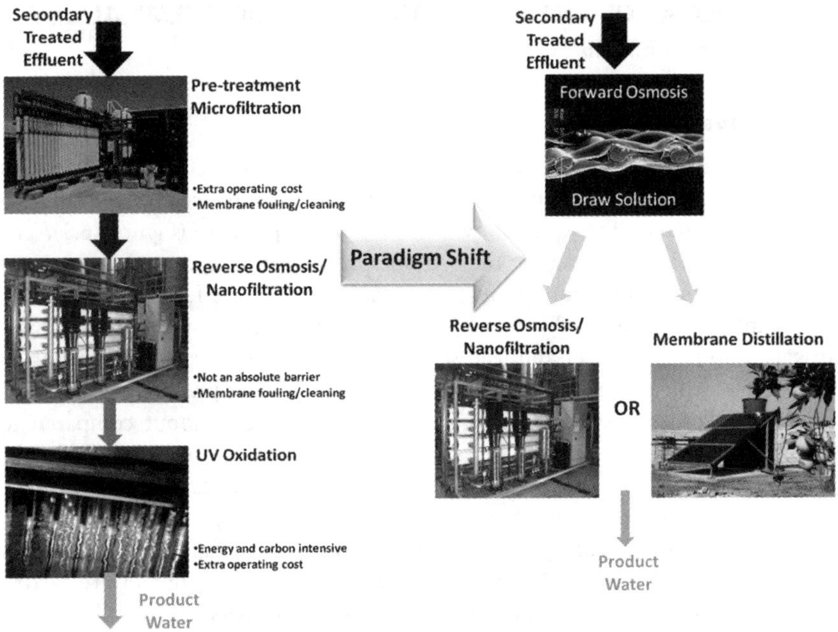

Figure 16.2. A schematic concept of a paradigm shift to FO membrane based processes in wastewater reclamation

removal of certain persistent emerging trace organic chemicals with low molecular weight (Asano et al. 2007). These energy-intensive post treatments substantially increase the energy and carbon footprint of wastewater reclamation (Rothausen and Conway 2011). In addition, membrane fouling by effluent organic matter is also a major hindrance to the NF and RO membrane filtration in wastewater reclamation. Current operational strategies to deal with membrane fouling include pre-treatment of feed water and cleaning of the fouled membrane, both of which would significantly impact the cost and operation of the membrane process. As a result, there is a growing need to design wastewater reclamation systems using alternative and simple processes that provide multiple barriers without substantially increasing the energy and carbon footprint.

As a resurgent membrane process, FO has the potential to be integrated with pressure- or thermal-driven membrane processes to improve or offer a sustainable alternative to wastewater reclamation. Successful development and deployment of such FO membrane based processes would produce a safe and economical paradigm shift in wastewater reclamation (Figure 16.2). In the following sections, we critically evaluate the technical feasibility and performance efficiency of FO for the rejection of emerging trace organic chemicals. Key aspects impacting the rejection are highlighted and elucidated: process parameters including the properties of membrane and draw solution and operating conditions; membrane fouling; and FO-based hybrid processes.

16.3 REMOVAL OF EMERGING TRACE ORGANIC CHEMICALS BY FO PROCESSES

16.3.1 Overview

For a stand-alone FO process, a water-permeable and salt-rejecting non-ideal FO membrane (i.e., less than 100% solute rejection) is placed between a feed solution and a concentrated draw solution with high osmotic pressure (Figure 16.3). The extraction of water is driven by the osmotic pressure difference and at the same time, it couples the rejection of salt and contaminants in the feed solution by the FO membrane. As a result, mass transfer in an FO process is bi-directional: forward diffusion of water and feed solute is combined with reverse diffusion of draw solute, which is unique to FO. FO utilising an osmotic pressure difference as the driving force substantially reduces the external energy input compared to pressure driven NF and RO processes; in addition, there is no applied hydraulic pressure and thus no cake layer compaction in FO, which yields greater membrane fouling resistance and higher membrane cleaning efficiency.

These inherent advantages make FO a promising candidate for advancement of wastewater treatment. Consequently, several early investigations were carried out to examine the effectiveness of FO in the treatments of landfill leachate (Herron 1997), anaerobic digester concentrate (Holloway et al. 2007), activated sludge solution (Cornelissen et al. 2008; Achilli et al. 2009), and domestic wastewater (Cath et al. 2005). Specifically, good removal efficiency of emerging trace organic chemicals by FO processes has been demonstrated. Cartinella et al. (2006) found a near complete rejection of three hormones by an FO process. Cath et al. (2010) reported the rejection of six pharmaceutical compounds ranging from 72% (salicylic acid) to more than 99% (diclofenac). More recently, a number of studies have delineated the rejection performance of emerging trace organic chemicals by an FO process. A comprehensive study on the removal of 23 emerging trace organic chemicals revealed that the rejection of charged trace organic chemicals was consistently above 80% whereas that of neutral emerging organic chemicals varied from 40 to 90% (Hancock et al. 2011). A similar observation was also reported by Valladares Linares et al. (2011) when they

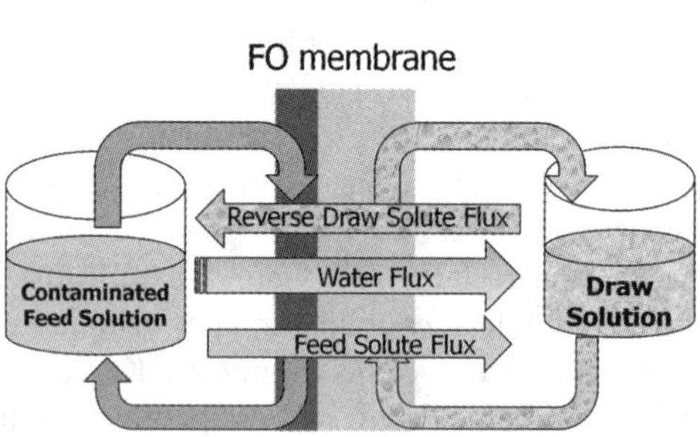

Figure 16.3. A schematic of FO and typical mass transfer phenomena in an FO process

examined the removal of 13 emerging trace organic chemicals. Alturki et al. (2013) proposed and elucidated rejection mechanisms for emerging trace organic chemicals by using a total of 40 emerging trace organic chemicals. They found that the rejection of charged trace organic chemical was governed by both electrostatic interaction and size exclusion, while that of neutral trace organic chemical was dominated by size exclusion.

Significant research and development of FO processes in wastewater treatment over more than a decade has substantially proven its technical feasibility and excellent rejection of emerging trace organic chemicals (Table 16.2). As a promising technology, FO diversifies and enhances the current portfolio of technology solutions to upgrade or reengineer existing wastewater infrastructure to a sustainable manner whilst also reducing its environmental footprint. However, the current understanding of rejection mechanisms of emerging trace organic chemicals in an FO process is still limited.

16.3.2 Impact of Process Parameters

16.3.2.1 Membrane Properties

Membrane properties significantly impact membrane transport behaviour, especially the water permeability and feed solute passage. An ideal membrane specifically tailored for FO operation should reject dissolved solutes, produce high permeate water flux, be compatible with the selected draw solution, and withstand the mechanical stresses generated by the operating conditions (Yip et al. 2010). However, commercially available membranes lack one or more of the above-mentioned characteristics, thereby inhibiting their application to an FO process. For example, a commercial asymmetric CTA membrane specifically designed for FO process is chemically stable only within a small pH range from 4 to 7 such that the membrane will hydrolyse when it is exposed to an ammonium bicarbonate draw solution. In addition, the CTA membrane has a relatively low water flux and salt rejection. Alternatively, a TFC polyamide RO membrane, as the golden standard for seawater desalination, generates extremely low water flux when used in an FO mode, resulting from severe internal concentration polarization (ICP). As a result, a flurry of membrane development work has been carried out to improve FO membrane properties focusing on minimizing ICP and achieving higher water permeate flux and solute rejection.

McCutcheon and Elimelech (2008) correlated water flux to the hydrophilicities of various support layer polymers and concluded that support layer hydrophilicity played a crucial role in water diffusion in an FO process. Sufficient wetting of the support layer improves the available pathways and sites for water transport, thereby enhancing water transport. This hypothesis has been carefully examined by fabricating the membrane support layer with a series of different hydrophilic polymers. Widjojo et al. (2011) fabricated a TFC membrane comprising a hydrophilic sulphonated polysulfone as its support layer and obtained a water flux of 21 $L/m^2 h$ for 2 M NaCl draw solution and deionised water feed.

Table 16.2. Summary of FO studies for wastewater treatment

		Operating conditions			*Performance*				
Year	*Membrane*	*Feed*	*Draw*	*Mode*	*Water flux* (L/m^2h)	*Reverse salt flux*	*Water recovery (%)*	*Rejection*	*Ref*
1997	CTA[a]	Landfill leachate (Benton County, Oregon)	Saturated NaCl solution (~6.2 M NaCl)	FO[b]	7.4	N.A.[c]	95	Final discharge in compliance with the *US National Pollutant Discharge Elimination System*	(Herron 1997)
2005	CTA	Synthetic wastewater	1.7 M NaCl	FO	6.5 (initial) to 6 (end)	0.3 g/L wastewater recovered	95	Urea rejection of 51.3%	(Cath et al. 2005)
2005	CTA	Synthetic wastewater	1–1.7 M NaCl	FO	0.8	N.A.	N.A.	Near complete rejection of urea	(Cath et al. 2005)
2006	CTA	Mixture of hygiene wastewater, humidity condensate and urine	Initial of 1.2 M NaCl to 0.34 M NaCl	FO	12.5 (initial) to 5.5 (end)	N.A.	70	Rejection of estrone and estradiol up to 95%	(Cartinella et al. 2006)
2007	CTA	Digested sludge	Initial of 1.2 M NaCl	FO	11(initial) to 5 (end)	N.A.	72	Rejections of ammonia, TKN[e] and orthophosphate were 88.0, 89.1 and 99.8%, respectively	(Holloway et al. 2007)

2009	CTA	RO concentrate with TDS[d] of 7.5 g/L	0.9 M NaCl	FO	12 (initial) to 4 (end)	N.A.	80	N.A.	(Martinetti et al. 2009)
2010	CTA	Secondary effluent feed	35 g/L synthetic seawater	FO	5.5 (initial) to 3.5 (end)	500 mg/L as TDS	63	Rejection: diclofenac >99%, gemfibrozil >80%, naproxen >90%, and salicylic acid >72%	(Cath et al. 2010)
2011	CTA	Synthetic wastewater	Red Sea seawater (~0.5 M NaCl)	FO	5.5 (initial) to 2 (end)	N.A.	N.A.	Moderate rejections for hydrophilic neutral compounds (44–95%), and hydrophobic neutral contaminants (48–92%); and high rejection for the hydrophilic ionic contaminants (96–99%)	(Valladares Linares et al. 2011)
2011	CTA	MBR permeate	0.5 M NaCl	FO	7.5 (initial) to 4 (end)	N.A.	N.A.	Almost all contaminants werehighly rejected	(Hancock et al. 2011)
2011	CTA	Synthetic wastewater	Varying concentration of NaCl	FO	13.8	N.A.	N.A.	Boron rejection of 40%	(Jin et al. 2011)

(Continued)

Table 16-2. Summary of FO studies for wastewater treatment *(Continued)*

		Operating conditions			*Performance*				
Year	*Membrane*	*Feed*	*Draw*	*Mode*	*Water flux* (L/m^{2}h)	*Reverse salt flux*	*Water recovery (%)*	*Rejection*	*Ref*
2012	CTA and TFC[f]	Synthetic wastewater	2 M NaCl	FO	11.8 and 29.3, respectively	N.A.	N.A.	Rejection: carbamazepine 95–96%, diclofenac 92–95%, ibuprofen 82–83%, naproxen 64–73%	(Jin et al. 2012)
2012	CTA	Synthetic wastewater	0.5 M NaCl	FO	5.4	4.28 g/m^{2}h	25	Rejection: bisphenol A 86%; diclofenac and triclosan 100%	(Xie et al. 2012a)
2012	CTA	Synthetic wastewater	0.5 M NaCl	FO	5.4	4.28 g/m^{2}h	25	Rejection: carbamazepine 90%; sulfamethoxazole 96%	(Xie et al. 2012b)
2012	CTA	Synthetic wastewater	2 M NaCl	FO	14.4	14.5 mg/m^{2}h	N.A.	Boron rejection of 50%	(Kim et al. 2012)

2013	CTA	Synthetic wastewater	0.5 and 2 M NaCl	FO	6 and 10, respectively	4 and 8.5 g/m^2h, respectively	50	Rejection of 13 charged compounds from 80 to 100%; rejection of 27 neutral compounds varied from 20 to 100%	(Alturki et al. 2013)
2013	CTA and TFC	Synthetic wastewater	0.5 M NaCl	FO	5.4 and 14.5, respectively	4.28 and 0.32 g/m^2h, respectively	25	Rejection of charged contaminants above 90%, rejection of neutral contaminants varied from 40 to 98%	(Xie et al. 2013)
2013	CTA	Drilling wastewater	1.3 M NaCl	FO	15 (initial) to 2.4 (end)	N.A.	80	All *USEPA primary drinking water regulated* constituents were below the maximum contamination level	(Hickenbottom et al. 2013)

(Continued)

Table 16-2. Summary of FO studies for wastewater treatment *(Continued)*

		Operating conditions			*Performance*				
Year	*Membrane*	*Feed*	*Draw*	*Mode*	*Water flux* (L/m^2h)	*Reverse salt flux*	*Water recovery (%)*	*Rejection*	*Ref*
2013	TFC	Frac flowback and produced waters from natural gas extraction operations	5.5–6.0 M NH_3/CO_2	FO	2.6	N.A.	64	Product water met the *water quality standards for surface discharge in the state of Pennsylvania*	(McGinnis et al. 2013)

[a]Cellulose triacetate
[b]FO mode where the active layer of the membrane faces feed solution
[c]Not available in the literature
[d]Total dissolved solid
[e]Total Kjehldahl nitrogen
[f]Thin-film composite

Han et al. (2012) selected the super hydrophilic sulphonated poly(ether ketone) polymer to fabricate membrane support layer, and yielded a low membrane structure parameter (101 μm), which is an indicator of potential for ICP; and water flux of 35 L/m^2 h for 2 M NaCl draw solution and deionised water feed. Huang et al. (2013) employed a intrinsically hydrophilic nylon 6,6 microfiltration membrane as support layer and obtained a similar water flux as standard commercial CTA membrane albeit that its membrane structure parameter was two-fold higher than the standard commercial CTA membrane.

It is noteworthy that because the support layer acts as a stagnant diffusive boundary layer (Phillip et al. 2010), ICP significantly reduces the effective osmotic pressure difference across the active layer and thus the water flux (McCutcheon and Elimelech 2006). As a result the optimal structure for the support layer of FO membrane will be thin, highly porous and provide a direct path from the draw solution to the active layer of the membrane. Typically, the support layer possesses a mixed structure where a thin sponge-like layer sits on top of highly porous macrovoids extending to the bottom surface (Tiraferri et al. 2011). This structural characteristic substantially reduces the membrane structural parameter and thus alleviates the ICP, thereby increasing the effective osmotic pressure difference across the membrane. This hypothesis was further examined by fabricating a support layer with high porosity and low tortuosity. Song et al. (2011) utilised an electro-spun polyethersulfone nanofiber as the support layer and reported a low membrane structural parameter of 106 μm and a high water flux of 37 L/m^2 h for 0.5 M NaCl draw solution and deionised water feed. Bui et al. (2011) fabricated a nanofiber supported-polyamide composite membrane that exhibited two to five times higher flux than a standard commercial CTA membrane; and attributed this excellent performance to the superior porosity and pore interconnectivity of the electro-spun nanofiber. In fact, the membrane performance will be notably enhanced by a support layer coupling a hydrophilic polymer and superior porosity. Bui and McCutcheon (2012) fabricated an effective support layer composed of a hydrophilic nanofiber by blending polyacrylonitrile and cellulose acetate via electrospinning. The intrinsically hydrophilic open pore structure with superior interconnectivity resulted in a low membrane structural parameter of 109 μm and a substantially higher water flux (2-3 times) than a standard commercial CTA membrane.

Despite some progress in the development of high performance membranes for FO applications, to date most studies investigating the rejection of emerging trace organic chemicals by an FO process have employed a standard asymmetric commercial CTA membrane (Table 1). Jin et al. (2012) compared rejections of four pharmaceuticals–carbamazepine, diclofenac, ibuprofen and naproxen–between a standard commercial CTA membrane and their homemade TFC polyamide membrane. They found that compared to the CTA membrane, the TFC polyamide membrane exhibited excellent overall performance, with high water flux, excellent pH stability and high rejections of all four pharmaceuticals investigated (>94%). In a recent study, Coday et al. (2013) investigated rejections of humic acid and a set of cations (Mg^{2+}, Li^{+} and K^{+}) and anions (SO_4^{2-}, Br^{-} and NO_3^{-})

in an FO process using one commercial CTA membrane and two commercial TFC polyamide membranes. They found that there was no discernible difference between these three membranes in rejection of humic acid and anions, and a better rejection of cations (Li^+ and K^+) by the CTA membrane compared to that of the two TFC membranes. As a result, it is worthwhile to systematically examine rejection performance of asymmetric CTA and polyamide TFC membranes and provide insights in the relationship between membrane properties and rejection of emerging trace organic chemicals in an FO process.

16.3.2.2 Draw Solution Properties

The draw solution, providing the driving force for separation, is a key component for successful FO applications. In general, an appropriate draw solution for FO applications should generate high osmotic pressure, minimize reverse diffusion and be compatible to suitable re-concentration processes (Achilli et al. 2010). The first two properties of this list are of particular interest to the rejection of emerging organic chemicals in an FO process.

Similar to the pressure-dependent water flux behaviour in pressure driven NF and RO membrane processes, an osmotic pressure generated by the draw solution also governs water permeate flux in an FO process. Water flux has been found to increase non-linearly with an increase in draw solution concentration (i.e., osmotic pressure) due to ICP (McCutcheon and Elimelech 2006). This increase in water flux resulted in an increase in feed NaCl rejection (McCutcheon et al. 2006). NaCl rejection increases with increasing water flux due to the “dilution effect”, which is consistent with the “dilution effect” observed in pressure driven NF and RO processes (Nghiem et al. 2004). In fact, Alturki et al. (2013) observed an increase in rejection for most neutral emerging trace organic chemicals when they compared their rejections at two water fluxes using 0.5 and 2 M NaCl draw solutions. This result implies that the well-established diffusion-solution mechanism used in pressure driven RO membranes is also application to the transport of emerging organic chemicals in FO.

Water flux is coupled to a reverse permeation of the draw solute through a non-ideal FO membrane (with less than 100% solute rejection). Recently, several studies have been conducted to understand this mechanism (Hancock and Cath 2009; Xie et al. 2012b) and to quantify this bi-directional mass transfer (Phillip et al. 2010; Hancock et al. 2011; Yong et al. 2012). This reverse transport of draw solute significantly impacts solute – solute interaction in an FO process and thus influences the rejection of feed solute. Xie et al. (2012a) observed a lower adsorption of hydrophobic trace organic chemicals and thus a higher rejection in FO than that in RO. This has been attributed to a phenomenon termed “retarded forward diffusion”. This occurs when reverse permeation of the draw solute hinders forward diffusion of feed solutes, thereby increasing the rejection of feed solutes (Figure 16.4). A similar “retarded forward diffusion” phenomenon was reported by Kim et al. (2012) when they examined boron rejection in FO using a set of draw solutions with different reverse draw solute fluxes. Alturki et al. (2013) attributed notably higher rejection of neutral emerging trace organic

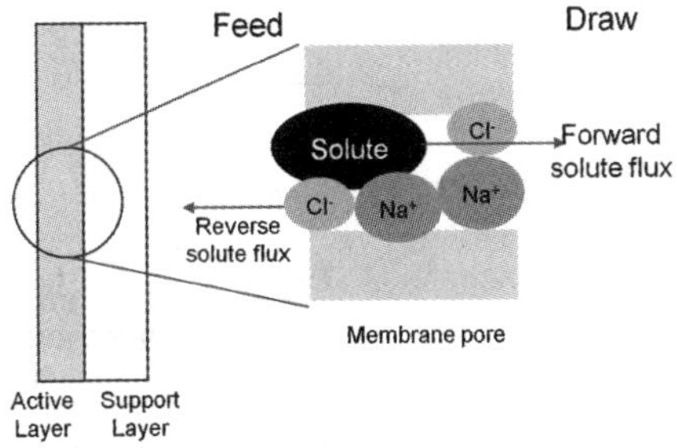

Figure 16.4. A schematic of retarded forward diffusion where the forward diffusion of feed neutral solute is hindered by the reverse transport of draw solutes, thereby leading to a higher rejection
SOURCE: Adapted from Xie et al. 2012a; reproduced with permission from Elsevier

chemicals in FO than that in RO to the "retarded forward diffusion" effect. As a result, retarded forward diffusion could be evoked to assess the rejection of neutral feed solutes by measuring the reverse draw solute flux in an FO process. This effect also has an important implication in the selection of a draw solution as there could be a trade-off between the rejection of feed solute and the loss of draw solute.

16.3.2.3 Operating Conditions

The impact of key operating conditions can be delineated, including feed solution pH, membrane orientation and solution temperature, on rejection of emerging trace organic chemicals in FO. These key operating conditions could either influence mass transfer (e.g., membrane orientation and solution temperature) in an FO process or alter the speciation of some feed emerging trace organic chemicals (e.g., ionic contaminants).

16.3.2.3.1 Feed Solution pH

Both membrane surface charge and speciation of ionic emerging trace organic chemicals vary with the feed solution pH, thereby influencing electrostatic repulsion (Valladares Linares et al. 2011). As a result, the rejection behaviour of ionic emerging trace organic chemicals can be highly pH-dependent. Xie et al. (2012b) observed that rejection of carbamazepine was relatively pH-independent, while that of sulfamethoxazole was strongly influenced by pH. In addition, it was interesting to note that the sigmoidal rejection curve of sulfamethoxazole as a function of feed pH matches the change in its speciation as an anionic or cationic compound (Figure 16.5). These two mirrored curves illustrate that electrostatic repulsion is a significant governing mechanism in the separation of sulfamethoxazole by FO. Jin et al. (2012) also highlighted pH-dependent rejections of naproxen and ibuprofen when they examined rejections of four pharmaceutical compounds by a CTA membrane at different pH values of 3, 6 and 8. However, in this study, they also reported a pH-independent rejection behaviour of all four pharmaceuticals by their homemade TFC membrane over the entire pH range

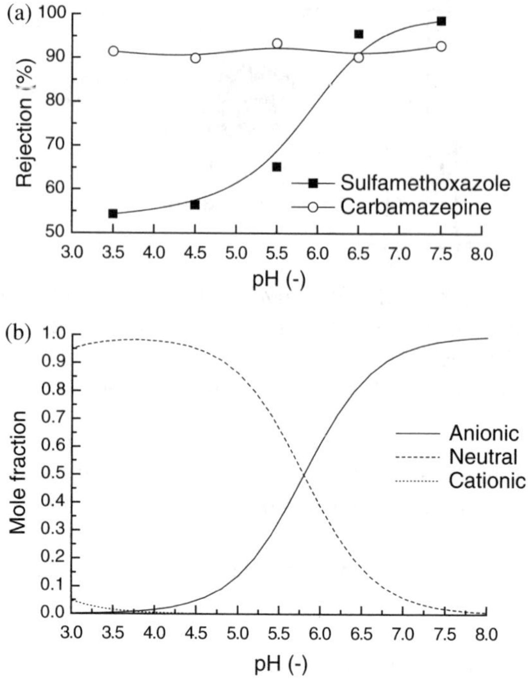

Figure 16.5. (Top) rejection of carbamazepine and sulfamethoxazole and (bottom) speciation of sulfamethoxazole as a function of feed pH
SOURCE: Adapted from Xie et al. 2012b; reproduced with permission from Elsevier

tested. This observation suggested that rejection of ionic emerging trace organic chemicals is also governed by steric hindrance. In a comprehensive investigation conducted by Alturki et al. (2013), they concluded that in addition to electrostatic hindrance, size exclusion can also play a major role in influencing rejection of ionic emerging trace organic chemicals in an FO process, as witnessed by the increased rejection with increasing molecular weight of ionic compound.

16.3.2.3.2 Membrane Orientation

An FO membrane can be operated in two different configurations, namely the normal FO mode and pressure retarded osmosis (PRO) mode. The former refers to a configuration where the active layer of the FO membrane is placed against the feed solution, while the latter refers to a configuration in which the active layer of the FO membrane is placed against the draw solution. Jin et al. (2011) through a modelling study showed that the boron flux in the PRO mode was higher than that in the FO mode. Xie et al. (2012a) reported a 20% lower rejection of carbamazepine in the PRO mode than that in the FO mode. Alturki et al. (2013) found lower rejection of charged and small molecular weight neutral emerging trace organic chemicals in PRO mode in comparison to the FO mode. In addition, rejection of a set of inorganic contaminants, calcium, boron and arsenate, was found to be lower in the PRO mode than that in the FO mode (Jin et al. 2012). The consistent lower rejection in the PRO mode reported in these studies is attributed to the concentrative ICP effect in the PRO mode that results in an elevated solute concentration gradient across the dense active layer of the membrane, thereby leading to a lower rejection. Furthermore, it is noteworthy that FO mode is favourable from a practical perspective, when the FO membrane is used to purify contaminated wastewater because of more severe and irreversible membrane fouling in the PRO mode than that in the FO mode (Mi and Elimelech 2008).

16.3.2.3.3 Feed and Draw Solution Temperatures

Temperature is an important factor governing mass transfer in membrane separation processes, including the FO process. In several practical applications of FO, there can be significant temporal and spatial variation in the temperature of feed solutions, such as secondary treated effluent or seawater. Similarly, draw solutions can be at higher temperatures than the feed solution as a result of thermal separation and recycling of the draw solution or using higher temperatures to increase the solubility of the draw solute. Such temperature variations could substantially impact the rejection of emerging trace organic chemicals by the FO process, as also observed in the NF and RO processes (Sharma and Chellam 2006; Arsuaga et al. 2008).

Several studies have examined the effect of temperature on the permeation of water (Zhao and Zou 2011; Phuntsho et al. 2012) and inorganic salts (Phuntsho et al. 2012) in the FO process. Generally, it was observed that water and salt permeabilities increased with increasing temperature in the FO process (McCutcheon and Elimelech 2006; Nayak and Rastogi 2010; Zhao and Zou 2011; Phuntsho et al. 2012). Recent studies have also focused on the impact of the temperature difference between the feed and draw solutions on water and draw solute permeation across FO membranes. Phuntsho et al. (2012) examined the water flux behaviour with feed and draw solutions of different temperature and found that water flux increased significantly by increasing draw solution temperature. You et al. (2012) proposed that the heat flux generated by the temperature difference between the feed and draw solutions could enhance the water flux due to the decrease in feed solution viscosity and the increase in water diffusivity.

Yet to date little attention has been paid to the effect of temperature and temperature difference between feed and draw solutions on the rejection of emerging trace organic chemicals in the feed solution, which is a critical aspect to the deployment of the FO process in wastewater reclamation. Recently, Xie et al. (2013a) reported that the rejection of neutral trace emerging organic chemicals decreased significantly as the feed solution temperature increased from 20 to 40°C whereas it increased when the feed and draw solution temperatures were 20 and 40°C, respectively. This enhanced rejection of neutral trace emerging organic chemicals and improved water flux have important implications for the integration of the FO process with a thermally-driven separation process, such as MD or conventional column distillation, for recovering the draw solutes.

16.3.3 Membrane Fouling

Fouling of water treatment membranes decreases water productivity, deteriorates permeate quality, and shortens membrane lifespan (Ang et al. 2006). While all conventional membrane processes are hampered by membrane fouling, as mentioned earlier, FO has been observed to inherently endure fouling compared to other systems that employ salt rejecting RO membranes (Cornelissen et al. 2008; Mi and Elimelech 2010).

Previous FO studies using a variety of foulants demonstrated a substantially stable water flux when the FO membranes were fouled with activated sludge (Achilli et al. 2009; Alturki et al. 2012; Zhang et al. 2012) and humic acid (Tang et al. 2010; Mi and Elimelech 2008; Parida and Ng 2013). More pronounced water flux decline was observed when alginate was used as a model foulant in the presence of divalent ions, such as calcium, thereby resulting in the formation of gel layer (Mi and Elimelech 2008; Zou et al. 2011). Calcium bridging was identified as the main reason for alginate fouling; however, water flux was significantly recovered after a simple physical cleaning step (Mi and Elimelech 2010), indicating the reversible characteristics of the fouling. While the factors governing the fouling behaviour vary from foulant to foulant, the low water flux decline in FO can be mainly attributed to the loose structure of the fouling layer formed without hydraulic pressure. On the other hand, reverse draw solute diffusion was found to increase the foulant-membrane interaction (She et al. 2012) or exacerbate cake-enhanced osmotic pressure within the formed fouling layer (Boo et al. 2012).

Effects of membrane fouling on the rejection of emerging trace organic chemicals have been investigated extensively in NF and RO processes. These studies suggest that membrane fouling influences the rejection of emerging trace organic chemicals via modification of membrane surface charge (Plakas et al. 2006; Xu et al. 2006), pore blockage (Nghiem and Hawkes 2007) or cake enhanced concentration polarization (Ng and Elimelech 2004; Vogel et al. 2010), thereby either improving or reducing their rejection. By drawing on these well-established mechanisms in NF and RO processes, several studies have also been initiated to shed light on the impact of membrane fouling on the rejection of emerging trace organic chemicals in FO (Table 16.3). Hancock et al. (2011) observed that rejection of emerging trace organic chemicals by the FO process substantially increased when the membrane was fouled by wastewater effluent in a pilot-scale setup. Valladares Linares et al. (2011) proposed that the fouling layer altered the charge and hydrophobicity of the FO membrane surface, thereby enhancing the rejection of ionic and neutral emerging trace organic chemicals. Jin et al. (2012) highlighted the enhanced membrane sieving effect by membrane fouling when they compared the rejections of boron and arsenate by an alginate-fouled FO membrane. Recently, Xie et al. (2013b) observed that permeation of carbamazepine and sulfamethoxazole decreased as the deposition of humic acid increased, which correlated well with the decrease in the membrane salt (NaCl) permeability coefficient. They concluded that the humic acid fouling layer hindered solute diffusion through the membrane pore and enhanced solute rejection by steric hindrance.

Table 16.3. Summary of impact of membrane fouling on rejection of feed contaminants in FO

	FO experimental conditions				
*Feed/[ref.]**	*Membrane*	*Draw*	*Scale*	*Rejection behaviour*	*Proposed mechanism*
WW** effluent/[1]	CTA	0.5 M synthetic sea salt	Pilot	Increased rejection for most emerging organic chemicals	Fouling layer separate and inhibit the interaction of contaminants with the membrane surface
WW effluent/[2]	CTA	Red Sea seawater (~0.5 M)	Bench	Increased rejection for hydrophilic ionic and hydrophobic neutral contaminants; 5% decrease in rejection of hydrophilic neutral contaminants	Higher negative charge and hydrophilicity of the fouled membrane; membrane swelling
Synthetic WW with alginate/[3]	CTA	0.5–5 M NaCl	Bench	Increased rejection of arsenate; no observable change in rejection of boron	Improved sieving effect by alginate fouling
Synthetic WW with humic acid & varying conc. of calcium/[4]	CTA	0.5 M NaCl	Bench	Decrease in permeation of carbamazepine and sulfamethoxazole	hydrated humic acid fouling layer hindered solute diffusion through the membrane pore and enhanced solute rejection by steric hindrance

*References: [1] Hancock et al. (2011); [2] Valladares Linares et al. (2011); [3] Jin et al. (2012); [4] Xie et al. (2013a)
**WW = wastewater

It is of practical interest to investigate the rejection behaviour of emerging trace organic chemicals after membrane physical cleaning. Obtaining such knowledge is of significant benefit in the consideration of the full scale deployment of FO in wastewater reclamation. Xie et al. (2013b) reported that membrane physical cleaning restored the permeation of carbamazepine and sulfamethoxazole as well as the reverse salt (NaCl) flux to those of the virgin (clean) membrane. However, more effort should be made to examine the impact of membrane cleaning on rejection of emerging contaminants.

16.3.4 FO-based Hybrid Systems for Wastewater Reclamation

Integration of FO with other processes offers an important alternative to wastewater treatment and reclamation. In order to produce freshwater, FO must be combined with existing processes. Such hybrid systems can provide an extra barrier to intercept emerging trace organic chemicals, which results in better quality product water. In addition, re-concentrated draw solution will close loop back to the FO process, thereby leading to a near-zero liquid discharge system. Furthermore, FO can be an excellent pre-treatment for fouling-susceptible membranes, such as RO membranes, in the treatment of low quality source water. As a result, successful development and deployment of FO-based hybrid systems have the potential to expand the availability of source water, lower technical barrier for implementation and thus improve the sustainability of freshwater supply (Table 16.4). To fulfil this concept, two major approaches have been proposed and examined: combination with pressure driven membrane processes including NF and RO; and FO used with thermally driven processes, such as column distillation and MD (Figure 16.6).

The feasibility of a combination of FO and RO process has been the subject of recent studies. A hybrid FO-RO system for seawater desalination was modelled based on bench-scale experimental data, the energy balance of which demonstrated a lower overall energy requirement for seawater desalination (Bamaga et al. 2011). Another modelling of an FO-RO hybrid system was conducted by Shaffer et al. (2012) whose results indicated that the system could achieve boron and chloride water quality requirements for agricultural irrigation while consuming less energy than a conventional two-pass reverse osmosis process. A 14-day operation of a lab scale FO-RO hybrid system using real Red Sea seawater draw solution and secondary wastewater effluent feed achieved high quality product water that was acceptable for water reclamation purposes with low energy consumption of 1.5 kWh/m^3 of product water in comparison with direct RO seawater desalination (Yangali-Quintanilla et al. 2011; Li et al. 2012). A 13-day test of a pilot scale FO-RO hybrid system for osmotic dilution and subsequent treatment of wastewater effluent highlighted a high quality of product water, nutrient and organic contaminant concentrations which were far below the detection limit (Cath et al. 2010). A more comprehensive bench- and pilot-scale FO-RO hybrid system demonstrated that this dual barrier treatment of impaired water could lead to more than 99% rejection of almost all emerging organic chemicals that were regulated in reclaimed water. In addition, utilising divalent

Table 16.4. Summary of FO-based hybrid systems

	Hybrid system description			Hybrid system performance			
Category	Hybrid system	Scale	Membrane	Water flux	Rejection	Energy consumption	Ref
Combination with pressure driven processes	FO-RO	Bench	CTA	Declined by 28% over 10-day operation	N.A.	~1.5 kWh/m^3 of product water	(Yangali-Quintanilla et al. 2011)
	FO-RO	Pilot	CTA	Declined from 8 to 5 L/m^2h during 40-day operation	more than 99% rejection of almost all emerging organic chemicals	N.A.	(Hancock et al. 2011)
	FO-RO	Pilot	CTA	Declined from 6 to 3 L/m^2h during 6-day operation	Concentrations of all emerging organic chemicals below detection limit in product water.	Economical water recovery of 63%	(Cath et al. 2010)
	FO-NF	Bench	CTA		TDS in product water less than 500 mg/L	N.A.	(Zhao et al. 2012)
	FO-NF	Bench	CTA	10 L/m^2h for both FO and NF processes	TDS in product water of 114 mg/L	Equivalent to seawater RO desalination	(Tan and Ng 2010)

(Continued)

Table 16-4. Summary of FO-based hybrid systems *(Continued)*

	Hybrid system description			*Hybrid system performance*			
Category	*Hybrid system*	*Scale*	*Membrane*	*Water flux*	*Rejection*	*Energy consumption*	*Ref*
	FO-NF	Bench	CTA	Water flux varied dependent on fertilizer draw solution	Nutrient concentrations in product water decreased, while most fertiliser draw solutions acceptable for direct fertiliser.	N.A.	(Phuntsho et al. 2013)
	FO-RO	Modelling	CTA	N.A.	Concentrations of boron and chloride in productwater meet quality requirements for agricultural irrigation	Overall energy consumption less than two-pass RO seawater desalination	(Shaffer et al. 2012)
	FO-RO	Modelling	CTA	7 (seawater feed) to 11(brackish water feed) L/m^2h	N.A.	~2.2 kJ per each liter of permeate water	(Bamaga et al. 2011)
Combination with thermally driven processes	FO-column distillation	Pilot	TFC	water flux of 2.6 L/m^2h with water recovery of 64%	Concentrated brine concentration of 180 g/L TDS, and product water with 300 mg/L TDS	275 kWh_{th}/m^3 of product water; 57% less thermal energy input than that by a conventional evaporator	(McGinnis et al. 2013)

	FO-MD	Bench	CTA	Water transfer rate of 0.1 L/h	BSA in feed solution concentrated from 1 g/L to 2.1 g/L in four hours.	N.A.	(Wang et al. 2011)
	FO-MD	Bench	CTA	Water flux varied from 8 to 25 L/m^2h	Acid orange 8 in the feed concentrated from 50 to 83 mg/L in two hours	N.A.	(Ge et al. 2012)
	FO-column distillation	Modelling	CTA	N.A.	N.A.	Energy savings compared to current technologies, on an equivalent work basis, range from 72 to 85%.	(McGinnis and Elimelech 2007)

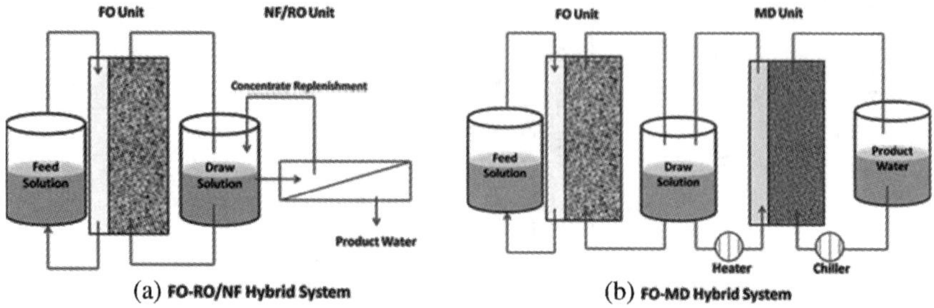

Figure 16.6. Closed loop hybrid systems (left) FO-NF/RO and (right) FO-MD

inorganic salts as draw solution in an FO process, such as $MgSO_4$ and Na_2SO_4, favours FO-NF hybrid system for desalination of brackish water (Zhao et al. 2012) and seawater (Tan and Ng 2010). In these FO-NF hybrid systems, due to the high rejection of divalent salts by NF membranes (above 98% for $MgSO_4$), the quality of product water can meet the drinking water guideline provided by the World Health Organisation (World Health 2002). Recently, Phuntsho et al. (2013) employed an FO-NF hybrid system using fertilizers as draw solution for brackish water desalination and they found that the FO-NF hybrid system was more advantageous in terms of reduced nutrient concentrations in the final product water and energy consumption. In fact, a life cycle assessment revealed that with the progress of development and commercialization of second-generation FO membrane, the environmental impact of FO-RO hybrid system could be 25% less than the state-of-the-art seawater RO; and the FO-NF hybrid system for both seawater and wastewater treatment could achieve similar levels of environmental impact (Hancock et al. 2012).

Effectiveness of an integrated FO and thermally driven process was first demonstrated by McGinnis and Elimelech (2007) when they utilised a multi-stage column distillation to re-concentrate thermolytic NH_3/CO_2 draw solution and to produce freshwater. The energy saving of the FO-multi-stage column distillation hybrid system on an equivalent work basis, ranged from 72 to 85%, in comparison to current technologies. Later, a pilot FO-multi-stage column distillation hybrid system using NH_3/CO_2 draw solution was employed to desalinate highly saline wastewater from natural gas extraction (McGinnis et al. 2013). The measured energy consumption of this pilot-scale hybrid system was 275 kWh/m^3 of product water, 57% less thermal energy input than that estimated for a conventional evaporator operated in a comparable single stage. On the other hand, MD, which is a thermally driven membrane process, has been widely recognised as a potential draw solution recovery process (Cath et al. 2005; Martinetti et al. 2009). In the MD process, solar thermal or low-grade heat can be utilised to increase the feed solution (i.e. the diluted draw solution of the FO process) temperature for the extraction of water vapour across a microporous membrane which then condenses to the liquid form. Several investigations have been conducted to demonstrate the technological feasibility of the FO-MD hybrid system. Wang et al. (2011) employed a bench scale FO-MD hybrid system to treat highly viscous protein solution where FO is employed for dewatering protein solutions while MD is used for draw solution recovery. Another bench scale FO-MD hybrid system was used to treat acid orange 8 solution (Ge et al. 2012) where the most efficient treatment was achieved when the water transfer rate of the FO system matched that of the MD system. However, neither of the above studies provided any details about their energy consumption. It is expected that renewable energy sources, such as solar and geothermal heat, wind energy, could facilitate these FO-based hybrid systems to further reduce their carbon and energy footprints.

Potential benefits of the FO-based hybrid systems extend beyond better product water quality and energy and cost savings. By using fewer treatment chemicals and prolonging the life of treatment equipment, these FO-based hybrid systems improve the sustainability of seawater desalination and wastewater reclamation that will be increasingly important contributors to solving global water supply challenges.

16.4 CONCLUDING REMARKS

This chapter provides the rationale for FO being a promising technology to remove emerging trace organic chemicals in the context of wastewater treatment and reclamation. The occurrence of emerging trace organic chemicals is known to be a challenge to many wastewater treatment and reclamation applications. Key factors governing the rejection of emerging trace organic chemicals by FO have been summarised and discussed. Developments of FO-based hybrid systems are highlighted as a significant alternative and a paradigm shift from conventional wastewater treatment and reclamation in the near future. The rejection behaviour of emerging trace organic chemicals by FO is significantly governed by membrane properties, draw solution properties, key operating conditions and membrane fouling. First, improvements in high performance membranes specifically designed for FO not only offer a reasonably high water flux via alleviation of any ICP effect, but also enhance the rejection of emerging trace organic chemicals; second, reverse transport of draw solutes through the membrane enhanced solute – solute interaction where the reverse diffusion of draw solutes hinder forward transport of feed solute, thereby increasing feed solute rejection; in addition, optimization of key operating conditions, including feed solution pH, membrane orientation and temperatures of feed and draw solutions, could further increase the rejection of emerging trace organic chemicals by FO. Furthermore, membrane fouling in FO, which has insignificant impact on water flux, notably altered rejection of emerging trace organic chemicals. Based on these state-of-the-art understandings of FO process, the development and deployment of FO-based hybrid systems will enable a paradigm shift to carbon-neutral wastewater treatment and reclamation processes. Resurgent of FO diversifies and advances technological solutions to wastewater treatment and reclamation. Better rejection of emerging trace organic chemicals by FO will satisfy ever more stringent water quality regulations; and lower energy and carbon footprint of FO-based hybrid system is clearly important from sustainability and cost considerations.

16.4 ABBREVIATIONS

CTA	cellulose triacetate
EDCs	endocrine disrupting chemicals
FO	forward osmosis
ICP	internal concentration polarization
MD	membrane distillation
NF	nanofiltration
PPCPs	pharmaceuticals and personal care products
RO	reverse osmosis
TDS	total dissolved solid
TFC	thin film composite
TKN	total Kjehldahl nitrogen
UV	ultra violet

References

Achilli, A., Cath, T. Y., and Childress, A. E. (2010). "Selection of inorganic-based draw solutions for forward osmosis applications." *Journal of Membrane Science*, 364(1–2), 233–241.

Achilli, A., Cath, T. Y., Marchand, E. A., and Childress, A. E. (2009). "The forward osmosis membrane bioreactor: A low fouling alternative to MBR processes." *Desalination*, 239(1-3), 10–21.

Al-Rifai, J. H., Gabelish, C. L., and Schäfer, A. I. (2007). "Occurrence of pharmaceutically active and non-steroidal estrogenic compounds in three different wastewater recycling schemes in Australia." *Chemosphere*, 69(5), 803–815.

Alturki, A., McDonald, J., Khan, S. J., Hai, F. I., Price, W. E., and Nghiem, L. D. (2012). "Performance of a novel osmotic membrane bioreactor (OMBR) system: Flux stability and removal of trace organics." *Bioresource Technology*, 113, 201–206.

Alturki, A. A., McDonald, J. A., Khan, S. J., Price, W. E., Nghiem, L. D., and Elimelech, M. (2013). "Removal of trace organic contaminants by the forward osmosis process." *Separation and Purification Technology*, 103, 258–266.

American Water Works, A., and James, E. (2011). "WATER REUSE FOR DRINKING WATER augmentation." *Water Quality & Treatment: A Handbook on Drinking Water*, McGraw Hill Professional, Access Engineering.

Ang, W. S., Lee, S., and Elimelech, M. (2006). "Chemical and physical aspects of cleaning of organic-fouled reverse osmosis membranes." *Journal of Membrane Science*, 272(1–2), 198–210.

Arsuaga, J. M., López-Muñoz, M. J., Aguado, J., and Sotto, A. (2008). "Temperature, pH and concentration effects on retention and transport of organic pollutants across thin-film composite nanofiltration membranes." *Desalination*, 221(1–3), 253–258.

Asano, T., Burton, F., and Leverenz, H. (2007). *Water Reuse: Issues, Technologies and Applications*, McGraw-Hill, New York.

Auriol, M., Filali-Meknassi, Y., Tyagi, R. D., Adams, C. D., and Surampalli, R. Y. (2006). "Endocrine disrupting compounds removal from wastewater, a new challenge." *Process Biochemistry*, 41(3), 525–539.

Bamaga, O. A., Yokochi, A., Zabara, B., and Babaqi, A. S. (2011). "Hybrid FO/RO desalination system: Preliminary assessment of osmotic energy recovery and designs of new FO membrane module configurations." *Desalination*, 268(1–3), 163–169.

Basile, T., Petrella, A., Petrella, M., Boghetich, G., Petruzzelli, V., Colasuonno, S., and Petruzzelli, D. (2011). "Review of Endocrine-Disrupting-Compound Removal Technologies in Water and Wastewater Treatment Plants: An EU Perspective." *Industrial & Engineering Chemistry Research*, 50(14), 8389–8401.

Boo, C., Lee, S., Elimelech, M., Meng, Z., and Hong, S. (2012). "Colloidal fouling in forward osmosis: Role of reverse salt diffusion." *Journal of Membrane Science*, 390–391, 277–284.

Bui, N.-N., Lind, M. L., Hoek, E. M. V., and McCutcheon, J. R. (2011). "Electrospun nanofiber supported thin film composite membranes for engineered osmosis." *Journal of Membrane Science*, 385–386(0), 10–19.

Bui, N.-N., and McCutcheon, J. R. (2012). "Hydrophilic Nanofibers as New Supports for Thin Film Composite Membranes for Engineered Osmosis." *Environmental Science & Technology*, 47(3), 1761–1769.

Carballa, M., Omil, F., Lema, J. M., Llompart, M. a., García-Jares, C., Rodríguez, I., Gómez, M., and Ternes, T. (2004). "Behavior of pharmaceuticals, cosmetics and hormones in a sewage treatment plant." *Water Research*, 38(12), 2918–2926.

Cartinella, J. L., Cath, T. Y., Flynn, M. T., Miller, G. C., Hunter, K. W., and Childress, A. E. (2006). "Removal of Natural Steroid Hormones from Wastewater Using Membrane Contactor Processes†." *Environmental Science & Technology*, 40(23), 7381–7386.

Cath, T. Y., Adams, D., and Childress, A. E. (2005). "Membrane contactor processes for wastewater reclamation in space: II. Combined direct osmosis, osmotic distillation, and membrane distillation for treatment of metabolic wastewater." *Journal of Membrane Science*, 257(1–2), 111–119.

Cath, T. Y., Gormly, S., Beaudry, E. G., Flynn, M. T., Adams, V. D., and Childress, A. E. (2005). "Membrane contactor processes for wastewater reclamation in space: Part I. Direct osmotic concentration as pretreatment for reverse osmosis." *Journal of Membrane Science*, 257(1–2), 85–98.

Cath, T. Y., Hancock, N. T., Lundin, C. D., Hoppe-Jones, C., and Drewes, J. E. (2010). "A multi-barrier osmotic dilution process for simultaneous desalination and purification of impaired water." *Journal of Membrane Science*, 362(1–2), 417–426.

Coday, B. D., Heil, D. M., Xu, P., and Cath, T. Y. (2013). "Effects of Transmembrane Hydraulic Pressure on Performance of Forward Osmosis Membranes." *Environmental Science & Technology*, 47(5), 2386–2393.

Cornelissen, E. R., Harmsen, D., de Korte, K. F., Ruiken, C. J., Qin, J.-J., Oo, H., and Wessels, L. P. (2008). "Membrane fouling and process performance of forward osmosis membranes on activated sludge." *Journal of Membrane Science*, 319(1–2), 158–168.

Council, N. R. (2012). *Water Reuse: Potential for Expanding the Nation's Water Supply Through Reuse of Municipal Wastewater*, The National Academies Press.

Cunningham, V. L., Binks, S. P., and Olson, M. J. (2009). "Human health risk assessment from the presence of human pharmaceuticals in the aquatic environment." *Regulatory Toxicology and Pharmacology*, 53(1), 39–45.

Drewes, J. E., Bellona, C., Oedekoven, M., Xu, P., Kim, T.-U., and Amy, G. (2005). "Rejection of wastewater-derived micropollutants in high-pressure membrane applications leading to indirect potable reuse." *Environmental Progress*, 24(4), 400–409.

Farré, M., Klöter, G., Petrovic, M., Alonso, M. C., de Alda, M. J. L., and Barceló, D. (2002). "Identification of toxic compounds in wastewater treatment plants during a field experiment." *Analytica Chimica Acta*, 456(1), 19–30.

Ge, Q., Wang, P., Wan, C., and Chung, T.-S. (2012). "Polyelectrolyte-promoted forward osmosis-membrane distillation (FO-MD) hybrid process for dye wastewater treatment." *Environmental Science & Technology*, 46(11), 6236–6243.

Han, G., Chung, T.-S., Toriida, M., and Tamai, S. (2012). "Thin-film composite forward osmosis membranes with novel hydrophilic supports for desalination." *Journal of Membrane Science*, 423–424(0), 543–555.

Hancock, N. T., Black, N. D., and Cath, T. Y. (2012). "A comparative life cycle assessment of hybrid osmotic dilution desalination and established seawater desalination and wastewater reclamation processes." *Water Research*, 46(4), 1145–1154.

Hancock, N. T., and Cath, T. Y. (2009). "Solute Coupled Diffusion in Osmotically Driven Membrane Processes." *Environmental Science & Technology*, 43(17), 6769–6775.

Hancock, N. T., Phillip, W. A., Elimelech, M., and Cath, T. Y. (2011). "Bidirectional Permeation of Electrolytes in Osmotically Driven Membrane Processes." *Environmental Science & Technology*, 45(24), 10642–10651.

Hancock, N. T., Xu, P., Heil, D. M., Bellona, C., and Cath, T. Y. (2011). "Comprehensive Bench- and Pilot-Scale Investigation of Trace Organic Compounds Rejection by Forward Osmosis." *Environmental Science & Technology*, 45(19), 8483–8490.

Herron, J. R. B., Edward G.Salter, Robert (1997). " Direct osmotic concentration contaminated water." *OSMOTEK*, INC.

Hickenbottom, K. L., Hancock, N. T., Hutchings, N. R., Appleton, E. W., Beaudry, E. G., Xu, P., and Cath, T. Y. (2013). "Forward osmosis treatment of drilling mud and fracturing wastewater from oil and gas operations." *Desalination*, 312(0), 60–66.

Holloway, R. W., Childress, A. E., Dennett, K. E., and Cath, T. Y. (2007). "Forward osmosis for concentration of anaerobic digester centrate." *Water Research*, 41(17), 4005–4014.

Hoover, L. A., Phillip, W. A., Tiraferri, A., Yip, N. Y., and Elimelech, M. (2011). "Forward with Osmosis: Emerging Applications for Greater Sustainability." *Environmental Science & Technology*, 45(23), 9824–9830.

Huang, L., Bui, N.-N., Meyering, M. T., Hamlin, T. J., and McCutcheon, J. R. (2013). "Novel Hydrophilic Nylon 6, 6 Microfiltration Membrane Supported Thin Film Composite Membranes for Engineered Osmosis." *Journal of Membrane Science*, (0).

Huerta-Fontela, M., Galceran, M. T., and Ventura, F. (2011). "Occurrence and removal of pharmaceuticals and hormones through drinking water treatment." *Water Research*, 45(3), 1432–1442.

Jin, X., Shan, J., Wang, C., Wei, J., and Tang, C. Y. (2012). "Rejection of pharmaceuticals by forward osmosis membranes." *Journal of Hazardous Materials*, 227–228, 55–61.

Jin, X., She, Q., Ang, X., and Tang, C. Y. (2012). "Removal of boron and arsenic by forward osmosis membrane: Influence of membrane orientation and organic fouling." *Journal of Membrane Science*, 389, 182–187.

Jin, X., Tang, C. Y., Gu, Y., She, Q., and Qi, S. (2011). "Boric Acid Permeation in Forward Osmosis Membrane Processes: Modeling, Experiments, and Implications." *Environmental Science & Technology*, null-null.

Jones, O. A., Lester, J. N., and Voulvoulis, N. (2005). "Pharmaceuticals: a threat to drinking water?" *Trends in Biotechnology*, 23(4), 163–167.

Kannan, K., Battula, S., Loganathan, B. G., Hong, C. S., Lam, W. H., Villeneuve, D. L., Sajwan, K., Giesy, J. P., and Aldous, K. M. (2003). "Trace Organic Contaminants, Including Toxaphene and Trifluralin, in Cotton Field Soils from Georgia and South Carolina, USA." *Archives of Environmental Contamination and Toxicology*, 45(1), 0030–0036.

Kasprzyk-Hordern, B., Dinsdale, R. M., and Guwy, A. J. (2009). "The removal of pharmaceuticals, personal care products, endocrine disruptors and illicit drugs during wastewater treatment and its impact on the quality of receiving waters." *Water Research*, 43(2), 363–380.

Kim, C., Lee, S., Shon, H. K., Elimelech, M., and Hong, S. (2012). "Boron transport in forward osmosis: Measurements, mechanisms, and comparison with reverse osmosis." *Journal of Membrane Science*, 419–420(0), 42–48.

Kimura, K., Toshima, S., Amy, G., and Watanabe, Y. (2004). "Rejection of neutral endocrine disrupting compounds (EDCs) and pharmaceutical active compounds (PhACs) by RO membranes." *Journal of Membrane Science*, 245(1–2), 71–78.

Kolpin, D. W., Furlong, E. T., Meyer, M. T., Thurman, E. M., Zaugg, S. D., Barber, L. B., and Buxton, H. T. (2002). "Pharmaceuticals, Hormones, and Other Organic Wastewater Contaminants in U.S. Streams, 1999-2000: A National Reconnaissance." *Environmental Science & Technology*, 36(6), 1202–1211.

Li, Z.-Y., Yangali-Quintanilla, V., Valladares-Linares, R., Li, Q., Zhan, T., and Amy, G. (2012). "Flux patterns and membrane fouling propensity during desalination of seawater by forward osmosis." *Water Research*, 46(1), 195–204.

Martinetti, C. R., Childress, A. E., and Cath, T. Y. (2009). "High recovery of concentrated RO brines using forward osmosis and membrane distillation." *Journal of Membrane Science*, 331(1–2), 31–39.

McCutcheon, J. R., and Elimelech, M. (2006). "Influence of concentrative and dilutive internal concentration polarization on flux behavior in forward osmosis." *Journal of Membrane Science*, 284(1–2), 237–247.

McCutcheon, J. R., and Elimelech, M. (2008). "Influence of membrane support layer hydrophobicity on water flux in osmotically driven membrane processes." *Journal of Membrane Science*, 318(1–2), 458–466.

McCutcheon, J. R., McGinnis, R. L., and Elimelech, M. (2005). "A novel ammonia—carbon dioxide forward (direct) osmosis desalination process." *Desalination*, 174(1), 1–11.

McCutcheon, J. R., McGinnis, R. L., and Elimelech, M. (2006). "Desalination by ammonia-carbon dioxide forward osmosis: Influence of draw and feed solution concentrations on process performance." *Journal of Membrane Science*, 278(1–2), 114–123.

McGinnis, R. L., and Elimelech, M. (2007). "Energy requirements of ammonia-carbon dioxide forward osmosis desalination." *Desalination*, 207(1–3), 370–382.

McGinnis, R. L., and Elimelech, M. (2007). "Energy requirements of ammonia-carbon dioxide forward osmosis desalination." *Desalination*, 207(1–3), 370–382.

McGinnis, R. L., Hancock, N. T., Nowosielski-Slepowron, M. S., and McGurgan, G. D. (2013). "Pilot demonstration of the NH3/CO2 forward osmosis desalination process on high salinity brines." *Desalination*, 312(0), 67–74.

Mi, B., and Elimelech, M. (2008). "Chemical and physical aspects of organic fouling of forward osmosis membranes." *Journal of Membrane Science*, 320(1–2), 292–302.

Mi, B., and Elimelech, M. (2010). "Organic fouling of forward osmosis membranes: Fouling reversibility and cleaning without chemical reagents." *Journal of Membrane Science*, 348(1–2), 337–345.

Miège, C., Choubert, J. M., Ribeiro, L., Eusèbe, M., and Coquery, M. (2009). "Fate of pharmaceuticals and personal care products in wastewater treatment plants—Conception of a database and first results." *Environmental Pollution*, 157(5), 1721–1726.

Nayak, C. A., and Rastogi, N. K. (2010). "Forward osmosis for the concentration of anthocyanin from Garcinia indica Choisy." *Separation and Purification Technology*, 71(2), 144–151.

Ng, H. Y., and Elimelech, M. (2004). "Influence of colloidal fouling on rejection of trace organic contaminants by reverse osmosis." *Journal of Membrane Science*, 244(1–2), 215–226.

Nghiem, L. D., and Hawkes, S. (2007). "Effects of membrane fouling on the nanofiltration of pharmaceutically active compounds (PhACs): Mechanisms and role of membrane pore size." *Separation and Purification Technology*, 57(1), 176–184.

Nghiem, L. D., Schäfer, A. I., and Elimelech, M. (2004). "Removal of Natural Hormones by Nanofiltration Membranes: Measurement, Modeling, and Mechanisms." *Environmental Science & Technology*, 38(6), 1888–1896.

Onda, K., Nakamura, Y., Takatoh, C., Miya, A., and Katsu, Y. (2003). "The behavior of estrogenic substances in the biological treatment process of sewage." *Water Sci. Technol.*, 47(9), 109–116.

Parida, V., and Ng, H. Y. (2013). "Forward osmosis organic fouling: Effects of organic loading, calcium and membrane orientation." *Desalination*, 312(0), 88–98.

Phillip, W. A., Yong, J. S., and Elimelech, M. (2010). "Reverse Draw Solute Permeation in Forward Osmosis: Modeling and Experiments." *Environmental Science & Technology*, 44(13), 5170–5176.

Phuntsho, S., Hong, S., Elimelech, M., and Shon, H. K. (2013). "Forward osmosis desalination of brackish groundwater: Meeting water quality requirements for fertigation by integrating nanofiltration." *Journal of Membrane Science*, 436(0), 1–15.

Phuntsho, S., Vigneswaran, S., Kandasamy, J., Hong, S., Lee, S., and Shon, H. K. (2012). "Influence of temperature and temperature difference in the performance of forward osmosis desalination process." *Journal of Membrane Science*, 415–416(0), 734–744.

Plakas, K. V., Karabelas, A. J., Wintgens, T., and Melin, T. (2006). "A study of selected herbicides retention by nanofiltration membranes—The role of organic fouling." *Journal of Membrane Science*, 284(1–2), 291–300.

Rijal, G. K., Zmuda, J. T., Gore, R., Abedin, Z., Granato, T., Kollias, L., and Lanyon, R. (2009). "Antibiotic resistant bacteria in wastewater processed by the Metropolitan Water Reclamation District of Greater Chicago system." *Water Science & Technology*, 59 (12), 2297–2304.

Rodgers-Gray, T. P., Jobling, S., Morris, S., Kelly, C., Kirby, S., Janbakhsh, A., Harries, J. E., Waldock, M. J., Sumpter, J. P., and Tyler, C. R. (2000). "Long-term temporal changes in the estrogenic composition of treated sewage effluent and its biological effects on fish." *Environmental Science & Technology*, 34(8), 1521–1528.

Rothausen, S. G. S. A., and Conway, D. (2011). "Greenhouse-gas emissions from energy use in the water sector." *Nature Clim. Change*, 1(4), 210–219.

Routledge, E. J., Sheahan, D., Desbrow, C., Brighty, G. C., Waldock, M., and Sumpter, J. P. (1998). "Identification of Estrogenic Chemicals in STW Effluent. 2. In Vivo Responses in Trout and Roach." *Environmental Science & Technology*, 32(11), 1559–1565.

Schwarzenbach, R. P., Escher, B. I., Fenner, K., Hofstetter, T. B., Johnson, C. A., von Gunten, U., and Wehrli, B. (2006). "The Challenge of Micropollutants in Aquatic Systems." *Science*, 313(5790), 1072–1077.

Shaffer, D. L., Yip, N. Y., Gilron, J., and Elimelech, M. (2012). "Seawater desalination for agriculture by integrated forward and reverse osmosis: Improved product water quality for potentially less energy." *Journal of Membrane Science*, 415–416, 1–8.

Sharma, R. R., and Chellam, S. (2006). "Temperature and concentration effects on electrolyte transport across porous thin-film composite nanofiltration membranes: Pore transport mechanisms and energetics of permeation." *Journal of Colloid and Interface Science*, 298(1), 327–340.

Sharpe, R. M., Fraser, H. M., Brougham, M. F. H., McKinnell, C., Morris, K. D., Kelnar, C. J. H., Wallace, W. H. B., and Walker, M. (2003). "Role of the neonatal period of pituitary-testicular activity in germ cell proliferation and differentiation in the primate testis." *Human Reproduction*, 18(10), 2110–2117.

She, Q., Jin, X., Li, Q., and Tang, C. Y. (2012). "Relating reverse and forward solute diffusion to membrane fouling in osmotically driven membrane processes." *Water Research*, 46(7), 2478–2486.

Snyder, S. A., Adham, S., Redding, A. M., Cannon, F. S., DeCarolis, J., Oppenheimer, J., Wert, E. C., and Yoon, Y. (2007). "Role of membranes and activated carbon in the removal of endocrine disruptors and pharmaceuticals." *Desalination*, 202(1–3), 156–181.

Snyder, S. A., Westerhoff, P., Yoon, Y., and Sedlak, D. L. (2003). "Pharmaceuticals, Personal Care Products, and Endocrine Disruptors in Water: Implications for the Water Industry." *Environmental Engineering Science*, 20(5), 449–469.

Song, X., Liu, Z., and Sun, D. D. (2011). "Nano Gives the Answer: Breaking the Bottleneck of Internal Concentration Polarization with a Nanofiber Composite Forward Osmosis Membrane for a High Water Production Rate." *Advanced Materials*, 23(29), 3256–3260.

Steinle-Darling, E., Litwiller, E., and Reinhard, M. (2010). "Effects of Sorption on the Rejection of Trace Organic Contaminants During Nanofiltration." *Environmental Science & Technology*, 44(7), 2592–2598.

Tan, C. H., and Ng, H. Y. (2010). "A novel hybrid forward osmosis - nanofiltration (FO-NF) process for seawater desalination: Draw solution selection and system configuration." *Desalination and Water Treatment*, 13(1–3), 356–361.

Tang, C. Y., She, Q., Lay, W. C. L., Wang, R., and Fane, A. G. (2010). "Coupled effects of internal concentration polarization and fouling on flux behavior of forward osmosis membranes during humic acid filtration." *Journal of Membrane Science*, 354(1–2), 123–133.

Tiraferri, A., Yip, N. Y., Phillip, W. A., Schiffman, J. D., and Elimelech, M. (2011). "Relating performance of thin-film composite forward osmosis membranes to support layer formation and structure." *Journal of Membrane Science*, 367(1–2), 340–352.

Valladares Linares, R., Yangali-Quintanilla, V., Li, Z., and Amy, G. (2011). "Rejection of micropollutants by clean and fouled forward osmosis membrane." *Water Research*, 45(20), 6737–6744.

Victorin, K., and Ahlborg, U. G. (1987). "Impact On Health of Chlorinated Dioxins and Other Trace Organic Emissions." *Waste management & research*, 5(1), 203–224.

Vogel, D., Simon, A., Alturki, A. A., Bilitewski, B., Price, W. E., and Nghiem, L. D. (2010). "Effects of fouling and scaling on the retention of trace organic contaminants by a nanofiltration membrane: The role of cake-enhanced concentration polarisation." *Separation and Purification Technology*, 73(2), 256–263.

Wang, K. Y., Teoh, M. M., Nugroho, A., and Chung, T.-S. (2011). "Integrated forward osmosis-membrane distillation (FO-MD) hybrid system for the concentration of protein solutions." *Chemical Engineering Science*, 66(11), 2421–2430.

Watkinson, A. J., Murby, E. J., and Costanzo, S. D. (2007). "Removal of antibiotics in conventional and advanced wastewater treatment: Implications for environmental discharge and wastewater recycling." *Water Research*, 41(18), 4164–4176.

Widjojo, N., Chung, T.-S., Weber, M., Maletzko, C., and Warzelhan, V. (2011). "The role of sulphonated polymer and macrovoid-free structure in the support layer for thin-film composite (TFC) forward osmosis (FO) membranes." *Journal of Membrane Science*, 383(1–2), 214–223.

World Health, O. (2002). *Guidelines for Drinking-Water Quality (Second Edition)*, World Health Organization, Albany, NY, USA.

Xie, M., Nghiem, L. D., Price, W. E., and Elimelech, M. (2012a). "Comparison of the removal of hydrophobic trace organic contaminants by forward osmosis and reverse osmosis." *Water Research*, 46(8), 2683–2692.

Xie, M., Price, W. E., and Nghiem, L. D. (2012b). "Rejection of pharmaceutically active compounds by forward osmosis: Role of solution pH and membrane orientation." *Separation and Purification Technology*, 93, 107–114.

Xie, M., Nghiem, L. D., Price, W. E., and Elimelech, M. (2013a). "Impact of humic acid fouling on membrane performance and transport of pharmaceutically active compounds in forward osmosis." *Water Research*, 47(13), 4567–4575.

Xie, M., Price, W. E., Nghiem, L. D., and Elimelech, M. (2013b). "Effects of feed and draw solution temperature and transmembrane temperature difference on the rejection of trace organic contaminants by forward osmosis." *Journal of Membrane Science*, 438(0), 57–64.

Xu, P., Drewes, J. E., Kim, T.-U., Bellona, C., and Amy, G. (2006). "Effect of membrane fouling on transport of organic contaminants in NF/RO membrane applications." *Journal of Membrane Science*, 279(1-2), 165–175.

Yangali-Quintanilla, V., Li, Z., Valladares, R., Li, Q., and Amy, G. (2011). "Indirect desalination of Red Sea water with forward osmosis and low pressure reverse osmosis for water reuse." *Desalination*, 280(1–3), 160–166.

Yip, N. Y., Tiraferri, A., Phillip, W. A., Schiffman, J. D., and Elimelech, M. (2010). "High Performance Thin-Film Composite Forward Osmosis Membrane." *Environmental Science & Technology*, 44(10), 3812–3818.

Yong, J. S., Phillip, W. A., and Elimelech, M. (2012). "Coupled reverse draw solute permeation and water flux in forward osmosis with neutral draw solutes." *Journal of Membrane Science*, 392–393, 9–17.

You, S.-J., Wang, X.-H., Zhong, M., Zhong, Y.-J., Yu, C., and Ren, N.-Q. (2012). "Temperature as a factor affecting transmembrane water flux in forward osmosis: Steady-state modeling and experimental validation." *Chemical Engineering Journal*, 198-199(0): 52–60.

Yu, C.-P., and Chu, K.-H. (2009). "Occurrence of pharmaceuticals and personal care products along the West Prong Little Pigeon River in east Tennessee, USA." *Chemosphere*, 75(10), 1281–1286.

Zhang, J., Loong, W. L. C., Chou, S., Tang, C., Wang, R., and Fane, A. G. (2012). "Membrane biofouling and scaling in forward osmosis membrane bioreactor." *Journal of Membrane Science*, 403–404(0), 8–14.

Zhao, S., and Zou, L. (2011). "Effects of working temperature on separation performance, membrane scaling and cleaning in forward osmosis desalination." *Desalination*, 278(1–3), 157–164.

Zhao, S., Zou, L., and Mulcahy, D. (2012). "Brackish water desalination by a hybrid forward osmosis-nanofiltration system using divalent draw solute." *Desalination*, 284(0), 175–181.

Zou, S., Gu, Y., Xiao, D., and Tang, C. Y. (2011). "The role of physical and chemical parameters on forward osmosis membrane fouling during algae separation." *Journal of Membrane Science*, 366(1-2), 356–362.

CHAPTER 17

Fertiliser Drawn Forward Osmosis Desalination for Fertigation

Sherub Phuntsho
Jung Eun Kim
Tahir Majeed
Fezeh Lotfi
Soleyman Sahebi
Myoung Jun Park
Ho Kyong Shon

This chapter describes the application of the forward osmosis (FO) process for the desalination of any saline water for irrigation purpose. Known as the fertiliser drawn forward osmosis (FDFO) process, the concept here is to use fertiliser solution as draw solution (DS) to extract water through a semi-permeable membrane by natural osmotic processes. The concentrated fertiliser solution finally becomes diluted which can then be applied directly for fertigation. The distinct feature of the FDFO process over other FO applications is that, the diluted fertiliser DS can be directly applied for fertigation or fertilised irrigation of crops as the fertilisers contain essential nutrients for the plants. Although the concept seems simple, there are certain limitations that need addressing. In this chapter, the concept of the FDFO desalination process is first explained and the potential of fertiliser solutions as DS evaluated. Some of the major limitations of the FDFO process are then introduced, and few options are suggested to overcome these limitations.

17.1 INTRODUCTION

The world population has already crossed 7 billion in 2011 (UNFPA 2011) and the United Nations (UN) estimates that it would reach 9 billion by 2050 (UN 2009a, b). Therefore, there is no doubt that, one of the most crucial challenges for the

21st century will be in dealing with the pressure from the significant increase in the demand for food supplies to meet this enormous population growth (Ward and Pulido-Velazquez 2008). However, food supplies would be affected by the water availability in the future as the competition for the already scarce fresh water resources increases between the water needed for food production and water needed for other uses such as for domestic consumption and industries. An optimum management of global water resources, therefore, presents a crucial challenge for the 21st century (Jury and Vaux 2005; McDonald et al. 2011). As per the current estimate, about one third of the world's population lives in water-stressed countries which may possibly rise to nearly two third by 2025 (Service 2006). Global warming and climate change due to anthropogenic activities have further created uncertainty regarding water availability and food productivity by altering the global hydrological cycle (McDonald et al. 2011).

About 60% of the food to feed the increased population will have to come from irrigated production (Plusquellec 2002). Although, water scarcity is becoming a significant issue all over the world, water use for agricultural purposes is not sustainable in many parts of the world (Jury and Vaux 2005). A typical example is the over allocation of river water from the Murray-Darling Basin for consumptive use which has seriously undermined the river ecosystem, the rich biodiversity that it supports (Goss 2003; MDBA 2010; Wentworth Group 2010). Fresh water resources are essential for consumptive use and support life on earth. However, the need to protect the environment while managing the very limited resources is becoming increasingly urgent especially as the water resources are increasingly becoming scarcer especially in drought stricken countries such as Australia.

Measures such as water conservation, repair of infrastructure, conservation and improvement of catchment areas and improvement of distribution system could help alleviate water stress to a certain degree. Nevertheless, such measures could only help improve the existing water sources and not create new water resources (Elimelech and Phillip 2011). Therefore, one of the popular water management policies especially in most of the developed countries and oil rich countries is to adopt technologies such as desalination that create new water resources for various water uses. In the face of climate change and the increasing global water crisis, the prospects of the scientific and technological solutions playing a crucial role are demanding and timely (Jury and Vaux 2005). Membrane based water purification technologies are increasingly being considered as a reliable options for providing water supplies both as main supply options or as alterative options to augment existing water supplies (Shannon et al. 2008; Geise et al. 2010). The current generation of membrane technologies, particularly reverse osmosis (RO) processes, have been significantly improved in terms of its efficiency and energy consumption. The constant improvement in the RO technology has therefore increased its scope for desalination and wastewater reclamation and reuse so as to provide alternate source of water to augment the existing fresh water supplies or to reduce pressure on limited available freshwater resources. Desalination technologies, in particular the RO process are therefore seen as a promising alternative in alleviating water scarcity in arid and densely populated regions of the world (Service 2006; McGinnis and Elimelech 2007).

While the performance of membrane technology in terms of energy need has significantly improved in the last few decades, it is still seen or remains as an energy intensive process, particularly for seawater RO desalination (SWRO) processes (McGinnis and Elimelech 2007; Greenlee et al. 2009). SWRO operates using very high hydraulic pressure ranging from 50 to 80 bar and hence requires high pressure pumps. To withstand high pressure pumps, high quality and high pressure rating materials are required for plant accessories such as pressure vessels for housing membrane modules, pipes and fittings which all increases the capital cost of the SWRO desalination plants. Fouling still proves to be a major challenge for membrane processes (Greenlee et al. 2009; Phuntsho et al. 2011a, 2011c). Although various strategies are adopted to reduce the impact of membrane fouling such as multiple pre-treatments, physical and chemical cleaning, most strategies involve additional processes that proportionately increases the capital and operational costs of the plant (Semiat 2008). Moreover, the law of thermodynamics sets a minimum. However, practically the most efficient RO desalination plant equipped with the energy recovery system has been reported to consume up to $\sim 3.2\,kWh/m^3$ excluding the energy consumed by the pre-treatment processes. This unit energy consumption for seawater RO desalination increases with recovery rates (Elimelech and Phillip 2011; Subramani et al. 2011). Since energy and climate change issues are strongly inter-related (Crowley 2000; Semiat 2008), addressing global water scarcity problems requires robust and innovative methods of purifying water at minimum energy and cost (McGinnis and Elimelech 2008; Shannon et al. 2008; Elimelech and Phillip 2011). Low energy and low cost desalination technologies could have a significant impact to humanity by utilising the unlimited saline water sources available on this planet.

17.2 FO: A POTENTIALLY LOW-ENERGY DESALINATION PROCESS

Forward osmosis (FO), an osmotically driven membrane process has lately drawn research interest for several applications, notably desalination for potable (Cath et al. 2009; Chung et al. 2012a) and non-potable (Hoover et al. 2011; Phuntsho et al. 2011b, 2012a) purposes, water treatment (Cath et al. 2006), wastewater treatment (Cath et al. 2005; Holloway et al. 2007), pharmaceutical applications (Cath et al. 2006), food processing (Garcia-Castello and McCutcheon 2011), osmotic power generation (Achilli and Childress 2010; Chung et al. 2012b; She et al. 2012; Yip and Elimelech 2012), and pre-treatment to RO desalination (Bamaga et al. 2009; Choi et al. 2009; Bamaga et al. 2011). All the research on the FO process is driven mainly by the need for lower energy technology for desalination as most conventional technologies are energy intensive.

In any FO process, a concentrated draw solution (DS) draws water through a semipermeable membrane from a feed solution (FS) such as saline water.

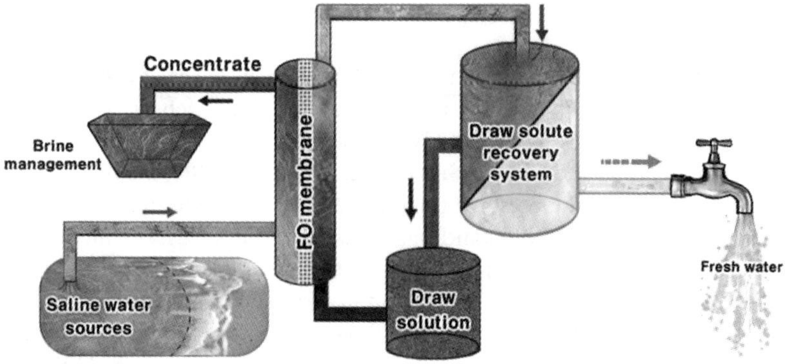

Figure 17.1. Schematic of the FO process for desalination of saline water for potable water supplies

The concentration or the osmotic pressure difference between the two solutions acts as a driving force that permeates the water through the membrane leaving the salt or any other pollutants behind. Therefore, the FO process relies on the natural osmotic process, does not require hydraulic pressure unlike the RO process and hence uses relatively lower energy. The water permeating from the FS dilutes the concentrated DS and exits as a diluted DS. Depending on the application, the diluted DS may be required to pass through some post-treatment processes to separate draw solutes from the pure water or in some cases use the diluted DS directly. Figure 17.1 shows a typical process layout diagram of the FO process for potable water application.

17.3 PRINCIPLE OF FDFO DESALINATION PROCESSES

Desalination using the natural osmotic process is a novel concept. However, its application for potable water using seawater or brackish water remains a significant challenge. In fact, the FO process simply converts th and e saline water feed into diluted DS, and the presence of draw solutes makes diluted DS unusable, and hence, the draw solutes and the pure water must be separated by an additional process (Phuntsho et al. 2012b, 2013a, 2014). Separation of diluted draw solutes from desalted water for recovery and regeneration is, however, not an easy task, requiring an additional processing unit and therefore consuming extra energy. So far, only few draw solutes have been found to be promising candidates for use as DS for potable water desalination, such as NH_3-CO_2 solution and magnetic nanoparticles. NH_3-CO_2 can be separated by heating at approximately 60°C, using low-grade heat (McCutcheon et al. 2005, 2006; McGinnis and Elimelech 2007).

The FO process is more suitable if the diluted DS after desalination can be directly used without further processing as its presence in the water could add value to the final product water (Kravath and Davis 1975; Kessler and Moody

1976; Hoover et al. 2011; Phuntsho et al. 2011b, 2012a). The FO process has been applied for nutritious drinks such as the hydration bags commercialised by the HTI (Stache 1989; Cath et al. 2006; Wallace et al. 2008; Chekli et al. 2012; Zhao et al. 2012b). A highly sugary solution is used as draw solution that draws water from the impaired water sources using CTA membranes and the diluted sugar can then be finally used directly as a nutritious drink. This approach does not require the post-treatment process to separating the draw solutes from the diluted DS. Similarly, in the FDFO process, a soluble fertiliser solution is used as DS and the diluted fertiliser draw solution after desalination containing the essential plant nutrients can be directly used for fertigation. Fertigation refers to the application of fertilisers (dissolved or suspended forms) to the crops with irrigation. Since fertilisers are extensively used in the modern agriculture system, the FDFO process would simply convert the saline water sources into nutrient-rich water for direct fertigation of crops using minimum energy. Such applications offer the advantage of not having to separate and recover draw solutes as in the case for potable water production, and therefore, the energy consumption will be significantly lower than the FO application where draw solute separation is required.

The concept of using fertiliser as draw solute was originally reported by Moody (Moody and Kessler 1976; Moody 1977; Moody and Kessler 1977). Although the potential for such a process is promising, the work on this particular concept did not receive enough attention until recently, mainly due to the lack of suitable membrane for the FO process (Phuntsho et al. 2011b; Phuntsho 2012; Phuntsho et al. 2012a, 2012b, 2013a). Figure 17.2 provides the general process layout of the FDFO desalination for fertigation. In the FO process, two different solutions are used on each side of the membrane: saline feed water on one side and highly concentrated fertiliser DS on the other side. The solutions are kept in contact with the membrane surfaces through a crossflow system that reduces the influence of concentration polarisation (CP) effects. Due to the osmotic pressure difference across the semi-permeable membrane, water flows from the FS (lower concentration) towards the highly concentrated fertiliser DS, in the process desalting the saline feed water. The fertiliser DS finally becomes diluted, and the extent of dilution depends on the concentration (or osmotic pressure) of the feed water. The final fertiliser solution can then be used directly for fertigation if it

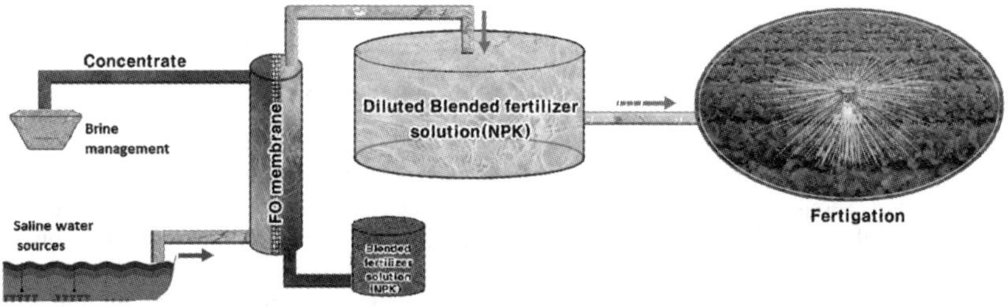

Figure 17.2. Schematic of the FDFO process for direct fertigation
SOURCE: Adapted from Phuntsho et al. 2012b; reproduced with permission from Elsevier

meets the water quality standards for irrigation. If the final fertiliser concentration exceeds the nutrient limit, then further dilution may be necessary before applying it for fertigation.

17.4 ADVANTAGES OF FDFO AND OPPORTUNITIES FOR SPECIFIC APPLICATIONS

17.4.1 Low Energy Desalination Processes

The driving force in the FO process is generated by the osmotic pressure difference between the two solutions across the semipermeable membrane. This process, therefore, does not involve hydraulic pressure to drive the water through the semipermeable membrane as in the case of the RO process. The only energy required for the FO process is for maintaining the crossflow conditions for the FS and DS in contact with the membrane surfaces so as to provide adequate velocity shear to minimise the CP effects that are intrinsic to any membrane filtration process.

The comparative energy requirement for different desalination technologies, are presented in Table 17.1. FO desalination process using NH_3-CO_2 as DS with draw solute recovery process is estimated to require comparatively lower energy than any other existing desalination technologies. The total energy required has been estimated at 0.84 kWh/m^3 that include 0.5 kWh/m^3 of energy for NH_3-CO_2 recovery and 0.24 kWh/m^3 electrical energy for the pumps. This total energy, when compared to other current desalination technologies on an equivalent work basis, can save between 72% and 85% of energy (McGinnis and Elimelech 2007). Although, direct comparison between the performances of NH_3-CO_2 and the

Table 17.1. Comparison of energy requirements for seawater desalination with the existing desalination technologies

Desalination Technology	*Total equivalent work energy (kWh/m^3)*
Multi stage flash (MSF) distillation	10-58[a], 5.66[b]
Multi effect distillation (MED)	6-58[a], 4.05[b]
MED-low temp/electrical	5-6.5[a], 3.21[b]
Reverse osmosis (RO)	4-6[b]
RO with energy recovery	3-4[a], 3.02[b], 2.1-3.2[c]
NH_3-CO_2 FO desalination[c]	0.84[b]

Adapted from [a]Semiat (2008), [b]McGinnis and Elimelech (2007) and [c]Elimelech and Phillip (2011); Subramani et al. (2011). The data for FO for direct fertigation has been adopted from [b]McGinnis and Elimelech (2007) by removing the energy required for draw solution separation by distillation process. [c]At the conditions of low temp, 1.5 M feed with draw solute recovery process. It includes energy for pumping of feed and draw solutions and pumping required during the distillation process. SOURCE: Adapted from Phuntsho et. al. (2012a); reproduced with permission from Elsevier fertiliser draw solutes cannot be made nevertheless, the figures in Table 17.1 provides a fair indication that, the energy required for FDFO desalination will also be significantly lower given that the draw solute recovery from the diluted draw solution is not necessary. The only energy required will be for the crossflow pumps for the FS and DS that does not require pressure. The energy for FDFO desalination for irrigation could be lower than the theoretical energy required based on limiting energy in thermodynamics for separating salt and water from seawater (Phuntsho et al. 2012a). In the RO process, the energy consumption increases with an increase in recovery rates however, for the FO process, recovery rates depend on the solubility and the osmotic pressure a draw solute can generate at maximum solubility, and therefore has no significant relation to external energy input (Phuntsho 2012; Phuntsho et al. 2014). Since FO desalination operates using very little energy, this technology could be easily powered by renewable energy, such as wind and solar energy, making FO technology with zero carbon footprint.

17.4.2 Direct Fertigation or Fertilised Irrigation

Fertilisers and water for irrigation are essential components for improving the agricultural productivity. Agriculture accounts for up to 80% of the total water consumption worldwide, and substantially more in the USA (Jury and Vaux 2005). In Australia, irrigation usage is up to 72% of the total water consumption (Khan 2008). Therefore, even small savings from agricultural water use through improved efficiency and also using unconventional water sources might make substantial quantities of conventional fresh water available for the community and the environment (Ward and Pulido-Velazquez 2008; McDonald et al. 2011). As freshwater resources are becoming scarcer every year, low energy desalination processes such as FDFO could be an effective method for augmenting water for irrigation and make more water available for other uses.

Fertigation is the application of fertiliser along with the irrigation water either in dissolved solution or in suspended form. FDFO-based fertigation has several advantages over the traditional method of applying water and fertilisers separately. The advantages include minimum loss of fertiliser from leaching, optimising the nutrient balance by supplying the nutrients directly to the root zone, control of nutrient concentration in the soil solution, savings in labour and energy, and flexibility in timing fertiliser application in relation to crop demand (Papadopoulos and Eliades 1987). FDFO-based desalination technology is also convenient for farms that adopt fertigation, as it can be easily integrated within their existing fertigation systems. Fertigation is an efficient and cost-effective way of supplying water and nutrients to the crops simultaneously, instead of conventional broadcast application of fertilisers (Hanson et al. 2006). Fertigation could also be advantageous for application in mixtures with other essential micronutrients, chemical pesticides (chemigation), and/or fungicides (fungigation), all in the correct or necessary proportions, thereby eliminating the need for a separate application modes for those chemicals (McBeath et al. 2007).

17.5 FERTILISER CHOICE AND PERFORMANCE OF FERTILISER DRAW SOLUTIONS

Different types of fertilisers are used for growing crops, and hence, it is important to understand what types of fertilisers are more suitable for the FDFO desalination process and how to screen and assess suitable fertiliser candidates for use as DS in the FDFO desalination process. The choice of fertiliser as DS would be guided by many factors, and the main factors include fertiliser economics and its performance as DS for the FDFO process. Fertiliser economics relates to the availability and the cost of the fertilisers. The fertiliser also should have physicochemical properties suitable for use as DS in the FO process. These properties include solubility, pH compatible to FO membrane used, types of species formed in the solution, osmotic pressure of the fertiliser solution, water extraction capacity of the fertiliser, etc. Finally, the choice also would be guided by the nutrient requirements such as its composition and concentrations for the particular target crop. The water extraction capacity refers to the volume of water a unit mass (in kg) of fertiliser can extract from a feed water of particular salinity, and is important in understanding the final nutrient concentration a particular type of fertiliser DS can result in the final FDFO product. The water extraction capacity of the fertiliser is also essential from the viewpoint of farming economics, because fertiliser that can extract a large volume of water is naturally desirable so that the cost of desalinated water remains lower. More on the water extraction capacity of the fertiliser draw solutes are discussed in Section 17.6.3.

A list of chemical fertilisers (synthetic and mineral) used for agriculture in many parts of the world is very long and impressive. However, only those fertilisers that are soluble and can generate osmotic pressure much higher than that of the FS are suitable for use as DS. The general properties of some of the selected fertiliser solutions are provided in Table 17.2. These fertilisers as commonly used in Australia and in many parts of the world are considered for assessment as DSs for the FDFO desalination process. Australia uses 5 to 6 million tonnes of fertiliser each year (one million tonnes of nitrogen, half a million tonnes of phosphorus and two hundred thousand tonnes of potassium) (FIFA 2009). The most-used single-nutrient fertilisers in Australia are single super phosphate and urea, followed by monoammonium phosphate (MAP) or $NH_4H_2PO_4$, diammonium phosphate (DAP) or $(NH_4)_2HPO_4$ and others.

Table 17.2 shows the types of species formed for each fertiliser and the osmotic pressure at 1.0 M solution based on the OLI Stream Analyser 3.2 streaming analysis. The total number of species formed is a colligative property of the solution, and is important because they contribute directly to the osmotic pressure of the solution that is responsible for generating the driving force. In addition, the species formed in the solution are also important because they affect the performance of the FO process, especially in terms of mass transfer across the membrane.

Table 17.2. Commonly used chemical fertilisers with their essential properties at 25°C

Fertiliser	*Chemical formula*	*MW*	*Max π (atm)**	*Species formed in 2.0 M solution at 25°C and 1.0 atm pressure*
Calcium nitrate	$Ca(NO_3)_2$	164.1	601.0	NO_3^- : 3.47 M, Ca^{2+} : 1.47 M, $CaNO_3^-$: 0.53 M, total = 5.47 M
Diammonium Phosphate	$(NH_4)_2HPO_4$	132.1	293.0	NH_4^+ : 3.94 M, HPO_4(ion) : 1.79 M, P_2O_7(ion) : 0.07 M, H_2PO_4(ion) : 0.02 M, HP_2O_7(ion) : 0.02 M.
Ammonium sulfate	$(NH_4)_2SO_4$	132.1	274.8	NH_4^+ : 3.07 M, SO_4^{2-} : 1.07 M, $NH_4SO_4^-$: 0.93 M
Ammonium chloride	NH_4Cl	53.5	356.0	NH_4^+ : 2 M Cl^- : 2 M
Monoammonium Phosphate	$NH_4H_2PO_4$	115.0	181.3	NH_4^- : 2.0 M, $H_2PO_4^-$: 1.76 M, $H_2P_2O_7$(ion) : 0.10 M, H_3PO_4(aq) : 0.02 M), HP_2O_7(ion) : 0.004 M,
Sodium nitrate	$NaNO_3$	85.0	417.9	Na^+ : 1.92 M, NO_3^- : 1.92 M, $NaNO_3$ (aq 0.08 M@2 M)
Potassium chloride	KCl	74.6	226.5	K^+ : 1.99 M, Cl^- : 1.99 M, KCl (aq.): 0.01 M @2 M,
Ammonium nitrate	NH_4NO_3	80.04	3346.1	NH_4^+ : 0.85 M, NO_3^- : 0.85 M, NH_4NO_3 : aq1.15 M
Potassium nitrate	KNO_3	101.1	101.4	K^+ : 2.0 M, NO_3^- : 2.0 M
Potassium dihydrogen phosphate	**KH_2PO_4	136.09	58.0	1.8 M K^+, 1.735 M H_2PO_4(ion), 0.026 $H_2P_2O_7$(ion), H_3PO_4 (aq 0.0064 M), HPO_4(ion)
Urea	$CO(NH_2)_2$	60.06	338.4	$CO(NH_2)_2$

*Osmotic pressure at maximum solubility.
**Data refers to maximum solubility at 1.8 M concentration.

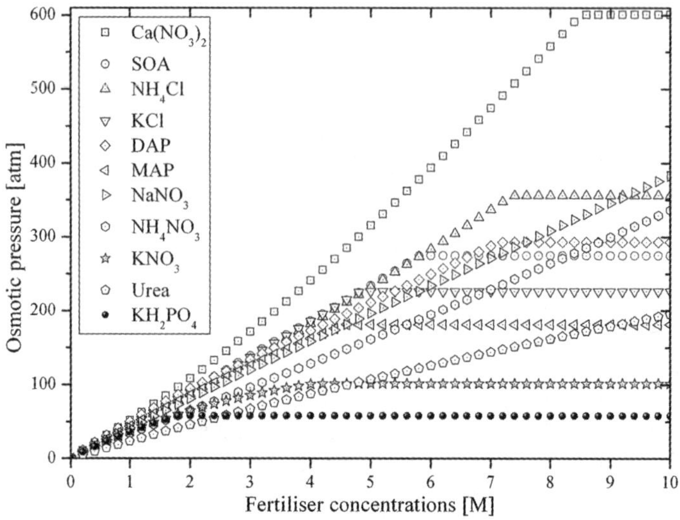

Figure 17.3. Osmotic pressure of the selected fertilisers commonly used around the world for agriculture. Osmotic pressure was predicted using OLI Stream Analyser 3.2

Most of the fertilisers listed in Table 17.2 generate charged species, with the exception of NH_4NO_3, $NaNO_3$, KCl, KH_2PO_4, $NH_4H_2PO_4$ and urea where some quantities of uncharged species are formed at higher concentrations. The osmotic pressure, being a colligative property depends only on the number of species formed in the solution and not on the type and nature of species formed. Figure 17.3 shows the variation of the osmotic pressure with the concentrations (in mol/L) of eleven selected fertiliser solutions at 25°C. $Ca(NO_3)_2$ generates the highest osmotic pressure of about 600 atm at its maximum solubility because the $Ca(NO_3)_2$ in solution forms the highest number of species compared to other fertilisers in solution. When a comparison is made at equal concentrations (e.g., at 1.0 M in Table 17.2), the next highest osmotic pressure generated is by DAP followed by SOA. The lowest osmotic pressure is observed for urea at 1.0 M; however, because of its very high solubility, it generates osmotic pressure in excess of 200 atm at concentrations higher than 10 M. These results indicate that all the selected fertilisers generate osmotic pressure much higher than seawater (~28 atm), indicating their suitability for use as a draw solute in the FDFO process.

The experimental pure water flux of each fertiliser draw solution as a function of its concentration in mol/L is presented in Figure 17.4. The water flux increased at higher molar fertiliser molar concentrations; however, the correlation between molar concentration and water fluxes was non-linear. In fact, a logarithmic correlation is generally observed between the DS concentration and the water flux for most types of draw solutes tested (Seppälä and Lampinen 2004; McCutcheon and Elimelech 2006; McCutcheon et al. 2006; Hancock and Cath 2009; Phuntsho et al. 2011b, 2013b, 2014). This means that, although the

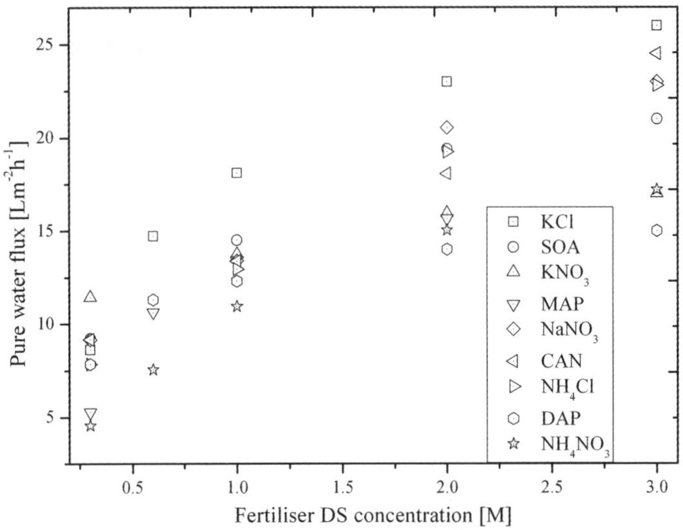

Figure 17.4. Pure water flux of the fertiliser DS as a function of molar concentrations. FS: DI water, Crossflow: 8.5 cm/s in counter-current mode, Temperature: 25°C. MAP: monoammonium phosphate, DAP: diammoium phosphate, SOA: ammonium sulphate.

water flux increased with the increase in DS concentrations, the increase in water flux at higher DS concentrations were not proportional to the increased osmotic pressure. At some point the water flux becomes almost flattening at high concentrations, which is resulted from the more severity of dilutive internal CP (ICP) effects at higher osmotic pressure. It is well known that, when a higher DS concentration is used, it increases the osmotic pressure difference or the driving force between the DS and the FS, and thus generates a higher water flux. This increased water flux is, however, only temporary because the increased water flux coming from the feed side of the membrane creates more dilution of the DS concentration at the membrane interface, which then reduces the driving force and hence the water flux. Since this dilutive phenomenon for the DS happens inside the membrane support layer, it is termed as dilutive ICP.

Tables 17.3 show the actual water flux, theoretical water flux and the performance ratio of the selected fertiliser DS (at a 2 M concentration) using DI water and brackish water or BW (0.3 M NaCl) as FS, respectively. The performance of each fertiliser DS varies considerably as do so with the other types of DS reported in the literature on the FO process (Cath et al. 2006; Achilli et al. 2010; Phuntsho et al. 2011b; Chekli et al. 2012; Ge et al. 2013). At similar molar concentrations, KCl DS generally showed the highest pure water flux at all concentrations except at the lowest concentration of 0.3 M where KNO_3 somehow showed the highest water flux. At 2 M concentrations, the next highest water flux was observed for $NaNO_3$ followed by $(NH_4)_2SO_4$ and NH_4Cl. $(NH_4)_2HPO_4$,

Table 17.3. Water fluxes and the performance ratio of fertiliser draw solutions (2 M) using either DI or 0.3 M NaCl as feed water. Membrane permeability coefficient $A = 1.48 \pm 0.108\ Lm^{-2}\,h^{-1}\,atm^{-1}$

	DI as feed water			*5 g/L NaCl as feed water*		
Fertiliser DS	*Actual J_w ($Lm^{-2}h^{-1}$)*	*Estimated J_w ($Lm^{-2}h^{-1}$)*	*PR (%)*	*Actual J_w ($Lm^{-2}h^{-1}$)*	*Estimated J_w ($Lm^{-2}h^{-1}$)*	*PR (%)*
KCl	22.81	138.40	16.48	14.86	116.6	12.7
$NaNO_3$	20.54	125.69	16.34	8.57	103.9	8.2
KNO_3	15.94	100.58	15.85	10.19	78.8	12.9
NH_4NO_3	15.04	100.58	14.95	7.79	78.8	9.9
NH_4Cl	19.25	135.92	14.16	14.52	114.1	12.7
SOA	19.41	142.74	13.60	12.20	120.9	10.1
MAP	15.66	133.75	11.71	7.63	111.9	6.8
$Ca(NO_3)_2$	18.08	168.16	10.75	11.76	146.3	8.0
DAP	14.01	147.24	9.52	8.07	125.4	6.4
KH_2PO_4	6.23	37.04	16.8	5.80	33.08	17.5
Urea	2.05	24.05	8.5	0.90	20.09	4.5

PR = performance ratio = $J_{w,actual}/J_{w,theoretical}$
MAP = Mono-ammonium phosphate
DAP = Diammonium hydrogen phosphate
SOURCE: Adapted from Phuntsho et al. (2012b); reproduced with permission from Elsevier

$NH_4H_2PO_4$ and NH_4NO_3 resulted in the lowest water flux amongst all of the selected fertilisers DS.

The water fluxes of all the fertiliser DS are much lower than the theoretical or estimated water flux based on the bulk osmotic pressure difference between the DS and the FS. The performance of DS is sometimes measured in terms of performance ratio (PR) which is defined as percentage ratio of the actual experimental water flux to theoretical flux. This ratio represents the percentage of the effective bulk osmotic pressure difference that is effectively available for generating water flux across the FO membrane (McCutcheon et al. 2006; Phuntsho et al. 2011b). The experimental water fluxes generally observed much lower than the theoretical or estimated fluxes and hence the PR is generally lower for all the draw solutes reported in the literature. Table 17.3 indicates that the maximum PR was only about 16% with DI as FS and 13% with BW as FS. This shows that only 16% or 13% of the bulk osmotic pressure difference between the DS and FS is available for effective generation of the driving force. This reduction in the effective driving force is mainly attributed to CP effects. The details on the types of CP effects that influences the FO process has been covered in other chapters.

Since the FDFO performances discussed in this chapter are mainly conducted in the FO mode of operation (i.e., the membrane support layer facing the DS and the active layer facing the FS), the lower water flux obtained in the experiment mainly due to concentrative external CP (ECP) and dilutive internal CP (ICP) effects.

When DI water is used as FS, only dilutive ICP is responsible for reducing the water flux. However, when NaCl is used as FS, both concentrative ECP and dilutive ICP are responsible for lowering the water flux. This is one of the main reasons why the PRO with the NaCl as FS is still lower than DI as FS. In fact, dilutive ICP (since ECP = 0 when DI water is used as FS) has been blamed as the major cause of lower water flux in the FO process because this CP phenomenon occurs inside the membrane support layer and cannot be simply mitigated by the hydrodynamic operational conditions (Gray et al. 2006; McCutcheon and Elimelech 2006; Achilli et al. 2010; Tang et al. 2010; Xu et al. 2010; Zou et al. 2011).

The maximum recovery rate at which the FDFO desalination process can be operated using brackish water (5,000 mg/L NaCl) and seawater (35,000 mg/L NaCl) is presented in Table 17.4. The calculations were made based on the maximum osmotic pressure that a particular fertiliser can generate in solution at its solubility limit. The movement of water across the membrane towards the DS will occur until the osmotic equilibrium is reached between the DS and the FS (Phuntsho et al. 2011b; Phuntsho 2012; Phuntsho et al. 2012b, 2014) irrespective of the rate of water transport across the membrane.

The data in Table 17.4 shows that the feed recovery rate of the FO process varies with the types of fertilisers used and the feed osmotic pressure. Theoretically, it is possible to achieve 100% feed recovery rate if the DS can generate osmotic pressure higher than the maximum solubility of the sodium chloride solution (6.15 M) in the FS with an osmotic pressure of 404 atm. For example, NH_4NO_3, $Ca(NO_3)_2$ and $NaNO_3$ all can easily generate osmotic pressure in excess of 404 atm and therefore, the use of these fertilisers as DS for FO desalination can theoretically achieve a 100% recovery rate. Other fertilisers, such as $(NH_4)_2SO_4$, can achieve about 94% and 87% recovery rates with brackish water and seawater, respectively. Most fertiliser DS listed in Table 17.4 can in fact achieve more than 90% recovery rate with brackish water except for KNO_3, with only about 85% due to its low solubility (4.03 M). The recovery rates with seawater for all fertilisers are higher than 80%, except with KNO_3.

However, it is important to note that 100% recovery is impractical; salt precipitation or crystallisation and membrane scaling could undermine the FO performance. Precipitation could occur earlier because in practice, the saline water from the natural sources can include many other dissolved elements such as Ca, Mg, etc., which have different solubility rates. Other issue such as increased pumping energy may also become significant due to increase in the viscosity of the feed water at higher concentrations.

17.6 LIMITATIONS AND OPTIONS FOR FDFO DESALINATION PROCESSES

Ideally, the FO process offers several novelties for desalination in comparison to conventional membrane processes such as RO. However, FO still faces several technological barriers, and thus, it has not been able to compete on a commercial

*Table 17.4. Theoretical recovery rates of the FDFO process using selected fertilisers as the DS**

Fertiliser	Max. solubility of DS	π at max solubility (atm)	NaCl at OE (M)	BGW5 (%)	BGW10 (%)	BGW20 (%)	BGW35 (%)
NH_4NO_3	101.9 M	3346.1	6.15	100	100	100	100
$Ca(NO_3)_2$	22.04 M	601.0	6.15	100	100	100	100
$NaNO_3$	10.95 M	417.9	6.15	100	100	100	100
NH_4Cl	7.35 M	356.0	5.59	98.5	96.9	93.9	89.3
Urea	19.65 M	338.4	5.38	98.4	96.8	93.6	88.8
DAP	7.13 M	293.0	4.82	98.2	96.4	92.9	87.6
SOA	5.8 M	274.8	4.58	98.1	96.3	92.5	86.9
KCl	4.82 M	226.5	3.94	97.8	95.7	91.3	84.8
MAP	4.56 M	181.3	3.29	97.4	94.8	89.6	81.8
KNO_3	4.03 M	101.4	2.02	95.7	91.5	83.1	70.3
KH_2PO_4	1.80 M	58.0	1.22	93.0	86.0	72.0	50.9

*Calculations were made based on the maximum osmotic pressure a fertiliser DS can generate at its highest solubility. The solubility and the osmotic pressure were predicted using OLI Stream Analyser 3.2.

scale with other desalination technologies. While some of the shortcomings elaborated below are applicable to the FO process in general, some are specific to FDFO desalination processes.

17.6.2 Issues of Salt Rejection and Reverse Movement of Draw Solutes

In the FO process, solutes are present on both sides of the membrane, and therefore, movement of solutes occurs in both the directions. Since none of the synthetic membranes are ideal semi-permeable membranes, the solute rejection is therefore not 100% (Phillip et al. 2010). The forward movement of feed solute is measured in terms of salt rejection rate as in any other pressure-based membrane process, and the reverse movement of draw solutes by diffusion is measured in terms of reverse solute flux (RSF). While the permeation of salute from the feed side towards the DS could occur because of both diffusive flux and convective force, however RSF occurs mainly because of diffusion since the convective force occurs in opposite direction to the RSF. In FDFO desalination, reverse movement of draw solutes is not only an economic loss, but also the presence of nutrients in the feed concentrate could make the concentrate management more complex as nitrogen (N) and phosphorous (P) can cause eutrophication and algal blooms in the receiving water (Hails 2002; Jickells 2005). If the feed rejection rates are low, the feed solute would permeate through the membrane and reach the diluted fertiliser solution. The presence of excess salts such as NaCl in the FDFO product water will also mean the potential to cause sodium and chlorine toxicity to the plants. It is therefore important that the DS with low reverse solute flux is identified and selected for application. However, one of the advantages of the FO process is that the FO process can achieve very high recovery rates without consuming significant additional energy (Martinetti et al. 2009) and therefore concentrate could be managed using simple methods such as evaporation ponds for salt recovery especially for inland FDFO desalination.

The feed salt rejection and RSF depends on both the membrane properties and the draw solute properties. The existing commercial CTA FO membrane has low working pH range, low pure water permeability and low salt rejection which are a cause of concern for the FO process (Yip et al. 2010). However, recently few new high performing polyamide-based thin film composite FO membranes have been commercialised which have not only higher water flux but also higher rejection and RSF compared to the CTA membrane.

In other studies (Achilli et al. 2010; Phuntsho et al. 2011b), the assessment of DS performance is usually done in terms of reverse solute flux or specific reverse solute flux (J_s/J_w or draw solute flux/water flux); however, a slightly different approach of assessment is adopted here. For fertilisation, the amount of essential nutrients present in the fertiliser is more important than the other non-essential elements or species present. Therefore, the reverse solute flux in FDFO is measured in terms of the loss of macronutrients (NPK) per unit volume of water

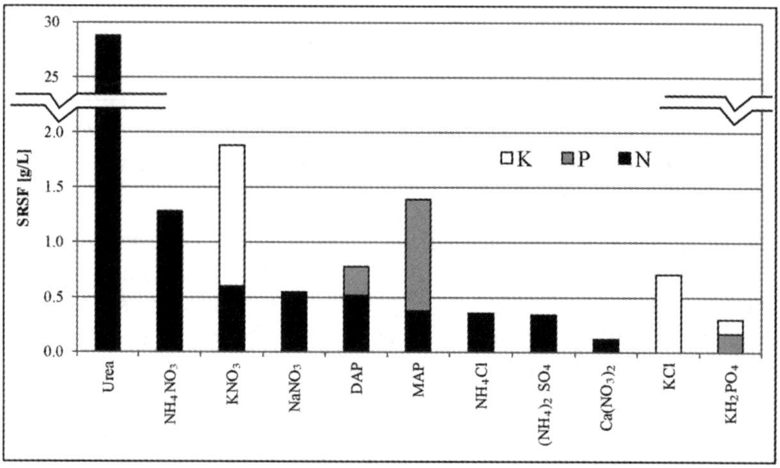

Figure 17.5. Reverse diffusion of draw solutes measured in terms of specific reverse solute flux (J_s/J_w) which is the ratio of reverse solute flux to the water flux of the selected fertilisers

extracted from the feed instead of the whole fertiliser compound (Phuntsho 2012; Phuntsho et al. 2012b). The NPK loss depends on the properties of the fertiliser salts in the solution such as their ionic charge, diffusion coefficient, valency, hydrated diameter, etc.

Figure 17.5 shows the comparative losses of NPK nutrients through reverse diffusion of different fertilier draw solutes. The results show that urea as a DS results in a very high loss of nutrient (29.2 g/L of N). High reverse solute flux of urea is due to its low molecular size, and also because urea remains neutral or uncharged in water solution. Uncharged species or low molecular weight compounds more readily diffuse through the membrane by diffusion (Hancock and Cath 2009). Besides high loss of nutrient by reverse diffusion, urea also has significantly lower water flux and, therefore, urea alone is not an ideal draw solute for the FDFO desalination process. Ammonium nitrate has the second highest loss of N after urea, probably because of the small hydrated diameter of the NH_4^+ and NO_3^- ions that could easily reverse diffuse through the FO membrane. Other monovalent fertiliser salts such as KNO_3, $NaNO_3$ and DAP all showed similar loss of N but lower compared to urea and NH_4NO_3.

However, fertilisers such as ammonium sulphate, MAP, DAP, and calcium nitrate, containing multivalent ions or ions with larger hydrated sizes (such as SO_4^{2+}, PO_4^{3+}, HPO_4^{2-}, Ca^{2+}) showed much lower reverse solute flux than the DS containing monovalent ions or ions with smaller hydrated size (Hancock and Cath 2009; Achilli et al. 2010; Phuntsho et al. 2011b). Calcium nitrate showed the lowest loss of nutrient by reverse diffusion (0.122 g/L of N) amongst all the fertilizer draw solutes in terms of nutrient concentration. In general, the ammonium compounds of sulphate and phosphate, and calcium nitrate containing monovalent elements, performed well in terms of SRSF since lower RSF is preferred for any FO process.

17.6.3 Challenges in Meeting Irrigation Water Quality Standards and their Options

The net transfer of water across the membrane cannot occur beyond the point of osmotic equilibrium between the DS and the FS. Without external intervention, DS cannot be diluted beyond the point of osmotic equilibrium (Phuntsho et al. 2014), and hence the point of osmotic equilibrium is dictated by osmotic pressure of the FS. The lowest concentration that the final diluted DS can achieve is equivalent to the initial feed concentration under counter-current crossflow mode and equivalent to the final feed concentration under co-current crossflow mode of FO operation. Based on the concept of osmotic equilibrium in the FO process, the following relationship provides an approximate volume of water a kilogram of draw solutes can extract:

$$V = \frac{1000}{M_w}\left[\frac{1}{C_{D,E}} - \frac{1}{C_{D,\max}}\right] \tag{17.1}$$

where $C_{D,E}$ is the molar concentration of the DS that generates equal bulk osmotic pressure (osmotic equilibrium condition) with the osmotic pressure of a FS; M_w is the molecular weight of the draw solute; and $C_{D,\max}$ is the maximum solubility of the draw solute. Eq. (17.1) indicates that the volume of water that a kilogram of solute can extract depends on the molecular weight of the draw solute used, the molar concentration of the DS at which it generates an equal osmotic pressure as the FS, and the maximum solubility of the draw solute. Eq. (17.1) also indicates that the water extraction capacity of the draw solute does not depend on the feed recovery rates when the FO process is operated under counter-current mode crossflow mode (Phuntsho et al. 2014).

Figure 17.6 shows the variation of water extraction capacity of different fertiliser draw solutes and NaCl as DS with the feed TDS (in g/L of NaCl) obtained using eq. (17.1) under counter-current crossflow direction of operation mode. When the FO process is operated under the counter-current crossflow direction, the water extraction capacity is not affected by the feed recovery rate, while under the co-current crossflow condition, the water extraction capacity decreases with increasing the feed recovery rate, even using the same FS concentrations (Phuntsho 2012, Phuntsho et al. 2014). The water extraction capacities of all the draw solutes decrease exponentially with an increase in feed TDS. Each fertiliser has its own unique property that makes certain fertilisers more suitable for high TDS feed while some are more suitable for feed with lower TDS (Phuntsho et al. 2014). While all the three fertilisers show a similar water extraction capacity at feed TDS lower than 5 g/L, $(NH_4)_2SO_4$ and $NH_4H_2PO_4$ could be more effective for higher TDS feed than the KNO_3 as DS. NaCl could be effective for all ranges of

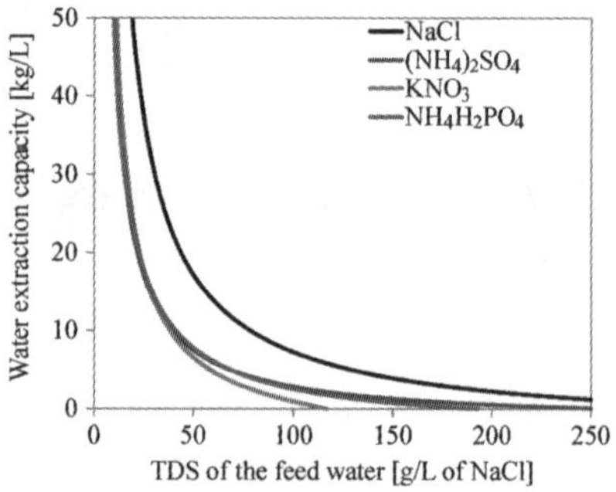

Figure 17.6. Variation of water extraction capacities of the draw solutes by FO processes at different feed TDS: (a) NaCl under different crossflow modes and (b) other selected draw solutes under counter-current crossflow mode

feed TDS, indicating the prospects of RO concentrate or brine for use as DS for a wide range of FO applications. NaCl, being the draw solute with the lowest M_w, exhibits the highest water extraction capacity amongst the four selected draw solutes.

Based on the water extraction capacity discussed earlier, the final expected nutrient concentrations of some of the selected commonly-used fertilisers when operated under different feed water TDS are calculated and presented in Table 17.5. These calculations were made based on the osmotic equilibrium between the osmotic pressures of DS and FS. The osmotic pressures were calculated using OLI Stream Analyzer software (OLI Inc.) It is clear from the table that there is a limit to which a particular type of DS could achieve its final concentration, and this will depend on the TDS or the osmotic pressure of the feed water. The lower the salinity or TDS of the feed water, the lower will be the final DS concentration of the fertiliser DS.

When NH_4NO_3 is used as the DS and seawater as the FS, the final NH_4NO_3 concentration after FO desalination would be 0.802 M, which contains an actual N concentration of 22.5 g/L. Likewise, $(NH_4)_2SO_4$, $Ca(NO_3)_2$ and DAP could contain 16.4, 15.1 and 14.5 g/L of N nutrient, respectively. The lowest N concentration was observed for MAP and NH_4NCl, with only little more than 8 g/L with seawater feed. The P concentration was also observed remained high, with 19.3 and 16.0 g/L for MAP and DAP, respectively. The potassium (K) nutrient was by far the highest amongst the major NPK nutrients, with 24.4 and 27.6 g/L for KCl and KNO_3, respectively. Although the required nutrient concentration for fertigation would vary depending on many factors such as types of crops to be irrigated, cropping seasons, soil nutrient conditions, etc. (Oliver and Barber 1966), the final concentrations obtained in Table 17.5 are too high for direct fertigation. For example, the required nutrient concentration varies from 50 to 200 mg/L for N, 12 to 60 mg/L for P, and 15 to 250 mg/L for K, depending on the types of crops and growing seasons (Phocaides 2007). Therefore, the data in Table 17.5

Table 17.5. Estimated nutrient concentrations (in mg/L) in the final FDFO desalination product water evaluated in terms of the three major nutrients concentrations NPK.

DS	BW5 N/P/K (mg/L)	BW10 N/P/K (mg/L)	BW20 N/P/K (mg/L)	BW35 N/P/K (mg/L)
MAP	1,200/2,660/0	2,420/5,360/0	4,900/10,800/0	8,900/19,700/0
$NaNO_3$	1,210	2,460/0/0	5,000	9,300
NH_4Cl	1,210	2,430/0/0	4,900	9,000
KNO_3	1,230/0/3,430	2,52/0/7,04	5,300/0/14,700	10,100/0/28,200
DAP	1,820/2,010/0	3,75/4,15/0	7,800/8,600/0	14,700/16,300/0
$Ca(NO_3)_2$	1,870	3,89/0/0	8,200	14,800
SOA	2,020	4,27/0/0	9,000	16,700
NH_4NO_3	2,670	5,66/0/0	12,100	23,000
Urea	4,500	8,99/0/0	18,200	33,300
KCl	0/0/3,360	0/0/6,78/	0/0/13,700	0/0/24,900
KH_2PO_4	0/2,720/3,440	0/5,60/7,07	0/11,800/14,800	0/22,700/28,700

MAP: Mono-ammonium phosphate
DAP: Diammonium hydrogen phosphate

with seawater as feed, indicate that a significant amount of water is required to further dilute the final fertiliser DS before fertigation.

Therefore, meeting the water quality in terms of the final nutrient concentrations for direct fertigation is one of the significant challenges of the FDFO desalination process. Since maintaining the required nutrient concentration is essential for fertigation, an additional process could be integrated with the FO unit. Although such a unit requires energy, it is nevertheless essential for maintaining a low final nutrient concentration and acceptability for direct fertigation. Therefore, such unit must have low energy consumption so that the total energy cost of the irrigation water remains comparatively lower than the conventional desalination processes making the desalination feasible for irrigation purpose. There are several possible options discussed below that could reduce the final nutrient concentrations.

17.6.3.1 Direct Dilution Method

Direct dilution of the final diluted fertiliser DS is one of the easiest options one can adopt for the FDFO process as shown in Figure 17.7(a). The feasibility of such option therefore would depend on many factors such as the dilution factor (diluted fertiliser solution to fresh water ratio) and more significantly, the availability of the fresh water itself. In those places where the fresh water sources are not available and where FDFO desalination is to be used as the main irrigation water source, such option will not be meaningful. However, where FDFO is to be

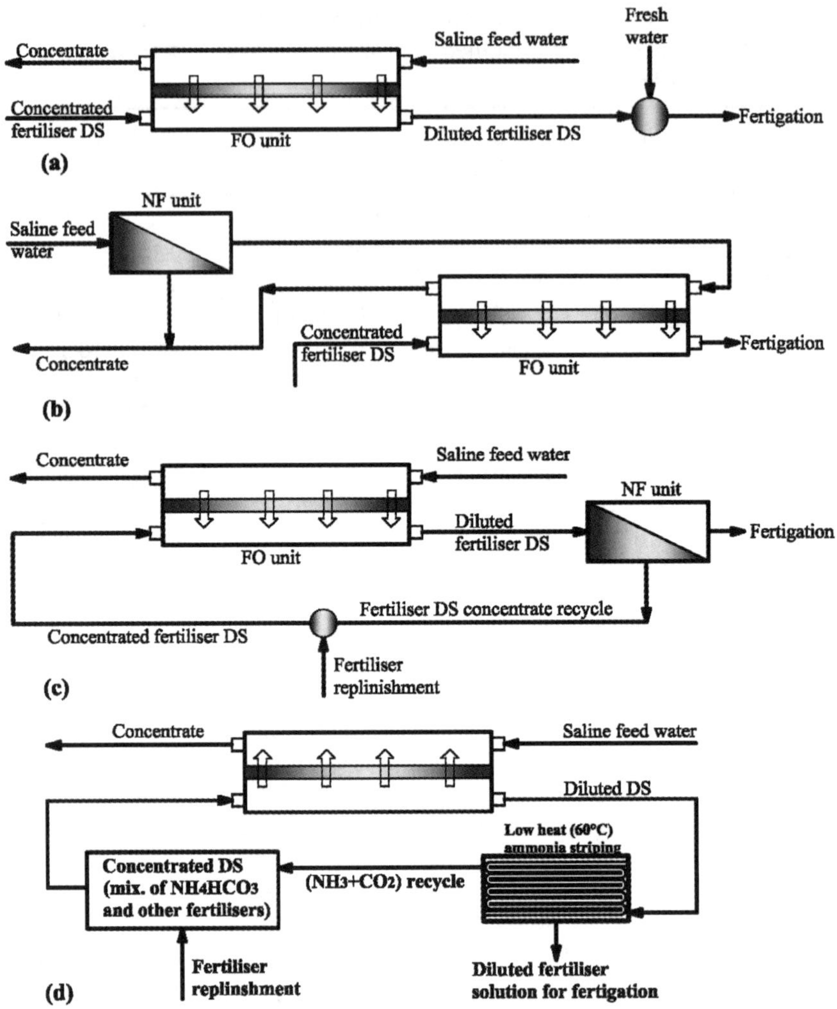

Figure 17.7. Various options for obtaining acceptable water quality for fertigation from the FDFO desalination process: (a) direct dilution using available fresh water sources; (b) using NF as a pre-treatment process to reduce feed TDS; (c) using NF as post-treatment process to remove the excess fertiliser solutes for recycling; and (d) using a DS containing mixture of non-volatile soluble fertiliser and NH_4HCO_3

used only as an additional source of water to supplement or augment the limited fresh water resources, this option is suitable.

17.6.3.2 Pre-treatment of Feed Water

The water extraction capacity and hence the final nutrient concentration in the diluted DS from the FDFO desalination process depends on the TDS of the feed water. Nanofiltration (NF) pre-treatment has been found very effective in reducing the TDS of the brackish groundwater (BGW). Multivalent ions such as Ca^{2+}, Mg^{2+}, SO_4^{2+}, etc. are characteristics of most brackish groundwater and hence can be quite effective in its removal including some percentage of monovalent ions. NF has been advantageously used as pre-treatment process to seawater RO desalination. NF has low operating pressure and energy cost, high flux, high retention of multivalent anion salts and relatively low investment and low operation and maintenance costs (Hassan et al. 1998; Lu et al. 2002; Van der Bruggen and Vandecasteele 2002). The reduced TDS after pre-treatment in turn can achieve much lower nutrient concentration that is acceptable for direct fertigation in the FDFO desalination process. The process layout for this option is presented Figure 17.7(b).

Most fertilisers were able to achieve nutrient concentrations significantly lower and quite closer to acceptable nutrient concentrations for direct fertigation when NF was used as pre-treatment of BGW before the FDFO process (Phuntsho et al. 2013a). The other important advantage of pre-treatment is that, NF pre-treatment significantly improves the feed water quality as most of the scaling and fouling components are removed. With high quality feed water, the FDFO process could to be operated in PRO mode of operation (support layer facing the pre-treated feed water instead of fertiliser DS) where water flux is usually observed significantly higher than in FO mode of operation (Phuntsho et al. 2013a).

17.6.3.3 Post-treatment Using Nanofiltration Processes

NF can also be used as post treatment process to reduce the fertiliser salt concentrations in the diluted fertiliser DS from the FDFO process as shown in Figure 17.7(c). Recovery and recycling of other draw solutes by the NF process have been investigated (Tan and Ng 2010; Zhao et al. 2012a). NF was found to be effective in reducing the fertiliser nutrient concentrations from the diluted fertiliser DS (Phuntsho 2012; Phuntsho et al. 2013a). NF has been found more effective for fertiliser DS containing multivalent ions particularly $(NH_4)_2SO_4$, $NH_4H_2PO_4$ and $(NH_4)_2HPO_4$ as they are highly rejected by negatively charged polyamide NF membrane surface. Although $Ca(NO_3)_2$ is a divalent ion as well, rejection by NF membrane is lower because of the affinity of the Ca^{2+} ions to absorb on the negatively charged membrane surface. Monovalent fertilisers have been observed to have lower rejection rates and hence the diluted fertiliser solution will have to be treated through second NF pass to achieve acceptable nutrient concentrations before fertigation. The NF concentrate can be recycled back as DS for the FDFO desalination process to extract more water from the feed water. The other advantage of the post-treatment is that, NF can operate more efficiently since the final diluted DS contains only dissolved fertiliser, and any foulants present in the feed water could have been removed during the FDFO process.

A comparative study on NF as a pre-treatment or post-treatment option indicates that NF post-treatment could be more advantageous in terms of process efficiency and the energy consumption because of the high quality feed received for the NF from the FDFO process.

17.6.3.4. Post-treatment by Thermal Distillation

Another potential post treatment alternative is to use fertilisers with thermolyte properties such as ammonium bicarbonate, ammonium carbonate and ammonium carbamate. In such case, a certain amount of thermolyte draw solutes from the final FDFO product water can be recovered and recycled for reuse of the DS. The DS can be prepared containing thermolyte solutions such as mixture of ammonium bicarbonate or ammonium carbonate or ammonium carbamate (McGinnis 2002; McCutcheon et al. 2005, 2006; McGinnis and Elimelech 2007) which can be further mixed with other essential fertilisers in specific proportions to meet specific crop requirements. These ammonium compounds would provide N nutrient while other macro nutrients such as P or K can be provided by mixing together other fertilisers. The final diluted DS containing a mixture of thermolyte solutions and other fertilisers after FDFO desalination can then be heated up to 60°C to recover certain percentage of ammonium compounds in the form of ammonia-carbon dioxide which can be regenerated and recycled back as concentrated DS to further extract water. This recovery concept was first developed by McGinnis (2002) and later reported in many other studies. Since the final product water still needs to contain certain amount of ammonia N nutrients, only certain fraction of the final diluted DS may be required to undergo ammonia-carbon dioxide recovery and regeneration process. The final product water after recovering and recycling ammonia-carbon dioxide will contain significantly lower nutrient concentration which can be directly used for fertigation. Since only a certain percentage of ammonium compound is required to be removed from the diluted DS for recovery and recycling, the diluted DS may not be required heating up to 60°C because ammonium bicarbonate decomposition starts at about 35°C (Gokel 2004), and therefore, the amount of energy required is expected to be much lower than recovering 100% as required for drinking water. Figure 17.7(d) shows the concept of the FDFO desalination process with ammonia-carbon dioxide recovery and recycling process.

17.6.3.5 Hybrid FO Systems

The other alternatives to achieve lower dilution of fertiliser DS from a FDFO desalination process are to use hybrid FO systems as shown in Figures 17.8 and 17.9. Figure 17.8 shows the conceptual process diagram of the FDFO desalination using impaired water as the source of water for further dilution. Wastewater effluent can be used for dilution of fertiliser solution, either directly if the effluent meets the irrigation water quality standards, or after further treatment by the FO process. The concept involves using a two-stage FO process to provide a multiple barrier for simultaneous wastewater treatment and desalination of seawater through osmotic dilution (Cath et al. 2010). The first stage FO process desalinates the saline water using fertiliser as the DS, and the diluted fertiliser DS then goes to the second stage FO process to extract more water from the impaired water source containing low TDS or osmotic pressure. The second stage FO process could offer dual advantages of treating wastewater effluent while at the same time, providing

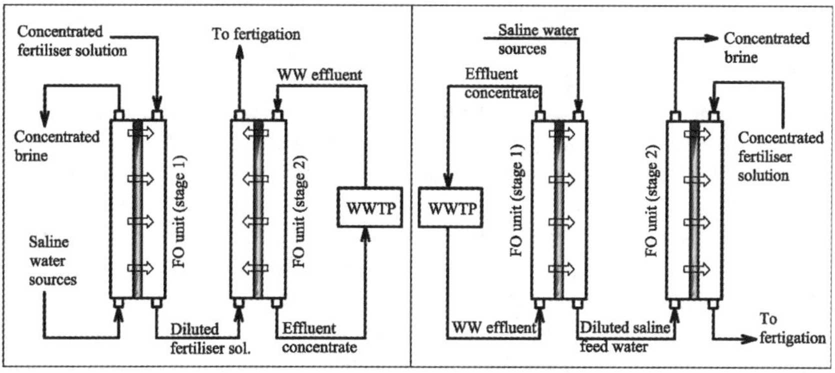

Figure 17.8. FDFO desalination process using 2 stage FO process with additional dilution water from a secondary wastewater effluent and limited river water source

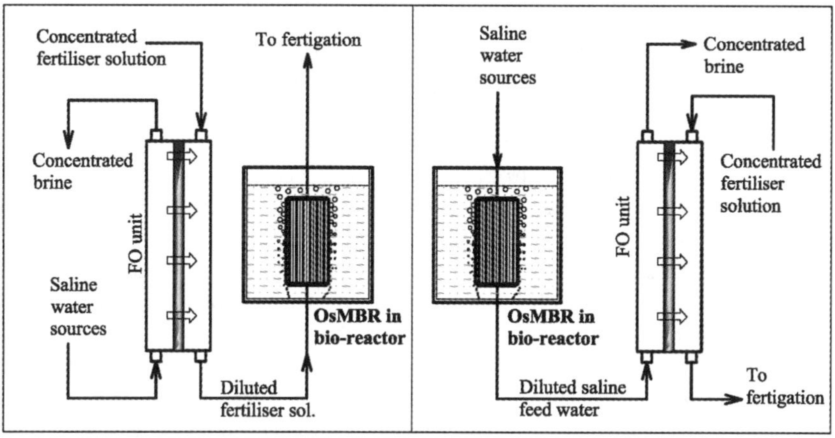

Figure 17.9. FDFO-OsMBR hybrid system for diluting the fertiliser concentration

further dilution to the fertiliser solution so that the final diluted fertiliser DS can be applied directly for fertigation. The alternative arrangement could be to use the saline water as DS in the first stage of the FO unit to extract water from the impaired water. The diluted saline water can then be used as the feed water in the second stage FO unit, with concentrated fertiliser as the DS. In both the process designs, the feed recovery rate could not only increase but also achieve a low final fertiliser concentration in the diluted DS. In the second option, the amount of fertiliser lost due to reverse solute flux would be less than the first option since the fertiliser is used as the DS only once, whereas in the first option, the fertiliser is used as the DS twice.

Similarly, FDFO can be integrated with the membrane bioreactor (MBR). The osmotic MBR (OMBR) has been recently studied (Achilli et al. 2009; Xiao et al. 2011). Since the water quality for irrigation is much lower than the drinking water, the potential application for OMBR with FDFO is quite significant. There are two possible combinations for the Osmotic MBR or OMBR with FDFO (Figure 17.9). The OMBR can be used as a post treatment process after the FDFO desalination process using the saline feed water. The diluted fertiliser solution is then further uses as a DS in the OMBR process. Since the salinity or the osmotic pressure of the MBR reactor is significantly lower than the osmotic pressure of the diluted fertiliser solution from the FDFO process, it still can draw water during the OsMBR process thereby further diluting the fertiliser solution. Alternatively, OsMBR can be used as a pre-treatment process in which the saline water can be used as the DS to extract from the wastewater by OMBR in the process diluting the saline water. This diluted saline water can be further used as feed water for the FDFO process in the process achieving much lower nutrient concentration in the final diluted fertiliser DS.

17.6.3.6 Blending of Fertilisers Containing Different Nutrients

The expected final NPK nutrient concentrations from a FDFO desalination process using a feed TDS of 5,000 mg/L NaCl and single fertiliser compound is presented in Figure 17.10. These expected final NPK nutrients are significantly higher than the allowable nutrient concentrations for any plants. One of the important observations noted from the final nutrient concentrations in Figure 17.9 is that those fertilisers containing a lower percentage of a particular nutrient element due to the presence of other elements in the compound result in lower final nutrient concentrations. While observing the final nutrient N concentrations of different fertiliser DS with the percentage composition of N in each fertilizer indicates a fairly good correlation between the percentage of N in different fertilisers and their N concentration in the final FDFO product water (Phuntsho 2012; Phuntsho et al. 2012b). This indicates that if the percentage of each fertiliser nutrient is reduced in the DS, the nutrient concentration in the final FDFO product water is also expected to be lower. One of the methods to reduce the percentage composition of the particular nutrient is by blending or mixing different types of fertiliser containing different nutrient species in a single DS.

Several different types of bleneded fertiliser solutions were prepared and tested as DS for the FDFO desalination process and found to be effective in reducing the concentration of a particular nutrient in the final diluted DS (Phuntsho 2012; Phuntsho et al 2012b). The presence of more numbers of different elemental species in the blended DS can provide adequate osmotic pressure at lower fertiliser concentration. Figure 17.10(b) shows that blending three different fertilisers ($NH_4H_2PO_4$, KCl and NH_4NO_3) to prepare a DS containing NPK nutrients could significantly reduce the concentration of each nutrient. The other advantage of blended fertiliser DS is that the blend ratio can be adjusted to meet the nutrient ratio required by the particular plant in a particular season and soil condition. Fertilisers can also be blended or mixed with other

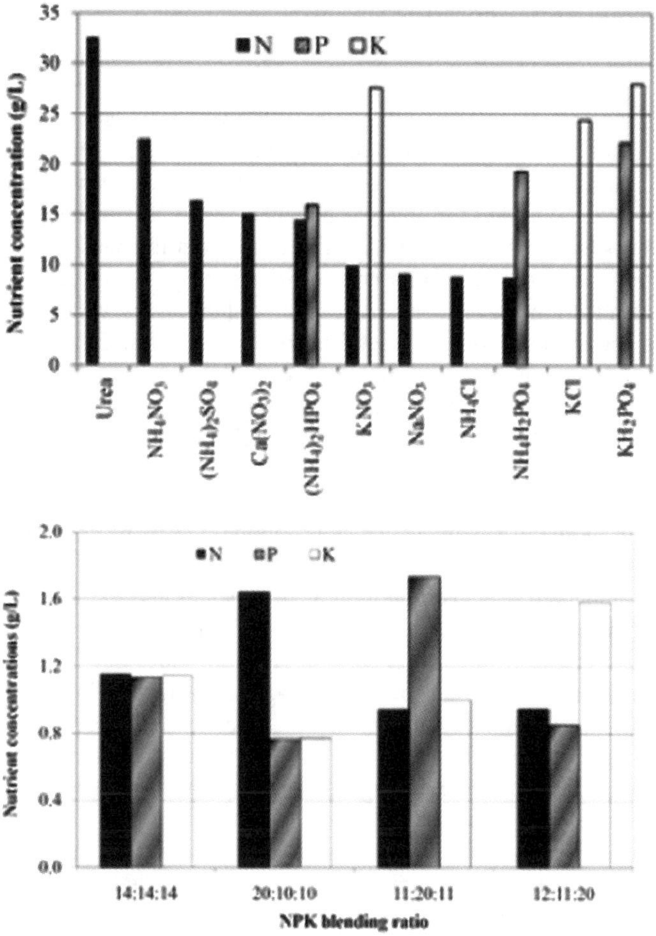

Figure 17.10. Expected final NPK nutrient concentrations of fertilizer solution after FO desalination or in the final FDFO product water using (a) brackish water as feedwater (5000 mg/L NaCl at 3.9 atm) and (b) MAP, KCl, and NH_4NO_3 blended in different ratios.

soluble agricultural chemicals such as pesticides/insecticides, fungicides, etc., to reduce the final nutrient concentration in the DS. The presence of other ionic or non-ionic species could increase the osmotic potential of the DS and hence lower the concentration of essential nutrient elements in the final DS. However, such mixtures should be compatible with the membranes as well as meet the fertigation requirements. Agricultural chemicals are often applied through an irrigation system, and as such, terms such as chemigation, fungigation, etc. are popular with the fertigation system.

17.7 CONCLUDING REMARKS

The FO process is a promising and emerging low-energy desalination technology. FO desalination for potable water, however, still suffers from lack of an ideal draw solute that can be easily recovered and regenerated without significant energy consumption. The novelty offered by the FO process has been advantageously applied in the desalination for non-potable purposes such as irrigation where the diluted draw solution (DS) can be used directly without the need for the recovery and regeneration process. In this chapter, the concept of fertiliser drawn FO (FDFO) desalination, its opportunities and its limitations were all discussed. Given that the fertiliser contains useful nutrients for plants, the diluted fertiliser draw solution after desalination can be directly applied for fertigation, thereby avoiding the need for separation and recovery of the DS. The energy required for fertiliser driven FDFO desalination has been estimated to be comparatively lower than the current state-of-the-art RO desalination technologies. Being a low energy process, the FDFO process should also be easily powered by the renewable energy, such as solar and wind energy, which exits abundant in many arid regions of the world.

Most soluble fertilisers can be used as draw solutions, and therefore extract water from saline water with very high recoveries. Since fertilisers are extensively used for agricultural production, FDFO desalination does not create additional environmental issues related to fertiliser usage. In fact, the FDFO desalination could add more value to irrigation water and, therefore, provide more opportunities for improving the efficiencies of water and fertiliser uses. Such technology could provide irrigation water through alternative sources such as saline groundwater and wastewater effluent.

The major limitations of the FDFO desalination process, however, are the challenge in meeting the irrigation water quality standards in terms of nutrient concentrations, limiting the direct use of FDFO product water for fertigation. Several options have been discussed that could be integrated with the FDFO desalination process in order to reduce the final nutrient concentration closer to acceptable limit. The options include integrating NF as either pre-treatment or post-treatment process, adopting hybrid two stage FDFO processes for both saline water and wastewater effluent, integrating FDFO with osmotic MBR, using mixture of thermolyte and fertiliser solutions and using blended fertilisers as draw solute.

17.8 ABBREVIATIONS/NOMENCLATURE

BGW	brackish groundwater
CP	concentration polarization
CTA	cellulose triacetate
DS	draw solution

ECP	external concentration polarisation
FO	forward osmosis
FDFO	fertiliser drawn forward osmosis
FS	feed solution
ICP	internal concentration polarisation
MAP	monoammonium phosphate
MBR	membrane bioreactor
NF	nanofiltration
OMBR	osmotic membrane bioreactor
PR	performance ratio
RSF	reverse solute flux
RO	reverse osmosis
SWRO	seawater reverse osmosis desalination

17.9 ACKNOWLEDGMENTS

This work is supported by the National Centre of Excellence in Desalination Australia (NCEDA), which is funded by the Australian Government through Water for the Future initiative.

References

Achilli, A., Cath, T. Y., and Childress, A. E. (2010). "Selection of inorganic-based draw solutions for forward osmosis applications." *Journal of Membrane Science*, 364(1–2), 233–241.

Achilli, A., Cath, T. Y., Marchand, E. A., and Childress, A. E. (2009). "The forward osmosis membrane bioreactor: A low fouling alternative to MBR processes." *Desalination*, 239(1–3), 10–21.

Achilli, A. and Childress, A. E. (2010). "Pressure retarded osmosis: From the vision of Sidney Loeb to the first prototype installation–Review." *Desalination*, 261(3), 205–211.

Bamaga, O. A., Yokochi, A., and Beaudry, E. G. (2009). "Application of forward osmosis in pretreatment of seawater for small reverse osmosis desalination units." *desalination and Water Treatment*, 5(2009), 183–191.

Bamaga, O. A., Yokochi, A., Zabara, B., and Babaqi, A. S. (2011). "Hybrid FO/RO desalination system: Preliminary assessment of osmotic energy recovery and designs of new FO membrane module configurations." *Desalination*, 268, 163–169.

Cath, T. Y., Adams, D., and Childress, A. E. (2005). "Membrane contactor processes for wastewater reclamation in space: II. Combined direct osmosis, osmotic distillation, and membrane distillation for treatment of metabolic wastewater." *Journal of Membrane Science*, 257(1–2), 111–119.

Cath, T. Y., Childress, A. E., and Elimelech, M. (2006). "Forward osmosis: Principles, applications, and recent developments: Review." *Journal of Membrane Science*, 281 (2006), 70–87.

Cath, T. Y., Drewes, J. E., and Lundin, C. D. (2009). A novel hybrid forward osmosis process for drinking water augmentation using impaired water and saline water sources. WERC and Water Research foundation.

Cath, T. Y., Hancock, N. T., Lundin, C. D., Hoppe-Jones, C., and Drewes, J. E. (2010). "A multi-barrier osmotic dilution process for simultaneous desalination and purification of impaired water." *Journal of Membrane Science*, 362(1–2), 417–426.

Chekli, L., Phuntsho, S., Vigneswaran, S., Kandasamy, J., and Chanan, A. (2012). "A review of draw solutes in forward osmosis process and their use in modern applications." *Desalination and Water Treatment*, 43(1–3), 167–184.

Choi, Y.-J., Choi, J.-S., Oh, H.-J., Lee, S., Yang, D. R., and Kim, J. H. (2009). "Toward a combined system of forward osmosis and reverse osmosis for seawater desalination." *Desalination*, 247(1–3), 239–246.

Chung, T.-S., Li, X., Ong, R. C., Ge, Q., Wang, H., and Han, G. (2012a). "Emerging forward osmosis (FO) technologies and challenges ahead for clean water and clean energy applications." *Current Opinion in Chemical Engineering*, 1(3), 246–257.

Chung, T.-S., Zhang, S., Wang, K. Y., Su, J., and Ling, M. M. (2012b). "Forward osmosis processes: Yesterday, today and tomorrow." *Desalination*, 287(2012), 78–81.

Crowley, T. J. (2000). "Causes of climate change over the past 1000 years." *Science*, 289 (5477), 270–277.

Elimelech, M., and Phillip, W. A. (2011). "The future of seawater desalination: Energy, technology, and the environment." *Science*, 333(6043), 712–717.

FIFA. (2009). *Fertiliser Industry environment report 2008*, Canberra ACT 2601, Australia.

Garcia-Castello, E. M., and McCutcheon, J. R. (2011). "Dewatering press liquor derived from orange production by forward osmosis." *Journal of Membrane Science*, 372(1–2), 97–101.

Ge, Q., Ling, M. and Chung, T.-S. (2013). "Draw solutions for forward osmosis processes: Developments, challenges, and prospects for the future." *Journal of Membrane Science*, 442(0), 225–237.

Geise, G. M., Lee, H.-S., Miller, D. J., Freeman, B. D., McGrath, J. E., and Paul, D. R. (2010). "Water purification by membranes: The role of polymer science." *Journal of Polymer Science Part B: Polymer Physics*, 48(15), 1685–1718.

Gokel, G. W. (2004). *Dean's Handbook of Organic Chemistry*, Second edition, pp 2.4–2.5.

Goss, K. F. (2003). "Environmental flows, river salinity and biodeversity conservation: managing trade-offs in the Murray-Darling basin." *Australian Journal of Botany*, 51, 619–625.

Gray, G. T., McCutcheon, J. R., and Elimelech, M. (2006). "Internal concentration polarization in forward osmosis: role of membrane orientation." *Desalination*, 197(1–3), 1–8.

Greenlee, L. F., Lawler, D. F., Freeman, B. D., Marrot, B., and Moulin, P. (2009). "Reverse osmosis desalination: Water sources, technology, and today's challenges." *Water Research*, 43(9), 2317–2348.

Hails, R. S. (2002). "Assessing the risks associated with new agricultural practices." *Nature*, 418(6898), 685–688.

Hancock, N. T. and Cath, T. Y. (2009). "Solute coupled diffusion in osmotically driven membrane processes." *Environmental Science & Technology*, 43(17), 6769–6775.

Hanson, B. R., Simunek, J., and Hopmans, J. W. (2006). "Evaluation of urea-ammonium-nitrate fertigation with drip irrigation using numerical modeling." *Agric. Water Manage.*, 86(1–2), 102–113.

Hassan, A. M., Al-Sofi, M. A. K., Al-Amoudi, A. S., Jamaluddin, A. T. M., Farooque, A. M., Rowaili, A., Dalvi, A. G. I., Kither, N. M., Mustafa, G. M., and Al-Tisan, I. A. R. (1998). "A new approach to membrane and thermal seawater desalination processes using nanofiltration membranes (Part 1)." *Desalination*, 118(1–3), 35–51.

Holloway, R. W., Childress, A. E., Dennett, K. E., and Cath, T. Y. (2007). "Forward osmosis for concentration of anaerobic digester centrate." *Water Research*, 41(2007), 4005–4014.

Hoover, L. A., Phillip, W. A., Tiraferri, A., Yip, N. Y., and Elimelech, M. (2011). "Forward with osmosis: Emerging applications for greater sustainability." *Environmental Science & Technology*, 45(23), 9824–9830.

Jickells, T. (2005). "External inputs as a contributor to eutrophication problems." *Journal of Sea Research*, 54(1), 58–69.

Jury, W. A. and Vaux, H. (2005). "The role of science in solving the world's emerging water problems." *Proceedings of the National Academy of Sciences of the United States of America*, 102(44), 15715–15720.

Kessler, J. O. and Moody, C. D. (1976). "Drinking water from sea water by forward osmosis." *Desalination*, 18(3), 297–306.

Khan, S. (2008). "Managing climate risks in Australia: options for water policy and irrigation management." *Australian journal of experimental Agriculture*, 48, 265–273.

Kravath, R. E. and Davis, J. A. (1975). "Desalination of seawater by direct osmosis." *Desalination*, 16(1975), 151–155.

Lu, X., Bian, X., and Shi, L. (2002). "Preparation and characterization of NF composite membrane." *Journal of Membrane Science*, 210(1), 3–11.

Martinetti, C. R., Childress, A. E., and Cath, T. Y. (2009). "High recovery of concentrated RO brines using forward osmosis and membrane distillation." *Journal of Membrane Science*, 331(1–2), 31–39.

McBeath, T. M., McLaughlin, M. J., Armstrong, R. D., Bell, M., Bolland, M. D. A., Conyers, M. K., Holloway, R. E., and Mason, S. D. (2007). "Predicting the response of wheat (Triticum aestivum L.) to liquid and granular phosphorus fertilisers in Australian soils." *Australian Journal of Soil Research*, 45(6), 448–458.

McCutcheon, J. R. and Elimelech, M. (2006). "Influence of concentrative and dilutive internal concentration polarization on flux behavior in forward osmosis." *Journal of Membrane Science*, 284(1–2), 237–247.

McCutcheon, J. R., McGinnis, R. L., and Elimelech, M. (2005). "A novel ammonia-carbon dioxide forward (direct) osmosis desalination process." *Desalination*, 174(2005), 1–11.

McCutcheon, J. R., McGinnis, R. L., and Elimelech, M. (2006). "Desalination by ammonia-carbon dioxide forward osmosis: Influence of draw and feed solution concentrations on process performance." *Journal of Membrane Science*, 278(2006), 114–123.

McDonald, R. I., Green, P., Balk, D., Fekete, B. M., Revenga, C., Todd, M., and Montgomery, M. (2011). "Urban growth, climate change, and freshwater availability." *Proceedings of the National Academy of Sciences*, 108(15), 6312–6317.

McGinnis, R. L. (2002). "Osmotic desalination process." US Patent, Pending. PCT/US02/02740 (2002).

McGinnis, R. L. and Elimelech, M. (2007). "Energy requirements of ammonia-carbon dioxide forward osmosis desalination." *Desalination*, 207(1–3), 370–382.

McGinnis, R. L. and Elimelech, M. (2008). "Global challenges in energy and water supply: The promise of engineered osmosis." *Environmental Science & Technology*, 42(23), 8625–8629.

MDBA. (2010). *Guide to the proposed Basin Plan: overview – Murray-Darling Basin*, Canberra, Murray-Darling Basin Authority. Vol: 1.

Moody, C. D. (1977). *Forward osmosis extractors: Theory, feasibility and design optimisation*, School of renewable natural resources. Arizona, The University of Arizona: 150.

Moody, C. D. and Kessler, J. O. (1976). "Forward osmosis extractors." *Desalination*, 18 (1976), 283–295.

Moody, C. D. and Kessler, J. O. (1977). *Reclamation of brackish waters for irrigation: optimization of process parameters for fertilizer-driven osmosis. U. S. NTIS*, PB Rep. Issue PB-277709/, 56 pp. Report 1977. Gov. Rep. Announce. Index (U. S.) 1978, 78(11), 126.

Oliver, S. and Barber, S. A. (1966). "An evaluation of the mechanisms governing the supply of Ca, Mg, K, and Na to soybean roots (Glycine max)1." *Soil Science Society of America Journal*, 30(1), 82–86.

Papadopoulos, I. and Eliades, G. (1987). "A fertigation system for experimental purposes." *Plant and soil*, 102, 141–143.

Phillip, W. A., Yong, J. S. and Elimelech, M. (2010). "Reverse draw solute permeation in forward osmosis: Modeling and experiments." *Environmental Science & Technology*, 44 (13), 5170–5176.

Phocaides, A. (2007). *Handbook on pressurized irrigation techniques*, Rome, Food and Agriculture Organization, UN.

Phuntsho, S., Listowski, A., Shon, H. K., Le-Clech, P., and Vigneswaran, S. (2011a). "Membrane autopsy of a 10 year old hollow fibre membrane from Sydney Olympic Park water reclamation plant." *Desalination*, 271(1–3), 241–247.

Phuntsho, S., Shon, H. K., Hong, S. K., Lee, S. Y., and Vigneswaran, S. (2011b). "A novel low energy fertilizer driven forward osmosis desalination for direct fertigation: evaluating the performance of fertilizer draw solutions." *Journal of Membrane Science*, 375(2011), 172–181.

Phuntsho, S., Shon, H. K., Vigneswaran, S., and Cho, J. (2011c). "Assessing membrane fouling potential of humic acid using flow field-flow fractionation." *J. Membr. Sci.*, 373(1–2), 64–73.

Phuntsho, S. (2012). *A Novel Fertiliser Drawn Forward Osmosis Desalination for Fertigation*, PhD Dissertation, University of Technology Sydney.

Phuntsho, S., Shon, H. K., Hong, S. K., Lee, S. Y., Vigneswaran, S., and Kandasamy, J. (2012a). "Fertiliser drawn forward osmosis desalination: the concept, performance and limitations for fertigation." *Rev. Environ. Sci. Biotechnol.*, 11(2), 147–168.

Phuntsho, S., Shon, H. K., Majeed, T., El Salibya, I., Vigneswarana, S., Kandasamy, J., Hong, S., and Leeb, S. (2012b). "Blended fertilisers as draw solutions for fertiliser drawn forward osmosis desalination." *Environmental Science & Technology*, 46(8), 4567–4575.

Phuntsho, S., Hong, S., Elimelech, M., and Shon, H. K. (2013a). "Forward osmosis desalination of brackish groundwater: Meeting water quality requirements for fertigation by integrating nanofiltration." *Journal of Membrane Science*, 436(0), 1–15.

Phuntsho, S., Mamisahebi, S., Majeed, T., Lotfi, F., Kim, J. E., and Shon, H. K. (2013b). "Assessing the major factors affecting the performances of forward osmosis and its implications on the desalination process." *Chemical Engineering Journal*, 231(2013), 484–496.

Phuntsho, S., Hong, S., Elimelech, M., and Shon, H. K. (2014). "Osmotic equilibrium in the forward osmosis process: Modelling, experiments and implications for process performance." *Journal of Membrane Science*, in press.

Plusquellec, H. (2002). "Is the daunting challenge of irrigation achievable?" *Irrigation and Drainage*, 51(3), 185–198.

Semiat, R. (2008). "Energy issues in desalination processes." *Environmental Science & Technology*, 42(22), 8193–8201.

Seppälä, A. and Lampinen, M. J. (2004). "On the non-linearity of osmotic flow." *Experimental Thermal and Fluid Science*, 28(4), 283–296.

Service, R. F. (2006). "Desalination freshens up." *Science*, 313(5790), 1088–1090.

Shannon, M. A., Bohn, P. W., Elimelech, M., Georgiadis, J. G., Marinas, B. J., and Mayes, A. M. (2008). "Science and technology for water purification in the coming decades." *Nature*, 452(7185), 301–310.

She, Q., Jin, X., and Tang, C. Y. (2012). "Osmotic power production from salinity gradient resource by pressure retarded osmosis: Effects of operating conditions and reverse solute diffusion." *Journal of Membrane Science*, 401–402(0), 262–273.

Stache, K. (1989). "Apparatus for transforming sea water, brackish water, polluted water or the like into a nutritious drink by means of osmosis." US Patent 4, 030.

Subramani, A., Badruzzaman, M., Oppenheimer, J., and Jacangelo, J. G. (2011). "Energy minimization strategies and renewable energy utilization for desalination: A review." *Water Research*, 45(5), 1907–1920.

Tan, C. H. and Ng, H. Y. (2010). "A novel hybrid forward osmosis – nanofiltration (FO-NF) process for seawater desalination: Draw solution selection and system configuration." *Desalination and Water Treatment*, 13(2010), 356–361.

Tang, C. Y., She, W., Lay, W. C. L., Wang, R., and Fane, A. G. (2010). "Coupled effects of internal concentration polarization and fouling on flux behavior of forward osmosis membranes during humic acid filtration." *Journal of Membrane Science*, 354(1–2), 123–133.

UN. (2009a). *Role of desalination in addressing water scarcity / Economic and Social Commission for Western Asia*, New York, United Nations.

UN. (2009b) "World population prospects: The 2008 revision." *Population Newsletter*, 87.

UNFPA. (2011). *State of world population 2011*, Information and External Relations Division, United Nations Population Fund.

Van der Bruggen, B., and Vandecasteele, C. (2002). "Modelling of the retention of uncharged molecules with nanofiltration." *Water Research*, 36(5), 1360–1368.

Wallace, M., Cui, Z., and Hankins, N. P. (2008). "A thermodynamic benchmark for assessing an emergency drinking water device based on forward osmosis." *Desalination*, 227(1–3), 34–45.

Ward, F. A. and Pulido-Velazquez, M. (2008). "Water conservation in irrigation can increase water use." *Proceedings of the National Academy of Sciences*, 105(47), 18215–18220.

Wentworth Group. (2010). "Sustainable diversion in the Murray-Darling basin: An analysis of the options for achieving a sustainable diversion limit in the Murray-Darling basin." Wentworth group of concerned scientists. June 2010.

Xiao, D., Tang, C. Y., Zhang, J., Lay, W. C. L., Wang, R., and Fane, A. G. (2011). "Modeling salt accumulation in osmotic membrane bioreactors: Implications for FO membrane selection and system operation." *Journal of Membrane Science*, 366(1–2), 314–324.

Xu, Y., Peng, X., Tang, C. Y., Fu, Q. S., and Nie, S. (2010). "Effect of draw solution concentration and operating conditions on forward osmosis and pressure retarded osmosis performance in a spiral wound module." *Journal of Membrane Science*, 348 (1–2), 298–309.

Yip, N. Y. and Elimelech, M. (2012). "Thermodynamic and energy efficiency analysis of power generation from natural salinity gradients by pressure retarded osmosis." *Environmental Science & Technology*, 46(9), 5230–5239.

Yip, N. Y., Tiraferri, A., Phillip, W. A., Schiffman, J. D., and Elimelech, M. (2010). "High performance thin-film composite forward osmosis membrane." *Environmental Science & Technology*, 44(10), 3812–3818.

Zhao, S., Zou, L., and Mulcahy, D. (2012a). "Brackish water desalination by a hybrid forward osmosis-nanofiltration system using divalent draw solute." *Desalination*, 284 (2012), 175–181.

Zhao, S., Zou, L., Tang, C. Y., and Mulcahy, D. (2012b). "Recent developments in forward osmosis: Opportunities and challenges." *Journal of Membrane Science*, 396(0), 1–21.

Zou, S., Gu, Y., Xiao, D., and Tang, C. Y. (2011). "The role of physical and chemical parameters on forward osmosis membrane fouling during algae separation." *Journal of Membrane Science*, 366(1–2), 356–362.

CHAPTER 18

Forward Osmosis Hybrid Processes for Mining Wastewater Treatment

R. Thiruvenkatachari
M. J. Francis
M. Cunnington
S. Su
C. Cote

In this chapter the integration of forward osmosis (FO) with other water treatment methods will be examined with a particular focus on the integration of FO and reverse osmosis (RO) technologies. The limitations of RO in relation to the quality of the feed water will be discussed; in particular for the treatment of mine-affected water. Different modes of integration will be presented, and the advantage and limitations of different hybrid configurations will be examined.

18.1 INTRODUCTION

Membrane processes have been used for the desalination of saline water since 1960 (Loeb and Sourirajan 1960). The most widely used process of these technologies is reverse osmosis (RO), whereby water is forced through a semi-permeable membrane by the application of a hydraulic pressure greater than the osmotic pressure of the feed solution. Forward osmosis (FO) has only been studied as a method of desalination for the last ten years (McCucheon et al. 2005). In FO pure water is 'drawn' from the feed solution across a semi-permeable membrane into a draw solution with a higher osmotic pressure than the feed, whereby the osmotic pressure difference between the feed and the draw solutions drives the process. One major drawback of the FO process is that the pure water must be recovered subsequently from the draw solution. As such, FO is not currently used as a stand-alone water treatment method on a commercial scale, although research is being carried out to find draw solutions from which pure water can be readily recovered (Kim et al. 2012; Hancock et al. 2013; Stone et al. 2013), including the use of fertilisers as draw solutions (Phuntsho et al. 2012). Research is also being carried out into the effect of membrane composition on the permeate flow and rejection of the FO process (Cath et al. 2006; Han et al. 2012; Setiawan et al. 2012).

18.1.1 Hybrid FO Systems

The bulk of FO systems used in water treatment are hybrid systems whereby FO is combined with another membrane treatment method, generally replacing either chemical pre-treatment, which this chapter will focus on, or it is used to reduce the volume of the waste stream as a post treatment step. FO membranes are stronger and more resistant to fouling than the finer membranes used in other membrane processes making them more suitable to be used for the pre-treatment of raw water (Lee et al. 2010; Hickenbottom et al. 2013; Richardson et al. 2013).

18.1.1.1 Pre-treatment Hybrid Systems

Hybrid water treatment systems have been investigated which combine Membrane Distillation (MD), Ultrafiltration (UF), Nanofiltration (NF) and RO with FO as a pre-treatment system. These systems each have advantages and limitations as to the water that can be treated and the draw solutions that can be selected.

Forward Osmosis-Membrane Distillation (FO-MD)

Traditionally, particularly in the food industry, liquids are concentrated through multi-stage vacuum evaporation which is a high energy process (Onsekizoglu 2012). MD was developed to reduce this cost. In MD, the feed is heated and run parallel to a hydrophobic membrane through which only water vapour can pass. On the other side of the membrane is a cold condensing surface or a sweep gas which collects the vapour. FO is used in combination with MD when pure solutions, such as proteins and pharmaceuticals, need to be concentrated without the use of evaporation. The feed solution concentrates and dilutes the draw solution. The draw solution, of higher osmotic pressure than the protein or pharmaceutical feed solution, is then reconcentrated using MD (Wang et al. 2011). Recently this technology has been used to treat a feed solution containing a coloured dye which models the ability of FO-MD to treat domestic wastewater (Ge et al. 2012). The major limitations of FO-MD is the limited range of draw solutions which can be used in tandem with MD, the energy required to heat the draw solution for reconcentration (generally supplied by waste heat) and the necessity of a condensing or sweep gas apparatus to collect the pure water by-product.

FO-Ultrafiltration (FO-UF)/Nanofiltration (FO-NF)

FO can also act as a pre-treatment process for UF or NF. Not only do FO membranes foul less frequently than UF or NF membranes but by diluting the draw solution, the hydraulic pressure required for UF and NF is reduced. However, combining FO with either UF or NF requires the selection of a draw solution containing large ions that cannot pass through the UF or NF membrane pores when regenerating the draw solution. Recent research has focussed on polymeric draw solutions such as polyacrylic acids, which can have high osmotic pressures and can be rejected by UF and NF due to the high molecular weight of the polymer (Ling and Chung 2011). The advantage of UF and NF over RO is that the processes run at lower hydraulic pressures (Fritzmann et al. 2007), however RO has a higher rejection rate than UF and NF and a wider range of possible draw solutions to select from for the hybrid processes (Tan and Ng 2010).

FO-Reverse Osmosis (FO-RO)

FO, also referred to as osmotic dilution, has been investigated as a replacement for the chemical pre-treatment of RO since 2009 (Choi et al. 2009). The system has two main advantages, firstly the FO unit dilutes the feed to the RO unit and lowers the required hydraulic pressure, which is the main operating cost of RO and secondly, FO replaces chemical pre-treatment which can be costly, particularly for remote systems. Experiments on FO-RO have focussed on lowering the energy cost (Quintanilla et al. 2011), theoretically modelling the process (Choi et al. 2010; Bamaga et al. 2011) and treating low osmotic pressure solutions like secondary or tertiary treated domestic wastewater (Cath et al. 2010). As the osmotic gradient is the driving force behind FO, this technology is well suited for low osmotic pressure feedwaters. However a combined FO-RO process would also be ideally suited to treating mine affected water, as the FO process would act as a low cost pre-treatment and screen for the raw mine water, protecting the RO membrane and lowering the operating cost of the RO unit. Furthermore, RO is the cheapest desalination method currently in use making it the most economic technology for the regeneration of the draw solution.

18.1.1.2 Post-treatment Hybrid Systems

FO has been used for several years as a post treatment method for the concentration of liquids such as fruit juice (Petrotos and Lazarides 2001), which reduces the cost of transportation, and also to concentrate sugar solutions (Garcia-Castello et al. 2009). Recently FO has been used as a post-treatment method to concentrate sludge slurries to reduce waste volumes (Holloway et al. 2007). One such system, which has been tested at pilot scale, uses a standard hybrid FO-RO configuration with the feed solution low osmotic pressure secondary or tertiary treated effluent and the draw solution sea water (Cath et al. 2009). The effluent dilutes the draw solution which is then treated with conventional RO to create drinking water. The reject from the RO unit is then used as a draw solution in a second pass FO unit to concentrate the feed waste water from the first FO unit further, making it a FO-RO-FO process. The RO reject is diluted and returned to the sea.

Another domestic wastewater treatment method that utilises FO technology is the Osmotic Membrane BioReactor (OMBR) system. In sewage treatment, the feed water is treated in three steps – removal of heavy solids by gravity separation (primary treatment), removal of dissolved or suspended biological matter by oxidation (secondary treatment) and then chemical treatment (tertiary treatment)

(Forster 2003). A membrane bioreactor system involves a membrane (UF or MF) being submerged into the secondary treatment aeration tank. The aeration stops the membrane from fouling and clean water passes through the pores of the membrane. The osmotic membrane bioreactor system replaces the UF or MF membrane with a FO membrane. The FO membrane has a higher rejection rate and is more resistant to fouling then the other membranes (Achilli et al. 2009). The osmotic membrane bioreactor system is similar to other hybrid FO-RO systems. The FO membrane is submerged in the bioreactor and a draw solution of higher osmotic pressure is passed on the outer side of the membrane. The FO process dilutes the draw solution and concentrates the activated sludge. The draw solution is then conventionally treated using RO to recover the water and regenerate the draw solution (Cornelissen et al. 2008). A summary of hybrid FO systems is presented in Table 18.1

Although FO hybrid systems, particularly FO-RO systems, are ideal for treating raw brackish mine affected water, Table 18.1 shows that the FO hybrid systems currently in use are mainly used to treat NaCl solutions or secondary/tertiary treated effluent. Furthermore, the hybrid system that does treat brackish water employs NF which is a lower energy water treatment method but does not have the same rejection rate as RO or the wide range of available draw solutions.

18.2 TREATMENT OF MINE AFFECTED WATER

In 2010–11 the mining industry accounted for 4% (540 GL) of water consumption (10% more compared to 2009–10), increasing its gross value by $243 million per GL of water consumed (ABS 2012). The coal industry is by far the largest water user in the mining sector because of the huge mass of product mined (fourth largest producer in the world with total production of raw black coal of 471Mt in 2009–10, according to the Australian Coal Association (http://www.australian-coal.com.au/coal-production.html). Water is an integral part of coal mining operations and uses approximately 200–300 L per tonne of saleable coal produced. This "water use" compensates for water entrained in sold product and rejects. However, many coal mines are located in areas experiencing significant climate variability where water can be scarce, or where water systems are fully utilised (e.g. Hunter Valley region and the Murray-Darling Basin). It can be difficult to obtain new water entitlements. Water used by the coal industry (inputs) is obtained from a variety of sources including allocation from bulk water infrastructure (third party water), groundwater and surface water (rainfall and runoff). Water is used onsite for various purposes including longwall dust suppression (in underground mines), coal washing, road dust suppression, washing of mine equipment and vehicles, human consumption,. Water is returned to the environment (output) after contact with mining or processing activities, mainly through evaporation. Releases to the creeks or rivers are rare, but permitted in some river

Table 18.1. Summary of hybrid FO water treatment systems

Configuration	*Water treated*	*Flux rate* (Lm^{-2}-h^{-1})	*Scale*	*Reference*
FO-MD	Acid orange 8 dye solution	15–40	Lab-Scale	Ge et al. 2012
FO-UF	NaCl solution	2–15	Lab-Scale	Ling and Chung 2011
FO-NF	NaCl solution	5–25	Lab-Scale	Tan and Ng 2010
FO-NF	Brackish lake water	8–14	Lab-Scale	Zhao et al. 2012
FO-RO	NaCl solution	2–3	Lab-Scale	Choi et al. 2009
FO-RO	Tap water	4–14	Lab-Scale	Bamaga et al. 2011
FO-RO	Secondary waste water effluent	1–5	Lab-Scale	Quintanilla et al. 2011
FO-RO	Secondary/Tertiary effluent	1–8	Lab-Scale	Cath et al. 2010
FO-RO	Anaerobic digester centrate	7–16	Lab-Scale	Holloway et al. 2007
FO-RO-FO	Secondary/Tertiary effluent	21.4	Pilot-Scale	Cath et al. 2009

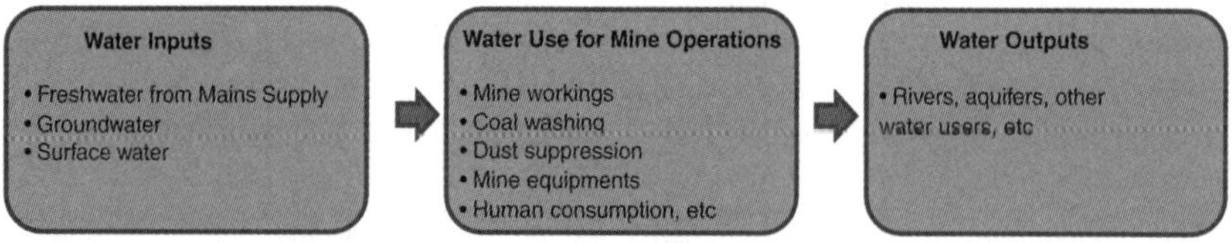

Figure 18.1. Water interactions with coal mining operation

systems, under highly controlled conditions. Figure 18.1 shows a diagrammatic representation of the water interactions in mining.

In the past, many coal mine operations managed water as an environmental issue, but today there is a paradigm shift towards viewing water as key business resource. Some coal mining companies have developed and adopted holistic water management principles and policies to ensure water is used efficiently and to maintain an appropriate water balance as well as minimising any potential environmental impacts from mine-affected water, in accordance to the water quality compliance regulations. Such measures are more likely to have a competitive, economic and reputational advantage. Guidance in identifying, mitigating and monitoring key water-related risks is also available (DRET 2008). Many of these risks are underpinned by the rapidly changing regulatory and economic environment, in the context of highly variable climatic conditions. The extreme variability in rainfall patterns makes water management for mines in Australia even more difficult. The most severe drought on record was witnessed during 1993–2008 and mines sites began to store sufficient water to provide adequate water supply. But this period was followed by above average rainfall resulting in flooding of mines in 2008 and more recently in 2010/11. Due to stringent water discharge regulations, most mines could not release any water and this led to many mines in Fitzroy Basin accumulating large volumes of water. The threat to the licence to operate mine sites due to non-compliant release of low quality mine-affected water and extreme variation in weather conditions (changes to rainfall, temperature and evaporation) are two main economic drivers to improve the approach to water management and to better control the risks.

During dry conditions, water management must focus on minimising dependence on high quality water. Adapting coal handling and preparation plants (CHPP) and dust suppression to use highly saline water reduces the volume of raw water that needs to be treated on mine sites. However, high quality water is still required for underground mining, and mine sites where the mine-affected water is hypersaline (typically with an electrical conductivity over 10000 μS/cm) require treatment to bring the water under the regulated discharge limits. Water sharing between sites and other industries also reduce the reliance on freshwater supply (Luba et al. 2006). Minimising the import of freshwater also has an advantage of minimising the introduction of additional water into the water balance, which will assist in controlling mine water inventory during wet season. During wet conditions, the build up of water on a mine site can occur at a faster rate than the discharge and so water management must focus on managing the water inventory by providing sufficient storage capacity and ensure monitoring and discharge conditions are met. However, augmenting storage capacity at an existing mine site can be hugely challenging (Cote et al. 2010). Adopting efficient treatment technologies on-site would minimise the risk of wet season run-offs and freshwater contamination and allow segregation into different qualities of water to enable greater water recycling.

18.2.1 Mine Water Characteristics

Characteristics of mine affected water vary from mine to mine and the possible impacts on local water sources vary according to local conditions. Water access and environmental performances including mine water discharge, is regulated through number of government agencies both Federal (DSEWPaC) and State-based (DEHP in Queensland, DLWC and DEP in NSW). A study conducted by the Qeensland-based (then) Department of Environment and Resource Management (DERM 2008) found that discharge quality limits and operating requirements for coal mine water discharges were inconsistent, and in the case of some coal mines, did not adequately protect downstream environments. The report's key recommendations included the need to improve the management of wastewater in mining activities, reduce the potential for cumulative impacts and improve water quality data. Most coal mines in the region now have highly regulated discharge conditions.

A recent study on the water quality from two representative mines from New South Wales and one from Queensland (Thiruvenkatachari 2011) has shown significant variation in the characteristics of mine water. Table 18.2 shows the concentration of some of the different parameters analysed from the various streams from the mine sites. These values are approximate concentration ranges obtained from various streams within the mine sites. The pH of the water sources

Table 18.2. Coal mine water quality characteristics

Parameters	*Units*	*Values*
pH	pH unit	3.5–8.9
Total Dissolved Solids	mg/L	700–20000
Suspended Solids	mg/L	5–50
Calcium	mg/L	20–500
Magnesium	mg/L	25–700
Sodium	mg/L	100–6000
Potassium	mg/L	10–120
Iron	mg/L	0–120
Chloride	mg/L	200–1300
Sulphate	mg/L	50–14000
Silica	mg/L	0–15
Bicarbonate Alkalinity	mg/L	5–250

ranged from acidic to basic in nature. Some streams showed high dissolved solids concentration and had the characteristics of scale formation.

Ongoing water management initiatives on site are helping to maximise the amount of water being reused and to minimise the volume discharged off site. Any discharges from site are undertaken in accordance with regulatory requirement. Part of the water management plan involves developing suitable water treatment facility to obtain more water for reuse and to enable water to be discharged off site without harming the environment.

18.2.2 Water Treatment Technologies

Mine affected waters are typically strongly acidic or alkaline and carry high concentrations of salts and trace metals and generally requires treatment before discharge into natural waterways. Mine water treatment technologies can be placed under different categories such as neutralisation of acid, removal of metals, desalination and removal of specific target compounds, as shown in Figure 18.2 (Niekerk et al. 2006).

An appropriate treatment process for a given site is selected based on the quality and quantity of mine-affected water, type of parameters that require removal/reduction, treated water quality objectives and capital and operating costs. Broadly, technologies have focussed on three main areas, namely lime neutralisation and chemical precipitation, desalination and passive treatment systems. Lime is the simple, low cost neutralising agent of choice in most applications. Lime sludges are heavy, low volume, easy to handle, and easy to clarify. Most metals contained in the sludge are as metal hydroxides which are insoluble and will not

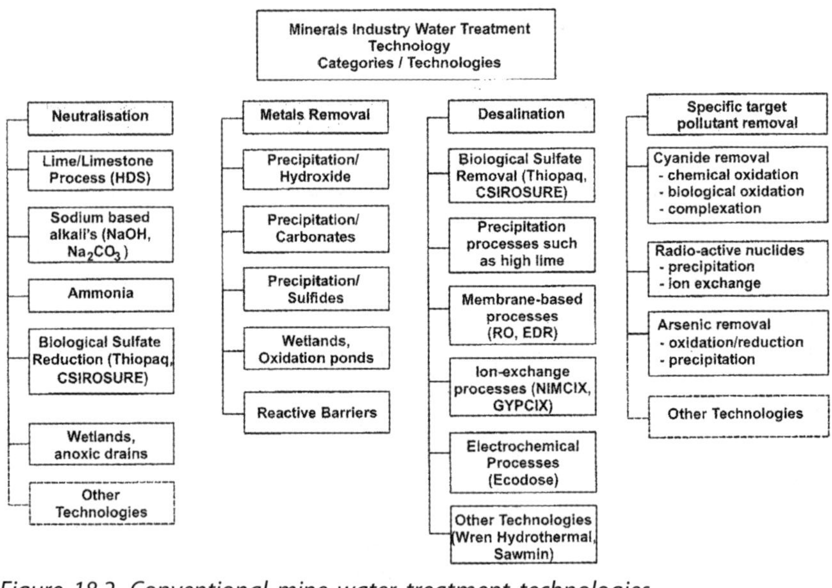

Figure 18.2. Conventional mine water treatment technologies

readily leach into the environment. Other commonly used chemical reagents are limestone, magnesium hydroxide, soda ash (sodium carbonate), caustic soda (sodium hydroxide) and in some cases ammonia. The main disadvantage in this process is they produce large amounts of sludge and the pH of the treated water needs to be readjusted to neutral values which requires large amount of chemical addition (Costello 2003; Nedved and Jansz 2006). In order to remove suspended solids in water, coagulants (such as inorganic iron and aluminium salts) and flocculants (anionic or cationic synthetic polymers) are often used. The most common passive treatment systems are sulphate reducing bacteria-based processes, anoxic limestone drains, constructed anaerobic and aerobic wetlands and biosorption (Watzlaf et al. 2004; Kuyucak 2006).

Currently, RO is a mature and one of the most commonly used desalination technologies in Australia (Hoang et al. 2009). Although membrane desalination processes are very expensive, due to a drop in the cost of membranes the cost of desalination has actually halved in the last five years, but is still not below the target value of $A500/ML (Moore 2009; Firth et al. 2002). Some mines reuse the RO treated water for their mine site operations and reduce the reliance on additional freshwater. However RO membranes can be very sensitive to fouling by various dissolved and undissolved constituents, particulate matter, salt precipitates, and microorganisms, particularly for mine affected water containing silica. It would require extensive and expensive pre-treatment to reduce membrane fouling and to ensure acceptable performance. The short life span of the membrane, membrane scaling, inability to achieve yield to design specifications, inconsistent output quality water are some of the problems generally encountered in the RO system for the treatment of mine impacted water (Davis 2009).

Mine affected water is pre-treated prior to the RO process and goes through elaborate procedures such as aeration, lime neutralisation where the pH is adjusted, coagulation and precipitation where suspended solids, excess lime, some precipitated metal elements and gypsum are allowed to settle, multimedia filters to further reduce the concentration of suspended solids, granular activated carbon filters to reduce the concentration of total organic carbon and microbiological activity, MF units to reduce colloidal material and water softener to reduce hardness and addition of antiscalents prior to RO treatment. Figure 18.3 shows the representation of pre-treatment train prior to RO process (Shao et al. 2009).

To our knowledge, FO has not been applied to the reuse of mine impacted water. As water is not forced through the membrane (osmotic pressure is the driving force, instead of hydraulic pressure), the FO system has fewer clogging or fouling problems. Application of an integrated FO-RO system for treating mine affected water would eliminate extensive pre-treatment requirements for conventional RO systems and has the potential to reduce the overall desalination cost. Through this process, the mine affected water is fed to the FO unit with minimal or no other pre-treatment and the water naturally permeates through the FO membrane to the draw solution, which is a higher salt concentration (osmotic pressure) than the feed mine water. As the treatment process proceeds, the draw solution becomes diluted. The diluted draw solution from this unit serves as the

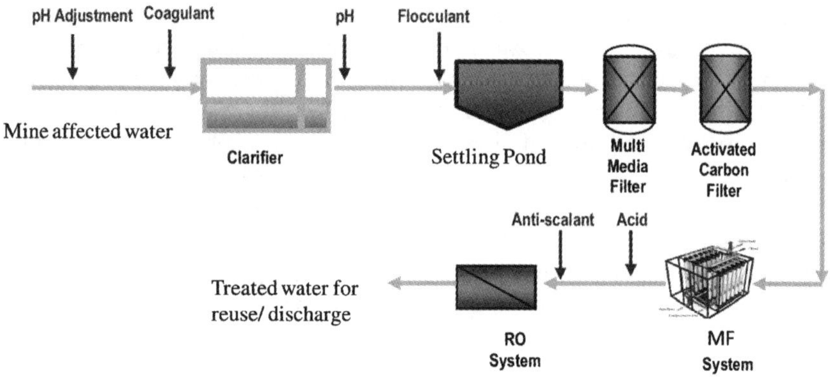

Figure 18.3. Mine affected water treatment using RO process with pre-treatment system

feed to the RO unit. The product water is the permeate from the RO unit, and the concentrate will be the draw solution which will be recirculated back into the FO unit, regenerating the draw solution.

To treat the mine affected water, once an appropriate FO draw solution (Kim et al. 2012; Ge et al. 2013; Stone et al. 2013) is selected then an suitable UF (Ling and Chung 2011), NF (Zhao et al. 2012) or RO membrane (Shaffer et al. 2013) technique can be selected to regenerate the draw solution and produce pure water from the draw solution. The ion size of the draw solution will determine whether ultra filtration, NF or RO can be selected. Although UF and NF require lower pressure than RO there are advantages to using RO as it has higher rejection and allows wider range of available draw solutions to choose.

18.3 INTEGRATED FO-RO SYSTEMS

Two modes of FO-RO integration are examined in detail in this chapter, both of which have advantages and disadvantages. The first, referred to as steady state operation, is shown in Figure 18.4. In this mode, the draw solution of a FO system is directly used as the feed water for a RO system. The permeate rate of the forward osmosis system is matched with the permeate rate of the RO system. This allows the FO system to act as a pre-treatment step and for steady state production of permeate to be produced. The draw solution/RO feed is kept at a constant volume through the re-addition of the reject stream. As the permeate rate from the RO unit is set to match the permeate rate from the FO unit, the concentration of the draw solution will remain constant. While maintaining the permeate rates, the hydraulic pressure in the RO system is decided by the concentration of the draw solution used to obtain the required osmotic gradient in FO and flux through FO for the given membrane characteristics and surface area.

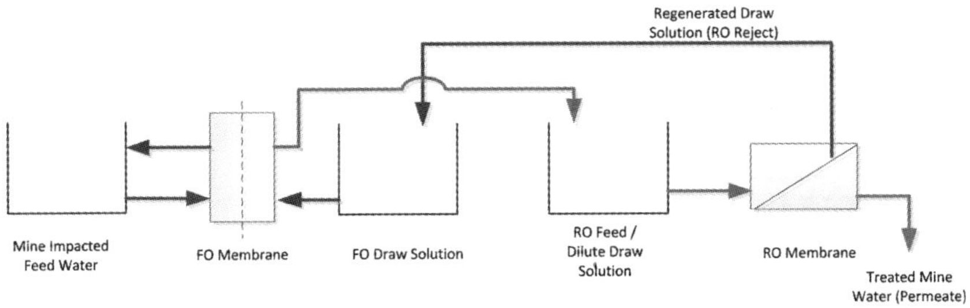

Figure 18.4. Steady state FO-RO system diagram

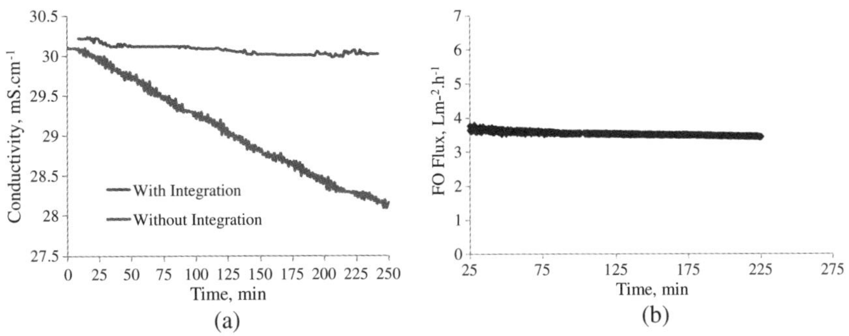

Figure 18.5. Steady state mode for hybrid FO-RO system. (a) FO draw conductivity with or without integration with RO. (b) Flux in FO when integrated with RO

The system will operate in a steady state operation and is only limited by the reduction in the osmotic gradient observed as the FO feed solution concentrates. This hybrid system also allows draw solutions to be selected which are easier to treat by RO, which have higher rejection due to larger molecular sizes and that have lower levels of back diffusion into the FO feed to reduce contamination. It also removes the need for chemical pre-treatment.

This system can overcome the problems of draw solution dilution, eliminate the need for chemical pre-treatment for the RO system and allows greater control on the quality and rate of permeate production through the selection of appropriate draw solutions. In standalone FO operation, the conductivity of the draw solution is gradually reduced due to dilution, thereby lowering the osmotic gradient. However, in the hybrid FO-RO process with steady state operation, the draw solution conductivity is able to be maintained constant as shown in Figure 18.5(a). With constant draw concentration, a more stable FO flux is able to be achieved (Figure 18.5(b)).

The second integration method is the reject management mode, shown in Figure 18.6. In this mode the FO and RO units are run independently with the reject stream from the RO unit being used to keep the concentration of the draw solution constant. In this mode the RO system pressure is independent of FO flux.

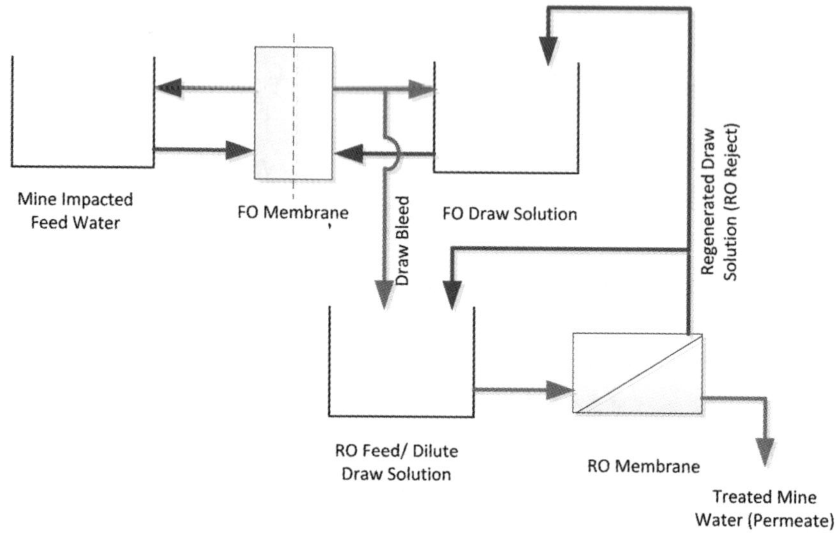

Figure 18.6. Reject management mode for hybrid FO-RO system diagram

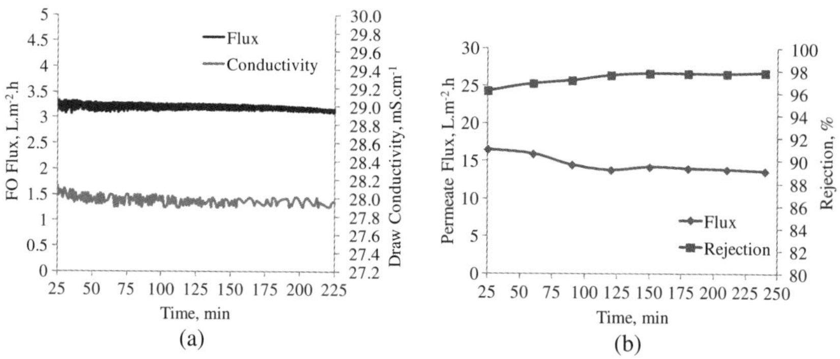

Figure 18.7. Reject management mode for hybrid FO-RO system. (a) Flux through FO and draw solution conductivity when integrated with RO. (b) Permeate water flux and salt rejection in RO

As the concentration of the RO feed increases, the quantity of RO reject added to FO draw, decreases. During this operation, the draw bleed is set to match the FO permeate flux and the reject return to FO draw, keeping the volume of the FO draw solution constant.

This system allows a higher rate of RO permeate to be produced than the steady state mode of operation as it is not limited by the FO flux (Figure 18.7(a)). The final treated mine water quality was about 600 μS/cm with RO rejection of the draw solution over 96% (Figure 18.7(b)). This method is also easier to tailor to specific mine water treatment requirements as the addition of RO reject is readily controllable. This configuration offers a greater flexibility in managing the reject volume.

The two modes adopted have their own advantages and disadvantages. The steady state mode can treat more mine affected water over a longer period of time but the RO permeate rate is limited by FO permeate rate. However this can be overcome by having FO system with higher effective membrane surface area, membrane with more permeability, or by having draw solutions with higher osmotic pressure. The reject management mode can deliver higher volume of RO permeate but is limited by the concentration of required draw solution based on the concentration of the mine affected water. The osmotic pressure of the mine affected water determines which of these two modes to be adopted. Both of these modes of integrated FO-RO systems would eliminate the need for extensive RO pre-treatment, potentially reducing the overall cost of desalination.

In some cases, two or more coal mine sites are in close proximity to each other and potentially have water streams with significantly different total dissolved solid concentrations. In such cases, water from one mine site can be used as the draw solution to treat the water from another site, without the need for any 'new' draw solution, as long as there is sufficient osmotic gradient between the waters from the mine sites. During the integrated FO-RO operation using two different mine waters, the RO is exposed to diluted mine affected water thereby reducing the cost of RO operation and the cost of any pre-treatment required to protect the membrane from fouling.

The current work at CSIRO investigating hybrid FO systems revolves around the study of a lab scale 3–5 LMH FO-RO system. The FO and RO systems were run individually, in steady state and reject management integration modes with the FO unit acting as a pre-treatment to the RO unit. Mine-affected water has been successfully treated using this system with reusable quality water produced as a final product. This unit has been used to evaluate a wide range of draw solutions including several natural products which have both high osmotic pressures and good rejection rates in post treatment. Furthermore, the unit has successfully utilised water from one mine site as the draw solution for mine-affected water from another site. Given the cost of chemicals and the water sharing agreements already in place between mine sites, this is a natural evolution of the technology.

Further to the success of the lab scale unit and the wide concentration range of mine-affected water that the system can treat, it is planned to take the hybrid FO-RO system to a pilot scale. The pilot unit will contain spiral wound FO and RO membranes and will produce about 1 m^3/day of reusable quality water per day using mine-affected water with a minimal pre-treatment system. The optimisation parameters obtained through this study using mine affected water, will provide valuable insight into the viability of this technology for mine site application. It is envisioned that the full scale FO unit with nominal capacity of 100 m^3/day would be integrated with the existing mine site RO system.

18.4 CONCLUDING REMARKS

Although FO is a promising and versatile water treatment technology, it is still not mature enough to be used in its own right. When FO is coupled with a mature treatment system like the RO, the hybrid process utilises the strengths of both technologies.

Application of the hybrid FO-RO system to mine affected water can potentially offer the industry a low cost desalination alternative. Two different integration modes of FO-RO systems were investigated to treat mine affected water. The results show that the FO and RO process can be successfully integrated to produce a reusable quality of treated mine water as permeate, while also able to continuously regenerate the FO draw solution. The hybrid system could potentially eliminate the extensive pre-treatment required for conventional RO process and thereby overall cost of desalination. The mine site application of hybrid FO-RO technology has yet to be investigated.

18.5 ABBREVIATIONS/NOMENCLATURE

FO	Forward Osmosis
MD	Membrane Distillation
MF	Microfiltration
NF	Nanofiltration
OMBR	Osmotic Membrane Bioreactor
RO	Reverse Osmosis
UF	Ultrafiltration

References

Achilli, A., Cath, T. Y., Marchand, E. A., and Childress, A. E. (2009). "The forward osmosis membrane bioreactor: A low fouling alternative to MBR processes." *Desalination*, 239, 10–21.

Australian Coal Association. (2013). Coal in Australia- Coal production, available at http://www.australiancoal.com.au/coal-production.html (accessed on July 10, 2013).

Australian Bureau of Statistics (ABS). (2012). *Water Account, Australia, 2010–11. Catalogue Number 4610,* Australian.

Bamaga, O. A., Yokoichi, A., Zabara, B., and Babaqi, A. S. (2011). "Preliminary assessment of osmotic energy recovery and designs of new FO membrane module configurations." *Desalination*, 268, 163–169.

Cath, T. Y., Drewes, J. E., and Lundin, C. D. (2009). *A novel hybrid forward osmosis process for drinking water augmentation using impaired water and saline water sources*, Water Research Foundation.

Cath, T. Y., Hancock, N. T., Lundin, C. D., Hoppe-Jones, C., and Drewes, J. E. (2010). "A multi-barrier osmotic dilution process for simultaneous desalination and purification of impaired water." *Journal of Membrane Science*, 362, 417–426.

Choi, J-S., Kim, H., Lee, S., Hwang, T-M., Oh, H., Yang, D. R., and Kim, J. H. (2010). "Theoretical investigation of hybrid desalination system combining reverseosmosis and forward osmosis." *Desalination and Water Treatment*, 15(1–3), 114–120.

Cath, T. Y., Childress, A. E., and Elimelech, M. (2006). "Forward osmosis: principles, applications, and recent developments." *J. Membr. Sci.*, 281, 70–87.

Choi, Y-J., Choi, J-S., Oh, H-J., Lee, S., Yang, D. R., and Kim, J. H. (2009). "Toward a combined system of forward osmosis and reverseosmosis for seawater desalination." *Desalination*, 247, 239–246.

Cornelissen, E. R., Harmsen, D., De Korte, K. F., Ruiken, C. J., and Qin, J-J. (2008). "Membrane fouling and process performance of forward osmosis membranes on activated sludge." *Journal of Membrane Science*, 319, 158–168.

Costello, C. (2003). *Acid mine drainage: Innovative treatment technologies*, U.S. Environmental Protection Agency, Office of Solid Waste and Emergency Response, Technology Innovation Office, Washington, D.C., USA.

Cote, C., Moran, C., Hedemann, C., and Koch, C. (2010). "Systems modelling for effective mine water management." *Environmental Modelling and Software*, 25, 1664–1671.

Department of Resource, Energy and Tourism (DRET) formerly, Department of Environment and Resource Management (DERM). (2008). *A study of the cumulative impacts on water quality of mining activities in the Fitzroy River Basin*, DERM Publication, New South Wales, Australia.

Davis, A. (2009). "More water at less cost: Austar mine." *Australian Journal of Mining*, January/February.

Firth, B., Taylor, R., Hannink, R., and O'Brien, M. (2002). *Remediation of saline water for coal mining*, Australian Coal Association Research Program (ACARP) Report Number C10013.

Forster, C. F. (2003). *Wastewater treatment and technology*, Thomas Telford, USA.

Fritzmann, C., Lowenberg, J., Wintgens, T., Melin, T. (2007). "State-of-the-art of reverseosmosis desalination." *Desalination*, 216, 1–76.

Ge, Q., Wang, P., Wan, C., and Chung, T-S. (2012). "Polyelectrolyte-promoted forward osmosis-membrane distillation (FO-MD) hybrid process for dye wastewater treatment." *Environmental Science and Technology*, 46, 6236–6243.

Garcia-Castello, E. M., McCutcheon, J. R., and Elimelech, M. (2009). "Performance evaluation of sucrose concentration using forward osmosis." *Journal of Membrane Science*, 338, 61–66.

Ge, Q., Ling, M., and Chung, T-S. (2013). "Draw solutions for forward osmosis processes: Developments, challenges, and prospects for the future," *Journal of Membrane Science*, 442, 225–237.

Han, G., Chung, T-S., Toriida, M., and Tamai, S. (2012). "Thin-film composite forward osmosis membranes with novel hydrophilic supports for desalination." *Journal of Membrane Science*, 423–424, 543–555.

Hancock, N. T., Xu, P., Roby, M. J., Gomez, J.D., and Cath, T. Y. (2013). "Towards direct potable reuse with forward osmosis: Technical assessment of long-term process performance at the pilot scale." *Journal of Membrane Science*, 445, 34–46.

Hickenbottom, K. L., Hancock, N. T., Hutchings, N. R., Appleton, E. W., Beaudry, E. G., Xu, P., and Cath, T. Y. (2013). "Forward osmosis treatment of drilling mud and fracturing wastewater from oil and gas operations." *Desalination*, 312, 60–66.

Hoang, M., Bolto, B., Haskard, C., Barron, O., Gray, S., and Leslie, G. (2009). *Desalination in Australia, CSIRO Water for a Healthy Country National Research Flagship Report*, CSIRO, Australia.

Holloway, R. W., Childress, A. E., Dennett, K. E., and Cath, T. Y. (2007). "Forward osmosis for concentration of anaerobic digester centrate." *Water Research*, 41, 4005–4014.

Kim, T-W., Kim, Y., Yun, C., Jang, C., Kim, W., and Park, S. (2012). "Systematic approach for draw solute selection and optimal system design for forward osmosis desalination." *Desalination*, 284, 253–260.

Kuyucak, N. (2006). "Selecting suitable methods for treating mining effluents." *Proc., Water in Mining 2006*, Brisbane, Queensland, Australia.

Lee, S., Boo, C., Elimelech, M., and Hong, S. (2010). "Comparison of fouling behaviour in forward osmosis (FO) and reverseosmosis (RO)." *Journal of Membrane Science*, 365, 34–39.

Ling, M. M., and Chung, T-S. (2011). "Desalination process using super hydrophilic nanoparticles via forward osmosis integrated with ultrafiltration regeneration." *Desalination*, 278, 194–202.

Loeb, S., and Sourirajan, S. (1960). *Sea water demineralization by means of a semipermeable membrane*, University of California, USA.

Luba, L., Jakman, M., Lefebvre, N., and Aseervatham, R. (2006). "Mining for water-partnering for sustainable water use in semi-arid regions." *Proc., Water in Mining Conference 2006*, pp 100–203, Brisbane, QLD.

McCutcheon, J. R., McGinnis, R. L., and Elimelech, M. (2005). "A novel ammonia—carbon dioxide forward (direct) osmosis desalination process." *Desalination*, 174(1), 1–11.

Moore, P. (2009). "Water, water everywhere?" *Mining Magazine*, July/Aug.

Nedved, M., and Jansz, J. (2006). "Wastewater pollution control in the Australian mining industry." *J. Cleaner Prod.*, 14, 1118–1120.

Niekerk, A. M., Wurster, A., and Cohen, D. (2006). "Technology advances in mine water treatment in Southern Africa over 20 years." *Proc., Water in Mining 2006*, Brisbane, Queensland, Australia.

Onsekizoglu, P. (2012). *Membrane Distillation: Principle, Advances, Limitations and Future Prospects in Food Industry*, Sina ZereshkiDr. (Ed.), Chapter 11, Distillation - Advances from Modeling to Applications.

Petrotos, K. B., and Lazarides, H. N. (2001). "Osmotic concentration of liquid foods." *Journal of Food Engineering*, 49(2–3), 201–206.

Phuntsho, S., Shon, H. K., Hong, S., Lee, S., Vigneswaran, S., and Kandasamy, J. (2012). "Fertiliser drawn forward osmosis desalination: the concept, performance and limitations for fertigation." *Rev. Environ. Sci. Biotechnol.*, 11, 147–168.

Quintanilla, V. Y., Li, Z., Valladares, R., Li, Q., and Amy, G. (2011). "Indirect desalination of Red Sea water with forward osmosis and low pressure reverseosmosis for water reuse." *Desalination*, 280, 160–166.

Richardson, T-M. J., Flynn, M. T., and Brozell, A. (2013). "zNano forward osmosis membrane for wastewater treatment processes." *Proc., 43rd International Conference on Environmental Systems*, American Institute of Aeronautics and Astronautics, USA.

Setiawan, L., Wang, R., Shi, L., Li, K., and Fane, A. G. (2012). "Novel dual-layer hollow fibre membranes applied for forward osmosis process." *Journal of Membrane Science*, 421–422, 238–246.

Shaffer, D. L., Arias Chavez, L. H., Ben-Sasson, M., Romero-Vargas Castrillón, S., Yip, N. Y., and Elimelech, M. (2013). "Desalination and Reuse of High-Salinity Shale Gas Produced Water: Drivers, Technologies, and Future Directions." *Environmental Science & Technology*, 47(17): 9569–9583.

Shao, E., Wei, J., Yo, A., and Levy, R. (2009). "Application of ultrafiltration and reverseosmosis for mine waste water reuse." *Proc., Water in Mining 2009*, Perth, WA.

Stone, M. L., Rae, C., Stewart, F. F, and Wilson, A.D. (2013). "Switchable polarity solvents as draw solutes for forward osmosis." *Desalination*, 312, 124–129.

Tan, C. H., and Ng, H. Y. (2010). "A Novel hybrid forward osmosis– nanofiltration (FO-NF) process for seawater desalination: draw solution selection and system configuration." *Desalination and Water Treatment*, 13(1–3), 356–361.

Thiruvenkatachari, R., Younes, M., and Su, S. (2011). "Coal minesite investigation of wastewater quality in Australia." *Desalination and Water Treatment*, 32, 357–364.

Wang, K. Y., Teoh, M. M., Nugroho, A., and Chung, T-S. (2011). "Integrated forward osmosis-membrane distillation (FO-MD) for the concentration of protein solutions." *Chemical Engineering Science*, 66, 2421–2430.

Watzlaf, G. R., Schroeder, K. T., Kleinmann, K. L. P., Kairies, C. L., and Nairn, R. W. (2004). *The Passive Treatment of Coal Mine Drainage*, NETL Report Number DOE/NETL-2004/1202.

Zhao, S., Zou, L., and Mulcahy, D. (2012). "Brackish water desalination by a hybrid forward osmosis-nanofiltration unit using divalent draw solution." *Desalination*, 284, 175–181.

CHAPTER 19

Pressure-Assisted Osmosis (PAO) for Water Purification

Sangho Lee
Pierre Le-Clech
Gaetan Blandin

19.1 INTRODUCTION

Forward osmosis (FO) is one of the main emerging water treatment technologies and has received interest as a promising alternative to conventional membrane technologies (Cath et al. 2006; Zhao et al. 2012). FO operation is driven by the osmotic pressure difference across a semi-permeable membrane. Compared with pressure-driven membrane processes, the FO process offers recognized advantages of potentially less operation energy, high rejection of many contaminants, and low membrane fouling tendency (Lee et al. 2010). FO has been considered for a range of potential applications such as wastewater treatment (Holloway et al. 2007; Cornelissen et al. 2008; Achilli et al. 2009), seawater/brackish desalination (McCutcheon et al. 2005; McCutcheon et al. 2006), and energy generation using salinity gradient power (Lee et al. 1981; Loeb 2002; Achilli et al. 2009; Achilli and Childress 2010; Yip et al. 2011; Kim and Elimelech 2012; She et al. 2012).

In FO processes, water flux through the membrane is inherently limited by a number of phenomena, including internal concentration polarization (ICP), external concentration polarization (ECP), membrane fouling, and reverse salt flux (RSF) (Cath et al. 2006; Mi and Elimelech 2008; Tang et al. 2010; She et al. 2012), which are strongly affected by current design and mode of operation (Blandin et al. 2013). Although recent improvements have been obtained in membrane materials and further modifications of the internal structure of the support layer, the water flux and productivity of FO are still lower than those of competing technologies such as reverse osmosis (RO) (Liu et al. 2009; Chanukya et al. 2013).

Recently, pressure-assisted osmosis (PAO) has been proposed as a novel method for FO operation to increase water flux (Lutchmiah et al. 2011; Yun et al. 2012; Blandin et al. 2013; Coday et al. 2013; Yun et al. 2013). PAO relies on the pressurization of the feed solution (FS) of the conventional FO system. The system takes advantage of this additional hydraulic driving force to enhance permeation

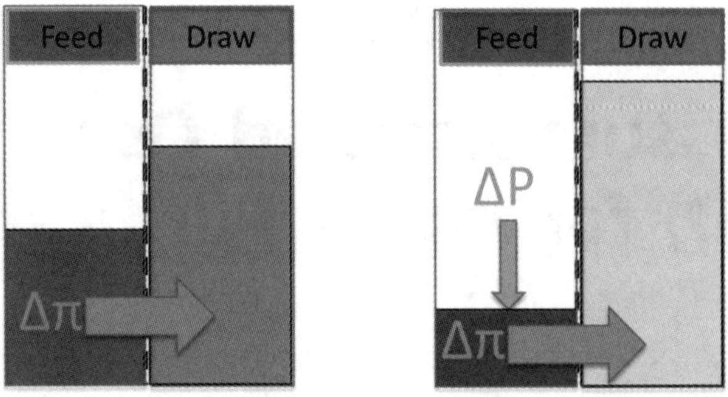

Figure 19.1. Basic concept for PAO (right) in contrast with FO (left)
SOURCE: Adapted from Blandin et al. 2013; reproduced with permission from Elsevier

flux (Figure 19.1). The hydraulic pressure on the feed side is expected to not only increase permeate water flux, but also reduce reverse solute flux. Accordingly, PAO has potential to tackle current limitations of FO and may be used in many application areas. Nevertheless, relatively little information is currently available on PAO design and operation.

This chapter presents a brief overview of the principles of PAO processes and the recent approaches aiming at the enhancement of water flux and reduction of reverse solute flux. The potential of PAO for desalination and wastewater reclamation is also highlighted.

19.2 PRINCIPLE OF PRESSURE-ASSISTED OSMOSIS (PAO)

19.2.1 Basic Equations

Two types of CP phenomena occur in osmotically-driven membrane processes; ICP and ECP [8]. Depending on the orientation of the membrane, the mechanisms for CP are different. When the active layer of the membrane is in contact with the draw solution (AL-DS orientation), the feed solution (FS) is concentrated within the support layer (concentrative ICP) and the draw solution is diluted in the active layer (dilutive ECP). If the active layer of the membrane is in contact with the FS (AL-FS orientation), the FS is concentrated in the active layer (concentrative ECP) and the DS is diluted in the support layer (dilutive ICP). Assuming that the solute rejection is sufficiently high, the following equations can be used to describe these phenomena (Phillip et al. 2010):

$$J_{w,FO} = A\left(\pi_{D,b} e^{-J_{w,FO}K} - \pi_{F,b} e^{-J_{w,FO}/k}\right) \quad \text{for AL-FS} \qquad (19.1)$$

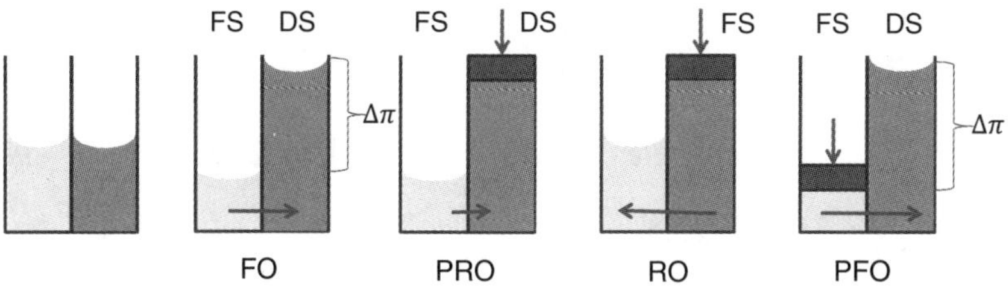

Figure 19.2. Water flows for FO, PRO, RO, and PAO (FS: feed solution; DS: draw solution)

$$J_{w,FO} = A\left(\pi_{D,b} e^{-J_{w,FO}/k} - \pi_{F,b} e^{J_{w,FO}K}\right) \quad \text{for AL-DS} \qquad (19.2)$$

where $J_{w,FO}$ is the water flux through the membrane in FO mode (i.e. water flux is driven by osmotic pressure); A is the water permeability of the membrane; $\pi_{D,b}$ is the osmotic pressure of the DS; $\pi_{F,b}$ is the osmotic pressure of the FS; k is the mass transfer coefficient in the boundary layer; and K is the solute resistance to diffusion in the support layer. On the other hand, the water flux in the RO process is described as:

$$J_{w,RO} = A(\Delta P - \Delta\pi) = A\left(\Delta P - \pi_{F,b} e^{J_{w,RO}/k}\right) \qquad (19.3)$$

where $J_{w,RO}$ is the water flux in RO mode (i.e. water flux is driven by external hydraulic pressure); ΔP is the applied hydraulic pressure difference (bar); and $\Delta\pi$ is the osmotic pressure difference (bar) between the feed and permeate across the active layer of the membrane. A concentrative ECP also exists in RO mode. Combining Eqs. (19.1)–(19.3), the generalized equations for PAO processes can be derived:

$$J_w = A\left(\pi_{D,b} e^{-J_w K} - \pi_{F,b} e^{J_w/k} + \Delta P\right) \quad \text{for AL-FS} \qquad (19.4)$$

$$J_w = A\left(\pi_{D,b} e^{-J_w/k} - \pi_{F,b} e^{J_w K} + \Delta P\right) \quad \text{for AL-DS} \qquad (19.5)$$

If ΔP is positive, this process is defined as PAO. If ΔP is negative and less than the osmotic driving force, this process is defined as the pressure retarded osmosis (PRO). Figure 19.2 illustrates water flows in FO, PRO, RO, and PAO.

As mentioned, CP plays an important role in any osmotically-driven membrane process. The effect of ECP can be estimated through the following equation defining k (Cath et al. 2006; Yun et al. 2013):

$$k = \frac{\text{Sh}D}{d_h} \qquad (19.6)$$

where Sh is the Sherwood number; and d_h is the hydraulic resistance. Sh is a function of the Reynolds number, Re, and the Schmidt number, Sc:

$$Sh = 1.85\left(ReSc\frac{d_h}{L}\right)^{0.33} \quad \text{for laminar flow} \tag{19.7}$$

$$Sh = 0.04Re^{0.75}Sc^{0.33} \quad \text{for turbulent flow} \tag{19.8}$$

where L is the channel length.

On the other hand, the effect of ICP is affected by K, which is given as (Phillip et al. 2010):

$$K = \frac{t\tau}{\varepsilon D} = \frac{S}{D} \tag{19.9}$$

where t is the membrane thickness; τ is the tortuosity; D is the diffusion coefficient of solute; and ε is the porosity. Here, $S = t\tau/\varepsilon$ is the structure parameter of the membrane and is inherently determined by the characteristics of membrane structure.

By replacing K with S/D in Eqs. (19.4) and (19.5), the final governing equations for PAO water flux are obtained as (Lee 2012; Yun et al. 2013):

$$J_{w,PFO} = A\left(\pi_{D,b}e^{\frac{-J_{w,PFO}S}{D}} - \pi_{F,b}e^{J_{w,PFO}/k} + \Delta P\right) \quad \text{for AL-FS} \tag{19.10}$$

$$J_{w,PFO} = A\left(\pi_{D,b}e^{-J_{w,PFO}/k} - \pi_{F,b}e^{\frac{J_{w,PFO}S}{D}} + \Delta P\right) \quad \text{for AL-DS} \tag{19.11}$$

In the FO process, salts can diffuse across the membrane from the highly concentrated DS to the FS (Boo et al. 2012). This phenomenon is known as RSF, which affects the water flux across the membrane and fouling propensity. When RSF is considered, the governing equations for PAO water flux may be rewritten as:

$$J_{w,PFO} = A\left(\frac{\pi_{D,b}e^{\frac{-J_{w,PFO}S}{D}} - \pi_{F,b}e^{J_{w,PFO}/k}}{1 - \frac{B}{J_{w,PFO}}\left(e^{\frac{-J_{w,PFO}S}{D}} - e^{J_{w,PFO}/k}\right)} + \Delta P\right) \quad \text{for AL-FS} \tag{19.12}$$

$$J_{w,PFO} = A\left(\frac{\pi_{D,b}e^{-J_{w,PFO}/k} - \pi_{F,b}e^{\frac{J_{w,PFO}S}{D}}}{1 - \frac{B}{J_{w,PFO}}\left(e^{J_{w,PFO}/k} - e^{\frac{J_{w,PFO}S}{D}}\right)} + \Delta P\right) \quad \text{for AL-DS} \tag{19.13}$$

The impact of RSF in PAO will be further discussed in section 3.2.

19.2.2 Effect of Pressure

The hydraulic pressure has both positive and negative effects on the PAO process. As pressure is applied on the feed side, the driving force for water transport through the membrane increases. Accordingly, the flux in PAO ($J_{w,PAO}$) is higher than that in FO ($J_{w,FO}$). Since ICP increases with water flux across the membrane, $J_{w,PAO}$ is expected to be, at best, the sum of fluxes resulting from osmotic pressure and hydraulic pressure (Eq. (19.14)). The effect of pressure on flux enhancement will be further discussed in Section 19.3.1.

$$J_{w,PFO} \leq J_{w,FO} + A\Delta P \tag{19.14}$$

Although S is assumed to be constant in Eq. (19.10) or (19.11), it may be affected by the hydraulic pressure. Pressurization of the membrane by the feed water is expected to reduce the surface porosity against fresh water. This could lead to reduced porosity and decreased interconnectivity between the pores, resulting in increased S values under elevated hydraulic pressure (Holt (2013).

19.2.3 Energy Consumption

The main advantage of using FO is that it operates at low or no hydraulic pressure and thus, it has potential for lower energy consumption than RO. Since PAO uses both hydraulic and osmotic pressures, it requires higher pressure than FO, but still, much lower pressure than RO. When the energy required for draw solute recovery is not needed/considered (i.e., fertilizer driven FO desalination (Phuntsho et al. 2011)), the energy consumption by PAO is expected to be larger than FO.

Also, the additional energy required by PAO may be counterbalanced by lower capital investment costs (i.e. lower membrane surface needed, and therefore lower number of vessels, piping, and circulating pumps) or additional savings in the subsequent process. For example if PAO is implemented instead of FO in a FO-RO hybrid process (see Section 19.4.2), additional water recovery from PAO will lead to further reduction in RO energy consumption.

19.3 RECENT APPROACHES TO PAO

The interest for PAO is very recent and, to date, only a few groups have developed research activities on the topic (Choi et al. 2009; Lutchmiah et al. 2011; Yun et al. 2012; Blandin et al. 2013; Coday et al. 2013; Yun et al. 2013). Therefore, a limited number of references are available, and most of the works conducted by the aforementionned groups are in progress. There have been three major topics in PAO research: i) flux enhancement; ii) reduction in RSF; and iii) fouling propensity.

19.3.1 Flux Enhancement

Based on Eqs. (19.10) and (19.11), water permeation flux is expected to increase when additional hydraulic pressure is applied on the feed side of the osmotic system. The first study on the effect of pressure on flux in PAO was conducted by Choi et al. (2009). According to their work, the water flux was found to increase from 0.5 to 1.4 L/m^2-h by applying hydraulic pressure up to 2.3 bar. Nevertheless, polyamide RO membranes for FO operation were used and very low flux was obtained.

Experimental works using CTA FO membrane (HTI, U.S.A.) were conducted by Yun et al. (2012). De-ionised water and 0.6 M of NaCl were used as FS and DS respectively. The results confirmed a significant increase of the permeation flux from 4.5 to 9.5 L/m^2-h for the conventional FO systems and PAO operated at 8 bar, respectively. It was also found that the characteristics of the FO membrane were important.

Another recent study by Blandin et al. (2013) has provided further insights on the increased performances offered by PAO systems. It was clearly demonstrated that hydraulic pressure acts positively on the permeation flux, but resulted, as expected, in more severe CP affecting the efficiency of the osmotic pressure. Moreover, as an indirect effect of the applied hydraulic pressure, membrane (HTI CTA) deformation was observed, leading to higher permeability of the filtration material. In a series of experiments based on deionized water and 35 g/L Red Sea salt as FS and DS, respectively, the specific contributions of hydraulic and osmotic pressures were measured for a range of feed pressures (Figure 19.3). Above 6 bar, the hydraulic pressure contribution became predominant, as a consequence of the important CP affecting the osmotic pressure efficiency. It has to be noted that, in similar operating conditions, but for the other membrane orientation (i.e. AL-DS), lower CP was observed. Under those conditions, and even at 6 bar, the osmotic pressure contribution remained predominant at 70%. It is expected that the development of new membrane with improved water permeability and limited CP will further improve PAO performances, thanks to a better use of both osmotic and hydraulic pressures.

Recent experimental works by Yun et al. (2013) have also highlighted the flux enhancement by hydraulic pressure in PAO. In their study, the flux increase in PAO was theoretically predicted using the models and compared with the experimental results. Figure 19.4 shows the results of their works. As the external pressure increases, the water flux increases from 4.0 to 8.0 L/m^2-h; this increase being smaller than the theoretical sum of fluxes in FO and RO modes (Eq. (19.12)). On the contrary, the model prediction based on Eq. (19.10) was found to closely match the experimental results.

An increase in flux by pressure was also observed in hollow fiber FO systems (Yun et al. 2012). The experiments were carried out using thin film composite hollow fiber FO membrane (Cheil Industry, Korea). Deionized water and 0.5 M of $MgCl_2$ were used as FS and DS, respectively. The flux was increased from 4 to 6.5 L/m^2-h, as the pressure increased by 2.5 bar (Figure 19.5).

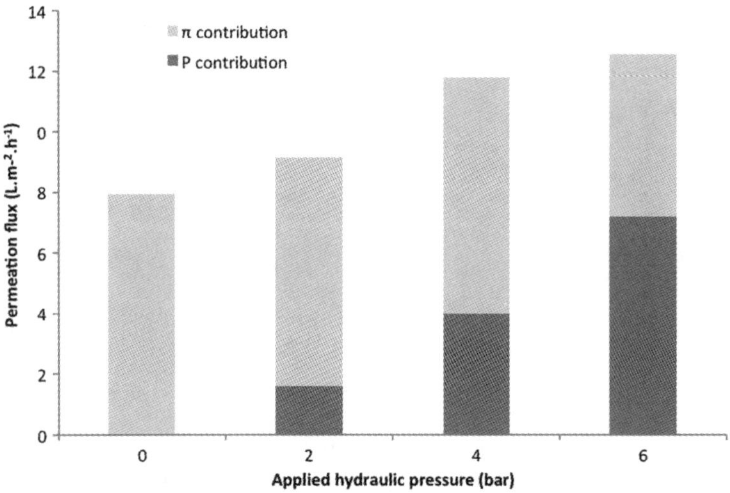

Figure 19.3. Hydraulic (P) and osmotic (π) pressures contribution to the permeate flux for different applied hydraulic pressures (CFV = 0.1 m · s^{-1}; T: 25°C; HTI CTA membrane; orientation: AL-FS; RO feed spacers on both feed and draw channels)

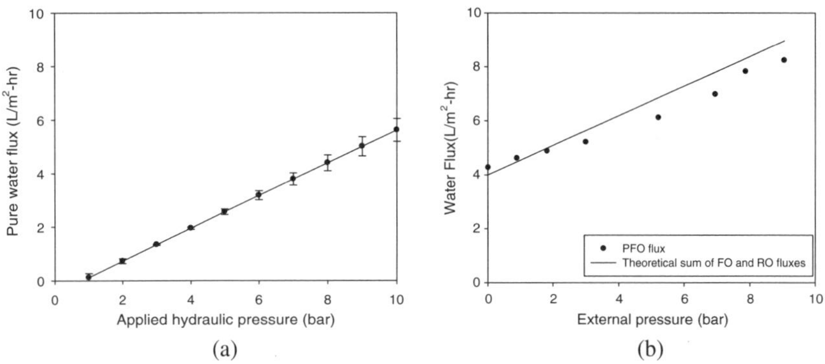

Figure 19.4. (a) Pure water flux of the FO membrane in RO mode. (b) Comparison of water flux in PAO compared with the sum of fluxes in FO and RO. The curve and symbols represent the water flux from the PAO test and the theoretical sum of FO flux and RO flux (based on Eq. (19.12)), respectively

SOURCE: Adapted from Yun et al. 2013; reproduced with permission from Taylor & Francis

In summary, PAO allows higher flux than FO by increasing the driving force for water transport but also increases ICP and limits the expected flux enhancement. Therefore, theoretical models seem to be useful to explore the optimum conditions for PAO operations.

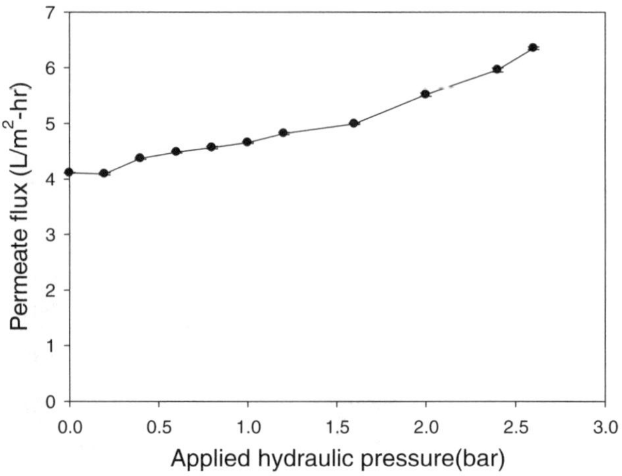

Figure 19.5. Effect of external hydraulic pressure on flux of hollow fiber FO membrane in PAO mode.

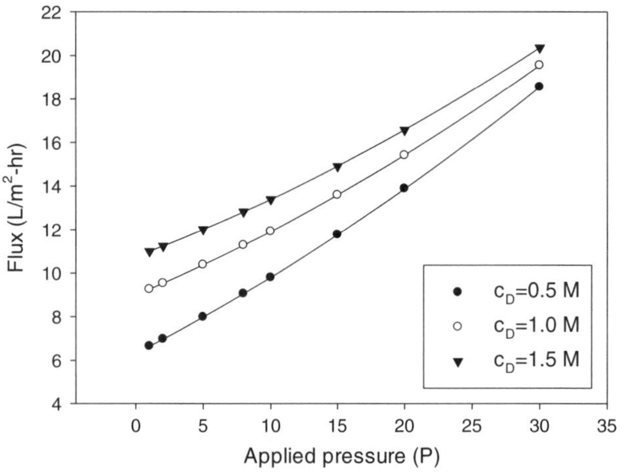

Figure 19.6. Simulation for the effect of external hydraulic pressure and DS concentration (C_D) on PAO flux

Figure 19.6 shows the simulation results by the model based on Eq. (19.10). The water flux was calculated as a function of hydraulic pressure and draw solute concentration. It is evident that the water flux increases with an increase in the pressure. Nevertheless, water flux is not linearly proportional to the pressure, due to the effect of ICP.

The ratio of PAO flux to FO flux, $J_{w,PAO}/J_{w,FO}$ is not constant. The equation for $J_{w,PAO}/J_{w,FO}$ can be obtained by rearranging Eqs. (19.1) and (19.10) in AL-FS orientation:

$$\frac{J_{w,PFO}}{J_{w,FO}} = \frac{\left(\pi_{D,b}e^{\frac{-J_{w,PFO}S}{D}} - \pi_{F,b}e^{J_{w,PFO}/k} + \Delta P\right)}{\left(\pi_{D,b}e^{\frac{-J_{w,FO}S}{D}} - \pi_{F,b}e^{J_{w,FO}/k}\right)} \tag{19.14}$$

For application in which $\pi_{F,b}$ is negligible compared to $\pi_{D,b}$, the following equation for $J_{w,PAO}$ can be derived:

$$J_{w,PFO} = \frac{J_{w,FO}\left(e^{-\frac{J_{w,PFO}S}{D}} + \frac{\Delta P}{\pi_{D,b}}\right)}{\left(e^{-\frac{J_{w,FO}S}{D}}\right)} \tag{19.15}$$

Eq. (19.15) indicates that the ratio of hydraulic pressure to osmotic pressure of DS ($\Delta P/\pi_{D,b}$) and structure parameter (S) are important factors affecting $J_{w,PAO}$. Rearranging Eq. (19.15), the ratio of RO flux to total PAO flux, $J_{w,P}/J_{w,PAO}$ is derived:

$$\frac{J_{w,P}}{J_{w,PFO}} = \frac{\frac{\Delta P}{\pi_{D,b}}}{\left(e^{\frac{-J_{w,PFO}S}{D}} + \frac{\Delta P}{\pi_{D,b}}\right)} \tag{19.16}$$

Figure 19.7 shows the simulation results by the model for $J_{w,PAO}$ and $J_{w,P}/J_{w,PAO}$. An increase in $\Delta P/\pi_{D,b}$ results in higher water flux for PAO operations, with a strong influence of the S value (Figure 19.7(a)). This implies that relatively high pressure (or $\Delta P/\pi_{D,b}$) and small S are required to obtain high flux in PAO. On the other hand, the $J_{w,P}/J_{w,PAO}$ is higher with large S values (Figure 19.7(b)), suggesting that the PAO operation becomes similar to FO or RO depending on S value. As S decreases, the importance of hydraulic pressure (or $J_{w,P}/J_{w,PAO}$) decreases, indicating that the PAO operation becomes comparable to FO. On the other hand, the importance of hydraulic pressure decreases with large S, indicating that the PAO is similar to RO operation.

19.3.2 Reduction in RSF

In the FO process, RSF and water permeation are both driven by the osmotic pressure difference across the membrane following diffusion mechanisms (Phillip et al. 2010)). Therefore, the potential improvement of FO performances by increase of DS concentration is expected to enhance the water permeation flux, but also to lead to more severe RSF, impacting negatively on the system. In the PAO system, according to the solution-diffusion equations, hydraulic pressure acts as an additional driving force only for water permeation but, unlike the osmotic pressure, it does not affect directly reverse salt permeation and therefore represents a way to overcome the RSF/water permeation trade-off.

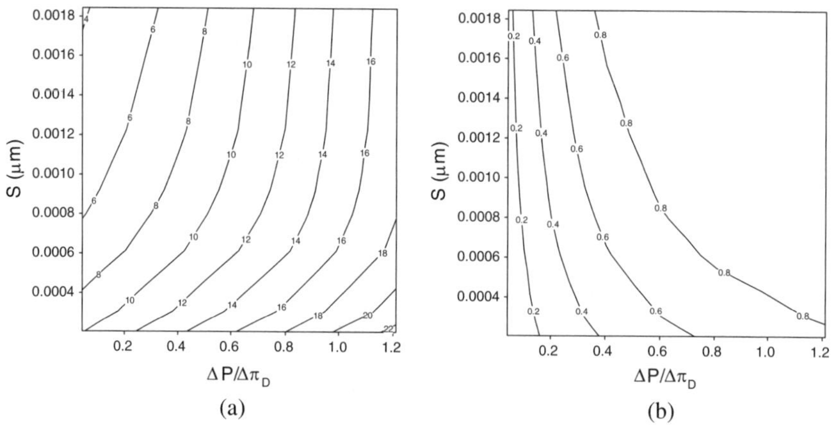

Figure 19.7. Contour diagrams of PAO flux ($J_{w,PAO}$ in L/m^2 hr) (a) and the ratio of RO flux to total PAO flux ($J_{w,P}/J_{w,PAO}$) (b) for different $\Delta P/\Delta\pi$ and S values

Moreover, as already discussed, the increased permeation flux is known to lead to more severe CP, which consequently decreases osmotic pressure difference across the membrane and therefore further limits the RSF. However, as a consequence of membrane deformation and its resulting higher permeability, increase in solute permeability has been observed when operating with the HTI CTA membrane. Since both salt and water fluxes being greater while increasing hydraulic pressure, the salt/water flux ratio (J_s/J_w) was used in order to effectively compare systems (Blandin et al. 2013)). Results obtained show that the salt/water flux ratio significantly decreased with hydraulic pressure (Figure 19.8). This phenomenon can be partly explained by comparatively higher water permeation increase and confirmed the interest of PAO as a driving force acting directly only on the water permeation.

Thus, the application of hydraulic pressure represents an improvement of the FO process not only by enhancing the water permeation, but also in limiting the negative impact of RSF, one of the major limitations related to the use of FO technology (Blandin et al. 2013).

19.3.3 Effect of Spacer

Feed and permeate spacers play an important role in hydraulic transfer mechanisms, especially in the context of spiral-wound module (Kim and Park (2011). Even in conventional FO practice, it is expected that hydraulic pressure is to be added on the draw and feed sides of the membrane to compensate the pressure drop along the channels. Moreover, in PAO configuration, the conjunction of pressurization of the feed and permeation of water to the draw is expected to significantly pressurize the draw channel (Kim and Park (2011). Therefore, the choice of spacer is predominant to properly support the membrane and to avoid

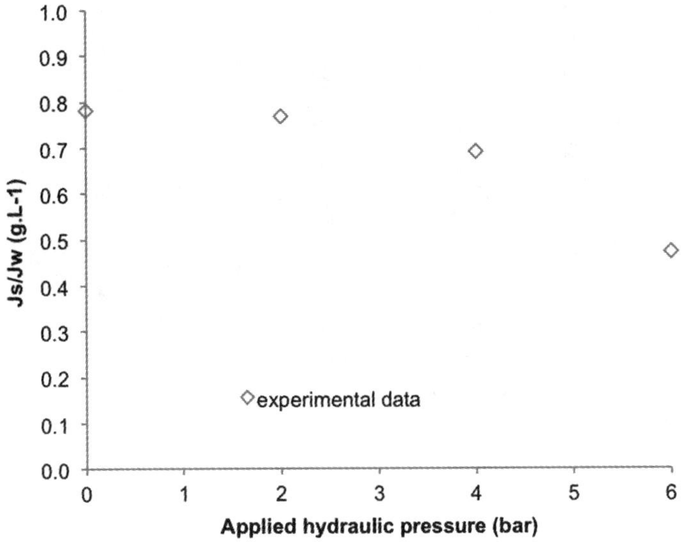

Figure 19.8. Salt/water flux ratio (Js/Jw) as function of the applied hydraulic pressure on the feed (CFV = 0.1 m · s^{-1}; T: 25°C; HTI CTA membrane; orientation: AL-FW; RO feed spacers on both feed and draw channels)

resistance in water transfer. RO mesh feed spacers are typically used in FO operation on both feed and draw channels.

In the context of PAO, and to better evaluate the combined effect of spacer and feed pressurization, comparative tests were performed in a RO cross flow setup using either typical RO mesh feed spacer or frit support as membrane support (Blandin et al. 2013). Results showed that HTI CTA membrane suffered from deformation due to the pressurization of the membrane on the spacer strands (Figure 19.9), as already described in the literature (Kim and Elimelech 2012; She et al. 2012). In the range of pressure tested (0–6 bar), it was also observed that a significant increase of permeability due to the membrane stretching occurred with no loss of membrane integrity. As a result, the choice of the spacer type and the potential resulting stretching of the membrane need to be better considered, assessed, and monitored, especially during long-term operation.

19.3.4 Fouling Propensity

Although the hydraulic pressure in PAO may increase the water flux, it may have adverse effect by accelerating membrane fouling. Although few studies have been completed for PAO systems, the effect of hydraulic pressure on membrane fouling in FO has been well assessed and contrasted with those obtained in RO (Lee et al. 2010)). According to this study, the structure of the fouling layers of FO and RO are quite different. A thin and compact fouling layer is formed in RO due to the

Figure 19.9. From left to right, FO membrane after RO setup water permeability test at 6 bar with RO feed spacer and frit support, respectively
Source: Blandin et al. 2013; reproduced with permission from Elsevier

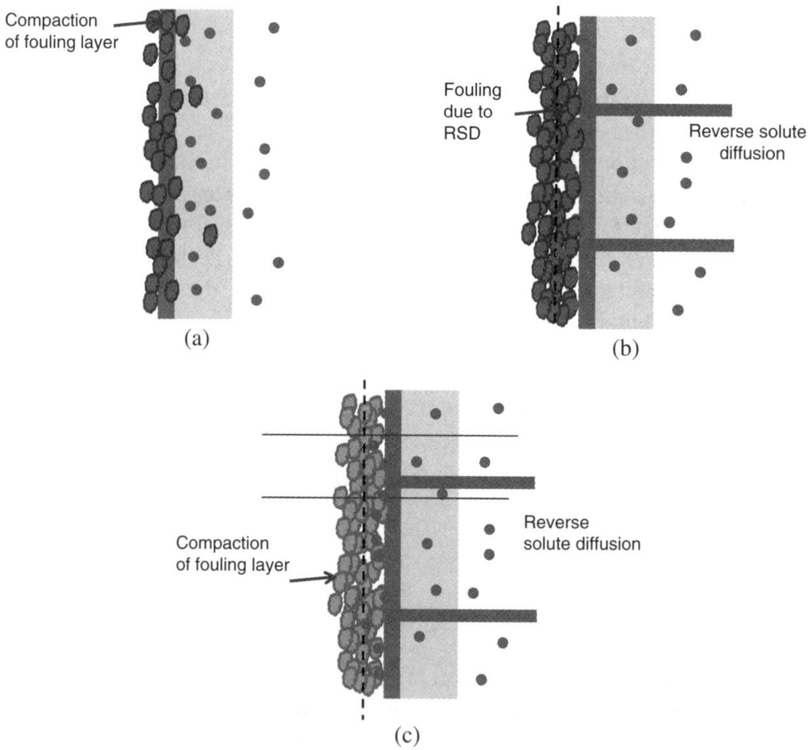

Figure 19.10. Proposed fouling mechanisms in (a) RO, (b) FO, and (c) PAO
SOURCE: Adapted from Yun et al. 2013; reproduced with permission from Taylor & Francis

use of high hydraulic pressure (Figure 19.10(a)). In FO, reverse diffusion of salt from the DS to the feed side exacerbates the cake-enhanced osmotic pressure within the fouling layer. The elevated osmotic pressure near the membrane surface on the feed side leads to substantial drop in the net osmotic driving force and, thus, to significant decline of permeate flux (Figure 19.10(b)).

In PAO, both fouling layer compaction and cake-enhanced osmotic pressure are expected to occur (Figure 19.10(c)). Depending on the relative importance of those two different effects, PAO may have higher or lower fouling propensity than FO. It is interesting to note that PAO is operated under lower pressure than RO and reduced RSF compared to FO. Accordingly, it is difficult to theoretically predict the fouling propensity in PAO. Unfortunately, very few studies were conducted to compare the fouling rates in PAO with those in FO or RO. In addition, it is difficult to draw definite conclusions yet. For instance, the fouling rates were found to be similar for FO, RO, and PAO, when operated under comparable conditions (Yun et al. 2013). Further systematic studies are required to understand the fouling propensity of PAO processes.

19.4 POSSIBLE APPLICATIONS

19.4.1 Seawater Desalination

FO is conceptually a low-energy desalination alternative to conventional evaporation or RO. Nevertheless, one of the limitations posed by FO-based desalination is low flux and thus low recovery of product water or extended membrane surface. To apply FO for seawater desalination, the water flux should be comparable to that of RO, which may require high concentration of DS, further membrane developments or alternatively by adopting the concept of PAO. The additional hydraulic pressure can increase both water flux and recovery by compensating a reduction in osmotic driving force efficiency. Figure 19.11 shows an example of osmotically-based desalination systems using PAO as a second stage unit process (Hong (2013). The high pressure pump increases the pressure of the feed stream to the PAO process. If the cost of draw solute recovery is substantial, the use of PAO may be effective to reduce such cost.

19.4.2 Osmotic Dilution of Seawater Using Reclaimed Wastewater

The osmotic dilution is a system that combines seawater desalination and wastewater reclamation (Cath et al. 2010; Boo et al. 2013; Hancock et al. 2013). In this system, FO uses a saline stream (i.e. seawater) as a DS to extract water from an impaired water source (i.e. reclaimed wastewater). The diluted seawater is further treated by RO. The advantages of this system include: i) lower energy required for seawater desalination; ii) multiple barriers against the contaminants for wastewater recycling; and iii) protection of RO membrane from foulants in the reclaimed wastewater.

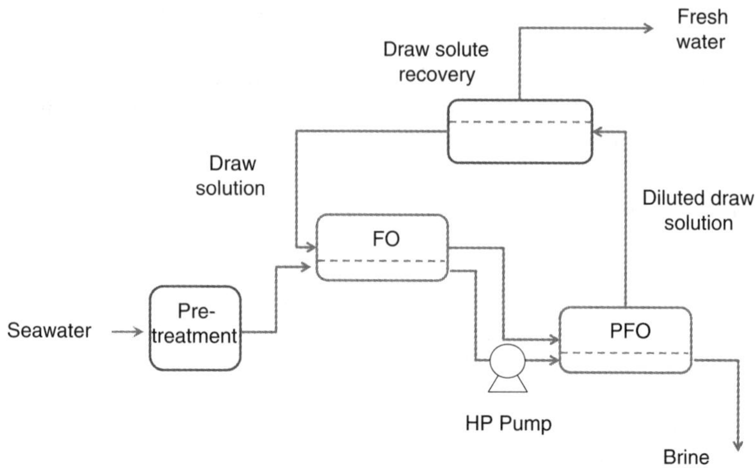

Figure 19.11. Schematic diagram of system using PAO for concentrating FO brine

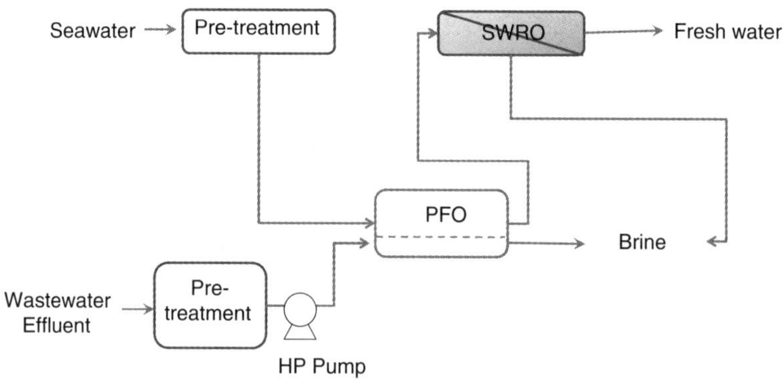

Figure 19.12. Schematic diagram of PAO-SWRO hybrid system for high permeate flux

Again, PAO may be used to increase water flux through the membrane. Figure 19.12 illustrates the basic concept of an osmotic dilution system using PAO. In combination with seawater osmotic pressure, the additional hydraulic pressure is used to exhibit relatively high flux in PAO. This allows not only high productivity of fresh water but also high recovery. In addition, the hydraulic pressure may be used to stabilize the system performance with respect to the changes in the feed water properties. As shown in Figure 19.13 (a), the water flux in FO decreases with the feed water temperature (i.e. 20% of flux loss from 20°C to 5°C). Applying hydraulic pressure allows the same water flux at lower feed water temperature without changing the concentration of DS (i.e. 12 bar of additional hydraulic pressure to obtain same water flux). As

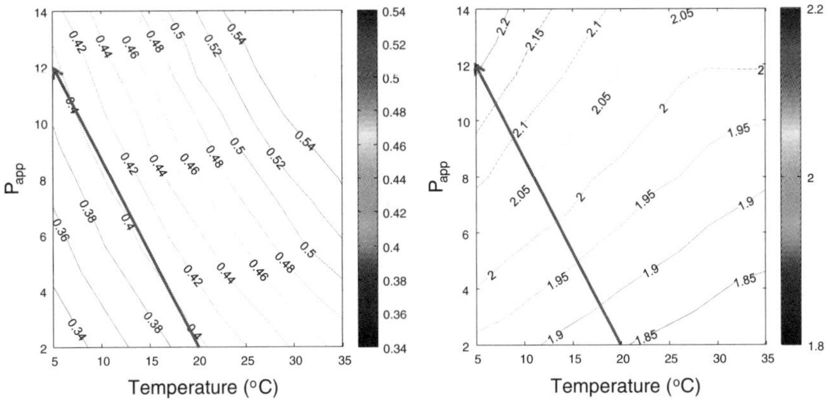

Figure 19.13. Contour diagrams of (a) normalized flux (J_w/J_{w0}) and (b) specific energy consumption (kWh/m^3) in PAO at different temperatures and applied pressures (P_{app})

illustrated in Figure 19.13(b), PAO requires additional energy, which may be counterbalanced with capital and operational costs for increasing draw solute concentration.

19.5 SUMMARY

Although FO is a promising technology in many water treatment applications, relatively low water flux and high RSF are technical limitations to be addressed. Various strategies have been suggested and the concept of PAO has been recently considered as an alternative to overcome actual FO operation limitations. PAO uses both osmotic and hydraulic pressures as driving forces to produce water. Flux enhancement in PAO has been confirmed in both experimental and theoretical studies. In PAO, the water flux increases non-linearly with the applied hydraulic pressure due to more severe ICP. The RSF in PAO has been also reported to be lower than that in FO, due to the effect of the hydraulic pressure. Accordingly, the application of hydraulic pressure represents an improvement of the FO process, not only by enhancing the water permeation but also in limiting the negative impact of RSF, one of the major limitations related to the use of FO technology.

However, the adverse effects of the pressure on the performance of membrane in PAO have also been reported. The selection of feed/draw spacers is predominant to properly support the membrane and to avoid resistance in water transfer. PAO membranes should be redesigned to have enough mechanical strength under intermediate or high pressure. Based on the early studies conducted on PAO fouling, the hydraulic pressure is also expected to accelerate membrane fouling and make the fouling layer more irreversible in PAO.

So far, two potential industrial applications of PAO have been seriously considered: seawater desalination and osmotic dilution of seawater using reclaimed wastewater. Due to its advantages of high water flux and lower RSF, the interest in PAO configurations is expected to significantly increase and to be considered for other applications. However, it should be emphasized that there are many technical issues related to PAO operation. Consequently, additional research on the design and operation of PAO systems should be pursued in parallel with the development and application of new membranes and spacers for PAO.

19.6 ABBREVIATIONS/NOMENCLATURE

$\pi_{D,b}$	Osmotic pressure of the DS
$\pi_{F,b}$	Osmotic pressure of the FS
A	Water permeability of the membrane
AL	Active layer
CFV	Cross flow velocity
CP	concentration polarization
CTA	cellulose triacetate
D	Diffusion coefficient of solute
d_h	Hydraulic resistance
DS	draw solution
ECP	External concentration polarization
FO	Forward osmosis
FS	feed solution
ICP	Internal concentration polarization
$J_{w,FO}$	Water flux in FO mode (i.e. driven by osmotic pressure),
$J_{w,RO}$	Water flux in RO mode (i.e. driven by external hydraulic pressure)
k	Mass transfer coefficient in the boundary layer
K	Solute resistance to diffusion in the support layer
L	Channel length
PAO	Pressure-assisted osmosis
PRO	Pressure retarded osmosis
Re	Reynolds number
RO	Reverse osmosis
RSF	reverse salt flux
S	Membrane structure parameter
Sc	Schmidt number
Sh	Sherwood number
t	Membrane thickness
ΔP	Applied hydraulic pressure difference
$\Delta\pi$	Osmotic pressure difference
ε	Membrane porosity
τ	Membrane tortuosity

References

Achilli, A., Cath, T. Y., and Childress, A. E. (2009). "Power generation with pressure retarded osmosis: An experimental and theoretical investigation." *Journal of Membrane Science*, 343, 1–2, 42–52.

Achilli, A., Cath, T. Y., Marchand, E. A., and Childress, A. E. (2009). "The forward osmosis membrane bioreactor: A low fouling alternative to MBR processes." *Desalination*, 239, 1–3, 10–21.

Achilli, A. and Childress, A. E. (2010). "Pressure retarded osmosis: From the vision of Sidney Loeb to the first prototype installation – Review." *Desalination*, 261, 3, 205–211.

Blandin, G., Verliefde, A. R. D., Tang, C. Y., Childress, A. E., and Le-Clech, P. (2013). "Validation of assisted forward osmosis (AFO) process: Impact of hydraulic pressure." *Journal of Membrane Science*, 447, 1–11.

Boo, C., Elimelech, M., and Hong, S. (2013). "Fouling control in a forward osmosis process integrating seawater desalination and wastewater reclamation." *Journal of Membrane Science*, 444, 0, 148–156.

Boo, C., Lee, S., Elimelech, M., Meng, Z., and Hong, S. (2012). "Colloidal fouling in forward osmosis: Role of reverse salt diffusion." *Journal of Membrane Science*, 390–391, 0, 277–284.

Cath, T. Y., Childress, A. E., and Elimelech, M. (2006). "Forward osmosis: Principles, applications, and recent developments." *Journal of Membrane Science*, 281, 1–2, 70–87.

Cath, T. Y., Hancock, N. T., Lundin, C. D., Hoppe-Jones, C., and Drewes, J. E. (2010). "A multi-barrier osmotic dilution process for simultaneous desalination and purification of impaired water." *Journal of Membrane Science*, 362, 1–2, 417–426.

Chanukya, B. S., Patil, S., and Rastogi, N. K. (2013). "Influence of concentration polarization on flux behavior in forward osmosis during desalination using ammonium bicarbonate." *Desalination*, 312, 0, 39–44.

Choi, Y.-J., Choi, J.-S., Oh, H.-J., Lee, S., Yang, D. R., and Kim, J. H. (2009). "Toward a combined system of forward osmosis and reverse osmosis for seawater desalination." *Desalination*, 247, 1–3, 239–246.

Coday, B. D., Heil, D. M., Xu, P., and Cath, T. Y. (2013). "Effects of transmembrane hydraulic pressure on performance of forward osmosis membranes." *Environmental Science and Technology*, 47, 2386–2393.

Cornelissen, E. R., Harmsen, D., de Korte, K. F., Ruiken, C. J., Qin, J. J., Oo, H., and Wessels, L. P. (2008). "Membrane fouling and process performance of forward osmosis membranes on activated sludge." *Journal of Membrane Science*, 319, 1–2, 158–168.

Hancock, N. T., Xu, P., Roby, M. J., Gomez, J. D., and Cath, T. Y. (2013). "Towards direct potable reuse with forward osmosis: Technical assessment of long-term process performance at the pilot scale." *Journal of Membrane Science*, 445, 0, 34–46.

Holloway, R. W., Childress, A. E., Dennett, K. E., and Cath, T. Y. (2007). "Forward osmosis for concentration of anaerobic digester centrate." *Water Research*, 41, 17, 4005–4014.

Holt, T. (2013). "Membranes for salinity power by PRO." *International FO/PRO summit 2013*, Seoul, Korea, WATEREX.

Hong, S. K. (2013). "Recent development of hollw fiber FO membrane for osmotic dilution process." Seoul, Korea, COEX, WATEREX.

Kim, Y. C., and Elimelech, M. (2012). "Adverse impact of feed channel spacers on the performance of pressure retarded osmosis." *Environmental science & technology*, 46(8), 4673–4681.

Kim, Y. C., and Park, S. J. (2011). "Experimental study of a 4040 spiral-wound forward-osmosis membrane module." *Environmental Science and Technology*, 45, 18, 7737–7745.

Lay, W. C. L., Zhang, J., Tang, C., Wang, R., Liu, Y., and Fane, A. G. (2012). "Factors affecting flux performance of forward osmosis systems." *Journal of Membrane Science*, 394–395, 0, 151–168.

Lee, K. L., Baker, R. W., and Lonsdale, H. K. (1981). "Membranes for power generation by pressure-retarded osmosis." *Journal of Membrane Science*, 8, 2, 141–171.

Lee, S. (2012). "Simulation and optimization of full-scale FO hybrid system for seawater desalination." *International FO/PRO Summit*, Seoul.

Lee, S., Boo, C., Elimelech, M., and Hong, S. (2010). "Comparison of fouling behavior in forward osmosis (FO) and reverse osmosis (RO)." *Journal of Membrane Science*, 365, 1–2, 34–39.

Liu, L., Wang, M., Wang, D., and Gao, C. (2009). "Current patents of forward osmosis membrane process." *Recent Patents on Chemical Engineering*, 2, 76–82.

Loeb, S. (2002). "Large-scale power production by pressure-retarded osmosis, using river water and sea water passing through spiral modules." *Desalination*, 143, 2, 115–122.

Lutchmiah, K., Cornelissen, E. R., Harmsen, D. J. H., Post, J. W., Lampi, K., Ramaekers, H., Rietveld, L. C., and Roest, K. (2011). "Water recovery from sewage using forward osmosis." *Water Science and Technology*, 64, 1443–1449.

McCutcheon, J. R., McGinnis, R. L., and Elimelech, M. (2005). "A novel ammonia – carbon dioxide forward (direct) osmosis desalination process." *Desalination*, 174, 1, 1–11.

McCutcheon, J. R., McGinnis, R. L., and Elimelech, M. (2006). "Desalination by ammonia-carbon dioxide forward osmosis: Influence of draw and feed solution concentrations on process performance." *Journal of Membrane Science*, 278, 1–2, 114–123.

Mi, B., and Elimelech, M. (2008). "Chemical and physical aspects of organic fouling of forward osmosis membranes." *Journal of Membrane Science*, 320, 1–2, 292–302.

Phillip, W. A., Yong, J. S., and Elimelech, M. (2010). "Reverse draw solute permeation in forward osmosis: modeling and experiments." *Environ. Sci. Technol.*, 44, 5170–5176.

Phuntsho, S., Shon, H. K., Hong, S., Lee, S., and Vigneswaran, S. (2011). "A novel low energy fertilizer driven forward osmosis desalination for direct fertigation: Evaluating the performance of fertilizer draw solutions." *Journal of Membrane Science*, 375, 1–2, 172–181.

Qin, J. J., Lay, W. C. L., and Kekre, K. A. (2012). "Recent developments and future challenges of forward osmosis for desalination: A review." *Desalination and Water Treatment*, 39, 1–3, 123–136.

She, Q., Jin, X., and Tang, C. Y. (2012). "Osmotic power production from salinity gradient resource by pressure retarded osmosis: Effects of operating conditions and reverse solute diffusion." *Journal of Membrane Science*, 401–402, 0, 262–273.

Tang, C. Y., She, Q., Lay, W. C. L., Wang, R., and Fane, A. G. (2010). "Coupled effects of internal concentration polarization and fouling on flux behavior of forward osmosis membranes during humic acid filtration." *Journal of Membrane Science*, 354, 1–2, 123–133.

Yip, N. Y., Tiraferri, A., Phillip, W. A., Schiffman, J. D., Hoover, L. A., Kim, Y. C., and Elimelech, M. (2011). "Thin-film composite pressure retarded osmosis membranes for sustainable power generation from salinity gradients." *Environmental science & technology*, 45, 10, 4360–4369.

Yun, T., Kim, Y.-J., Lee, S., Hong, S., and Kim, G. I. (2013). "Flux behavior and membrane fouling in pressure-assisted forward osmosis." *Desalination and Water Treatment*, in press.

Yun, T., Kim, Y.-J., Lee, S., Kim, H., and Hong, S. (2012). "Effect of applied pressure on water flux in forward osmosis: Membrane types and operation conditions." *The 7th Conference of Aseanian Membrane Society (AMS7)*, Busan Korea, AMS7.

Yun, T., Park, S.-M., Kim, Y., Lee, S., and Kim, G. I. (2012). "Effect of external pressure and feed/draw solution velocity on flux characteristics in hollow fiber forward osmosis membrane." *International Desalination Workshop*, Jeju, Korea.

Zhao, S., Zou, L., Tang, C. Y., and Mulcahy, D. (2012). "Recent developments in forward osmosis: Opportunities and challenges." *Journal of Membrane Science*, 396, 0, 1–21.

CHAPTER 20

Membrane Development for Pressure Retarded Osmosis

X. Li
G. Han
T. S. Chung

Salinity gradient energy (or osmotic power) has huge potential as a sustainable blue energy source. This chapter focuses on membrane development for osmotic power generation via pressure retarded osmosis processes. The concept of osmotic power generation and the fundamentals of pressure retarded osmosis (PRO) are presented in terms of osmotic processes, water and salt permeabilities, concentration polarization and power density. In addition, preferential types of PRO membranes and some recent exciting developments are reviewed.

20.1 RENEWABLE AND BLUE SALINITY-GRADIENT ENERGY

The explosive increase in energy demand and the shrinking reserves of fossil fuels have magnified the worldwide search for alternative energy sources nowadays. The global trend toward environmental sustainability has shifted the future power production from conventional fuels and internal combustion engines to renewable and environmentally friendly energy in order to reduce the emission of greenhouse gases. Salinity-gradient energy is one kind of renewable blue energy, which uses the Gibbs energy retrieved from the mixing of two salty solutions with different concentrations. This potential energy can be extracted in terms of electrical energy via membrane based pressure-retarded osmosis (PRO) and reverse electrodialysis (RED). Recently, much more attentions have been gained from both academia and industry in salinity-gradient energy generation via these two processes. As shown in Figure 20.1, the number of publications from the SciFinder database on salinity gradient energy is soaring in recent years. Theoretically, a large amount of energy can be harvested, but the problem is "how to effectively acquire it"? The membrane-based PRO technique may provide one of the answers.

Theoretically, salinity gradient energy is available worldwide where salty solutions of different salinity mix. For example, a large amount of energy can be

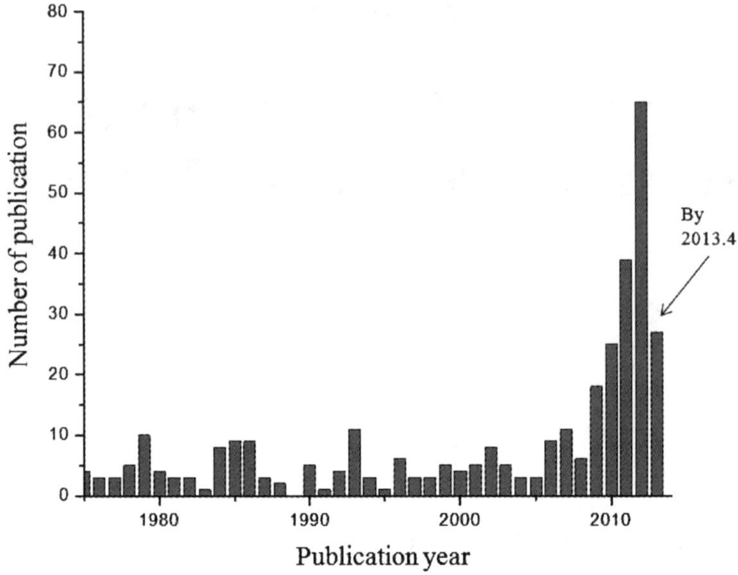

Figure 20.1. Number of publications on harvesting salinity gradient energy including pressure retarded osmosis and reverse electro-dialysis from the late 1970s

produced when fresh river water flows into the sea. The average osmotic pressure of seawater is 28 bar equivalent to the pressure of a water head of 286 m. The extent of such energy in the ocean is estimated to be 1750–2000 TWh per year (Pattle 1954; Wick and Schmitt 1977).

Not only can seawater and river water be used as feed streams for osmotic power generation, but also industrial plants such as reverse osmosis (RO) plants that discharge substantial volumes of water with a high salinity. When RO retentate is used as the draw solution instead of seawater, the salinity gradient between RO retentate and river water is much greater than that between seawater and river water (about 7.9–8.5 vs. 3.5 wt%). The former can result in a much higher osmotic pressure (about 70–77 vs. 28 bar) and osmotic energy than the latter. Consequently, the operation pressure may become 20–38 bar depending on the membrane properties and produce a much higher power density. In addition, if osmotic power generation and RO plants are successfully integrated, not only can it make seawater desalination less energy dependent and more sustainable, but also significantly alleviate the disposal and environmental issues of waste RO retentate. Furthermore, since the RO retentate has been well pre-treated in its previous processes, the use of RO retentate may significantly reduce membrane fouling in the high pressure compartment. As a result, the integration will save some of expensive pre-treatment costs originally required for seawater before entering the PRO power generator.

After the early works in the 1970s, the progress on the study of salinity gradient energy generation was slow. Up to the present day, commercialization

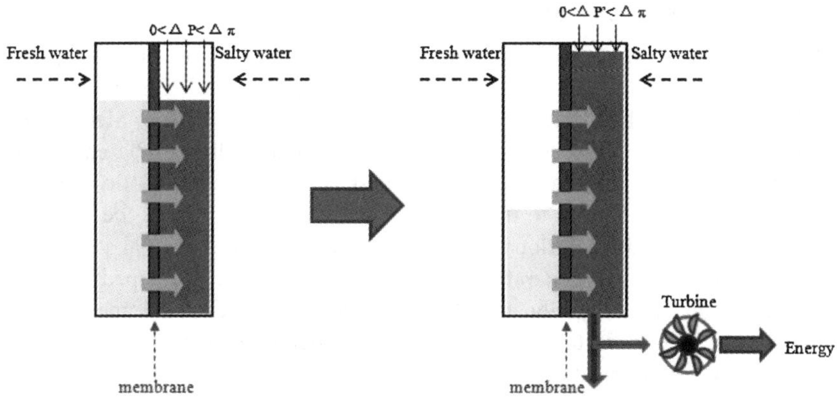

Figure 20.2. Illustration of pressure retarded osmosis for power generation

and industrial exploitation are still limited. Currently, two membrane-based approaches are applied to harvest the salinity gradient energy into electricity. They are (1) pressure retarded osmosis (PRO) and (2) reverse electrodialysis (RED). Even though they are fundamentally different, the major challenge of PRO and RED is the molecular design of the membranes partitioning the low and high salinity solutions which are the hearts of both technologies. In this chapter, we will only focus on PRO for harvesting salinity gradient energy.

20.2 FUNDAMENTALS OF PRESSURE RETARDED OSMOSIS (PRO)

There are four major components involved in the PRO process; namely, (1) a salty feed solution comprising a high osmotic pressure (referred to as "draw solution" thereafter), (2) a feed solution (i.e., usually fresh water) containing a lower osmotic pressure, (3) a semipermeable membrane working as a barrier between the two feeds and (4) a hydro-turbine converting the hydraulic pressure to electricity. Due to the salinity gradient (i.e., osmotic pressure difference) between the two feeds, water transports across the membrane naturally from the low salinity one to the high salinity one and build up the hydraulic pressure in the compartment of the high salinity one. Since water transports from a low pressure chamber to a high pressure one due to the osmotic gradient across the membrane, this process is called as "pressure retarded osmosis" by Loeb in 1975. Figure 20.2 illustrates the concept of pressure retarded osmosis for power generation.

Conceptually, PRO is an osmotic process between forward osmosis (FO) and reverse osmosis (RO). Although the hydraulic pressure on the saltwater is partially retarding the water flow through the membrane, water still permeates from the freshwater to the saltwater because the trans-membrane hydraulic pressure is lower than the osmotic pressure difference across the membrane. Since the salty water in the high pressure compartment continuously gains water in volume, it is able to produce extra work by pushing the turbine and generating electricity.

20.2.1 Development of PRO

After Norman's (1974) proposal on water salination as a source of energy, Loeb (Loeb 1976; Loeb et al. 1976; Loeb and Mehta 1979; Loeb et al. 1997; Loeb 1998; Loeb 2002) and Metha (Metha and Loeb 1978; Metha and Loeb 1979; Mehta 1982) were the pioneers in exploring membrane processes for salinity-gradient energy generation. Loeb and Metha conducted the first experiments of osmotic power via PRO processes. In addition to utilizing seawater and river water as the feed pair, they used high salinity solutions such as Dead Sea water and normal seawater as the feed pair for power generation. Later, Lee et al. (1981) developed the basic theoretical model to describe the PRO performance of a membrane from its osmosis and RO characteristics. However, the energy generation from salinity gradient was economically infeasible at that time because of ineffective PRO membranes.

Statkraft, a Norwegian power company, initiated serious research on osmotic power since 1997 and built the first prototype plant in 2009 by mixing river water and seawater across a semi-permeable membrane under the PRO mode. It demonstrated the potential of salinity gradient for energy generation. However, the early prototype plant used conventional cellulose acetate membranes with power densities of less than $1.5\,W/m^2$ which was well below the economically feasible value of $5\,W/m^2$ for flat sheet or $3\,W/m^2$ for hollow fiber membranes (Gerstandt et al. 2008; Thorsen and Holt 2009). In addition, the early generation membrane was operated only at about 6 bars because of membrane limitations. As a result, the percentage of water in the high-pressure brine chamber to the hydro-turbine could only be run at about 20%, which was below the preferred value of about 33% (Gerstandt et al. 2008; Skilhagen et al. 2008).

20.2.2 Development of PRO Membranes

Theoretically, FO and PRO membranes share similar osmotic mechanism and both require highly selective membranes with high water fluxes and low salt fluxes. Therefore, recent advancements in forward osmosis (FO) membranes (Wang KY et al. 2009; Yang et al. 2009; Hausman et al. 2010; Wang KY et al. 2010; Wang R et al. 2010; Zhang et al. 2010; Qiu et al. 2011; Widjojo et al. 2011; Li et al. 2012; Wang KY et al. 2012) and the commercialization of FO membranes by Hydration Technology Innovations Inc (HTI) would facilitate the development of PRO membranes. However, this proposition is only partially true because pioneering works on PRO membranes imply that PRO membranes have much stricter requirements on water flux, salt permeability and robustness than FO membranes because the former is operated under high trans-membrane pressures, while the latter is under no or negligible pressures. Therefore, different strategies must be implemented to design both membranes.

20.2.3 Basic Energy Calculations for PRO Processes

In terms of energy production, the specific power or power density (W) is defined as the osmotic energy output per unit membrane area, which can be calculated by the product of trans-membrane hydraulic pressure (ΔP) and water permeating flux (J_w) across the membrane (Lee et al. 1981; Gerstandt et al. 2008):

$$W = J_w \times \Delta P = A \times (\Delta\pi - \Delta P) \times \Delta P \tag{20.1}$$

W is a major performance indicator of the PRO membrane because it determines the required amount of membrane area and the size of a PRO plant for a given capacity of energy production. Mathematically, by differentiating Eq. (20.1) with respect to ΔP, the maximum power density can be obtained when the hydrostatic pressure difference is equal to the half of the osmotic pressure difference, $\Delta\pi/2$, across the membrane. Therefore, the optimal operation pressure of the saltwater compartment is about 13.5 bar for seawater. This value may be higher for other highly concentrated salty water if the membrane is mechanically strong. Substituting $\Delta\pi/2$ into Eq. (20.1) yields:

$$W_{\max} = A\frac{\Delta\pi^2}{4} \tag{20.2}$$

Therefore, the maximum power in a PRO system is directly proportional to the water permeability coefficient A, and also proportional to the square of the osmotic pressure difference.

20.2.4 Concentration Polarization in PRO

Theoretically, the water permeation flux (J_w) across an ideal semi-permeable thin film membrane that allows water passage but fully rejects all other solutes or ions can be related to water permeability A, effective osmotic pressure difference $\Delta\pi_m$ and trans-membrane hydraulic pressure difference ΔP as follows:

$$J_w = A(\Delta\pi_m - \Delta p) \tag{20.3}$$

However, when a non-ideal semi-permeable membrane is operated in the PRO process, salt is transported from the saltwater side to the freshwater side due to its concentration gradient. Since salt permeates in the opposite direction of water flux, the salt flux is called as reverse salt flux J_s which can be described as:

$$J_s = B(C_{D,m} - C_{F,m}) \tag{20.4}$$

where B is the salt permeability coefficient of the membrane active layer; and $C_{D,m}$ and $C_{F,m}$ are the solute concentrations at the interfaces of the active layer facing the draw and feed solutions, respectively (Achilli and Childress 2010; Wang KY et al. 2010; Su and Chung 2011; Wang KY et al. 2012; Zhao et al. 2012).

The salt transport is also limited by (1) external concentration polarization (ECP) due to the stagnant layers caused by reduced mixing on the membrane surface and (2) internal concentration polarization (ICP) due to resistance against salt transport in the porous support layer. Therefore, salt gradients exist at membrane outer boundaries as well as inside the membrane support that lower

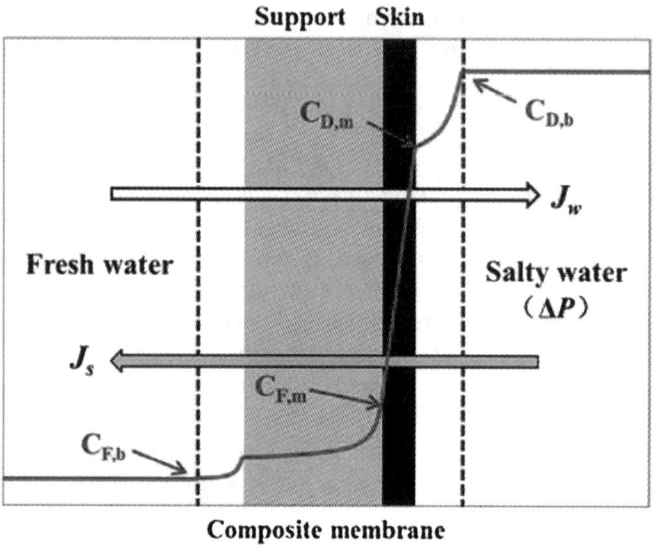

Figure 20.3. Schematic representations of the osmotic profiles of a PRO membrane. $C_{D,b}$ is the solute concentration of the bulk saltwater, $C_{D,m}$ the solute concentration at the dense top layer of the membrane, $C_{F,m}$ the solute concentration inside the membrane between the dense top-layer and the porous support, $C_{F,b}$ the solute concentration in the bulk freshwater

the effective osmotic driving force for water transport across the membranes. Figure 20.3 shows the transport of water and salt through a semi-permeable membrane synthesized on a porous support for PRO processes. ECP can be significantly lowered by intensified mixing with the aid of high flow rates on the membrane surface, while ICP cannot be completely eliminated but somewhat reduced because it happens within the porous support layer. As a result, the effective osmotic pressure across the selective layer of the membrane drops from $(\pi_{D,b} - \pi_{F,b})$ to $(\pi_{D,m} - \pi_{F,m})$. One must increase this effective driving force in order to increase the PRO performance.

Mathematically, the reverse salt flux is contributed by two components, salt diffusion due to concentration gradient and convective flow arising from water flux as follows (Lee et al. 1981; Su and Chung 2011; Achilli and Childress 2010; Zhang et al. 2013):

$$J_s = \frac{D\varepsilon}{\tau}\frac{dC(x)}{dx} - J_w C(x) \tag{20.5}$$

where ε is the porosity of the support layer, and τ is the tortuosity. At steady state, Eqs. (20.4) and (20.5) should be equal:

$$B(C_{D,m} - C_{F,m}) = \frac{D\varepsilon}{\tau}\frac{dC(x)}{dx} - J_w C(x) \tag{20.6}$$

Integration of Eq. (20.6) over the support layer thickness l results in the following equation:

$$J_w = A\frac{\pi_{D,b}\exp(-\frac{J_w}{k}) - \pi_{F,b}\exp(\frac{J_w S_t}{D})}{1+\frac{B}{J_w}(\exp(\frac{J_w S_t}{D})-1)} - A\Delta P \quad (20.7)$$

where $\pi_{D,b}$ is the bulk osmotic pressure of the salty water; $\pi_{F,b}$ is the bulk osmotic pressure of the fresh water; and S_t is the structural parameter:

$$S_t = \frac{\tau l}{\varepsilon} \quad (20.8)$$

The salt reverse flux can then be expressed as a function of J_w using van't Hoff factor i (Su and Chung 2011; She et al. 2012):

$$J_s = \frac{B}{iRT}\left(\frac{J_w}{A} + \Delta P\right) \quad (20.9)$$

Mathematically, the ECP effect could be expressed as follows (McCutcheon and Elimelech 2006; Su and Chung 2011):

$$\pi_{D,m} = \pi_{D,b}\exp\left(-\frac{J_w}{k}\right) \quad (20.10)$$

where k is the mass transport coefficient which has the following relationship with process conditions:

$$k = \frac{ShD}{d_h} \quad (20.11)$$

where D is the solute diffusion coefficient, d_h is the hydraulic diameter; and Sh is the Sherwood number of a laminar flow in a rectangular channel given by:

$$Sh = 1.85\left(ReSc\frac{d_h}{L}\right)^{0.33} \quad (20.12)$$

where Re is the Reynolds number; Sc is the Schmidt number; and L is the length of the channel.

20.3 MEMBRANE FOR PRESSURE RETARDED OSMOSIS (PRO)

20.3.1 Early PRO Works Using RO and FO Membranes

Loeb and his co-workers studied PRO using commercially available RO hollow fiber membranes for seawater desalination (Loeb 1976; Loeb et al. 1976). The membranes possessed a typical asymmetric structure with an aromatic polyamide selective layer on top of a porous hollow fiber support. These RO membranes could withstand a hydraulic pressure up to 75 atm but had a relatively low water permeation flux. As a result, it had a low power density. Jellinek and Masuda (1981) employed flat-sheet cellulose triacetate (CTA) membranes for PRO experiments. The maximum hydraulic pressure applied in their studies was about 17 atm, and the maximum power was about 0.26 W. Mehta (1982) investigated several types of RO membranes for PRO applications. Reduction in water permeation coefficients were observed in PRO processes but no permanent damage was found to these membranes after tests. Nevertheless, all aforementioned PRO studies showed power densities far below expectation, which were due to the severe ICP occurring inside the RO membranes (McCutcheon and Elimelech 2008). The thick support layer in RO membranes retarded the free diffusion of ions and thus significantly reduced the effective driving force.

In the last two decades, in addition to the flat-sheet cellulose triacetate (CTA) membranes commercialized from HTI (Figure 20.4), various FO membranes have been developed (McCutcheon and Elemelech 2006; Wang KY et al. 2009; Hausman et al. 2010; Wang KY et al. 2010; Wang R et al. 2010; Zhang et al. 2010; Li et al. 2012; Zhao et al. 2012). Some of them had been tested for PRO, but most of them exhibited very low power densities because membranes failed at very low hydraulic pressures. One of the best early generation FO membranes showed a power density of 10.6 W/m^2 using 1 M NaCl seawater brine and 40 mM NaCl wastewater as feeds, but their membrane could only withstand a hydraulic pressure of less than 10 bar (Chou et al. 2012).

In the case of HTI membranes, various PRO data have been reported. The CTA membranes are embedded in woven supports with an overall thickness of only around 50 mm, as shown in Figure 20.4. Achilli and his coworkers

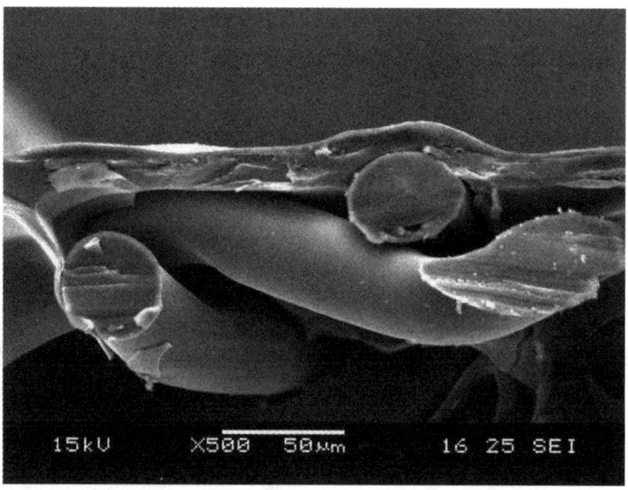

Figure 20.4. A SEM image of HTI's FO membrane. A cellulose triacetate layer is laid over a polyester mesh

(Achilli et al. 2009) reported that the HTI flat-sheet FO membrane had a maximum power density of 2.7 W/m^2 at 9.7 bar using 35 g/L sodium chloride in the salty water compartment. Xu et al. (2010) tested HTI commercial FO spiral-wound modules for PRO and found the maximum power density of about 0.5 W/m^2 using synthetic seawater as the draw solution. She et al. (2012) studied three different HTI CTA membranes and obtained a peak power density of about 4 W/m^2 utilizing a 1 M NaCl draw solution. All of them are below the economically feasible value of 5 W/m^2 set by Statkraft for flat sheet membranes (Gerstandt et al. 2008; Thorsen and Holt 2009).

Clearly, breakthroughs on membrane materials and membrane design are urgently needed to increase membrane power density. Materials with enhanced water transport properties and membranes with superior semi-permeable characteristics and robust mechanical properties become absolutely necessary to harvest osmotic power from salinity gradient systems.

20.3.2 Recently Developed PRO Membranes

20.3.2.1 Thin-film Composite Membranes

Thin-film composite (TFC) membranes comprising an aromatic polyamide selective skin sitting on a microporous support have dominated RO and NF membranes since it was invented in the 1970s (Cadotte 1977; Petersen 1993). The commercially available TFC-RO membranes consist of an asymmetric structure; namely, (1) a thin polyamide selective layer, (2) a porous intermediate layer for easy interfacial polymerization on top of it and better interactions with underneath non-woven fabrics, and (3) a layer of non-woven fabrics layer to withstand high pressures. The thin polyamide layer is synthesized via an in situ interfacial polymerization reaction between an aqueous poly-functional amine solution and a poly-functional acyl chloride dissolved in an apolar organic solvent. Because of (1) the solubility preference of monomers in two different phases, (2) immiscibility between the organic and aqueous phases, and (3) extremely rapid reaction, the interfacial polymerization generally occurs on the organic side near the interface and produces a defect-free ultrathin polyamide film. Comparing to the conventional asymmetric membranes made by one step phase inversion process, the TFC membranes have superior advantages such as higher water permeability, greater solute rejection, and availability of various moieties for the interfacial polymerization. Figure 20.5 shows images of a typical TFC membrane. The "ridge-and-valley" structure is the characteristic morphology of polyamide layers synthesized from the in situ interfacial polymerization.

Yip et al. (2010) and Wang R et al. (2010) are the pioneers in fabricating TFC flat-sheet and hollow fiber membranes, respectively, for FO applications. Widjojo et al. (2011) and Wang KY et al. (2012) advanced the membrane fabrication and FO performance of TFC FO membranes by employing hydrophilic supports. This is due to the fact that a hydrophobic substrate makes it difficult to be fully wetted by water, thus aggravating the internal concentration polarization (ICP) effects and decreasing water flux. Some of these TFC FO membranes have been tested for

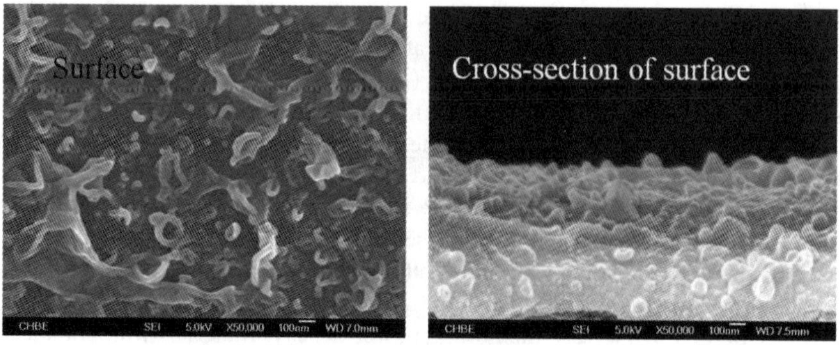

Figure 20.5. Surface SEM images of a typical TFC membrane

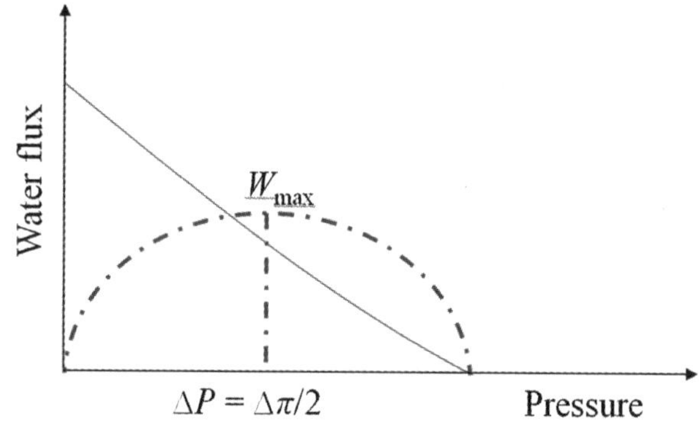

Figure 20.6. Power density and flux in PRO process

PRO applications. Yip et al. (2011) extrapolated their data and predicted a peak power density of 9.21 W/m^2 using seawater and river water as feeds. However, no experimental data was presented. Chou et al. (2012) reported their TFC hollow fiber membranes for PRO applications and showed a power density of 10.6 W/m^2 using 1 M NaCl as the draw solution. However, the membranes collapsed at 10 bar. Clearly, traditional TFC FO membranes must be designed differently for PRO applications.

20.3.2.2 Reinforcements on Membrane Supports

As expressed by Eqs. 20.1 and 20.2, the power density is a product of the hydraulic pressure difference across the membrane (ΔP) and water permeation flux (J_w), and the ideal maximum power density occurs at $\Delta P = \Delta\pi/2$ (Figure 20.6) and is proportional to $(\Delta\pi/2)^2$. Therefore, desirable PRO membranes must have the ability to withstand high hydraulic pressures in order to maximize the power density. Taking seawater and river water as feed streams, the optimum applied pressure is about 13 bar. This pressure would be increased to about 34 bar if model RO retentate (i.e., 79 g/L NaCl, twice the concentration of seawater) is used as the

Table 20.1. Mechanical properties of PAN supports

Membrane ID	*Elongation at break (%)*	*Tensile strength (MPa)*	*Young's modulus (MPa)*
PAN-10%	4.1 ± 0.6	2.4 ± 0.3	135.2 ± 10.7
PAN-15%	9.4 ± 2.7	3.7 ± 0.7	182.4 ± 13.7
PAN-18%	12.1 ± 3.5	4.2 ± 0.6	194.3 ± 17.2
PAN-22%	14.0 ± 3.7	4.5 ± 0.4	203.5 ± 18.1

SOURCE: Zhang et al. (2013); reproduced with permission from Elsevier

draw solution. Under this condition, most FO membranes would collapse in the high-pressure brine chamber.

Conventional FO membranes would exhibit compaction and deformation inevitably when subjected to high pressure PRO processes over a period of time, resulting in a densified porous support layer and a defective dense layer, and thus deteriorate PRO performance. Several approaches have been proposed to overcome this complicated issue.

Zhang et al. (2013) demonstrated that one can improve the mechanical properties of membrane supports by increasing polymer concentration in casting dopes and enhance water flux by post solvent treatments of the TFC layer. As a consequence, the resultant TFC membranes showed much higher power density. Polyacrylonitrile (PAN) porous support membranes were used in their study. Since Young's modulus, tensile strength and elongation at break are typical parameters that characterize membrane mechanical properties, Table 20.1 shows these properties increasing with an increase in polymer concentration in casting solutions. For example, the support made from 22 wt% PAN has a tensile strength around 1.5 times of the one from 10 wt% PAN, while the elongation at break shows an even more dramatic increase of almost 3 times. As a result, the toughness of the PAN support increases with increasing PAN concentration as toughness is a product of strain and stress. Meanwhile, the Young's modulus increases to around 1.5 times as well.

The TFC membranes synthesized on these PAN supports show initial water fluxes ($\Delta P = 0$ bar) of 26 $\mathrm{L m^{-2} h^{1}}$ (LMH), ~22 LMH, ~19 LMH, and ~17 LMH from supports made from 10, 15, 18 and 22 wt% PAN, respectively, when testing under the PRO mode using a 3.5 wt% NaCl solution as the model seawater and DI water as the feed. The flux decline can be understood due to smaller pore sizes, lower porosity and higher tortuosity of supports made from higher polymer concentrations. However, the maximum power density calculated from Eq. (20.1) shows a reversed trend with the initial water flux trend against the PAN concentration. It increases from about 0 to 0.88 $\mathrm{W/m^2}$. This interesting phenomenon arises from the fact that the burst pressure increases with an increase in PAN concentration as shown in Figure 20.7 and the theoretic maximum power density is proportional to ΔP.

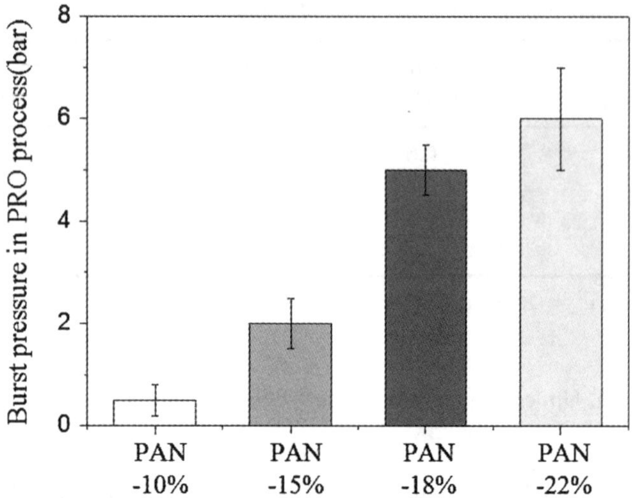

Figure 20.7. Burst pressures of TFC membranes synthesized on PAN supports from different polymer concentrations
SOURCE: Adapted from Zhang et al. 2013; reproduced with permission from Elsevier

The burst pressure in their study was defined as the hydraulic pressure where water flux in the PRO process starts to flow from the model seawater across the membrane to the feed water due to the irreversible membrane damage. At the burst pressure, membranes usually show a pure water permeability of around 5 to 10 times larger than the original value and a significant enhanced salt permeability. By increasing the PAN concentration in the casting solution, much stronger supports are obtained, which reduce the possibility of membrane damage and hence increase the burst pressure. Since a higher PAN concentration increases the burst pressure from 0.5 bar to 6 bar and since power density is proportional to both water flux and hydraulic pressure difference, power density increases with PAN content due to the rapidly raised burst pressure of the substrates.

Li et al. (2013) fabricated TFC PRO membranes and investigated the reinforcement of the membrane supports made of Torlon® polyamide-imide. Torlon® (hereafter referred to as PAI) is a commercially available polymer with properties from both polyamide and polyimide polymers. Figure 20.8 shows the effect of membrane morphology on membrane deformation and compaction. When subjected to high pressures, serious deformation can be found in the membrane with straight and long finger-like macrovoids. In comparison, the membrane with a fully sponge-like structure which was cast from the same polymer solution of the former has much less compaction. Clearly, under the same hydraulic pressure, the straight and long finger-like macrovoid makes an adverse contribution to the mechanical strength of a TFC PRO membrane (Peng et al. 2008).

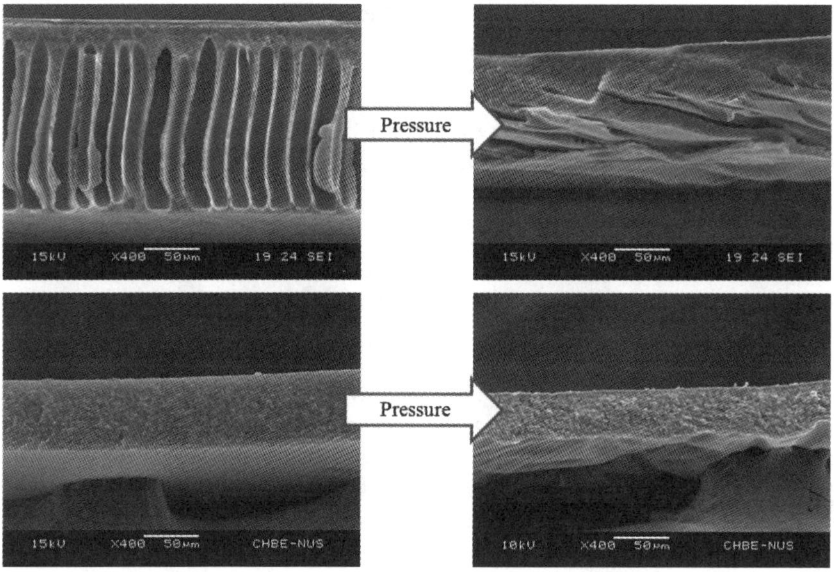

Figure 20.8. Comparison of membrane deformation before and after compression from different original morphologies

SOURCE: Adapted from Li et al. 2013; reproduced with permission from Elsevier

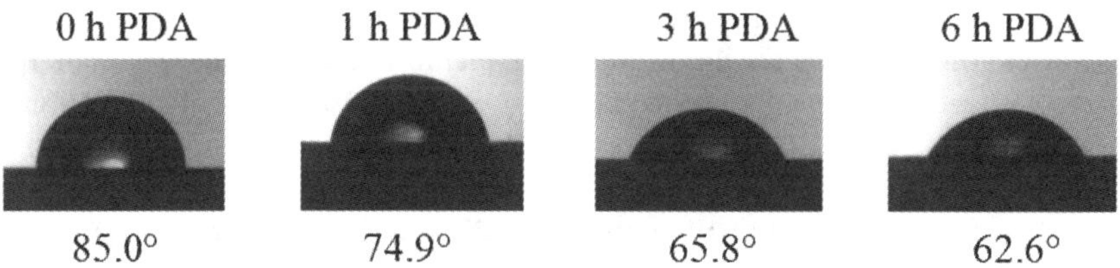

Figure 20.9. Evolution of contact angle for the surface of PDA modified PAI supports

SOURCE: Adapted from Li et al. 2013; reproduced with permission from Elsevier

Li et al., (2013) also found that a polydopamine (PDA) modification to the PAI support can benefit the PRO membrane in terms of both hydrophilicity and mechanical strength. Figure 20.9 illustrates the evolution of water contact angle of PAI membranes as a function of the duration of PDA modification. In a weak alkaline solution, dopamine molecules start to self-polymerize and keeps reacting with other dopamine molecules. Gradually, a thin PDA layer is formed on top of the PAI membrane and alters the nature of membrane surface from hydrophobic into hydrophilic. Therefore, the water contact angle decreases with increasing modification time and then reaches a plateau after a long immersion. Clearly, the hydrophilicity of PDA modified PAI substrates can be effectively manipulated by controlling the modification time. This tunability provides a convenient platform for the fabrication of desirable substrates for TFC membranes.

Table 20.2. Changes in mechanical properties of PAI membranes after PDA modification

	Elongation at break (%)	Tensile strength (MPa)	Young's modulus (MPa)
Control membrane	6.3 ± 1.8	3.33 ± 0.56	194.2 ± 43.8
PDA modified membrane	6.0 ± 0.7	6.07 ± 0.23	259.8 ± 6.2

SOURCE: Li et al. (2013); reproduced with permission from Elsevier

In addition, the PDA modification reinforces the mechanical strength of PAI supports. Table 20.2 summarizes the measured elongation at break, tensile strength, and Young's modulus of the control membrane, and PDA modified membranes. It reveals that the PDA modification can reinforce membrane's tensile strength, modules and toughness. The tensile strength of the PDA modified membrane is around 1.8 times of the control one, the Young's modulus increases to around 1.3 times as well.

Figure 20.10 compares the responses of the control and PDA modified membranes under a hydraulic pressure of 14 bar. The PDA modified membrane was prepared by immersing the control membrane in a PDA solution for 3 hours. Their water permeations decline as a function of compression time. Fast reductions in pure water permeability (PWP) are observed for both membranes. However, the PWP of the PDA modified membrane decreases less than the control membrane throughout the entire test. It reaches a plateau about 10.0% of its initial PWP value that nevertheless doubles the plateau value of the control membrane. When PDA is in contact with PAI membranes, the carbonyl groups of PAI chains may react with amine groups of PDA and a cross-linking via amide linkages is induced. Consequently, the cross-linking reaction enhances chain rigidity and inhibits intra-segmental and inter-segmental mobility (Xiao et al. 2011). As a result, in addition to exhibit better mechanical strength and stability, the PDA modified membrane not only mitigates membrane deformation under high hydraulic pressures but also sustains membrane performance in PRO processes. Figure 20.11 compares their burst pressures in the PRO process. At the burst pressure, the thin film polyamide layer possibly undergoes a deformation and defects are created. The PDA modified membrane is able to withstand a hydraulic pressure of 11 bar which is much higher than the control membrane (i.e., 7 bar). These results confirm the positive effect of PDA modification to the PAI support on PRO performance.

20.3.2.3 Effects of Solvent Treatments on TFC Membranes

It has been known that methanol and ethanol can swell up polymer chains and increase membrane permeability (Shao et al. 2004; Tin et al. 2004; Zuo et al. 2012). By immersing the TFC membrane that was coated by PDA into ethanol for 2 days,

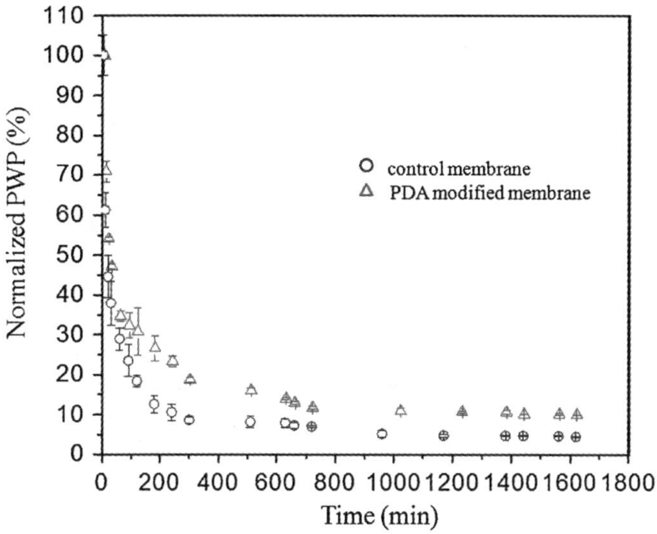

Figure 20.10. Reduction in pure water permeability (PWP) as a function of time for the control and PDA modified PAI membranes
SOURCE: Adapted from Li et al. 2013; reproduced with permission from Elsevier

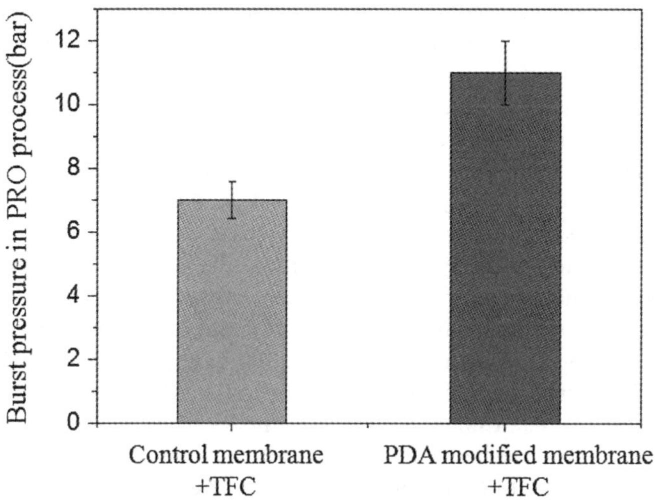

Figure 20.11. Burst pressures of the control TFC and PDA modified TFC membranes
SOURCE: Adapted from Li et al. 2013; reproduced with permission from Elsevier

the initial water flux of the resultant membrane increases from 20 to 38 LMH in PRO tests, as shown in Figure 20.12. Moreover, the mechanical stability of the TFC membrane is also enhanced after the ethanol treatment. The membrane still works at 10 bar and shows a water flux of 6 LMH in PRO tests. As a result, the maximum power density has doubled than the original TFC membrane.

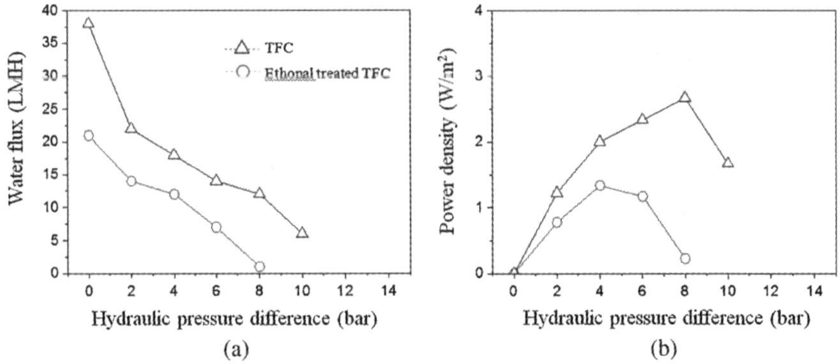

Figure 20.12. The PRO performance versus different hydraulic pressures of the PAN 22 wt% substrate-based TFC membranes

SOURCE: Adapted from Zhang et al. 2013; reproduced with permission from Elsevier

This is due to the fact that ethanol not only extracts the non-reacted monomers from the TFC membranes, but may also help remove the short and probably defective polyamide chains (Zhang et al. 2013). It may even take away a small portion of the components from the PAN support. Such extractions contribute to a clean and more open space within the polyamide layer, resulting in higher water permeability. The removal of short polyamide chains might also minimize the weak points in the skin and increase the overall mechanical stability.

Apart from the chemical changes, ethanol swells up the TFC membrane and improves its wettability. An isolated polyamide skin layer swells by 6% in water (Freger 2004), while ethanol can induce a greater swelling because it participates with both hydrogen bonding and interactions with nonpolar moieties of polyamide due to its lower polarity compared to water. As a result, a higher water permeation rate is obtained due to weak chain–chain interactions, enlarged chain–chain distance and enhanced chain flexibility in the polyamide skin layer.

Positron annihilation lifetime spectroscopy (PALS) has been employed to examine the asymmetric layer structure of both integrated and composite membranes (Tung et al. 2009; Zhang et al. 2011; Li et al. 2012). The membrane immersed in ethanol has a bigger free volume in the skin layer than the original one. By fitting the data using the three-layer model in the VEPFIT program, the thickness of the skin layer could be obtained and shown in Table 20.3. The original polyamide layer has a thickness of 203 ± 42 nm, while the thickness is reduced to 175 ± 17 nm after the ethanol immersion. The 13.8% percentage of thickness decrease is due to the fact that extraction outweighs swelling in terms of selective layer thickness. The enhanced water flux of the ethanol-treated membrane in the PRO process is therefore explained by a larger free volume and a smaller TFC thickness, as revealed by PALS.

However, a further extension of the immersion time in ethanol from 2 days to 3 days leads to a decrease in water flux in the PRO process. For TFC/PAN membranes, the salt permeability dramatically increases to 1.0 LMH even though

Table 20.3. Thickness of skin layer obtained by PALS

	TFC membrane	*Ethanol treated TFC membrane*
Boundary length of the selective layer analyzed by PALS	203 ± 42	175 ± 17

SOURCE: Zhang et al. (2013); reproduced with permission from Elsevier

water permeability also increases to 4.0 LMH/bar (Zhang et al. 2013). As indicated by Eq. (20.7), water permeability A and salt permeability B have opposite effects in determining water flux. A large water permeability favors a high water flux in PRO, while a large B causes a high salt reverse flux across the membrane which not only reduces the effective driving force but also eventually lowers the water flux. The ideal condition to maximize water flux is to increase A but decrease B. Nevertheless, the real situation is governed by the tradeoff relationship between permeability and selectivity where an increasing A is usually accompanied with an increasing B (Geise et al. 2011; Yip et al. 2011). Hence, a delicate balance between A and B is strategically important to achieve a high water flux.

Similar phenomena were reported on TFC/PAI membranes (Li et al. 2013). The water permeability A of the ethanol treated membrane increases to 1.4 times of the un-treated one, while the salt permeability increases by 1.9 times. In the PRO tests, the ethanol treated membrane obtains an initial water flux of 44 LMH at 0 bar comparing to 25 LMH of the untreated one. As a result, the membrane has the maximum power density of 2.84 W/m^2 occurring at 6 bar (Figure 20.13).

20.3.3 Current State-of-the-art PRO Membranes

Based on aforementioned results, membranes for high pressure PRO applications should have a high water permeability (A), low salt permeability (B), small structure parameter (S_t) but sufficient high mechanical strength. Recently, several breakthroughs on TFC PRO membranes have been made for osmotic power generation.

In case of flat-sheet membranes, Han et al. (2013a) designed TFC membranes on top of a porous polyimide membrane support. As shown in Figure 20.14, the support layer shows a fully sponge-like structure with a small structure parameter and excellent mechanical robustness, while the polyamide selective layer was chemically modified using novel post-fabrication procedures to achieve desired water permeability. By carefully controlling the reaction parameters, three batches of TFC-PRO membranes with different mass transport properties were prepared. Figure 20.15 summarizes their water and salt permeability. Compared to the pristine TFC membrane, the TFC membrane modified with 200 ppm NaOCl (TFC200) with a mild post-treatment exhibits an approximate 4-fold increase in A to 5.30 $L\,m^{-2}\,h^{-1}\,bar^{-1}$ and a big B value of 2.00 $L\,m^{-2}\,h^{-1}$. When further increasing

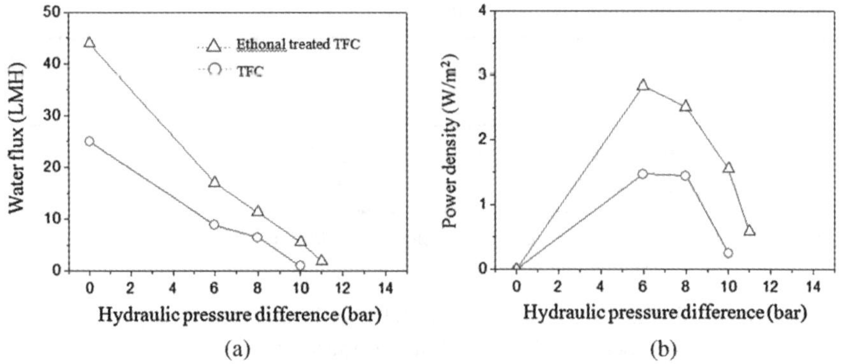

Figure 20.13. The PRO performance versus hydraulic pressure of the PAI substrate-based TFC membranes

SOURCE: Adapted from Li et al. 2013; reproduced with permission from Elsevier

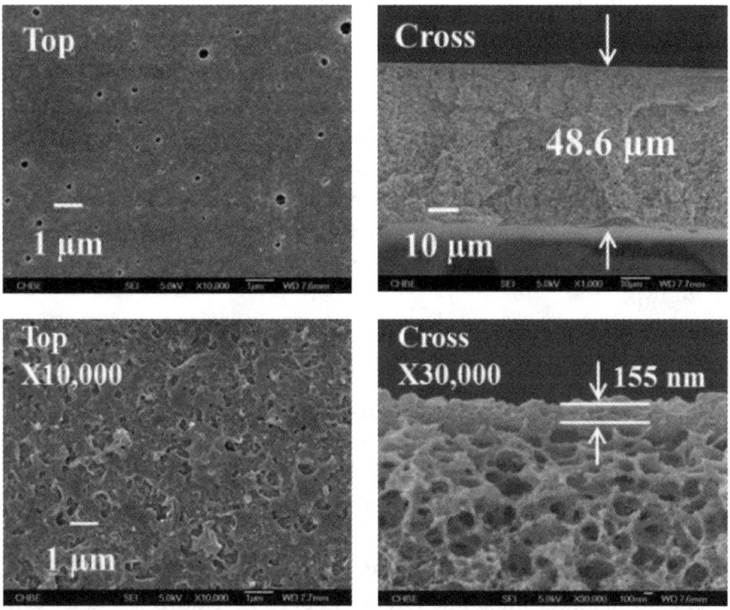

Figure 20.14. SEM images of the hand-cast polyimide membrane support (top) and TFC PRO membrane (bottom)

SOURCE: Adapted from Han et al. 2013a; reproduced with permission from Elsevier

the degree of modification, the TFC membrane modified with 600 ppm NaOCl (TFC600) shows the highest average A and B values of 10.03 $\mathrm{L\,m^{-2}\,h^{-1}\,bar^{-1}}$ and 5.40 $\mathrm{L\,m^{-2}\,h^{-1}\,bar^{-1}}$, respectively. Interestingly, these membranes still have salt rejections (R) of above 80% against NaCl when conducting RO tests at 2 bar.

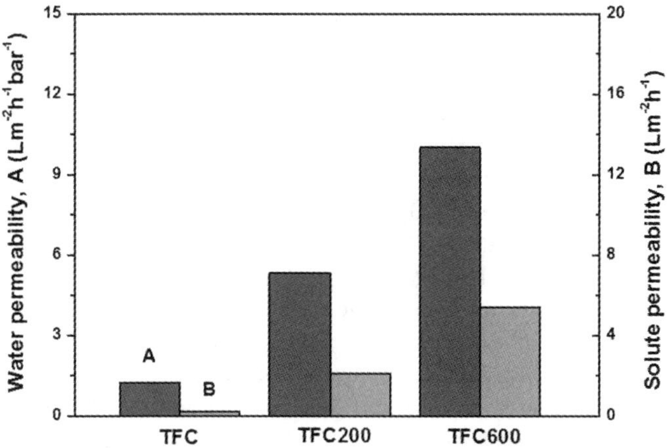

Figure 20.15. Transport properties (A and B) of TFC, TFC200 and TFC600 membranes (All tests were done at 1–4 bar using a 200 ppm NaCl solution)
SOURCE: Adapted from Han et al. 2013a; reproduced with permission from Elsevier

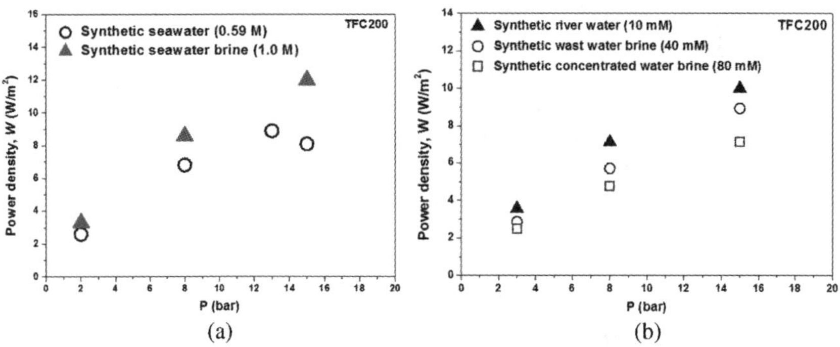

Figure 20.16. Power density of the TFC200 membrane vs. trans-membrane pressure as a function of NaCl concentration; (a) Draw solutions: seawater brine (1 M NaCl) and synthetic seawater (0.59 M NaCl), feed solution: deionized water; (b) Draw solution: seawater brine (1 M NaCl), feed solution: varying from river water to waste water brine and concentrated water brine, and temperature: 25°C
SOURCE: Adapted from Han et al. 2013a; reproduced with permission from Elsevier

The newly developed TFC200 PRO membranes show the best performance which not only exhibit an excellent water permeability ($A = 5.30\ \mathrm{L m^{-2} h^{-1} bar^{-1}}$) and membrane robust, but also overcome the bottlenecks of low power density. Under lab-scale PRO power generation tests, the membranes can withstand trans-membrane hydraulic pressures up 15 bar and exhibit a maximum power density ranging from 7 to 12 $\mathrm{W/m^2}$ using various synthesized seawater and brine as draw solutions (see Figure 20.16). The impressive mechanical stability and attractive power density suggest the great practicability of the newly developed composite membranes for harvesting osmotic energy via PRO process.

Comparing to flat sheet membranes, membranes with a hollow fiber configuration are of great interest because of their high packing density and ease of module fabrication. Most importantly, hollow fiber modules may not require spacers between the membranes (She et al. 2012; Kim and Elimelech 2012; Sivertsen et al. 2013). Not only could this minimize membrane deformation and structure parameter enhancement owing to unavoidable spacer-membrane interactions under high-pressure PRO operations, but also eliminate the extra energy consumption for water transport through woven or non-woven spacers. However, the development of PRO hollow fiber membranes is still in its infancy.

Recently, three advancements have been made. Fu et al. (2013) developed mixed matrix hollow fiber membranes consisting of a PBI (polybenzimidazole)/POSS (polyhedral oligomeric silsesquioxane) outer layer and a PAN/PVP (polyvinylpyrrolidone) inner layer for osmotic power generation. Even though the fiber has a relatively low power density of 2.47 W/m^2 at 7 bar using 1.0 M NaCl as the draw solution, but the dense-selective layer is located at the outer layer. This is the first hollow fiber with an outer dense-selective layer for PRO applications. Zhang et al. (2013) invented a TFC hollow fiber comprising substrates of high asymmetry and high porosity with a small and narrow pore size distribution. The TFC membrane produces a maximum power density of 24.0 W/m^2 at 20.0 bar by using 1 M NaCl as the concentrated brine. This PRO performance surpasses all polymeric membranes in literatures.

Han et al. also developed new design strategies to molecularly construct highly robust TFC PRO hollow fiber membranes with exciting power density for osmotic power generation (Han et al. 2013b). The developed TFC-PRO membranes consist of a selective polyamide skin formed in the lumen side of well-constructed Matrimid hollow fiber supports via interfacial polymerization (see Figure 20.17). The hollow fiber membranes can withstand trans-membrane pressures up to 16 bar and exhibit a peak power density as high as 12 W/m^2 and 10 W/m^2 using seawater brine (1.0 M NaCl) as the draw solution and river water and waste water brine as the feed, respectively (see Figure 20.18). The newly developed TFC hollow fiber membranes show great potential to be applied in PRO processes for osmotic power harvesting.

20.4 CONCLUDING REMARKS

Since most conventional FO membranes are designed for no or low-pressure operation environments, currently available FO membranes are likely to be damaged under this high pressure condition. High pressure PRO experiments must be conducted in order to obtain the real membrane power density. Most theoretical extrapolations based from water flux vs. pressure relationship and the initial water flux to estimate power density are misleading because the real power density deviates a lot from the predicted one. Although an increase in membrane thickness and polymer concentration during casting or spinning may improve membrane's mechanical strengths, it may also result in a lower water flux and

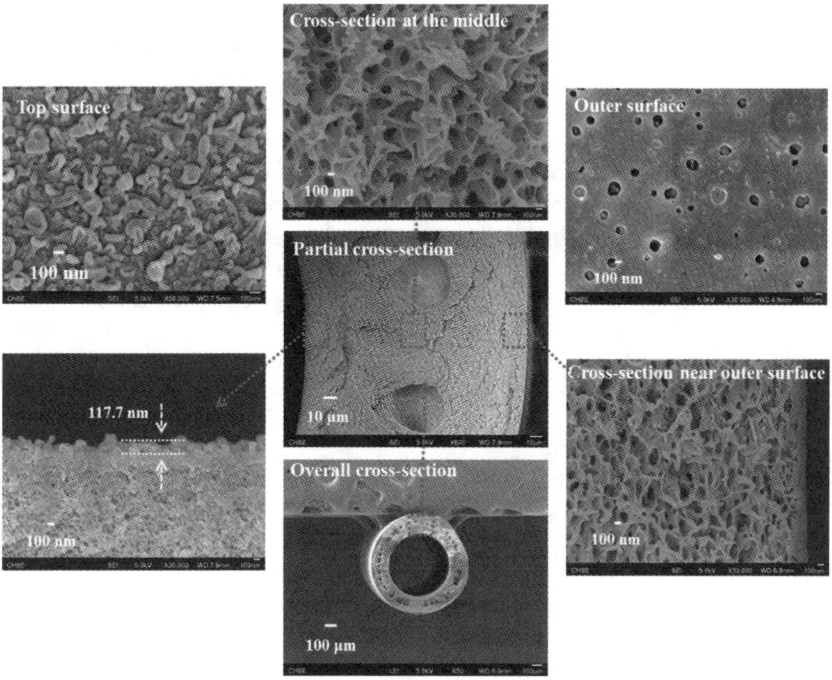

Figure 20.17. SEM micrographs of different bulk and surface morphologies of the hollow fiber supports

SOURCE: Adapted from Han et al. 2013b; reproduced with permission from Elsevier

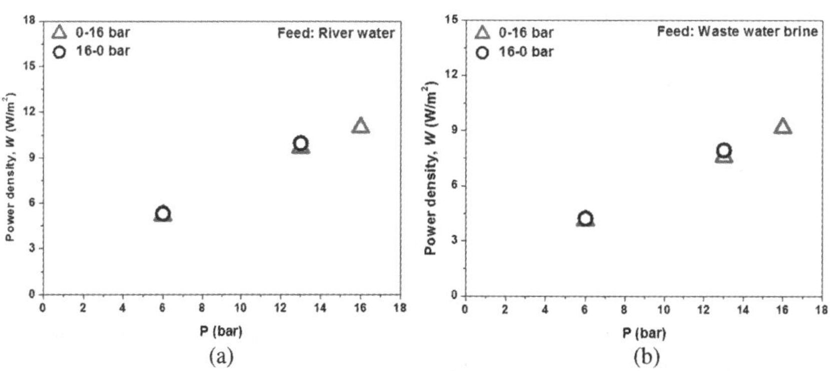

Figure 20.18. Power density of TFC PRO hollow fiber membranes using seawater brine (1 M NaCl) as the draw solution, and river water and waste water brine was feed solutions

SOURCE: Adapted from Han et al. 2013b; reproduced with permission from Elsevier power density. Therefore, a delicate balance among dope formulation, spinning and casting conditions, and membrane morphology must be searched for the fabrication of high performance PRO membranes. One must also consider membrane's physicochemical properties in the wet state as well as their changes under tensile, elongation, compression, and bending stresses. Based on the current works, PRO membranes must be much stronger and more robust than conventional FO membranes without compromising other performance characteristics such as high water flux, high solute rejection, low salt permeability and small structural parameter (i.e., low concentration polarization). Fouling, module fabrication and system design should also be investigated in the future.

20.5 ABBREVIATIONS/NOMENCLATURE

CTA	cellulose triacetate
ECP	external concentration polarization
FO	forward osmosis
HTI	Hydration Technology Innovations Inc
ICP	internal concentration polarization
PAI	polyamide-imide
PALS	Positron annihilation lifetime spectroscopy
PAN	Polyacrylonitrile
PBI	polybenzimidazole
PDA	polydopamine
POSS	polyhedral oligomeric silsesquioxane
PRO	pressure retarded osmosis
PWP	pure water permeability
PVP	Polyvinylpyrrolidone
RED	reverse electrodialysis
RO	reverse osmosis
SEM	Scanning electron microscopy
TFC	Thin-film composite

20.6 ACKNOWLEDGMENTS

This work was funded under the project entitled "Membrane development for osmotic power generation, Part 1. Materials development and membrane fabrication" (1102-IRIS-11-01) and NUS grant number of R-279-000-381-279. This research grant is supported by the Singapore National Research Foundation under its Environmental & Water Technologies Strategic Research Programme and administered by the Environment & Water Industry Programme Office (EWI) of the PUB. Thanks are due to Dr. S. Zhang for her useful suggestions.

References

Achilli, A., Cath, T. Y., and Childress, A. E. (2009). "Power generation with pressure retarded osmosis: an experimental and theoretical investigation." *J. Membr. Sci.*, 343, 42–52.

Achilli, A., and Childress, A. E. (2010). "Pressure retarded osmosis: from the vision of Sidney Loeb to the first prototype installation–review." *Desalination*, 261, 205–211.

Cadotte, J. E. (1977). "Reverse osmosis membrane." US patent 4039440.

Chou, S., Wang, R., Shi, L., She, Q., Tang, C. Y., and Fane, A. G. (2012). "Thin-film composite hollow fiber membranes for pressure retarded osmosis (PRO) process with high power density." *J. Membr. Sci.*, 389, 25–33.

Freger, V. (2004). "Swelling and morphology of the skin layer of polyamide composite membranes: an atomic force microscopy study." *Environ. Sci. Technol.*, 38, 3168–3175.

Fu, F. J., Zhang, S., Sun, S. P., Wang, K. Y., and Chung, T. S. (2013). "POSS-containing delamination-free dual-layer hollow fiber membranes for forward osmosis and osmotic power generation". *J. Membr. Sci.*, 443, 144–155.

Geise, G. M., Park, H. B., Sagle, A. C., Freeman, B. D., and McGrath, J. E. (2011). "Water permeability and water/salt selectivity tradeoff in polymers for desalination." *J. Membr. Sci.*, 369, 130–138.

Gerstandt, K., Peinemann, K. V., Skilhagen, S. E., Thorsen, T., and Holt, T. (2008). "Membrane processes in energy supply for an osmotic power plant." *Desalination*, 224, 64–70.

Han, G., Zhang, S., Li, X., Widjojo, N., and Chung, T. S. (2012). "Thin film composite forward osmosis membranes based on polydopamine modified polysulfone substrates with enhancements in both water flux and salt rejection." *Chem. Eng. Sci.*, 80, 219–231.

Han, G., Zhang, S., Li, X., and Chung, T. S. (2013a). "High performance thin film composite pressure retarded osmosis (PRO) membranes for renewable salinity-gradient energy generation." *J. Membr. Sci.*, 440, 108–121.

Han, G., Wang, P., and Chung, T. S. (2013b). "Highly robust thin-film composite pressure retarded osmosis (PRO) hollow fiber membranes with outstanding power density for renewable salinity-gradient energy generation." *Environ. Sci. Technol.*, dx.doi.org/10.1021/es4013917.

Hausman, R., Digman, B., Escobar, I. C., Coleman, M., and Chung, T. S. (2010). "Functionalization of polybenzimidizole membranes to impart negative charge and hydrophilicity." *J. Membr. Sci.*, 363, 195–203.

Jellinek, H. H., and Masuda, H. (1981). "Osmo-power. Theory and performance of an osmo-power pilot plant." *Ocean Engineering*, 8, 103–128.

Kim, Y. C., and Elimelech, M. (2012). "Adverse impact of feed channel spacers on the performance of pressure retarded osmosis." *Environ. Sci. Technol.*, 46, 4673–4681.

Kimura, K., Amy, G., Drewes, J., Heberer, T., Kim, T. U., and Watanabe, Y. (2003). "Rejection of organic micropollutants (disinfection by-products, endocrine disrupting compounds, and pharmaceutically active compounds) by NF/RO membranes." *J. Membr. Sci.*, 227, 113–121.

Lee, K. L., Baker, R. W., and Lonsdale, H. K. (1981). "Membranes for power-generation by pressure-retarded osmosis." *J. Membr. Sci.*, 8, 141–171.

Li, X., Wang, K. Y., Helmer, B., and Chung, T. S. (2012). "Thin-film composite membranes and formation mechanism of thin-film layers on hydrophilic cellulose acetate propionate substrates for forward osmosis processes." *Ind. Eng. Chem. Res.*, 51, 10039–10050.

Li, X., Zhang, S., Fu, F. J., and Chung, T. S. (2013). "Deformation and reinforcement of thin-film composite (TFC) polyamide-imide (PAI) membranes for osmotic power generation." *J. Membr. Sci.*, 434, 204–217.

Loeb, S. (1975). "Osmotic power plants." *Science*, 189, 654–655.

Loeb, S. (1976). "Production of energy from concentrated brines by pressure-retarded osmosis. I. Preliminary technical and economic correlations." *J. Membr. Sci.*, 1, 49–63.

Loeb, S., Van Hessen, F., and Shahaf, D. (1976). "Production of energy from concentrated brines by pressure-retarded osmosis, II. Experimental results and projected energy costs." *J. Membr. Sci.*, 1, 249–269.

Loeb, S., and Mehta, G. D. (1979). "A two coefficient water transport equation for pressureretarded osmosis." *J. Membr. Sci.*, 4, 351–362.

Loeb, S., Titelman, L., Korngold, E., and Freiman, J. (1997). "Effect of porous support fabric on osmosis through a Loeb-Sourirajan type asymmetric membrane." *J. Membr. Sci.*, 129, 243–249.

Loeb, S. (1998). "Energy production at the Dead sea by pressure-retarded osmosis: challenge or chimera." *Desalination*, 120, 247–262.

Loeb, S. (2001). "One hundred and thirty benign and renewable megawatts from Great Salt Lake? The possibilities of hydroelectric power by pressure-retarded osmosis." *Desalination*, 141, 85–91.

Loeb, S. (2002). "Large-scale power production by pressure-retarded osmosis, using river water and sea water passing through spiral modules." *Desalination*, 143, 115–122.

Mehta, G. D., and Loeb, S. (1978). "Internal polarization in the porous substructure of a semipermeable membrane under pressure-retarded osmosis." *J. Membr. Sci.*, 4, 261–265.

Mehta, G. D., and Loeb, S. (1979). "Performance of permasep b-9 and b-10 membranes in various osmotic regions and at high osmotic pressures." *J. Membr. Sci.*, 4, 335–349.

Mehta, G. D. (1982). "Further results on the performance of present-day osmotic membranes in various osmotic regions." *J. Membr. Sci.*, 10, 3–19.

McCutcheon, J. R., and Elimelech, M. (2006). "Influence of concentrative and dilutive internal concentration polarization on flux behavior in forward osmosis." *J. Membr. Sci.*, 284, 237–247.

McCutcheon, J. R., and Elimelech, M. (2008). "Influence of membrane support layer hydrophobicity on water flux in osmotically driven membrane processes." *J. Membr. Sci.*, 318, 458–466.

Norman, R. S. (1974). "Water salination: a source of energy." *Science*, 186, 350–352.

Pattle, R. E. (1954). "Production of electric power by mixing fresh and salt water in the hydroelectric pile." *Nature*, 174, 660.

Peng, N., Chung, T. S., and Wang, K. Y. (2008). "Macrovoid evolution and critical factors to form macrovoid-free hollow fiber membranes." *J. Membr. Sci.*, 318, 363–372.

Petersen, R. J. (1993). "Composite reverse-osmosis and nanofiltration membranes." *J. Membr. Sci.*, 83, 81–150.

Qiu, C., Qi, S., and Tang, C. Y. (2011). "Synthesis of high flux forward osmosis membranes by chemically crosslinked layer-by-layer polyelectrolytes." *J. Membr. Sci.*, 381, 74–80.

Shao, L., Chung, T. S., Goh, S. H., and Pramoda, K. P. (2004). "Transport properties of crosslinked polyimide membranes induced by different generations of diaminobutane (DAB) dendrimers." *J. Membr. Sci.*, 238, 153–163.

She, Q., Jin, X., and Tang, C. Y. (2012). "Osmotic power production from salinity gradient resource by pressure retarded osmosis: effects of operating conditions and reverse solute diffusion." *J. Membr. Sci.*, 401, 262–273.

Skilhagen, S. E., Dugstad, J. E., and Aaberg, R. J. (2008). "Osmotic power- power production based on the osmotic pressure difference between waters with varying salt gradients." *Desalination*, 220, 476–482.

Sivertsen, E., Holt, T., Thelin, W., and Brekke, G. (2012). "Modelling mass transport in hollow fibre membranes used for pressure retarded osmosis." *J. Membr. Sci.*, 417–418, 69–79.

Sivertsen, E., Holt, T., Thelin, W., and Brekke, G. (2013). "Pressure retarded osmosis efficiency for different hollow fibre membrane module flow configurations." *Desalination*, 312, 107–123.

Su, J. C., and Chung, T. S. (2011). "Experimental and theoretical study of sublayer structure and its effect on concentration polarization and membrane performance in FO processes." *J. Membr. Sci.*, 376, 214–224.

Tin, P. S., Chung, T. S., and Hill, A. J. (2004). "Advanced fabrication of carbon molecular sieve membranes by nonsolvent pretreatment of precursor polymers." *Ind. Eng. Chem. Res.*, 43, 6476–6483.

Thorsen, T., and Holt, T. (2009). "The potential for power production from salinity gradients by pressure retarded osmosis." *J. Membr. Sci.*, 335, 103–110.

Tung, K. L., Jean, Y. C., Nanda, D., Lee, K. R., Hung, W. S., Lo, C. H., and Lai, J. Y. (2009). "Characterization of multilayer nanofiltration membranes using positron annihilation spectroscopy." *J. Membr. Sci.*, 343, 147–156.

Wang, K. Y., Yang, Q., Chung, T. S., and Rajagopalan, R. (2009). "Enhanced forward osmosis from chemically modified polybenzimidazole (PBI) nanofiltration hollow fiber membranes with a thin wall." *Chem. Eng. Sci.*, 64, 1577–1584.

Wang, K. Y., Ong, R. C., and Chung, T. S. (2010). "Double-skinned forward osmosis membranes for reducing internal concentration polarization within the porous sublayer." *Ind. Eng. Chem. Res.*, 49, 4824–4831.

Wang, K. Y., Chung, T. S., and Amy, G. (2012). "Developing thin-film-composite forward osmosis membranes on the PES/SPSf substrate through interfacial polymerization." *AIChE J.*, 58, 770–781.

Wang, R., Shi, L., Tang, C. Y., Chou, S., Qiu, C., and Fane, A. G. (2010). "Characterization of novel forward osmosis hollow fiber membranes." *J. Membr. Sci.*, 355, 158–167.

Wick, G. L., and Schmitt, W. R. (1977). "Prospects for renewable energy from the sea." *Marine Technol. Soc. J.*, 11, 16–21.

Widjojo, N., Chung, T. S., Weber, M., Maletzko, C., and Warzelhan, V. (2011). "The role of sulphonated polymer and macrovoid-free structure in the support layer for thin-film composite (TFC) forward osmosis (FO) membranes." *J. Membr. Sci.*, 383, 214–223.

Yang, Q., Wang, K. Y., and Chung, T. S. (2009). "Dual-layer hollow fibers with enhanced flux as novel forward osmosis membranes for water reclamation." *Environ. Sci. Technol.*, 43, 2800–2805.

Yip, N. Y., Tiraferri, A., Phillip, W. A., Schiffman, J. D., and Elimelech, M. (2010). "High performance thin-film composite forward osmosis membrane." *Environ. Sci. Technol.*, 44, 3812–3818.

Yip, N. Y., Tiraferri, A., Phillip, W. A., Schiffman, J. D., Hoover, L. A., Kim, Y. C., and Elimelech, M. (2011). "Thin-film composite pressure retarded osmosis membranes for sustainable power generation from salinity gradients." *Environ. Sci. Technol.*, 45, 4360–4369.

Yip, N. Y., and Elimelech, M. (2011). "Performance limiting effects in power generation from salinity gradients by pressure retarded osmosis." *Environ. Sci. Technol.*, 45, 10273–10282.

Xiao, Y. C., and Chung, T. S. (2011). "Grafting thermally labile molecules on cross-linkable polyimide to design membrane materials for natural gas purification and CO2 capture." *Energy Environ. Sci.*, 4, 201–208.

Xu, Y., Peng, X., Tang, C. Y., Fu, Q. S., and Nie, S. (2010). "Effect of draw solution concentration and operating conditions on forward osmosis and pressure retarded osmosis performance in a spiral wound module." *J. Membr. Sci.*, 348, 298–309.

Zhang, S., Wang, K. Y., Chung, T. S., Chen, H., Jean, Y. C., and Amy, G. (2010). "Well-constructed cellulose acetate membranes for forward osmosis: Minimized internal concentration polarization with an ultra-thin selective layer." *J. Membr. Sci.*, 360, 522–535.

Zhang, S., Wang, K. Y., Chung, T. S., Chen, H. M., and Jean, Y. C. (2011). "Molecular design of the cellulose ester-based forward osmosis membranes for desalination." *Chem. Eng. Sci.*, 66, 2008–2018.

Zhang, S., Fu, F. J., and Chung, T. S. (2013). "Substrate modifications and alcohol treatment on thin film composite membranes for osmotic power." *Chem. Eng. Sci.*, 87, 40–50.

Zhao, S., Zou, L., Tang, C. Y., and Mulcahy, D. (2012). "Recent developments in forward osmosis: opportunities and challenges." *J. Membr. Sci.*, 396, 1–21.

Zuo, J., Wang, Y., Sun, S. P., and Chung, T. S. (2012). "Molecular design of thin film composite (TFC) hollow fiber membranes for isopropanol dehydration via pervaporation." *J. Membr. Sci.*, 405–406, 123–133.

Index

Page numbers followed by *e, f,* and *t* indicate equations, figures, and tables, respectively.

About the Editors

Ho Kyong Shon, Ph.D., is an ARC Future Fellow and Associate Professor in the School of Civil and Environmental Engineering at the University of Technology Sydney (UTS). He was awarded a Ph.D. from UTS in 2006. After acquiring his Ph. D., he was employed as a UTS Chancellor's Postdoctoral Fellow for 3 years and then worked as a Lecturer at UTS; a position he held from 2009 to 2011. His research interests are membrane-based desalination and nanotechnology for water purification. Since 1998, Dr. Shon has made significant contributions to the understanding of physico-chemical water treatment processes. The initial stage of his research career focused on understanding membrane fouling mechanisms of organic/inorganic materials for wastewater reuse. He has been recently involved in extensive experience of membrane-based desalination processes through the National Centre of Excellence Desalination in Australia, Australia Research Council Discovery Projects, European Desalination project, SeaHERO Desalination project and the GMVP desalination project. He has so far published over 180 refereed journal articles including five patents and fifteen book chapters. He is the Editor of *Water Science and Technology* (IWA publishing), the Associate Editor in *Journal of Hazardous, Toxic, and Radioactive Waste* (American Society of Civil Engineers) and the guest Editor of *Desalination and Water Treatment.*

Sherub Phuntsho, Ph.D., is a research fellow at the School of Civil and Environmental Engineering under the Faculty of Engineering and Information Technology of the University of Technology Sydney (UTS). He received his Ph.D. in Environmental Engineering in 2013 and Master of Engineering in 2006 from UTS. He obtained his Bachelors in Civil Engineering from SVRCET (now NIT Surat, South Gujarat, India) in 2000 under the Government of India competitive scholarship. He has engaged in a number of research works at UTS and most recently relating to membrane processes for water purifications. Currently his major research work includes the development of fertilizer drawn forward osmosis (FDFO), a potentially low energy and lost membrane process for fertigation. Dr. Phuntsho has authored over 40 technical papers including 4 book chapters and over 40 peer-reviewed journal articles. He also regularly reviews research manuscripts for the *Journal of Membrane Science* and *Environmental Science and Technology*. Earlier, Dr. Phuntsho worked for a number of years at the Department of Urban Development and Engineering Services (DUDES) under the Ministry of Works and Human Settlements (MOWHS) where he was involved in a number of urban development projects funded by the Royal Government of Bhutan and Danish International Development Agency or DANIDA between 2001 and 2008.

Tian C. Zhang, Ph.D., P.E., D.WRE, BCEE, F.ASCE, A.EASA, is a Professor in the department of Civil Engineering at the University of Nebraska-Lincoln (UNL), USA. He received his Ph.D. in environmental engineering from the University of Cincinnati in 1994. He joined the UNL faculty in August 1994. Professor Zhang teaches courses related to water/wastewater treatment, remediation of hazardous wastes, and non-point pollution control. Professor Zhang's research involves fundamentals and applications of nanotechnology and conventional technology for water, wastewater, and stormwater treatment and management, remediation of contaminated environments, and detection/control of emerging contaminants in the environment. Professor Zhang has published more than 75 peer-reviewed journal papers, 50 book chapters, and 9 books since 1994. Professor Zhang is a member of American Association for the Advancement of Science (AAAS), the Water Environmental Federation (WEF), and Association of Environmental Engineering and Science Professors (AEESP). Professor Zhang is a Diplomate of Water Resources Engineer (D.WRE) of the American Academy of Water Resources Engineers, and Board Certified Environmental Engineers (BCEE) of the American Academy of Environmental Engineers, Fellow of American Society of Civil Engineers (F.ASCE), and Academician of European Academy of Sciences and Arts. Professor Zhang is the Associate Editor of *Journal of Environmental Engineering* (since 2007), *Journal of Hazardous, Toxic, and Radioactive Waste* (since 2006), and *Water Environment Research* (since 2008). He has been a registered professional engineer in Nebraska, USA since 2000.

Rao Y. Surampalli, Ph.D., P.E., Dist.M.ASCE, is President and Chief Executive Officer of the Global Institute for Energy, Environment and Sustainability. He was with the U.S. Environmental Protection Agency (USEPA) for 29 years and retired as an Engineer Director in October, 2014. He received his M.S. and Ph.D. degrees in Environmental Engineering from Oklahoma State University and Iowa State University, respectively. He is a Registered Professional Engineer in the branches of Civil and Environmental Engineering, and also a Board Certified Environmental Engineer (BCEE) of the American Academy of Environmental Engineers (AAEE) and a Diplomate of the American Academy of Water Resources Engineers (DWRE). He is an Adjunct Professor in seven universities and distinguished/honorary visiting professor in six universities. Currently, he serves, or has served on 66 national and international committees, review panels, or advisory boards including the ASCE National Committee on Energy, Environment and Water Policy. He also served as President of Civil Engineering Certification (CEC), Inc., that was started by ASCE for Board Certification of various specialties within civil engineering. He is a Distinguished Engineering Alumnus of both the Oklahoma State and Iowa State Universities, and is an elected Fellow of the American Association for the Advancement of Science, an elected Member of the European Academy of Sciences and Arts, an elected Member of the Russian Academy of Engineering, an elected Fellow of the Water Environment Federation and International Water Association, and a Distinguished Member of the American Society of Civil Engineers. He also is Editor-in-Chief of the ASCE

Journal of Hazardous, Toxic and Radioactive Waste, past Vice-Chair of Editorial Board of *Water Environment Research Journal*, and serves on the editorial boards of eight other refereed environmental journals. He has authored over 600 technical publications in journals and conference proceedings, including 15 patents, 16 books, 110 book chapters, and presented over 110 plenary/keynote or invited presentations around the world.